West Nile Encephalitis Virus Infection

Emerging Infectious Diseases of the 21st Century

Series Editor: I.W. Fong

Professor of Medicine, University of Toronto
Infectious Diseases, St. Michael's Hospital

Recent volumes in this series:

MALARIA: GENETIC AND EVOLUTIONARY ASPECTS
Edited by Krishna R. Dronamraju and Paolo Arese

INFECTIONS AND THE CARDIOVASCULAR SYSTEM:
New Perspectives
Edited by I.W. Fong

REEMERGENCE OF ESTABLISHED PATHOGENS IN THE 21ST CENTURY
Edited by I.W. Fong and Karl Drlica

BIOTERRORISM AND INFECTIOUS AGENTS:
A New Dilemma for the 21ST Century
Edited by I.W. Fong and Ken Alibek

MOLECULAR PARADIGMS OF INFECTIOUS DISEASE:
A Bacterial Perspective
Edited by Cheryl A. Nickerson and Michael J. Schurr

NEW AND EVOLVING INFECTIONS OF THE 21ST CENTURY
Edited by I.W. Fong and Ken Alibek

ANTIMICROBIAL RESISTANCE AND IMPLICATIONS FOR THE 21ST CENTURY
Edited by I.W. Fong and Karl Drlica

WEST NILE ENCEPHALITIS VIRUS INFECTION
Edited by Michael S. Diamond

A Continuation Order Plan is available for this series. A continuation order will bring delivery of each new volume immediately upon publication. Volumes are billed only upon actual shipment. For further information please contact the publisher.

Michael S. Diamond

West Nile Encephalitis Virus Infection

Viral Pathogenesis and the Host Immune Response

Springer

Editor
Michael S. Diamond
Departments of Medicine, Molecular Microbiology,
Pathology & Immunology
Washington University School of Medicine
St. Louis, MO, USA

ISBN: 978-0-387-79839-4 e-ISBN: 978-0-387-79840-0
DOI: 10.1007/978-0-387-79840-0

Library of Congress Control Number: 2008940591

Printed on acid-free paper

springer.com

Preface

West Nile virus is a neurotropic flavivirus that has emerged globally as a primary cause of viral encephalitis. Infection of humans and other vertebrate animals is associated with a febrile illness that can progress to a lethal encephalitis or flaccid paralysis syndrome. Its appearance in the Western Hemisphere in 1999 and the corresponding increase in global disease burden over the last decade have been accompanied by intensive study, including the entry of many scientists into the field. Breakthroughs have been made in understanding the unique transmission pattern between the vector and the multiple avian and mammalian hosts and targets. Studies in mammalian systems have dissected the viral and host factors that determine the pathogenesis and outcome of West Nile virus infection. On the basis of these experiments, progress has been made on the identification of genetic factors that predispose to severe human disease. Thus, in a remarkably short period of time, insight has been gained on a wide variety of disciplines related to West Nile virus biology.

The aim of this book was to assemble an up-to-date and cutting-edge anthology from the leading experts in the field. The chapters are balanced by submissions from newcomers who have made significant recent contributions with those from established investigators who have dedicated their careers to the study of West Nile virus. The topics are directed at the biology of West Nile virus, and cover ecology, vertebrate biology, epidemiology, clinical disease, pathogenesis, host immune response, structural biology, immune evasion, and progress on the development of vaccines and therapeutics. Nonetheless, because it belongs to a family of clinically relevant arthropod-borne human pathogens including dengue virus, Japanese encephalitis virus, yellow fever virus, and tick-borne encephalitis virus, many of the paradigms established for West Nile virus will be relevant for the transmission and pathogenesis of these viruses. Reciprocally, advances with other flaviviruses have influenced our understanding of the West Nile pathogenesis. As such, in some sections, chapters address West Nile virus biology in the context of findings with other pathogenic flaviviruses.

While the topics in this book are focused on West Nile virus, they are broad in scope ranging from understanding vector transmission patterns to the dynamics of structural transitions of proteins on the surface of the virion. As such, it is hoped that this book will have broad use for readers from a variety of backgrounds including clinical infectious disease, epidemiology, virology, immunology, and vector biology. The almost daily discoveries by investigators in this field make this virus an exciting and evolving area of study, and encourage the entry of new talented individuals and ideas. As an editor, I am deeply indebted and grateful to the dedication and perseverance of colleagues who pursue West Nile virus biology with passion, rigor, integrity, and collegiality. Indeed, the significant scientific progress in a relatively short period in the field has prompted optimism on the development, implementation, and targeting of vaccines and therapeutics that control West Nile virus disease in nature.

Contents

1. Global Epidemiology of West Nile Virus

Lyle R. Petersen

2. West Nile Virus: Molecular Epidemiology and Diversity

Gregory D. Ebel and Laura D. Kramer

3. Vector Biology and West Nile Virus

Carol D. Blair

4. Clinical Manifestations of Neurological Disease

J. David Beckham and Kenneth L. Tyler

5. Molecular Biology of West Nile Virus

Margo A. Brinton

8. Innate Immune Responses to West Nile Virus Infection

Alvaro Arjona and Erol Fikrig

9. Mechanisms of Complement Regulation of Infection by Flaviviruses

Erin Mehlhop and Michael S. Diamond

10. Antibody-Mediated Neutralization of West Nile Virus: Factors that Govern Neutralization Potency

Christiane A. Jost and Theodore C. Pierson

11. Structural Basis of Antibody Protection Against West Nile Virus

Richard J. Kuhn

12. Molecular Mechanisms of Flaviviral Membrane Fusion

Yorgo Modis and Vinod Nayak

13. $CD4^+$ and $CD8^+$ T-Cell Immune Responses in West Nile Virus Infection

Mario Lobigs, Arno Müllbacher, and Matthias Regner

14. Enhanced Antigen Processing or Immune Evasion? West Nile Virus and the Induction of Immune Recognition Molecules

Nicholas J.C. King, Ariane Davison, Daniel R. Getts, David Ping Lu, Meghann Teague Getts, Amanda Yeung, James K. Peterson, and Alison M. Kesson

Contributors

Alvaro Arjona
Section of Infectious Diseases – Department of Internal Medicine
Yale University School of Medicine
New Haven, CT, USA

Alan D.T. Barrett
Institute for Human Infections and Immunity
Sealy Center for Vaccine Development
Center for Biodefense and Emerging Infectious Diseases, and University of Texas Medical Branch
Galveston, TX, USA

David W.C. Beasley
Institute for Human Infections and Immunity
Sealy Center for Vaccine Development
Center for Biodefense and Emerging Infectious Diseases, and University of Texas Medical Branch
Galveston, TX, USA

J. David Beckham
Division of Infectious Diseases
University of Colorado at Denver Health Sciences Center, and Denver Veterans Affairs Medical Center
Denver, CO, USA

Carol D. Blair
Colorado State University
Fort Collins, CO, USA

Margo A. Brinton
Department of Biology
Georgia State University
Atlanta, GA, USA

David C. Chang
School of Molecular and Microbial Sciences
University of Queensland
St. Lucia, QLD, Australia

Ariane Davison
Department of Pathology
School of Medical Sciences
Bosch Institute, University of Sydney
Sydney, Australia

Michael S. Diamond
Departments of Medicine
Molecular Microbiology
Pathology & Immunology
Washington University School of Medicine
St. Louis, MO, USA

Gregory D. Ebel
Health Sciences Center School
University of New Mexico
Albuquerque, NM, USA

Jared Evans
Fox Chase Cancer Center
Philadelphia, PA, USA

Erol Fikrig
Section of Infectious Diseases – Department of Internal Medicine
Yale University School of Medicine
New Haven, CT, USA

Daniel R. Getts
Department of Pathology
School of Medical Sciences
Bosch Institute
Bosch Institute, University of Sydney
Sydney, Australia

Meghann Teague Getts
Department of Pathology
School of Medical Sciences
Bosch Institute, University of Sydney
Sydney, Australia

Jaap Goudsmit
Crucell Holland BV
Leiden, The Netherlands

Roy A. Hall
School of Molecular and Microbial Sciences
University of Queensland
St. Lucia, QLD, Australia

Christiane A. Jost
Laboratory of Viral Diseases
Viral Pathogenesis Section
National Institutes of Health
Bethesda, MD, USA

Alison M. Kesson
Department of Paediatrics and Child Health
Faculty of Medicine
The Children's Hospital at Westmead
University of Sydney
Sydney, Australia

Nicholas J.C. King
Department of Pathology
University of Sydney
Sydney, Australia

Alexander A. Khromykh
School of Molecular and Microbial Sciences
University of Queensland
St. Lucia, QLD, Australia

Laura D. Kramer
Arbovirus Laboratories, Wadsworth Center
State University of New York at Albany
New York State Department of Health
Slingerlands, NY, USA

John de Kruif
Crucell Holland BV
Leiden, The Netherlands

Richard J. Kuhn
Department of Biological Sciences
Markey Center for Structural Biology
Purdue University
West Lafayette, IN, USA

Mario Lobigs
Division of Immunology and Genetics
John Curtin School of Medical Research
The Australian National University
Canberra, Australia

David Ping Lu
Department of Pathology
School of Medical Sciences
Bosch Institute, University of Sydney
Sydney, Australia

Erin Mehlhop
Department of Pathology & Immunology
Washington University School of Medicine
St. Louis, MO, USA

Yorgo Modis
Department of Molecular Biophysics & Biochemistry
New Haven, CT, USA

Arno Müllbacher
Division of Immunology and Genetics
John Curtin School of Medical Research
The Australian National University
Canberra, Australia

Vinod Nayak
Department of Molecular Biophysics & Biochemistry
New Haven, CT, USA

Lyle R. Petersen
Division of Vector-Borne Infectious Diseases
Centers for Disease Control and Prevention
Fort Collins, CO, USA

James K. Peterson
Department of Mathematical Sciences and Biology
Clemson University
Clemson, SC, USA

Theodore C. Pierson
Laboratory of Viral Diseases
Viral Pathogenesis Section
National Institutes of Health
Bethesda, MD, USA

Matthias Regner
Division of Immunology and Genetics
John Curtin School of Medical Research
The Australian National University
Canberra, Australia

John T. Roehrig
Arboviral Diseases Branch
Division of Vector-Borne Infectious Diseases
National Center for Zoonotic
Vector-Borne, and Enteric Diseases
Coordinating Center for Infectious Diseases
Centers for Disease Control and Prevention
Fort Collins, CO, USA

Melanie A. Samuel
Department of Molecular Microbiology
Washington University School of Medicine
St. Louis, MO, USA

Christoph Seeger
Fox Chase Cancer Center
Philadelphia, PA, USA

Pei-Yong Shi
Wadsworth Center
New York State Department of Health, and Department of Biomedical Sciences
School of Public Health
State University of New York
Albany, NY, USA

Robert B. Tesh
Department of Pathology
Center for Biodefense and Emerging Infectious Diseases
University of Texas Medical Branch
Galveston, TX, USA

Mark Throsby
Crucell Holland BV
Leiden, The Netherlands

Kenneth L. Tyler
University of Colorado at Denver Health Sciences Center, and Denver Veterans Affairs Medical Center
Denver, CO, USA

Shu-Yuan Xiao
Department of Pathology
Center for Biodefense and Emerging Infectious Diseases
University of Texas Medical Branch
Galveston, TX, USA

Amanda Yeung
Department of Pathology
School of Medical Sciences
Bosch Institute, University of Sydney
Sydney, Australia

1. Global Epidemiology of West Nile Virus

Lyle R. Petersen

Abstract

West Nile virus (WNV) exists in a bird–mosquito–bird cycle, with humans and horses as incidental hosts. The global epidemiology follows several patterns (1) widespread enzootic transmission throughout tropical Africa, the Caribbean, Central America, and northern South America without significant human or equine morbidity; (2) periodic human and equine outbreaks in the Mediterranean Basin, Russia, and South Africa separated by years of low-level enzootic activity; (3) sporadic human disease in India; (4) little or no WNV enzootic activity in Southeast and East Asia and no human cases; (5) sporadic human disease and small outbreaks in Australia; and (6) repeated annual outbreaks in North America. The biology underlying these patterns is poorly understood, but probably results from a complex interaction of climate and weather, mosquito vectors, avian hosts, viral strain, other circulating flaviviruses, and other undefined factors. The virus was first detected in the Americas in New York City in 1999 and spread to Argentina within 7 years. The high infection incidence in North America has resulted in new modes of transmission, such as blood transfusion and organ donation. Older age and certain immunosuppressive conditions confer a high risk of neuroinvasive disease after infection.

Keywords

West Nile virus, epidemiology, flavivirus

1 Introduction

For 60 years after its initial isolation from the blood of a febrile woman in the West Nile district of Uganda in 1937 (Smithburn et al., 1940), WNV remained an occasional cause of febrile illness in Africa, the Middle East, parts of Europe and the former Soviet Union, South Asia, and Australia. It emerged from obscurity in the 1990s following large outbreaks of unusual severity in Algeria (1994 and 1997), Romania (1996), Tunisia (1997), Russia (1999), United States (1999), Israel (2000), and Sudan (2002) (Mackenzie et al., 2004). Following its recognition in New York City in 1999 (Nash et al., 2001), the virus spread rapidly throughout the New World, reaching Argentina by 2006 (Morales et al., 2006).

WNV is a member of the Japanese encephalitis serocomplex, which contains three other medically important flaviviruses: Japanese encephalitis, St Louis encephalitis, and Murray Valley encephalitis. Before WNV emerged in the Americas, each of these viruses had relatively unique geographic distributions (Fig. 1). Phylogenetic analysis of the complete genome sequence or of the E protein gene sequence divides strains of WNV into two genetic lineages (I and II) (Lanciotti et al., 1999, 2002). Major human outbreaks of WNV have been associated only with lineage I WNVs. Lineage II WNVs are maintained in enzootic cycles primarily in Africa and are not associated with human or animal outbreaks. Kunjin virus in Australia, although originally classified as a separate virus in the Japanese encephalitis serocomplex, is now recognized as a subtype lineage I WNV.

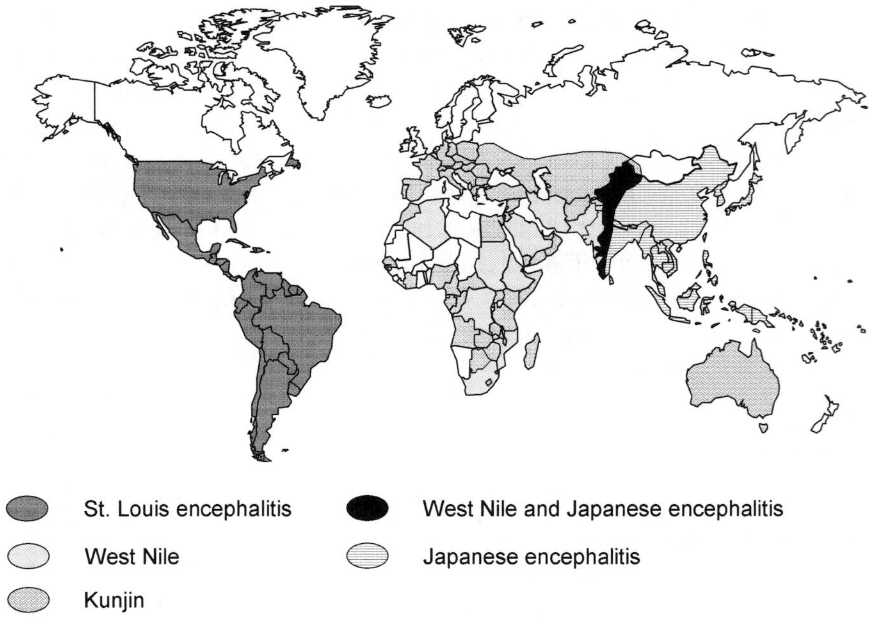

Figure 1. Geographic distributions of West Nile, St Louis encephalitis, Japanese encephalitis, and Kunjin viruses before 1999.

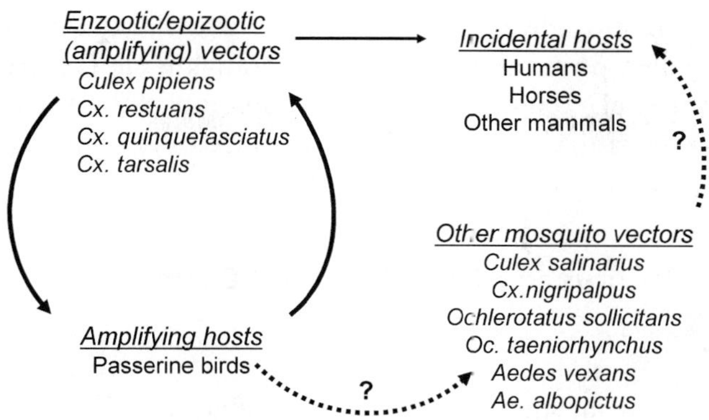

Figure. 2. WNV transmission cycle. WNV is transmitted in a cycle between birds and *Culex* mosquitoes. *Cx.* pipiens is an important vector in the northern United States and Canada. *Culex quinquefasciatus* is important in the southern United States; whereas, *Cx. tarsalis* is important in the western United States. In addition, *Cx. nigripalpus* is a vector in Florida. Other mosquito species may be "bridge" vectors to horses, humans, and other "dead end" hosts, which typically do not develop high viremia and do not participate in the transmission cycle. However, bridge vectors likely have a minor role compared to enzootic vectors in viral transmission to "dead end" hosts.

Birds are the primary amplifying hosts, and the virus is maintained in a bird–mosquito–bird cycle primarily involving *Culex* spp. mosquitoes (Hubalek and Halouzka 1999) (Fig. 2). *Culex* spp. mosquitoes are important for their role in overwintering the virus in temperate climates (Murgue et al., 2001a). The amplification cycle begins when infected overwintering mosquitoes emerge in spring and infect the birds. Amplification within the bird–mosquito–bird cycle continues until late summer and fall, when *Culex* mosquitoes begin diapause (a reduction of physiologic activity in which development is arrested) and rarely blood feed. Humans and other vertebrates, such as horses, do not develop a high-titer viremia and, thus, play a minor role in the transmission cycle.

Approximately 20–30% of persons develop illness after infection with lineage I strains currently circulating in North America (Mostashari et al., 2001; Brown et al., 2007). The typical incubation period ranges from 2 to 14 days, although longer incubation periods have been observed among the immunosuppressed (Pealer et al., 2003). The usual clinical presentation, called West Nile fever, is characterized by acute onset of fever, headache, malaise, back pain, myalgias, anorexia, and fatigue. Eye pain, pharyngitis, nausea, vomiting, diarrhea, abdominal pain, and rash can also occur. Neuroinvasive disease, which occurs in approximately 1 in 140 infected persons, can present as encephalitis, meningitis, or flaccid paralysis, singly or in combination. Other neurological manifestations include tremor, myoclonus, and Parkinsonian features such as rigidity, postural instability, and bradykinesia (Pepperell et al., 2003; Sejvar et al., 2003). Other infrequent complications include rhabdomyolysis, hepatitis, pancreatitis, central diabetes insipidus, stiff person syndrome, and hemorrhagic manifestations.

2 West Nile Virus in Africa, Asia, and Europe

2.1 Africa

The major mosquito vector in Africa and the Middle East is *Cx. univittatus*, with other *Culex* species important in some areas (Hubalek and Halouzka, 1999; Solomon, 2004). Serological studies in humans conducted in several African countries indicate considerable human WNV infection incidence, yet severe illness is uncommon (Murgue et al., 2002). Egyptian studies conducted in the 1950s demonstrated WNV antibodies in more than 60% of the human population (Hurlbut et al., 1956). Similar results were obtained in 1979, with prevalence ranging from 74% in Upper Egypt to 28% in Cairo (Darwish and Ibrahim, 1971). Although WNV has been occasionally isolated from the blood of febrile children in Egypt, neuroinvasive disease has remained uncommon. Similarly, WNV antibodies could be identified in half of the horses studied in Egypt, yet equine disease is uncommon (Schmidt and Elmansoury, 1963).

Elsewhere in northern Africa, 5% of serum samples collected in Tunisia in the late 1960s, mostly from children, had WNV antibodies, and in northern Senegal in 1989, a 98% WNV antibody seroprevalence was noted, although seroprevalence was lower elsewhere in the country (Murgue et al., 2002). In Kenya, WNV antibody prevalence among humans in different regions ranged from 3 to 65% in surveys conducted from 1966 to 1968. Sporadic human cases were reported from the Central African Republic in 1983 and 1993, although human serologic surveys conducted in 1966 and 1967 indicated that antibody prevalence was very low. In South Africa, 13–24% of samples collected from healthy adults between 1961 and 1965 had WNV antibodies. Seroprevalence up to 45% was reported from Madagascar (Murgue et al., 2002). Although serological data suggests that WNV exposure in Africa is frequent, these data must be interpreted cautiously because of differences in testing methodology and the considerable serological cross-reactivity among the flaviviruses.

In contrast to the apparent frequent exposure to WNV throughout much of Africa, outbreaks are uncommon, particularly in the tropics.

Epidemics occurred in South Africa in 1974 and in 1983/1984 following heavy rains and above normal temperatures in early summer. Despite an estimated 18,000 cases in 1974, neuroinvasive disease or mortality was not recorded (Jupp, 2001). More recently, a new pattern of outbreaks of unusual severity in northern Africa has occurred. An outbreak of approximately 50 human cases with neurological disease occurred in Algeria in 1994, which was followed in 1997 by an outbreak involving 173 patients (Triki et al., 2001). In Sudan in 2002, at least 31 cases of encephalitis occurred during an outbreak in children (Depoortere et al., 2004). In addition, 94 and 7 equines developed severe neurological disease during outbreaks in Morocco in 1996 and 2003, respectively (Schuffenecker et al., 2005). These outbreaks were associated with viral strains of apparent increased virulence and were closely related to the strains that caused large human outbreaks in Romania, Russia, Israel, and the United States (Lanciotti et al., 1999; Schuffenecker et al., 2005).

2.2 Middle East, Russia, Asia, and Australia

In Asia, *Cx. tritaeniorhynchus*, *Cx. vishnui*, and *Cx. quinquefasciatus* are predominant vectors (Hubalek and Halouzka 1999). The first reported WNV human epidemic occurred in Israel in 1951–1952 (Bernkopf et al., 1953), and was followed by a series of human outbreaks throughout the 1950s and in 1980. Illnesses were generally mild, although neuroinvasive disease was prominent during an outbreak among the elderly in 1957 (Murgue et al., 2002). There was little subsequent WNV activity in Israel until 1997 and 1998 when WNV was identified from dying migrating storks from Europe as well as other bird species, including domestic geese (Malkinson and Banet, 2002; Malkinson et al., 2002). Two human fatalities occurred in 1999, followed by an outbreak with 417 cases and 35 deaths the following year (Chowers et al., 2001; Weinberger et al., 2001). As noted in later sections of this chapter, the WNV strain introduced into Israel may have originated in central Europe and is closely related to viruses causing outbreaks of neuroinvasive disease in Africa, Europe, Russia, and North America (Lanciotti et al., 2002).

WNV illness is infrequent elsewhere in the Middle East and South Asia. A human serologic survey conducted in Iran in 1967 indicated that about 25% had WNV antibodies; however, human illness in that country has not been reported (Naficy and Saidi, 1970). Serological evidence of WNV antibodies in humans in India was first identified in 1952. WNV was identified since in mosquitoes and in ill humans in several areas (Paul et al., 1970; George et al., 1984). A study of samples collected throughout India from 1992 to 2001 identified 88 ill persons with WNV-specific antibodies, suggesting that WNV-related illness incidence may be higher than that currently recognized (Thakare et al., 2002).

Since 1963, WNV has been isolated from ticks, birds, and mosquitoes in the southern area of European Russia and western Siberia, and in adjacent republics of the former Soviet Union. Serological surveys of healthy humans indicated up to 8% anti-WNV IgG antibody seroprevalence (Platonov, 2001). Yet outbreaks were uncommon until 1999 when a large outbreak of severe neurological disease involved 318 cases and 40 deaths in Volgograd (Platonov et al., 2001). This outbreak occurred during an unusually hot and dry summer. The virus that caused the 1999 outbreak was genetically related to 1996 Romanian outbreak (Lanciotti et al., 2002). WNV has rarely been isolated from East Asia and human cases are extremely uncommon.

Kunjin virus, a lineage I WNV variant, is endemic throughout most of tropical Australia and eastern Queensland. Documented illness is rare, and cases occur in infrequent small outbreaks or sporadically. Illness is typically mild, non-encephalogenic, and non-life threatening.

2.3 Europe

Although little information exists regarding WNV mosquito vectors in Europe, *Cx. pipiens* and *Cx. modestus* are recognized as important vectors (Hubalek and Halouzka, 1999). In contrast to the North American experience, avian infection in Europe is generally asymptomatic, probably reflecting a long co-evolution of virus and hosts in the Old World. Nevertheless, infection with a lineage I strain identical to that imported into New York City in 1999 resulted in clinical illness in white storks (*Cicona ciconia*) migrating from central Europe through Israel in 1998 (Malkinson et al., 2002). In 2003, a WNV strain most closely genetically linked to the Israel 1998 and New York 1999 strains caused an outbreak of encephalitis in Hungarian geese (*Anser anser domesticus*), leading to the hypothesis that these strains originated in central Europe (Bakonyi et al., 2006). Interestingly, in 2004 and 2005, several deaths in goshawks (*Accipiter gentilis*) and a sparrow hawk (*Accipiter nisus*) from the same region of Hungary were attributed to a central African lineage II virus, which is the first report of a lineage II virus outside of Africa (Erdelyi et al., 2007). Another novel WNV, named Ravensberg virus, was isolated from mosquitoes in 1997 and 1999 at the Czech Republic–Austria border (Bakonyi et al., 2005). These observations suggest relatively frequent transport by migrating birds of WNV strains between Europe and Africa.

Since the 1960s WNV isolates have been obtained infrequently in southern and central Europe from mosquitoes, humans, birds, and horses, although serological surveys conducted in humans, birds, and other animals suggested more widespread viral exposure (Hubalek and Halouzka, 1999; Linke et al., 2007). In addition, serological evidence of WNV exposure has been noted in resident bird species in the United Kingdom (Buckley et al., 2003). Human or equine illness in Europe has been sporadic, with isolated human cases of WN fever identified in France, Spain, Romania, Belarus, and Czechland (Hubalek and Halouzka, 1999). An outbreak in the Camargue region of southern France from 1962 to 1965 resulted in 15 virologically confirmed human cases with one death and approximately 80 equine cases with 25–30% mortality (Murgue et al., 2001a). No further equine outbreaks of this magnitude were noted until an outbreak with 14 cases occurred in Italy in 1998 and another with 76 cases in the Camargue region in 2000 (Murgue et al., 2001b; Autorino et al., 2002). Four humans with WNV-related illness were noted in the Camargue region in 2003, which represented the first human WNV-related illnesses documented there since 1965 (Del Giudice et al., 2004). The first large human outbreak in Europe occurred in Romania with 352 cases of neuroinvasive disease in 1996 (Tsai et al., 1998). Like the Volograd outbreak, this outbreak occurred during exceptionally hot and dry weather.

3 West Nile Virus in the Americas

3.1 United States

The virus was first detected in North America during a human outbreak of meningitis and encephalitis and an accompanying epizootic in birds in 1999 in the New York City area (Nash et al., 2001). This outbreak also occurred during a period of abnormally hot weather. While the means of importation remained unknown, sequencing of isolates (NY99) implicated the Middle East, or possibly central Europe, as the virus' likely origin (Lanciotti et al., 1999, 2002; Bakonyi et al., 2006). Others have suggested viral importation via ships traveling from Black Sea ports (Lvov et al., 2004).

3.1.1 Mosquitoes and Vertebrates

In the United States, more than 60 mosquito species have been infected with WNV. WNV and St Louis encephalitis virus, a related Japanese encephalitis serogroup virus native to the Americas, share the same maintenance vectors (*Cx. pipiens* and *Cx. restuans* in the Northern United States and Canada, *Cx. quinquefasciatus* in the southern United States, and *Cx. tarsalis* in the western United States and Canada (Centers for Disease Control and Prevention, 2002d; Solomon, 2004). *Cx. nigripalpus* may be important in Florida (Blackmore et al., 2003).

Lethal WNV infection has been documented in more than 320 resident and exotic avian species in the United States (Centers for Disease Control and Prevention, 2007a), with significant declines in populations of highly susceptible species, such as American crows and other corvids (crows, ravens, and jays) (LaDeau et al., 2007). The high virulence to American crows from the NY99 WNV strain compared to two other lineage I WNV strains (Kunjin and a strain isolated in Kenya [KEN-3829]) was confirmed in the laboratory (Brault et al., 2004) and was shown to be due to a single nucleotide change in the NS3 gene (Brault et al., 2007).

A total of 24,213 equine cases of WNV disease were reported from 1999 to 2006 in the United States. Incidence peaked in 2002 at 14,539 and markedly decreased after a WNV vaccine became available in 2003. Approximately 10% of experimentally infected horses develop clinical illness (Bunning et al., 2001). Clinical series indicate a 20–40% mortality among clinically ill horses (Cantile et al., 2000; Porter et al., 2003; Weese et al., 2003; Salazar et al., 2004; Ward et al., 2004). Age, vaccination status, inability to rise, and female gender are associated with the risk of death (Salazar et al., 2004). Viremia in horses is low and of short duration; thus, horses are unlikely to serve as important amplifying hosts for WNV (Bunning et al., 2001, 2002).

WNV has been isolated or identified serologically in many native and imported vertebrate species in North America, including bats, wolves, eastern fox and gray squirrels, chipmunks, sheep, alligators, alpacas, black bears, macaques, a Suffolk ewe, reindeer, dogs, monkeys and baboons, raccoons, skunks, and opossums. While some of these species experience viremia sufficient to infect mosquitoes, their role, if any, in virus maintenance or amplification remains undefined.

3.1.2 Human Incidence and Distribution

The initial outbreak in New York City in 1999 caused 62 cases and 7 deaths in Queens and surrounding areas (Nash et al., 2001). Despite extensive geographic spread of WNV from 1999 to 2001, human disease was infrequent, with only 149 cases reported in the United States (Marfin et al., 2001; Petersen and Hayes, 2004) (Table 1; Fig. 3). However, in 2002 and 2003, multistate outbreaks in the Midwest resulted in more than 2,800 reported cases of neuroinvasive disease each year (O'Leary et al., 2004;

Table 1. Reported West Nile virus neuroinvasive disease cases, total number of reported cases (includes neuroinvasive disease, WN fever, and clinically unspecified cases), and WNV-related deaths in humans, United States, 1999–2007

Year	WNND[a]	Total	Deaths
1999	59	62	7
2000	19	21	2
2001	64	66	9
2002	2,946	4,156	284
2003	2,866	9,862	264
2004	1,148	2,539	100
2005	1,309	3,000	119
2006	1,495	4,269	177
2007	1,227	3,630	124
Total	11,123	27,605	1,086

[a] *WNND* West Nile virus neuroinvasive disease.

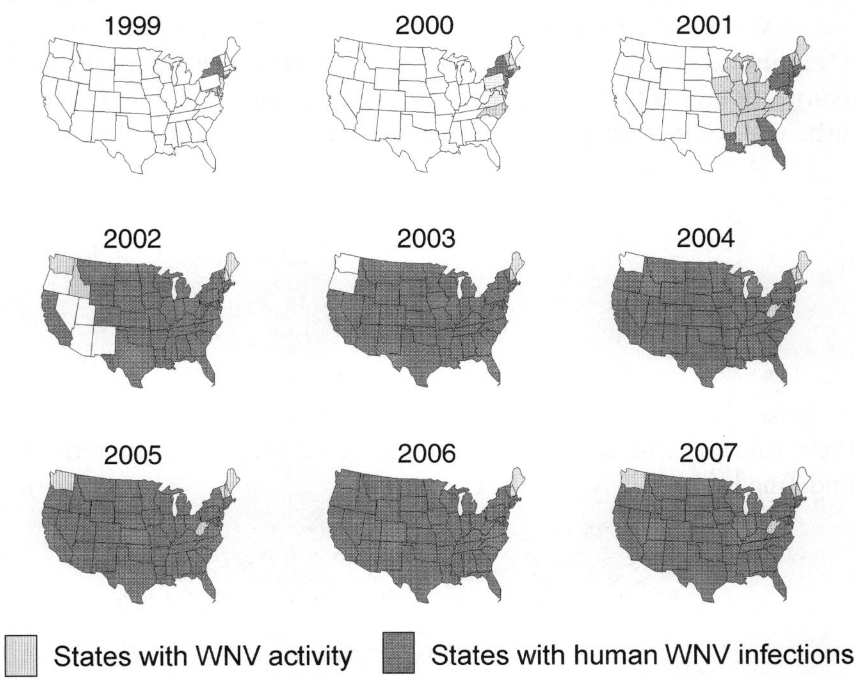

Figure 3. WNV activity in the United States, 1999–2007. States with any WNV activity (detected in mosquitoes, birds, equines, or other vertebrates) and states with human cases are indicated.

Petersen and Hayes, 2004). Lower incidence of WNV neuroinvasive disease occurred from 2004 to 2007. The case fatality rate among persons with neuroinvasive disease is approximately 10%.

Serological surveys indicate that even in areas experiencing outbreaks, less than 10% of the population is infected with WNV (Tsai et al., 1998; Centers for Disease Control and Prevention, 2001; Mostashari et al., 2001; Petersen and Hayes 2004; Schellenberg et al., 2006). Based on the 9,906 reported cases of WNV neuroinvasive disease from 1999 to 2006 and the proportions of infected persons who develop WNV neuroinvasive disease, an estimated 1.5 million persons were infected with the WNV in the United States through 2006 (Petersen and Hayes, 2004).

Human WNV infection incidence increases in early summer and peaks in August or early September (Fig. 4). Within large regional WNV epidemics in the United States, human disease incidence varies markedly from county to county, suggesting the importance of local ecological conditions. A follow-up study of viremic blood donors and uninfected controls showed that males and persons living in a rural area were approximately 1.4 times and 3.4 times more likely to have WNV infection than females or persons living in surburban or urban locations (Orton et al., 2006). In Ontario, Canada, persons living in rural areas were approximately six times more likely than urban dwellers to have WNV antibodies (Schellenberg et al., 2006). Farming (Miramontes et al., 2006), increased environmental temperatures from April to October (Miramontes et al., 2006), vegetation abundance in an urban area (Brownstein et al., 2002; Ruiz et al., 2007), and living in an

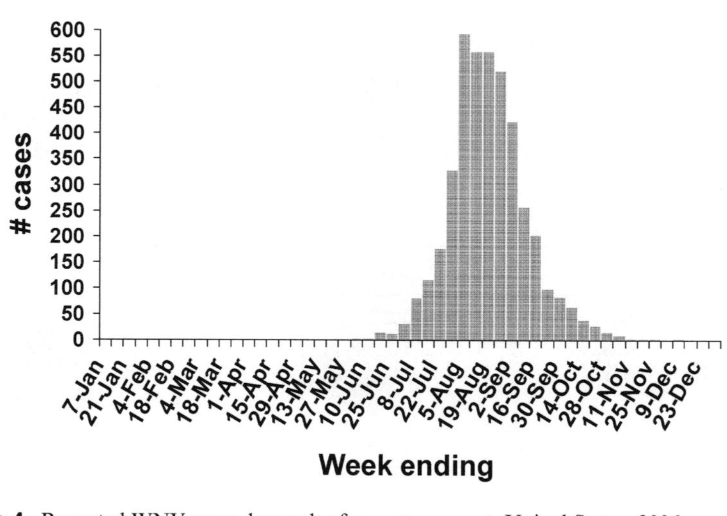

Figure 4. Reported WNV cases, by week of symptom onset, United States, 2006.

inner suburb (Ruiz et al., 2007) have been linked to increased WNV infection incidence.

3.2 Canada

The epidemiology of WNV in Canada largely resembles that of the northern United States. WNV was first detected in Canada in southern Ontario in 2001 and by 2002 had spread to Manitoba, Quebec, Nova Scotia, and Saskatchewan (Drebot et al., 2003). The first cases of human WNV disease and a large outbreak in Canada occurred in 2002 in Quebec and Ontario (Table 2). Subsequent spread of the virus has extended as far east as New Brunswick and west into Alberta. Large outbreaks also occurred in 2004 and 2007.

3.3 Latin America and the Caribbean

WNV was first detected south of the United States border in 2001, when a resident of the Cayman Islands developed WNV encephalitis (Centers for Disease Control and Prevention, 2002e). Subsequently, serologic studies in birds and horses suggested that WNV has circulated (first year of detection indicated) in Argentina (2006) (Morales et al., 2006), Colombia (2004) (Mattar, 2005), Cuba (2004) (Dupuis et al., 2005), Dominican Republic (2002) (Komar et al., 2003), El Salvador (2005) (Cruz et al., 2005), Guadeloupe (2002) (Quirin et al., 2004; Lefrancois et al., 2005), Guatemala (2003) (Morales-Betoulle et al., 2006), Haiti (2004) (Beatty et al., 2007), Jamaica (2002) (Dupuis et al., 2003), Mexico (2002) (Blitvich et al., 2003; Estrada-Franco et al., 2003; Fernandez-Salas et al., 2003;

Table 2. Reported West Nile virus disease neuroinvasive disease cases and total number of reported cases (includes neuroinvasive disease, WN fever, and clinically unspecified cases) in humans, Canada, 2002–2007

Year	WNND[a]	Total
2002	259	414
2003	217	1,481
2004	13	25
2005	49	225
2006	38	151
2007	216	2,215
Total	792	4,360

[a] *WNND* West Nile virus neuroinvasive disease.

Lorono-Pino et al., 2003; Marlenee et al., 2004), Puerto Rico (2002) (Dupuis et al., 2005), and Venezuela (2004) (Bosch et al., 2007). However, viral isolations have been infrequent and documented avian and equine morbidity and mortality have been extremely uncommon despite surprisingly high WNV antibody seroprevalence in some areas. Documented equine morbidity in Mexico has mostly occurred in its northern states. Unfortunately, serological cross-reactivity with other circulating flaviviruses in Latin America and the Caribbean, such as St Louis encephalitis virus, has complicated the interpretation of serological tests from that region. Documented human infections have been limited to a few patients in Cayman Islands (Centers for Disease Control and Prevention, 2002d), Cuba (Pupo et al., 2006), Haiti (Beatty et al., 2007), and northern Mexico (Elizondo-Quiroga et al., 2005).

Genetic sequencing of an isolate from a dead raven in Tabasco State, Mexico, in 2003 (Estrada-Franco et al., 2003) clustered most closely with strains collected in Florida and Louisiana in 2001 and 2002, while strains collected in Baja California, Sonora, and Tamaulipas in 2003 and 2004 clustered with strains from Louisiana, Texas, and Arizona collected those same years (Deardorff et al., 2006). These results suggest at least two independent WNV introductions into Mexico, but must be interpreted cautiously given the few genetic sequences available for study. The isolate from the Common Raven had a unique E-156 Pro residue, which ablates the N-linked glycosylation site found in most North American strains (Beasley et al., 2004); however, other isolates from the Americas do not share this residue.

The discrepancy between the serologic evidence indicating widespread WNV circulation in the Caribbean, Central and South America, and Mexico and the lack of substantial avian, equine, or human morbidity remains a mystery. Possible causes include false-positive serologic test results due to circulation of other flaviviruses, circulation of attenuated strains, protection resulting from previous flavivirus exposure modulating disease expression, and suboptimal disease surveillance.

4 Clinical Epidemiology

4.1 Risk Factors Associated with Human Disease

Serologic surveys in Romania, New York City, and Canada (Tsai et al., 1998; Mostashari et al., 2001; Schellenberg et al., 2006) as well as blood donor screening data (Kleinman et al., 2005) indicate that WNV infection incidence is constant across all age groups during outbreaks in naïve populations. Little is known about risk factors for development of WN fever after infection. Follow-up of viremic blood donors indicated that the risk of developing WN fever decreased with age and increased with virus concentration at the time of donation (Brown et al., 2007). Another study showed that yellow fever and Japanese encephalitis vaccination provided no protection for development of WN fever following subsequent mosquito-borne WNV infection, and, in fact, may have increased the risk (Johnson et al., 2005).

In the United States, the incidence of neuroinvasive disease increases approximately 1.5-fold for each decade of life, resulting in a risk approximately 30 times greater for a person 80–90 years old compared to a child younger than 10 years old (O'Leary et al., 2004). Thus, age is by far the most important host risk factor for the development of neuroinvasive disease after infection.

Organ transplant recipients are at extreme risk of developing neuroinvasive disease after WNV infection. A seroprevalence study carried out in patients of Canadian outpatient transplant clinics following a WNV epidemic in 2002, indicated that the risk of neuroinvasive disease following infection was 40% (Kumar et al., 2004a,b). Individual case reports or small case series have described WNV neuroinvasive disease in transplant recipients of kidney (Armali et al., 2003; Hardinger et al., 2003; Iwamoto et al., 2003; Cushing et al., 2004; DeSalvo et al., 2004; Kleinschmidt-DeMasters et al., 2004; Kumar et al., 2004b; Smith et al., 2004; Wadei et al., 2004), kidney/pancreas (Kleinschmidt-DeMasters et al., 2004; Ravindra et al., 2004), liver (Iwamoto et al., 2003; Pealer et al., 2003; Kleinschmidt-DeMasters et al., 2004; Kumar et al., 2004b), lung (Kleinschmidt-DeMasters et al., 2004), heart (Iwamoto et al., 2003; Kumar et al., 2004b), bone marrow (Hiatt et al., 2003; Pealer et al., 2003; Robertson et al., 2004), pancreatic islets (Barshes et al., 2006), and stem cells (Hong et al., 2003; Pealer et al., 2003; Kleinschmidt-DeMasters et al., 2004; Reddy et al., 2004). In a series of 11 transplant patients with WNV neuroinvasive disease, two died and three developed significant residual neurological deficits (Kleinschmidt-DeMasters et al., 2004).

A case–control study showed that cancer and chemotherapy increased the risk of developing meningitis or encephalitis at least sixfold (Patnaik et al., 2006). Eleven percent of patients inoculated with WNV as a cancer treatment in the early 1950s developed encephalitis, sometimes accompanied by paralysis, involuntary twitching, and cogwheel rigidity (Southam and Moore, 1954). The risk of adverse outcomes was related to underlying illness severity, particularly among patients with hematological malignancies (Southam and Moore, 1954). Neuroinvasive disease has been reported in patients with other immunocompromising conditions including common variable immunodeficiency (Alonto et al., 2003), infliximab (monoclonal antibody that inhibits tumor necrosis factor alpha) treatment for rheumatoid arthritis (Batsis and Phy, 2005; Chan-Tack and Forrest, 2006) and in five persons with human immunodeficiency virus infection (Torno et al., 2007). However, case reports do not allow for determining whether these conditions were predisposing factors.

Other studies have suggested that older age, alcohol abuse, diabetes, hypertension, and history of cardiovascular disease were risk factors for WN encephalitis (Bode et al., 2006; Murray et al., 2006), and that the black race, chronic renal disease, hepatitis C virus, and immunosuppression were risk factors for death from WNV infection (Murray et al., 2006). Another large study indicated that hypertension, diabetes, heart disease, kidney disease, cancer, and chemotherapy were risk factors for encephalitis (Patnaik et al., 2006). One study suggested that persons homozygous for a deletion in the CCR5 gene were at increased risk for symptomatic infection and death from WNV (Glass et al., 2006).

4.2 Transmission Modes

Nearly all human infections of WNV result from mosquito bites. Transmission also has been described via transfused blood (Pealer et al., 2003), transplanted organs (Iwamoto et al., 2003; Centers for Disease Control and Prevention, 2005), transplacental transmission (Centers for Disease Control and Prevention, 2002a), occupational transmission via percutaneous exposure (Centers for Disease Control and Prevention, 2002b) and inhalation (Nir, 1959), conjunctival exposure (Fonseca et al., 2005), and in a dialysis center by unidentified means (Centers for Disease Control and Prevention, 2004). Transmission via breast milk is also likely (Centers for Disease Control and Prevention, 2002c), although this mode of transmission appears uncommon (Hinckley et al., 2007). An outbreak of WNV infection occurred among turkey farm workers; however, the means of transmission in that setting was unknown (Centers for Disease Control and Prevention, 2003).

Transfusion-associated WNV transmission was proven during the 2002 US WNV epidemic when 23 transfusion recipients became infected after receipt of platelets, red blood cells, or fresh frozen plasma from 16 viremic blood donors (Harrington et al., 2003; Pealer et al., 2003).

Mathematical models indicated that the risk of transfusion-associated WNV transmission during the 2002 epidemic ranged from 2.1 to 4.7 per 10,000 donors in high incidence states (Biggerstaff and Petersen, 2003).

Since 2003, the United States and Canadian blood supplies have been screened for WNV using nucleic acid amplification (NAT) tests. Blood centers conduct NAT testing on minipools of 6–16 specimens, depending on the test kit manufacturer. From 2003 to 2007, screening identified approximately 2,000 NAT-positive blood donors, with as many as 1 in 150 donors being positive in some outbreak areas (Busch et al., 2005; Stramer et al., 2005). However, not all of these donations are infectious, as those from donors lacking WNV IgM antibody appear for less capable of transmitting WNV (Pealer et al., 2003; Centers for Disease Control and Prevention, 2007b). Universal blood donation screening has not eliminated WNV transfusion transmission. Through 2006, nine "breakthrough" transmissions have occurred from donations without WNV IgM antibody and with virus levels below the limit of detection by minipool screening (Petersen and Epstein, 2005; Centers for Disease Control and Prevention, 2007b).

In 2002, transmission via donated organs was first documented when four recipients of organs from a common donor developed WNV infection (Iwamoto et al., 2003). Serum from the day of organ harvest was positive for WNV by NAT and culture. A second transmission occurred in 2005 in which three of four organ recipients developed WNV infection (Centers for Disease Control and Prevention, 2005). Serum from the day of organ harvest was positive for WNV-specific IgG and IgM antibodies, but was negative for WNV RNA, suggesting that transmission can occur from virus sequestered in organs in the absence of detectable viremia in serum.

A causal relationship between WNV and fetal abnormalities has not been proven. A woman who had WNV encephalitis during the 27th week of pregnancy and delivered a term infant with chorioretinitis and laboratory evidence of congenitally acquired WNV infection (Centers for Disease Control and Prevention, 2002a). However, follow-up of 72 live infants born to women infected with WNV during pregnancy found that none had conclusive laboratory evidence of congenital WNV infection (O'Leary et al., 2006). Three infants born to women with symptomatic WNV infection within 3 weeks of delivery had symptomatic WNV disease at or shortly after delivery (O'Leary et al., 2006). Another study also failed to find differences between infants of WNV IgG-seropositive and -seronegative mothers with respect to birth outcomes (Paisley et al., 2006).

5 Summary

West Nile virus (WNV) exists in a bird–mosquito–bird cycle; humans and horses are incidental hosts. The global epidemiology follows several patterns (1) widespread enzootic transmission throughout tropical Africa, the Caribbean, Central America, and northern South America without significant human or equine morbidity; (2) periodic human and equine outbreaks in the Mediterranean Basin, Russia, and South Africa separated by years of low-level enzootic activity; (3) sporadic human disease in India; (4) little or no WNV enzootic activity in Southeast and East Asia and no human cases; (5) sporadic human disease and small outbreaks in Australia; and (6) repeated annual outbreaks in North America. The biology underlying these patterns is poorly understood, but probably results from a complex interaction of climate and weather, mosquito vectors, avian hosts, viral strain, other circulating flaviviruses, and other undefined factors. The virus was first detected in the Americas in New York City in 1999 and spread to Argentina within 7 years. The high infection incidence in North America has resulted in new modes of transmission, such as blood transfusion and organ donation. Older age and certain immunosuppressive conditions confer a high risk of neuroinvasive disease after infection.

References

Alonto, A. M., Aronoff, D. M., and Malani, P. N. 2003. West Nile virus meningitis in patient with common variable immunodeficiency. Emerg Infect Dis 9:1353–1354

Armali, Z., Ramadan, R., Chlebowski, A., and Azzam, Z. S. 2003. West Nile meningo-encephalitis infection in a kidney transplant recipient. Transplant Proc 35:2935–2936

Autorino, G. L., Battisti, A., Deubel, V., Ferrari, G., Forletta, R., Giovannini, A., Lelli, R., Murri, S., and Scicluna, M. T. 2002. West Nile virus epidemic in horses, Tuscany region, Italy. Emerg Infect Dis 8:1372–1378

Bakonyi, T., Hubalek, Z., Rudolf, I., and Nowotny, N. 2005. Novel flavivirus or new lineage of West Nile virus, central Europe. Emerg Infect Dis 11:225–231

Bakonyi, T., Ivanics, E., Erdelyi, K., Ursu, K., Ferenczi, E., Weissenbock, H., and Nowotny, N. 2006. Lineage 1 and 2 strains of encephalitic West Nile virus, central Europe. Emerg Infect Dis 12:618–623

Barshes, N. R., Agee, E. E., Zgabay, T., Brunicardi, F. C., Goss, J. A., and DeBakey, M. E. 2006. West Nile virus encephalopathy following pancreatic islet transplantation. Am J Transplant 6:3037

Batsis, J. A., and Phy, M. P. 2005. West Nile virus meningitis in a chronic immunosuppressed patient with rheumatoid arthritis. Clin Rheumatol 24:548–550

Beasley, D. W., Davis, C. T., Estrada-Franco, J., Navarro-Lopez, R., Campomanes-Cortes, A., Tesh, R. B., Weaver, S. C., and Barrett, A. D. 2004. Genome sequence and attenuating mutations in West Nile virus isolate from Mexico. Emerg Infect Dis 10:2221–2224

Beatty, M. E., Hunsperger, E., Long, E., Schurch, J., Jain, S., Colindres, R., Lerebours, G., Bernard, Y. M., Dobbins, J. G., Brown, M., and Clark, G. G. 2007. Mosquitoborne infections after Hurricane Jeanne, Haiti, 2004. Emerg Infect Dis 13:308–310

Bernkopf, H., Levine, S., and Nerson, R. 1953. Isolation of West Nile virus in Israel. J Infect Dis 93:207–218

Biggerstaff, B. J., and Petersen, L. R. 2003. Estimated risk of transmission of the West Nile virus through blood transfusion in the US, 2002. Transfusion 43:1007–1017

Blackmore, C. G., Stark, L. M., Jeter, W. C., Oliveri, R. L., Brooks, R. G., Conti, L. A., and Wiersma, S. T. 2003. Surveillance results from the first West Nile virus transmission season in Florida, 2001. Am J Trop Med Hyg 69:141–150

Blitvich, B. J., Fernandez-Salas, I., Contreras-Cordero, J. F., Marlenee, N. L., Gonzalez-Rojas, J. I., Komar, N., Gubler, D. J., Calisher, C. H., and Beaty, B. J. 2003. Serologic evidence of West Nile virus infection in horses, Coahuila State, Mexico. Emerg Infect Dis 9:853–856

Bode, A. V., Sejvar, J. J., Pape, W. J., Campbell, G. L., and Marfin, A. A. 2006. West Nile virus disease: A descriptive study of 228 patients hospitalized in a 4-county region of Colorado in 2003. Clin Infect Dis 42:1234–1240

Bosch, I., Herrera, F., Navarro, J. C., Lentino, M., Dupuis, A., Maffei, J., Jones, M., Fernandez, E., Perez, N., Perez-Eman, J., Guimaraes, A. E., Barrera, R., Valero, N., Ruiz, J., Velasquez, G., Martinez, J., Comach, G., Komar, N., Spielman, A., and Kramer, L. 2007. West Nile virus, Venezuela. Emerg Infect Dis 13:651–653

Brault, A. C., Langevin, S. A., Bowen, R. A., Panella, N. A., Biggerstaff, B. J., Miller, B. R., and Komar, N. 2004. Differential virulence of West Nile strains for American crows. Emerg Infect Dis 10:2161–2168

Brault, A. C., Huang, C. Y., Langevin, S. A., Kinney, R. M., Bowen, R. A., Ramey, W. N., Panella, N. A., Holmes, E. C., Powers, A. M., and Miller, B. R. 2007. A single positively selected West Nile viral mutation confers increased virogenesis in American crows. Nat Genet 39:1162–1166

Brown, J. A., Factor, D. L., Tkachenko, N., Templeton, S. M., Crall, N. D., Pape, W. J., Bauer, M. J., Ambruso, D., Dickey, W. C., and Marfin, A. A. 2007. West Nile viremic blood donors and risk factors for subsequent West Nile fever. Vector Borne Zoonotic Dis 7(4):479–488

Brownstein, J. S., Rosen, H., Purdy, D., Miller, J. R., Merlino, M., Mostashari, F., and Fish, D. 2002. Spatial analysis of West Nile virus: Rapid risk assessment of an introduced vector-borne zoonosis. Vector Borne Zoonotic Dis 2:157–164

Buckley, A., Dawson, A., Moss, S. R., Hinsley, S. A., Bellamy, P. E., and Gould, E. A. 2003. Serological evidence of West Nile virus, Usutu virus and Sindbis virus infection of birds in the UK. J Gen Virol 84:2807–2817

Bunning, M. L., Bowen, R. A., Cropp, B., Sullivan, K., Davis, B., Komar, N., Godsey, M., Baker, D., Hettler, D., Holmes, D., and Mitchell, C. J. 2001. Experimental infection of horses with West Nile virus and their potential to infect mosquitoes and serve as amplifying hosts. Ann N Y Acad Sci 951:338–339

Bunning, M. L., Bowen, R. A., Cropp, C. B., Sullivan, K. G., Davis, B. S., Komar, N., Godsey, M. S., Baker, D., Hettler, D. L., Holmes, D. A., Biggerstaff, B. J., and Mitchell, C. J. 2002. Experimental infection of horses with West Nile virus. Emerg Infect Dis 8:380–386

Busch, M. P., Caglioti, S., Robertson, E. F., McAuley, J. D., Tobler, L. H., Kamel, H., Linnen, J. M., Shyamala, V., Tomasulo, P., and Kleinman, S. H. 2005. Screening the blood supply for West Nile virus RNA by nucleic acid amplification testing. N Engl J Med 353:460–467

Cantile, C., Di Guardo, G., Eleni, C., and Arispici, M. 2000. Clinical and neuropathological features of West Nile virus equine encephalomyelitis in Italy. Equine Vet J 32:31–35

Centers for Disease Control and Prevention. 2001. Serosurveys for West Nile virus infection–New York and Connecticut counties, 2000. MMWR Morb Mortal Wkly Rep 50:37–39

Centers for Disease Control and Prevention. 2002a. Intrauterine West Nile virus infection – New York, 2002. MMWR Morb Mortal Wkly Rep 51:1135–1136

Centers for Disease Control and Prevention. 2002b. Laboratory-acquired West Nile virus infections – United States, 2002. MMWR Morb Mortal Wkly Rep 51:1133–1135

Centers for Disease Control and Prevention. 2002c. Possible West Nile virus transmission to an infant through breast-feeding – Michigan, 2002. MMWR Morb Mortal Wkly Rep 51:877–878

Centers for Disease Control and Prevention. 2002d. Provisional surveillance summary of the West Nile virus epidemic – United States, January–November 2002. MMWR Morb Mortal Wkly Rep 51:1129–1133

Centers for Disease Control and Prevention. 2002e. West Nile Virus activity – United States, 2001. MMWR Morb Mortal Wkly Rep 51:497–501

Centers for Disease Control and Prevention. 2003. West Nile virus infection among turkey breeder farm workers – Wisconsin, 2002. MMWR Morb Mortal Wkly Rep 52:1017–1019

Centers for Disease Control and Prevention. 2004. Possible dialysis-related West Nile virus transmission – Georgia, 2003. MMWR Morb Mortal Wkly Rep 53:738–739

Centers for Disease Control and Prevention. 2005. West Nile virus infections in organ transplant recipients – New York and Pennsylvania, August–September, 2005. MMWR Morb Mortal Wkly Rep 54:1021–1023

Centers for Disease Control and Prevention. 2007a. West Nile virus activity – United States, 2006. MMWR Morb Mortal Wkly Rep 56:556–559

Centers for Disease Control and Prevention. 2007b. West Nile virus transmission through blood transfusion – South Dakota, 2006. MMWR Morb Mortal Wkly Rep 56:76–79

Chan-Tack, K. M., and Forrest, G. 2006. West Nile virus meningoencephalitis and acute flaccid paralysis after infliximab treatment. J Rheumatol 33:191–192

Chowers, M. Y., Lang, R., Nassar, F., Ben-David, D., Giladi, M., Rubinshtein, E., Itzhaki, A., Mishal, J., Siegman-Igra, Y., Kitzes, R., Pick, N., Landau, Z., Wolf, D., Bin, H., Mendelson, E., Pitlik, S. D., and Weinberger, M. 2001. Clinical characteristics of the West Nile fever outbreak, Israel, 2000. Emerg Infect Dis 7:675–678

Cruz, L., Cardenas, V. M., Abarca, M., Rodriguez, T., Reyna, R. F., Serpas, M. V., Fontaine, R. E., Beasley, D. W., Da Rosa, A. P., Weaver, S. C., Tesh, R. B., Powers, A. M., and Suarez-Rangel, G. 2005. Short report: Serological evidence of West Nile virus activity in El Salvador. Am J Trop Med Hyg 72:612–615

Cushing, M. M., Brat, D. J., Mosunjac, M. I., Hennigar, R. A., Jernigan, D. B., Lanciotti, R., Petersen, L. R., Goldsmith, C., Rollin, P. E., Shieh, W. J., Guarner, J., and Zaki, S. R. 2004. Fatal West Nile virus encephalitis in a renal transplant recipient. Am J Clin Pathol 121:26–31

Darwish, M. A., and Ibrahim, A. H. 1971. Survey for antibodies to arboviruses in Egyptian sera. I. West Nile virus antihemagglutinins in human and animal sera. J Egypt Public Health Assoc 46:61–70

Deardorff, E., Estrada-Franco, J., Brault, A. C., Navarro-Lopez, R., Campomanes-Cortes, A., Paz-Ramirez, P., Solis-Hernandez, M., Ramey, W. N., Davis, C. T., Beasley, D. W., Tesh, R. B., Barrett, A. D., and Weaver, S. C. 2006. Introductions of West Nile virus strains to Mexico. Emerg Infect Dis 12:314–318

Del Giudice, P., Schuffenecker, I., Vandenbos, F., Counillon, E., and Zeller, H. 2004. Human West Nile virus, France. Emerg Infect Dis 10:1885–1886

Depoortere, E., Kavle, J., Keus, K., Zeller, H., Murri, S., and Legros, D. 2004. Outbreak of West Nile virus causing severe neurological involvement in children, Nuba Mountains, Sudan, 2002. Trop Med Int Health 9:730–736

DeSalvo, D., Roy-Chaudhury, P., Peddi, R., Merchen, T., Konijetti, K., Gupta, M., Boardman, R., Rogers, C., Buell, J., Hanaway, M., Broderick, J., Smith, R., and Woodle, E. S. 2004. West Nile virus encephalitis in organ transplant recipients: Another high-risk group for meningoencephalitis and death. Transplantation 77:466–469

Drebot, M. A., Lindsay, R., Barker, I. K., Buck, P. A., Fearon, M., Hunter, F., Sockett, P., and Artsob, H. 2003. West Nile virus surveillance and diagnostics: A Canadian perspective. Can J Infect Dis 14:105–114

Dupuis, A. P., Marra, P. P., and Kramer, L. D. 2003. Serologic evidence of West Nile virus transmission, Jamaica, West Indies. Emerg Infect Dis 9:860–863

Dupuis, A. P., 2nd, Marra, P. P., Reitsma, R., Jones, M. J., Louie, K. L., and Kramer, L. D. 2005. Serologic evidence for West Nile virus transmission in Puerto Rico and Cuba. Am J Trop Med Hyg 73:474–476

Elizondo-Quiroga, D., Davis, C. T., Fernandez-Salas, I., Escobar-Lopez, R., Velasco Olmos, D., Soto Gastalum, L. C., Aviles Acosta, M., Elizondo-Quiroga, A., Gonzalez-Rojas, J. I., Contreras Cordero, J. F., Guzman, H., Travassos da Rosa, A., Blitvich, B. J., Barrett, A. D., Beaty, B. J., and Tesh, R. B. 2005. West Nile virus isolation in human and mosquitoes, Mexico. Emerg Infect Dis 11:1449–1452

Erdelyi, K., Ursu, K., Ferenczi, E., Szeredi, L., Ratz, F., Skare, J., and Bakonyi, T. 2007. Clinical and pathologic features of lineage 2 West Nile virus infections in birds of prey in Hungary. Vector Borne Zoonotic Dis 7:181–188

Estrada-Franco, J. G., Navarro-Lopez, R., Beasley, D. W., Coffey, L., Carrara, A. S., Travassos da Rosa, A., Clements, T., Wang, E., Ludwig, G. V., Cortes, A. C., Ramirez, P. P., Tesh, R. B., Barrett, A. D., and Weaver, S. C. 2003. West Nile virus in Mexico: Evidence of widespread circulation since July 2002. Emerg Infect Dis 9:1604–1607

Fernandez-Salas, I., Contreras-Cordero, J. F., Blitvich, B. J., Gonzalez-Rojas, J. I., Cavazos-Alvarez, A., Marlenee, N. L., Elizondo-Quiroga, A., Lorono-Pino, M. A., Gubler, D. J., Cropp, B. C., Calisher, C. H., and Beaty, B. J. 2003. Serologic evidence of West Nile virus infection in birds, Tamaulipas State, Mexico. Vector Borne Zoonotic Dis 3:209–213

Fonseca, K., Prince, G. D., Bratvold, J., Fox, J. D., Pybus, M., Preksaitis, J. K., and Tilley, P. 2005. West Nile virus infection and conjunctival exposure. Emerg Infect Dis 11:1648–1649

George, S., Gourie-Devi, M., Rao, J. A., Prasad, S. R., and Pavri, K. M. 1984. Isolation of West Nile virus from the brains of children who had died of encephalitis. Bull World Health Org 62:879–882

Glass, W. G., McDermott, D. H., Lim, J. K., Lekhong, S., Yu, S. F., Frank, W. A., Pape, J., Cheshier, R. C., and Murphy, P. M. 2006. CCR5 deficiency increases risk of symptomatic West Nile virus infection. J Exp Med 203:35–40

Hardinger, K. L., Miller, B., Storch, G. A., Desai, N. M., and Brennan, D. C. 2003. West Nile virus-associated meningoencephalitis in two chronically immunosuppressed renal transplant recipients. Am J Transplant 3:1312–1315

Harrington, T., Kuehnert, M. J., Kamel, H., Lanciotti, R. S., Hand, S., Currier, M., Chamberland, M. E., Petersen, L. R., and Marfin, A. A. 2003. West Nile virus infection transmitted by blood transfusion. Transfusion 43:1018–1022

Hiatt, B., DesJardin, L., Carter, T., Gingrich, R., Thompson, C., and de Magalhaes-Silverman, M. 2003. A fatal case of West Nile virus infection in a bone marrow transplant recipient. Clin Infect Dis 37:e129–e131

Hinckley, A. F., O'Leary, D. R., and Hayes, E. B. 2007. Transmission of West Nile virus through human breast milk seems to be rare. Pediatrics 119:e666–e671

Hong, D. S., Jacobson, K. L., Raad, II, de Lima, M., Anderlini, P., Fuller, G. N., Ippoliti, C., Cool, R. M., Leeds, N. E., Narvios, A., Han, X. Y., Padula, A., Champlin, R. E., and Hosing, C. 2003. West Nile encephalitis in 2 hematopoietic stem cell transplant recipients: Case series and literature review. Clin Infect Dis 37:1044–1049

Hubalek, Z., and Halouzka, J. 1999. West Nile fever – a reemerging mosquito-borne viral disease in Europe. Emerg Infect Dis 5:643–650

Hurlbut, H. S., Rizk, F., Taylor, R. M., and Work, T. H. 1956. A study of the ecology of West Nile virus in Egypt. Am J Trop Med Hyg 5:579–620

Iwamoto, M., Jernigan, D. B., Guasch, A., Trepka, M. J., Blackmore, C. G., Hellinger, W. C., Pham, S. M., Zaki, S., Lanciotti, R. S., Lance-Parker, S. E., DiazGranados, C. A., Winquist, A. G., Perlino, C. A., Wiersma, S., Hillyer, K. L., Goodman, J. L., Marfin, A. A., Chamberland, M. E., and Petersen, L. R. 2003. Transmission of West Nile virus from an organ donor to four transplant recipients. N Engl J Med 348:2196–2203

Johnson, B. W., Kosoy, O., Martin, D. A., Noga, A. J., Russell, B. J., Johnson, A. A., and Petersen, L. R. 2005. West Nile virus infection and serologic response among persons previously vaccinated against yellow fever and Japanese encephalitis viruses. Vector Borne Zoonotic Dis 5:137–145

Jupp, P. G. 2001. The ecology of West Nile virus in South Africa and the occurrence of outbreaks in humans. Ann N Y Acad Sci 951:143–152

Kleinman, S., Glynn, S. A., Busch, M., Todd, D., Powell, L., Pietrelli, L., Nemo, G., Schreiber, G., Bianco, C., and Katz, L. 2005. The 2003 West Nile virus United States epidemic: The America's Blood Centers experience. Transfusion 45:469–479

Kleinschmidt-DeMasters, B. K., Marder, B. A., Levi, M. E., Laird, S. P., McNutt, J. T., Escott, E. J., Everson, G. T., and Tyler, K. L. 2004. Naturally acquired West Nile virus encephalomyelitis in transplant recipients: Clinical, laboratory, diagnostic, and neuropathological features. Arch Neurol 61:1210–1220

Komar, O., Robbins, M. B., Klenk, K., Blitvich, B. J., Marlenee, N. L., Burkhalter, K. L., Gubler, D. J., Gonzalvez, G., Pena, C. J., Peterson, A. T., and Komar, N. 2003. West Nile virus transmission in resident birds, Dominican Republic. Emerg Infect Dis 9:1299–1302

Kumar, D., Drebot, M. A., Wong, S. J., Lim, G., Artsob, H., Buck, P., and Humar, A. 2004a. A seroprevalence study of West Nile virus infection in solid organ transplant recipients. Am J Transplant 4:1883–1888

Kumar, D., Prasad, G. V., Zaltzman, J., Levy, G. A., and Humar, A. 2004b. Community-acquired West Nile virus infection in solid-organ transplant recipients. Transplantation 77:399–402

LaDeau, S. L., Kilpatrick, A. M., and Marra, P. P. 2007. West Nile virus emergence and large-scale declines of North American bird populations. Nature 447:710–713

Lanciotti, R. S., Roehrig, J. T., Deubel, V., Smith, J., Parker, M., Steele, K., Crise, B., Volpe, K. E., Crabtree, M. B., Scherret, J. H., Hall, R. A., MacKenzie, J. S., Cropp, C. B., Panigrahy, B., Ostlund, E., Schmitt, B., Malkinson, M., Banet, C., Weissman, J., Komar, N., Savage, H. M., Stone, W., McNamara, T., and Gubler, D. J. 1999. Origin of the West Nile virus responsible for an outbreak of encephalitis in the northeastern United States. Science 286:2333–2337

Lanciotti, R. S., Ebel, G. D., Deubel, V., Kerst, A. J., Murri, S., Meyer, R., Bowen, M., McKinney, N., Morrill, W. E., Crabtree, M. B., Kramer, L. D., and Roehrig, J. T. 2002. Complete genome sequences and phylogenetic analysis of West Nile virus strains isolated from the United States, Europe, and the Middle East. Virology 298:96–105

Lefrancois, T., Blitvich, B. J., Pradel, J., Molia, S., Vachiery, N., Pallavicini, G., Marlenee, N. L., Zientara, S., Petitclerc, M., and Martinez, D. 2005. West Nile virus surveillance, Guadeloupe, 2003–2004. Emerg Infect Dis 11:1100–1103

Linke, S., Niedrig, M., Kaiser, A., Ellerbrok, H., Muller, K., Muller, T., Conraths, F. J., Muhle, R. U., Schmidt, D., Koppen, U., Bairlein, F., Berthold, P., and Pauli, G. 2007. Serologic evidence of West Nile virus infections in wild birds captured in Germany. Am J Trop Med Hyg 77:358–364

Lorono-Pino, M. A., Blitvich, B. J., Farfan-Ale, J. A., Puerto, F. I., Blanco, J. M., Marlenee, N. L., Rosado-Paredes, E. P., Garcia-Rejon, J. E., Gubler, D. J., Calisher, C. H., and Beaty, B. J. 2003. Serologic evidence of West Nile virus infection in horses, Yucatan State, Mexico. Emerg Infect Dis 9:857–859

Lvov, D. K., Butenko, A. M., Gromashevsky, V. L., Kovtunov, A. I., Prilipov, A. G., Kinney, R., Aristova, V. A., Dzharkenov, A. F., Samokhvalov, E. I., Savage, H. M., Shchelkanov, M. Y., Galkina, I. V., Deryabin, P. G., Gubler, D. J., Kulikova, L. N., Alkhovsky, S. K., Moskvina, T. M., Zlobina, L. V., Sadykova, G. K., Shatalov, A. G., Lvov, D. N., Usachev, V. E., and Voronina, A. G. 2004. West Nile virus and other zoonotic viruses in Russia: Examples of emerging–reemerging situations. Arch Virol Suppl 85–96

Mackenzie, J. S., Gubler, D. J., and Petersen, L. R. 2004. Emerging flaviviruses: The spread and resurgence of Japanese encephalitis, West Nile and dengue viruses. Nat Med 10:S98–109

Malkinson, M., and Banet, C. 2002. The role of birds in the ecology of West Nile virus in Europe and Africa. Curr Top Microbiol Immunol 267:309–322

Malkinson, M., Banet, C., Weisman, Y., Pokamunski, S., King, R., Drouet, M. T., and Deubel, V. 2002. Introduction of West Nile virus in the Middle East by migrating white storks. Emerg Infect Dis 8:392–397

Marfin, A. A., Petersen, L. R., Eidson, M., Miller, J., Hadler, J., Farello, C., Werner, B., Campbell, G. L., Layton, M., Smith, P., Bresnitz, E., Cartter, M., Scaletta, J., Obiri, G., Bunning, M., Craven, R. C., Roehrig, J. T., Julian, K. G., Hinten, S. R., and Gubler, D. J. 2001. Widespread West Nile virus activity, eastern United States, 2000. Emerg Infect Dis 7:730–735

Marlenee, N. L., Lorono-Pino, M. A., Beaty, B. J., and Blitvich, B. J. 2004. Detection of antibodies to West Nile and Saint Louis encephalitis viruses in horses. Salud Publica Mexico 46:373–375

Mattar, S., Edwards, E., Laguado, J., Gonzales, M., Alvarez, J., Komar, N. 2005. West Nile virus antibodies in Colombian horses. Emerg Infect Dis 11:1496–1497

Miramontes, R., Jr., Lafferty, W. E., Lind, B. K., and Oberle, M. W. 2006. Is agricultural activity linked to the incidence of human West Nile virus? Am J Prev Med 30:160–163

Morales, M. A., Barrandeguy, M., Fabbri, C., Garcia, J. B., Vissani, A., Trono, K., Gutierrez, G., Pigretti, S., Menchaca, H., Garrido, N., Taylor, N., Fernandez, F., Levis, S., and Enria, D. 2006. West Nile virus isolation from equines in Argentina, 2006. Emerg Infect Dis 12:1559–1561

Morales-Betoulle, M. E., Morales, H., Blitvich, B. J., Powers, A. M., Davis, E. A., Klein, R., and Cordon-Rosales, C. 2006. West Nile virus in horses, Guatemala. Emerg Infect Dis 12:1038–1039

Mostashari, F., Bunning, M. L., Kitsutani, P. T., Singer, D. A., Nash, D., Cooper, M. J., Katz, N., Liljebjelke, K. A., Biggerstaff, B. J., Fine, A. D., Layton, M. C., Mullin, S. M., Johnson, A. J., Martin, D. A., Hayes, E. B., and Campbell, G. L. 2001. Epidemic West Nile encephalitis, New York, 1999: Results of a household-based seroepidemiological survey. Lancet 358:261–264

Murgue, B., Murri, S., Triki, H., Deubel, V., and Zeller, H. G. 2001a. West Nile in the Mediterranean basin: 1950–2000. Ann N Y Acad Sci 951:117–126

Murgue, B., Murri, S., Zientara, S., Durand, B., Durand, J. P., and Zeller, H. 2001b. West Nile outbreak in horses in southern France, 2000: The return after 35 years. Emerg Infect Dis 7:692–696

Murgue, B., Zeller, H., and Deubel, V. 2002. The ecology and epidemiology of West Nile virus in Africa, Europe and Asia. Curr Top Microbiol Immunol 267:195–221

Murray, K., Baraniuk, S., Resnick, M., Arafat, R., Kilborn, C., Cain, K., Shallenberger, R., York, T. L., Martinez, D., Hellums, J. S., Hellums, D., Malkoff, M., Elgawley, N., McNeely, W., Khuwaja, S. A., and Tesh, R. B. 2006. Risk factors for encephalitis and death from West Nile virus infection. Epidemiol Infect 134:1325–1332

Naficy, K., and Saidi, S. 1970. Serological survey on viral antibodies in Iran. Trop Geogr Med 22:183–188

Nash, D., Mostashari, F., Fine, A., Miller, J., O'Leary, D., Murray, K., Huang, A., Rosenberg, A., Greenberg, A., Sherman, M., Wong, S., and Layton, M. 2001. The outbreak of West Nile virus infection in the New York City area in 1999. N Engl J Med 344:1807–1814

Nir, Y. D. 1959. Airborne West Nile virus infection. Am J Trop Med Hyg 8:537–539

O'Leary, D. R., Marfin, A. A., Montgomery, S. P., Kipp, A. M., Lehman, J. A., Biggerstaff, B. J., Elko, V. L., Collins, P. D., Jones, J. E., and Campbell, G. L. 2004. The epidemic of West Nile virus in the United States, 2002. Vector Borne Zoonotic Dis 4:61–70

O'Leary, D. R., Kuhn, S., Kniss, K. L., Hinckley, A. F., Rasmussen, S. A., Pape, W. J., Kightlinger, L. K., Beecham, B. D., Miller, T. K., Neitzel, D. F., Michaels, S. R., Campbell, G. L., Lanciotti, R. S., and Hayes, E. B. 2006. Birth outcomes following West Nile virus infection of pregnant women in the United States: 2003–2004. Pediatrics 117:e537–e545

Orton, S. L., Stramer, S. L., and Dodd, R. Y. 2006. Self-reported symptoms associated with West Nile virus infection in RNA-positive blood donors. Transfusion 46:272–277

Paisley, J. E., Hinckley, A. F., O', D. R., Kramer, W. C., Lanciotti, R. S., Campbell, G. L., and Hayes, E. B. 2006. West Nile virus infection among pregnant women in a northern Colorado community, 2003 to 2004. Pediatrics 117:814–820

Patnaik, J. L., Harmon, H., and Vogt, R. L. 2006. Follow-up of 2003 human West Nile virus infections, Denver, Colorado. Emerg Infect Dis 12:1129–1131

Paul, S. D., Murthy, D. P., and Das, M. 1970. Isolation of West Nile virus from a human case of febrile illness. Indian J Med Res 58:1177–1179

Pealer, L. N., Marfin, A. A., Petersen, L. R., Lanciotti, R. S., Page, P. L., Stramer, S. L., Stobierski, M. G., Signs, K., Newman, B., Kapoor, H., Goodman, J. L., and Chamberland, M. E. 2003. Transmission of West Nile virus through blood transfusion in the United States in 2002. N Engl J Med 349:1236–1245

Pepperell, C., Rau, N., Krajden, S., Kern, R., Humar, A., Mederski, B., Simor, A., Low, D. E., McGeer, A., Mazzulli, T., Burton, J., Jaigobin, C., Fearon, M., Artsob, H., Drebot, M. A., Halliday, W., and Brunton, J. 2003. West Nile virus infection in 2002: Morbidity and mortality among patients admitted to hospital in southcentral Ontario. Can Med Assoc J 168:1399–1405

Petersen, L. R., and Epstein, J. S. 2005. Problem solved? West Nile virus and transfusion safety. N Engl J Med 353:516–517

Petersen, L. R., and Hayes, E. B. 2004. Westward ho? – The spread of West Nile virus. N Engl J Med 351:2257–2259

Platonov, A. E. 2001. West Nile encephalitis in Russia 1999–2001: Were we ready? Are we ready? Ann N Y Acad Sci 951:102–116

Platonov, A. E., Shipulin, G. A., Shipulina, O. Y., Tyutyunnik, E. N., Frolochkina, T. I., Lanciotti, R. S., Yazyshina, S., Platonova, O. V., Obukhov, I. L., Zhukov, A. N., Vengerov, Y. Y., and Pokrovskii, V. I. 2001. Outbreak of West Nile virus infection, Volgograd Region, Russia, 1999. Emerg Infect Dis 7:128–132

Porter, M. B., Long, M. T., Getman, L. M., Giguere, S., MacKay, R. J., Lester, G. D., Alleman, A. R., Wamsley, H. L., Franklin, R. P., Jacks, S., Buergelt, C. D., and Detrisac, C. J. 2003. West Nile virus encephalomyelitis in horses: 46 cases (2001). J Am Vet Med Assoc 222:1241–1247

Pupo, M., Guzman, M. G., Fernandez, R., Llop, A., Dickinson, F. O., Perez, D., Cruz, R., Gonzalez, T., Estevez, G., Gonzalez, H., Santos, P., Kouri, G., Andonova, M., Lindsay, R., Artsob, H., and Drebot, M. 2006. West Nile virus infection in humans and horses, Cuba. Emerg Infect Dis 12:1022–1024

Quirin, R., Salas, M., Zientara, S., Zeller, H., Labie, J., Murri, S., Lefrancois, T., Petitclerc, M., and Martinez, D. 2004. West Nile virus, Guadeloupe. Emerg Infect Dis 10:706–708

Ravindra, K. V., Freifeld, A. G., Kalil, A. C., Mercer, D. F., Grant, W. J., Botha, J. F., Wrenshall, L. E., and Stevens, R. B. 2004. West Nile virus-associated encephalitis in recipients of renal and pancreas transplants: Case series and literature review. Clin Infect Dis 38:1257–1260

Reddy, P., Davenport, R., Ratanatharathorn, V., Reynolds, C., Silver, S., Ayash, L., Ferrara, J. L., and Uberti, J. P. 2004. West Nile virus encephalitis causing fatal CNS toxicity after hematopoietic stem cell transplantation. Bone Marrow Transplant 33:109–112

Robertson, K. B., Barron, M. A., and Nieto, Y. 2004. West Nile virus infection in bone marrow transplant patients. Bone Marrow Transplant 34:823–824

Ruiz, M. O., Walker, E. D., Foster, E. S., Haramis, L. D., and Kitron, U. D. 2007. Association of West Nile virus illness and urban landscapes in Chicago and Detroit. Int J Health Geogr 6:10

Salazar, P., Traub-Dargatz, J. L., Morley, P. S., Wilmot, D. D., Steffen, D. J., Cunningham, W. E., and Salman, M. D. 2004. Outcome of equids with clinical signs of West Nile virus infection and factors associated with death. J Am Vet Med Assoc 225:267–274

Schellenberg, T. L., Anderson, M. E., Drebot, M. A., Vooght, M. T., Findlater, A. R., Curry, P. S., Campbell, C. A., and Osei, W. D. 2006. Seroprevalence of West Nile virus in Saskatchewan's Five Hills Health Region, 2003. Can J Public Health 97:369–373

Schmidt, J. R., and Elmansoury, H. K. 1963. Natural and experimental infection of Egyptian equines with West Nile virus. Ann Trop Med Parasitol 57:415–427

Schuffenecker, I., Peyrefitte, C. N., el Harrak, M., Murri, S., Leblond, A., and Zeller, H. G. 2005. West Nile virus in Morocco, 2003. Emerg Infect Dis 11:306–309

Sejvar, J. J., Haddad, M. B., Tierney, B. C., Campbell, G. L., Marfin, A. A., Van Gerpen, J. A., Fleischauer, A., Leis, A. A., Stokic, D. S., and Petersen, L. R. 2003. Neurologic manifestations and outcome of West Nile virus infection. J Am Med Assoc 290:511–515

Smith, R. D., Konoplev, S., DeCourten-Myers, G., and Brown, T. 2004. West Nile virus encephalitis with myositis and orchitis. Hum Pathol 35:254–258

Smithburn, K. C., Hughes, T. P., Burke, A. W., and Paul, J. H. 1940. A neurotropic virus isolated from the blood of a native of Uganda. Am J Trop Med Hyg 20:470–492

Solomon, T. 2004. Flavivirus encephalitis. N Engl J Med 351:370–378

Southam, C. M., and Moore, A. E. 1954. Induced virus infections in man by the Egypt isolates of West Nile virus. Am J Trop Med Hyg 3:19–50

Stramer, S. L., Fang, C. T., Foster, G. A., Wagner, A. G., Brodsky, J. P., and Dodd, R. Y. 2005. West Nile virus among blood donors in the United States, 2003 and 2004. N Engl J Med 353:451–459

Thakare, J. P., Rao, T. L., and Padbidri, V. S. 2002. Prevalence of West Nile virus infection in India. Southeast Asian J Trop Med Public Health 33:801–805

Torno, M., Vollmer, M., and Beck, C. K. 2007. West Nile virus infection presenting as acute flaccid paralysis in an HIV-infected patient: A case report and review of the literature. Neurology 68:e5–e7

Triki, H., Murri, S., Le Guenno, B., Bahri, O., Hili, K., Sidhom, M., and Dellagi, K. 2001. West Nile viral meningo-encephalitis in Tunisia. Med Trop (Mars) 61:487–490

Tsai, T. F., Popovici, F., Cernescu, C., Campbell, G. L., and Nedelcu, N. I. 1998. West Nile encephalitis epidemic in southeastern Romania. Lancet 352:767–771

Wadei, H., Alangaden, G. H., Sillix, D. H., El-Amm, J. M., Gruber, S. A., West, M. S., Granger, D. K., Garnick, J., Chandrasekar, P., Migdal, S. D., and Haririan, A. 2004. West Nile virus encephalitis: An emerging disease in renal transplant recipients. Clin Transplant 18:753–758

Ward, M. P., Levy, M., Thacker, H. L., Ash, M., Norman, S. K., Moore, G. E., and Webb, P. W. 2004. Investigation of an outbreak of encephalomyelitis caused by West Nile virus in 136 horses. J Am Vet Med Assoc 225:84–89

Weese, J. S., Baird, J. D., DeLay, J., Kenney, D. G., Staempfli, H. R., Viel, L., Parent, J., Smith-Maxie, L., and Poma, R. 2003. West Nile virus encephalomyelitis in horses in Ontario: 28 cases. Can Vet J 44:469–473

Weinberger, M., Pitlik, S. D., Gandacu, D., Lang, R., Nassar, F., Ben David, D., Rubinstein, E., Izthaki, A., Mishal, J., Kitzes, R., Siegman-Igra, Y., Giladi, M., Pick, N., Mendelson, E., Bin, H., and Shohat, T. 2001. West Nile fever outbreak, Israel, 2000: Epidemiologic aspects. Emerg Infect Dis 7:686–691

2. West Nile Virus: Molecular Epidemiology and Diversity

Gregory D. Ebel and Laura D. Kramer

Abstract

West Nile virus (WNV) is a member of the Japanese encephalitis (JE) antigenic complex in the family *Flaviviridae*, genus *Flavivirus*. Since the mid-1990s the incidence of West Nile neuroinvasive disease has increased and the geographic range of the virus has expanded significantly. This collection of widely geographically distributed genotypes has facilitated research on the molecular epidemiology and genetic variation of the virus, and important population parameters such as the mutation rate and natural selective forces. Novel insights have been gained regarding the evolution of a mosquito-borne zoonotic virus on a regional and intercontinental scale. This chapter reviews recent literature on WNV genetic diversity, and discusses research on the phenotypic impact of this diversity on enzootic transmission of the virus and pathogenesis in vertebrates. Also addressed are analyses of viral fitness and intrahost genetic diversity, as well as population genetic determinants that facilitate perpetuation and adaptation of WNV in new environments. Future challenges, including re-analysis of relationships of JE serogroup viruses, and significance of genetic diversity, are posed for consideration.

Keywords

West Nile, Flavivirus, evolution, selection, population dynamics, genetic diversity, molecular epidemiology

1 Overview of WNV Genetic Diversity at a Global Scale

West Nile virus (WNV) is a member of the Japanese encephalitis virush (JEV) antigenic complex in the family *Flaviviridae*, genus *Flavivirus*. Flaviviruses are found on every continent except Antarctica (Gould et al., 2003), but none is as widely distributed as is WNV.

The virus was first detected in the West Nile district of Uganda, Africa, in 1937 (Smithburn et al., 1940) and since then has been isolated from specimens collected in Europe (Hubalek and Halouzka, 1999), Asia (Zeller and Schuffenecker, 2004) and Australia (MacKenzie et al., 1994). Since the mid-1990s the incidence of West Nile neuroinvasive disease has increased and the geographic range of WNV has expanded, with outbreaks occurring in Russia and Europe (Murgue et al., 2002), and the introduction of the virus into the Americas (Lanciotti et al., 1999; Hayes and Gubler, 2005). As the range of the virus has expanded, so has the interest in genetic and antigenic variation as investigators have sought to understand the relationships between geographically disparate WNV-like viruses, and to determine the pattern(s) and impact(s) of viral evolution on a regional and intercontinental scale.

The genetic and antigenic diversity of WNV at the global scale has been addressed for several decades, with rapid recent increases in the level of analytic detail, sophistication and the number of recognized lineages of WNV. Prior to the availability of extensive genetic data on members of the JE serogroup, cross neutralization studies demonstrated that WNV, Kunjin (KUNV) and Koutango (KOUV) viruses are closely related antigenically (Calisher et al., 1989; DeMadrid and Porterfield, 1974). WNV was separated into three largely geographically distinct groups (Africa, India, and Europe and the Middle East) using polyclonal sera and monoclonal antibodies (Blackburn et al., 1987; Hammam et al., 1965). Phylogenetic analyses based on nucleotide sequencing of an approximately 1 kb fragment of the 3′ terminus of the NS5 coding sequence of many of these viruses confirmed the earlier findings, defining genotypes that agreed with the previously proposed antigenic groupings (Kuno et al., 1998). A study undertaken shortly after the introduction of WNV into North America analyzed complete genomes and included several individual WNV strains alongside prototype viruses (Lanciotti et al., 2002). This work proposed two main WNV lineages. Lineage 1 includes three sublineages: (a) distributed in Africa, the Middle East, Europe and the Americas; (b) found in Australia is also known as KUNV; and (c) strains isolated in India. Lineage 2 is generally confined to sub-Saharan Africa and may ocasionally cause outbreaks of encephalitis (Petersen and Roehrig, 2001). There is at least one report of lineages 1 and 2 strains isolated from the same region, i.e., Hungary, in 2003–2004. In this case, lineage 1 virus was associated with encephalitis in geese (Glavits et al., 2005) and mild encephalitis and meningitis in humans (Ferenczi et al., 2005) in Hungary in 2003, followed one year later by isolation of an encephalic lineage 2 virus from a goshawk (*Accipiter gentilis*) in the same region (Bakonyi et al., 2006).

KOUV has not been included in recent phylogenetic studies, but its relationship to the described lineages 1 and 2, was demonstrated antigenically (Varelas-Wesley and Calisher, 1982; Calisher et al., 1989). Charrel and colleagues (Charrel et al., 2003) revisited the phylogeny of Old World WNV strains, including partial coding sequences of KOUV and determined that this agent is a "distant variant" of WNV, in support of the previous antigenic studies. Data presented by these authors did not attempt to incorporate KOUV into the lineage structure proposed by Lanciotti et al. (Lanciotti et al., 2002), but the collected data from genetic and antigenic studies suggest that it is both an independent genetic lineage and an antigenic subtype. Subsequent publications have proposed additional WNV lineages, including "Rabensburg virus," or WNV lineage 3 (Bakonyi et al., 2005), a Russian strain that has been assigned to lineage 4 (Lvov et al., 2004), and a series of Indian strains that comprises lineage 5 (Bondre et al., 2007). KOUV would then be lineage 6.

Although the multiple WNV lineages proposed on the basis of phylogenetic studies are well supported by nucleotide sequence data, considerable confusion remains regarding the proper classification of these lineages as new virus species or WNV subtypes. Various authors have proposed criteria based on nucleotide distances. For example, Kuno and colleagues (Kuno et al., 1998) proposed that > 84% pairwise sequence identity at the nucleotide level be considered the quantitative genetic criteria for inclusion within a species, while Charrel and colleagues proposed a value of > ~79% for inclusion within the species WNV (Charrel et al., 2003). Applying either of these criteria, KOUV (with ~76% identity) and Rabensburg virus (with ~77% identity) would not be classified as WNV, despite demonstrated but limited serologic cross-reactivities (Bakonyi et al., 2005). In light of the increasing number of proposed lineages, and the increasing importance of the JEV serogroup from a public health perspective, an antigenic reexamination of these agents using high quality standardized immune reagents is clearly warranted.

2 Molecular Epidemiology of WNV in the Americas

2.1 Evidence for a Single Point Introduction

The introduction of WNV into North America in 1999 provided a unique opportunity to observe prospectively the outcome of the introduction of an exotic pathogen into a naïve environment. Although the agent was initially misidentified as St. Louis Encephalitis virus (Asnis et al., 2000), it was rapidly recognized on the basis of phylogenetic analyses and monoclonal antibody binding to be a strain of WNV that was most closely related to a strain that had been isolated in Israel the previous year (Lanciotti et al., 1999). Although the epidemiology and epizootiology of the initial outbreak were consistent with a single recent introduction, it was unclear whether the virus had been introduced in the same year it was initially detected, or whether it had been present at undetectable levels prior to its recognition. The first population study of WNV since its introduction into North America examined envelope (E) coding sequences from eleven WNV strains collected from mosquitoes and birds in New York during the 2000 transmission season (Ebel et al., 2001). This work reported extreme genetic conservation among the strains sampled, with all strains, including a strain collected in 1999, having greater than 99% identity. Some geographic patterns were noted, with a C to U mutation at nucleotide position 1974 occurring in four of five strains collected on Staten Island, suggesting that molecular epidemiologic studies of WNV might be informative in tracking the spread of WNV in North America. From these studies it was concluded that the WNV epidemic/epizootic ongoing in the northeastern US was the result of a single point introduction followed by primary expansion during 1999, overwintering and secondary expansion during 2000 (Ebel et al., 2001). A second population study that included 82 WNV strains confirmed the close genetic relationships of circulating WNV and presented stronger evidence of geographic clustering of strains carrying particular mutations, in this case a C to U mutation at genome position 858. Therefore, early studies of WNV presented data consistent with a single introduction of virus in 1999, and raised the possibility that molecular epidemiologic studies would provide insights into its patterns of perpetuation and dispersal in the Americas.

2.2 Genetic Conservation and Diversification During Colonization

Genetic conservation is typical of most arboviruses (Weaver, 2006; Pisano and Tolou, 1996); however geographic isolation and/or population bottlenecks may lead to diversification, or the virus may adapt to local host populations. The initial studies of WNV collected in the Northeastern US suggested a pattern of conservation, at least during the first two years after introduction (Lanciotti et al., 2002; Ebel et al., 2001). The first study describing WNV sequences collected from a distant point from that of WNV introduction in the Northeastern

United States, included strains collected in Texas (Beasley et al., 2003). Analysis of these strains confirmed the high degree of genetic homogeneity in WNV, reporting a maximum of 0.35% divergence from the NY99 strain. Notably two "sequence subtypes" were detected, one of which differed from NY99 by a U to C substitution at genome position 1442 that resulted in the substitution of an Ala for a Val residue at amino acid position 159 of the envelope protein – a position that had shown some sequence variation among strains collected in Africa and Europe (Lanciotti et al., 2002). It was thus suggested that the amino acid substitution was likely to be a result of genetic drift that occurred during the westward migration of WNV. A further analysis of additional strains collected in Texas established the presence of two clades, one defined by the V to A substitution at E159, and the other defined by five mutations in the PrM-E coding sequences. These clades clustered geographically, with one group of sequences (V to A, E159) clustering in Houston, and the other along a coastal region to the southeast of the main metropolitan area (Davis et al., 2005). WNV isolates in New York also showed evidence of the emergence of strains carrying the mutation resulting in the V to A substitution at E159; this clade was termed "WN02" because it was first noted in New York State during transmission season of the year 2002 (Ebel et al., 2004). The proportion of strains collected in the US that belong to the WN02 genotype increased from 2001 to 2003, suggesting a selective advantage of WN02. WN02 is now the dominant WNV genotype in North America (Davis et al., 2005; Snappin et al., 2007), with the NY99 genotype apparently persisting below the limit of detection of current molecular genetic studies, if at all. Indeed, experimental studies demonstrated that at least two *Culex* species mosquitoes have decreased extrinsic incubation periods for WN02 strains (Ebel et al., 2004; Moudy et al., 2007). Other genetic variants also have been noted. Attenuated strains have been isolated from infected mosquitoes and birds in 2003 in Texas (Davis et al., 2004) and from a phenotypically mixed population isolated from an American crow collected in New York (Jia et al., 2007). Attenuated populations, however, tend to persist in nature for relatively short periods. For example, the attenuated strains detected in Texas in 2003 were not detected in subsequent years (A.D.T. Barrett, personal communication). Thus, during colonization of North America, WNV has remained relatively genetically homogeneous, with the frequent appearance of ephemeral genetic variants, and the rarer emergence of variants (i.e., V159A) that become firmly established.

2.3 Insights into WNV Population Dynamics

Phylogenetic analyses have shown that the most divergent strains of WNV are of African origin (lineage 2), suggesting that the lineage 1 strains probably evolved from lineage 2 ancestors. Since WNV is known to be transported by migratory birds to new locations where competent *Culex* species mosquitoes are available to act as vectors (Malkinson and Banet, 2002), it seems likely that ancestral WNVs dispersed from Africa by this mechanism. WNV in Australia (KUN) and India represent two distinct clades within lineage I, suggesting that after establishment of the virus in a new location, either genetic drift or adaptation to local host populations led to viral genetic change.

The molecular epidemiologic studies that have been conducted since the introduction of WNV into North America, coupled with intensive and extensive WNV surveillance, have facilitated studies examining the population dynamics of WNV as it colonized and adapted to the naïve environment of North America. These studies have shown that the effective population size of WNV (i.e., the effective number of infections, $N_{e\tau}$) increased dramatically in the United States during 2002–2003 (Snappin et al., 2007) (Fig. 1a, b), coincident with a marked spike in the number of human cases (Hayes and Gubler, 2005) (Table 1). This increase in activity is widely attributed

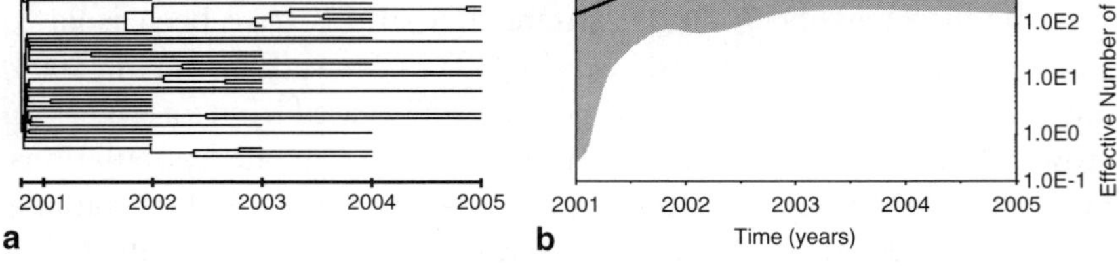

Figure 1. **a** Maximum a posteriori phylogenetic tree of 110 WN02 genotype viruses from samples obtained during the period from 2001 to 2005. For all branches, the times assigned to each tip correspond to the dates of sampling. **b** Bayesian skyline plot for the WN02 genotype. The *bold line* represents the median estimate of the effective number of infections through time, with the 95% HPD values shown in the *shaded area*. The effective number of infections, a measure of relative genetic diversity, is given as $N_e\tau$, where N_e is the effective population size and τ is the generation time. (Snappin et al., 2007, with permission by the American Society for Microbiology.)

Table 1. Human West Nile disease cases 1999–2007[a] United States.

Year	ND[b]	Fever	Unspecified symptoms	Total cases	Deaths
1999	59	3	0	62	7
2000	19	2	0	21	2
2001	64	2	0	66	9
2002	2,946	1,160	50	4,156	284
2003	2,860	6,830	166	9,856	264
2004	1,142	1,269	128	2,539	100
2005	1,294	1,607	99	3,000	119
2006	1,459	2,616	194	4,269	177
2007	1,227	2,350	53	3,630	117
Total	10,973	15,704	696	27,373	1,060

[a]Reported to CDC, as of December 21, 2007 (http://www.cdc.gov/ncidod/dvbid/westnile/)
[b]Neuroinvasive disease including meningitis, encephalitis, acute flaccid paralysis

to the westward expansion of the WNV epizootic to the Rocky Mountains and the midcenter of the US, but increases in activity were also noted during this period in regions such as New York, where WNV activity was already well established (Lukacik et al., 2006). Peak prevalence of WNV in North America was reached in 2003, after which both the WNV population and the number of human cases remained fairly constant. Notably, the rapid population growth of WNV during 2002–2003 was coincident with the V159A mutation becoming fixed in North American WNV populations. The evolutionary rate of WNV has been estimated, and is approximately 5×10^{-4} substitutions per site per year (Snappin et al., 2007; Bertolotti et al., 2007). Several investigators have demonstrated increases in pairwise viral nucleotide diversity over time (Bertolotti et al., 2007; Ebel et al., 2004) suggesting an expanding pool of available WNV genotypes in nature (Fig. 2). Coupled with an increasing rate of anthropogenic environmental change, this pool of WNV genotypes seems likely to insure continued virus perpetuation in North America.

2.4 Sampling Bias and Methodological Issues: Impact on Conclusions

Observational studies are frequently plagued by questions regarding the sampling strategy and criteria for strain inclusion. Specifically, it is difficult to determine whether molecular epidemiologic studies are biased because the investigator has focused sampling on a particular type of host, a specific location, etc. For example,

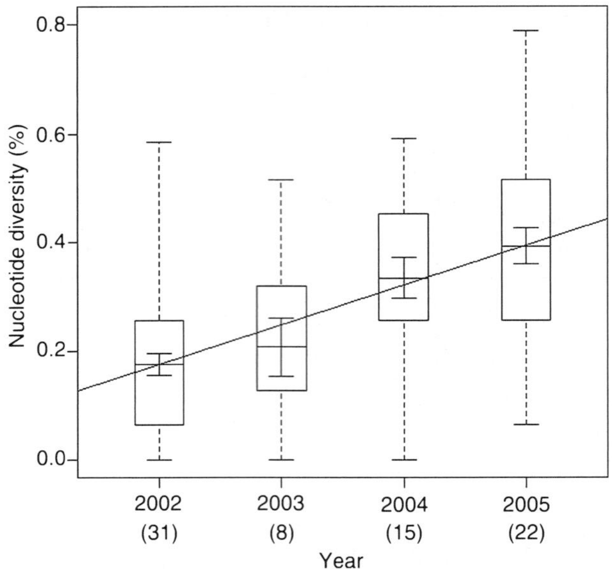

Figure 2. Genetic diversity in WNV from the Midwestern USA, 2002–2005. Genetic diversity is expressed as nucleotide diversity (Nei, 1987), or mean proportion of nucleotide differences among sequences within each year, corrected for multiple substitutions using a model of molecular evolution derived from a hierarchical likelihood ratio test approach implemented using the computer program Modeltest, version 3.7 (Posada and Crandall, 2001). *Lines* and *small error bars* indicate means and standard errors, respectively. Boxes indicate third quartiles, and large error bars indicate ranges. The *solid line* is a least squares regression line indicating the rate of increase of viral genetic diversity over time ($R2 = 0.3626$, $t = 21.69$, $P < 0.0001$). Sample sizes for each year are indicated in parentheses (Bertolotti et al., 2007, with permission from Elsevier Limited).

dead birds serve as excellent sentinels of WNV activity in North America. However, sequences drawn from dead bird samples may represent only the most pathogenic strains in a focus. One early study (Ebel et al., 2001) attempted to address this issue by including samples form both mosquitoes and birds. No signature mutations were observed in this or other studies (Anderson et al., 2001; Beasley et al., 2003) suggesting that both mosquito- and avian-derived sequences represent an essentially random sample of the WNV strains present in a sampling site, thereby increasing confidence that sampling strategy has not significantly biased sequence-based studies of WNV population biology. However, no study has yet included a sufficient number of complete genome sequences from divergent hosts and applied a sufficiently powerful analytic method (i.e., to detect subtle "signatures") to rule out the possibility that bias has been introduced. Importantly, several groups of investigators, using a wide array of WNV sequence sets derived from mosquitoes, birds and humans (Herring et al., 2007) have come to similar conclusions regarding the basic population dynamics of WNV.

2.5 Implications for WNV Pathogenesis

Viral genetics impact pathogenesis. The genetic variation that exists in naturally occurring WNV populations is thus clearly of central importance in determining the disease burden that the virus places on birds, horses and human beings. Several virally-encoded determinants of pathogenesis are variable in natural populations. One well characterized determinant is the glycosylation state of the envelope (E) protein. Most US strains isolated to date are glycosylated at position 154 in the E protein, as are many of the WNV strains responsible for more severe outbreaks of WN disease. Interestingly, this glycosylation site is relatively well conserved among other flaviviruses (Ryman et al., 1997). Studies using reverse genetics have shown that glycosylation of the E protein leads to a more neuroinvasive phenotype in mice than non-glycosylated strains, although viruses lacking E protein glycosylation retain neurovirulence (Beasley et al., 2005; Shirato et al., 2004). E protein glycosylation also appears to play a role in attachment and entry of WNV into host cells, with non-glycosylated viruses attaching to mosquito cells at rates up to 30-fold more efficiently than glycosylated viruses *in vitro*. Overall replication rates of glycosylated and non-glycosylated viruses, however, were similar (Hanna et al., 2005). In seeming contrast to these results, deletion of the glycosylation site in the WNV E protein and the homologous site in tick-borne encephalitis virus resulted in substantially decreased viral particle release from mammalian cells (Goto et al., 2005; Hanna et al., 2005). WNV isolated from *Anopheles maculipennis* mosquitoes in 1971 in Portugal, associated with an equine epizootic, was non-glycosylated (Parreira et al., 2007). Serosurvey indicated a low level of transmission of this non-glycosylated variant, suggesting that E protein glycosylation may be associated with widespread transmission of WNV.

Additional determinants of pathogenesis are also variable in nature. Mutations in the NS2a and prM genes in members of the viral swarm of an American crow isolate led to attenuation in infection of mosquitoes and birds, but the variants reverted to a wild type phenotype after infection of these hosts (Jia et al., 2007). Recently, a single amino acid residue at position 249 of the NS3 helicase was implicated in WNV pathogenesis in American Crows (Brault et al., 2004, 2007), but not house sparrows (Langevin et al., 2005). Nonetheless, this substitution of a proline for other amino acids at position 249 of the NS3 has been observed in many outbreaks throughout the world and appears to be subject to positive selection (Brault et al., 2007). Further, mouse attenuated, small plaque, temperature sensitive WNV variants were detected in Texas in 2003 (Davis et al., 2004). Studies of mouse neuroinvasiveness and neurovirulence indicated that several of these isolates were attenuated in neuroinvasiveness, but all were neurovirulent. Variable nucleotide changes were detected, suggesting that more than one set of nucleotide changes may lead to attenuation. Interestingly, while several WNV strains have been isolated that are decreased in their ability to produce disease in selected *in vivo* laboratory models, none has yet been detected that is significantly more pathogenic than the NY99 strain of WNV. This suggests that this strain represents a maximum in the pathogenic potential for this virus, from which reductions are significantly easier to achieve compared with increases.

3 Ecology and Phylogeny: WNV Adaptation in the Western Hemisphere

3.1 General Considerations

Arboviruses perpetuate in nature by replication in alternating hosts, i.e., blood-feeding arthropods and vertebrates. Their population dynamics are therefore dependent on those of at least two other species. Mosquito populations may be unstable due to temperature and rainfall patterns that vary both within and between seasons. Similarly, bird populations vary in abundance due to several factors, including seasonal migration and roosting behaviors. Generally, however, perpetuation of mosquito-transmitted infections depends on the vectorial capacity (VC) of host mosquito populations. Vectorial capacity is essentially an entomological restatement of the "basic reproductive rate" (R_0) of a pathogen, referring to the number of secondary infections expected to occur from the introduction of a single infection in a naïve population. An equation formalizing VC was described by MacDonald (1961) and later modified by others. One of these, described by Black and Moore (1996), provides a useful platform for rational examination of selective forces that may shape WNV (and other arboviruses). This formula is

$$VC = \frac{ma^2 p^n b}{-In(p)}$$

where VC is vectorial capacity (R_0); m is vector density in relation to the host; a is the probability that a vector feeds on a host in 1 day (i.e., the host preference index × feeding frequency); p is the probability that a vector survives one day; n is the duration of the extrinsic incubation period (EIP) in days; b is vector competence (the proportion of vectors ingesting an infective meal that are later able to transmit the infection; and $1/(-\ln(p)$ is the duration of the vector's life in days after surviving the EIP. This equation demonstrates that the abundance (m) and vector competence (b) of mosquito populations would impact the reproductive rate of WNV linearly and thus relatively weakly. In contrast, host feeding (a), vector longevity (p) and EIP (n) would impact R_0 much more powerfully (e.g., as a square or exponent.) It seems to follow that virus infectivity for mosquitoes, which would be incorporated into VC as b, would be of relatively minor importance relative to viral factors such as the speed of dissemination from the midgut that would impact the duration of the EIP, which would influence VC as n. Thus, natural selection might favor a poorly infectious but rapidly disseminating virus over a highly infectious virus that disseminates slowly. Similar predictions might be made about viral influences on other mosquito-associated factors such as host preference, survivorship, etc.

Arboviruses that exist in temperate environments also must adapt to the seasonal activity of their hosts. WNV, which is generally believed to have originated in Sub-Saharan Africa, readily survives the harsh winters in the northeastern and north central US. This seems to be facilitated by vertically infected, diapausing adult female mosquitoes (Nasci et al., 2001). Experimental studies with flaviviruses demonstrate that the rates of successful vertical transmission are extremely low [reviewed in (Rosen, 1987)]. Although some studies of WNV using intrathoracically inoculated *Culex tarsalis* indicate minimum filial infection rates as high as 6.9 (Goddard et al., 2003), the relevance of these findings to the natural transmission cycle, where mosquitoes are infected perorally, is unclear. While it is conceivable that overwintering following vertical transmission could create a population bottleneck wherein genetic drift might become important, the success of WNV over several years in temperate North America suggests that bottlenecks may in fact be rare and demonstrates the phenotypic robustness of this virus.

3.2 Adaptation to Mosquito Hosts

The general considerations discussed above suggest that, in theory, WNV would be expected to rapidly maximize the efficiency of the mosquito phase of its life cycle. Strong evidence now has been presented that this has occurred. Phenotypic studies were undertaken comparing WNV strains belonging to the NY99 and WN02 genotypes. These studies demonstrated that strains belonging to the WN02 genotype were transmitted more efficiently, and transmission was evident two days earlier than with those belonging to the NY99 genotype (Ebel et al., 2004; Moudy et al., 2007). This reduction in EIP results in increased vectorial capacity of WN02-infected mosquitoes relative to those infected with NY99. In subsequent studies evaluating a range of extrinsic incubation temperatures, the difference in transmission between the two viral genotypes increased with temperature and days since feeding, with a 2% advantage at day 4 and a 3% advantage at day 6 at 20°C, but a 10% advantage at day 4 and a 14% advantage at day 6 at 30°C (Kilpatrick, 2008). This advantage in transmissibility has led to WN02 apparently displacing NY99-like viruses; WN02 is currently dominant and apparently stable throughout North America (Davis et al., 2005). Therefore, the pattern of relative evolutionary stasis that characterized the early years of the WNV outbreak in North America was punctuated by a period of rapid evolution, where a newly emergent genotype (WN02) rapidly spread throughout North America, displaced other WNV clades and achieved peak prevalence (Snappin et al., 2007).

3.3 Adaptation to Avian Hosts

Significant avian mortality has been a notable, but not entirely unique, feature of WNV in North America as avian mortality has been noted elsewhere (Kramer and Bernard, 2001; Ladeau et al., 2007; Bin et al., 2001). It might have been predicted that the virus and host populations would co-evolve toward a less pathogenic interaction in a manner similar to myxomavirus following introduction to Australia (Anderson and May, 1982). In the case of WNV, however, this expectation has several conceptual flaws. First, significant avian mortality is not as universal as is typically stated. In the neotropics, for example, little mortality has been noted, possibly due to cross protection from other *Culex* transmitted flaviviruses circulating in the region such as St. Louis encephalitis, Ilheus, or Rocio viruses. Further, experimental infections of numerous avian species have shown significant intra- and inter-specific variability in response to infection and even more variation between avian families (Komar et al., 2003). Therefore, it seems unlikely that avian mortality acts strongly to shape WNV populations because it is at best an inconsistent phenomenon. Moreover, there is not yet clear evidence of viral adaptation to birds in North America, or that birds have yet developed resistance to WNV. Intriguingly, Brault et al. demonstrated that a single amino acid substitution in the WNV NS3 protein that resulted in increased viremia in the American crow (Brault et al., 2007) was positively selected. This finding suggests that Corvids are of unique and paramount importance in the worldwide emergence of WNV, and would seem to challenge most existing models of the transmission dynamics of arthropod-borne viruses. Comparative infection studies with Old World and New World passeriform birds and WNV strains and with native and tropical passeriform birds with WNV isolated in the US and in tropical Americas will be helpful in clarifying whether adaptation to birds occurs in nature, and more clearly define the role of these hosts in shaping WNV population structure.

4 Evolutionary Mechanisms in West Nile Virus

Molecular epidemiologic and experimental studies on genetically defined WNV strains, described above, have clearly established that the virus has undergone evolutionary change in the eight years since it was introduced into North America. It therefore is valuable to examine in detail the evolutionary mechanisms that have led to this change in order to understand the extraordinary success of WNV in adapting to a new environment. Accordingly, recent studies have investigated mechanisms that may be important in shaping WNV populations. These include various types of natural selection, intrahost population dynamics, host switching and viral fitness. Importantly, and as revealed through some of the studies published to date, it is extremely difficult to isolate these processes from one another because they appear to be related. For example, natural selection and infection of a particular host might simultaneously affect intrahost population dynamics and viral fitness, leading to difficulty in interpreting results. Nonetheless, some consistent results are beginning to emerge from this increasingly complex literature.

Several studies have sought to determine the impact of divergent hosts such as mosquitoes and birds, or host cell types, on WNV. These studies have included field (Jerzak et al., 2005), and *in vitro* (Ciota et al., 2007b, c) and *in vivo* (Jerzak et al., 2007) laboratory passage studies.

They showed that WNV displays high levels of within-host genetic diversity suggestive of quasispecies structure, and that WNV sequences from mosquitoes or mosquito cell culture were more genetically diverse than were sequences from vertebrates or vertebrate cell culture. *In vitro* passage studies showed that WNV can achieve much more significant adaptations to a mosquito cell line, as compared to a vertebrate cell line following sequential passage which bypassed the alternate cell type (Ciota et al., 2007c). These studies also demonstrated that no replicative cost is accrued in other hosts when WNV becomes highly adapted to a single cell type (Ciota et al., 2007a, c). Analysis of patterns of synonymous and nonsynonymous variation of *in vivo* passed WNV revealed strong purifying selection in vertebrate-passed WNV, but not in mosquito-passed WNV. Analysis of field-collected WNV from both hosts also suggests significant purifying selection. Collectively, the results from fitness studies and measures of natural selection suggest that WNV evolutionary dynamics are largely limited by reliance on birds for perpetuation in nature. The strength of purifying selection is set by birds, but infection of mosquitoes appears to provide the genetic diversity upon which natural selection acts.

5 Summary and Future Studies

Publications addressing WNV molecular epidemiology and genetic diversity have increased significantly in recent years. These studies have made several important contributions to the field. First, they have built upon extensive previous literature, demonstrating that WNV is a collection of widely geographically distributed genotypes that comprise some of the most successful arthropod-borne viruses known. This genetic diversity has facilitated molecular epidemiologic studies that have documented intercontinental movement of the virus and increased our understanding of the rate and mechanisms of an intracontinental spread. Second, important population parameters such as the mutation rate and demographic patterns of WNV have been described, and natural selective forces quantified. These studies have established that the evolutionary forces impacting WNV are similar to other arthropod-borne RNA viruses. Third, analysis of nucleotide sequences and the phenotypes of field collected viruses have identified coding sequences that are functionally important in facilitating both enzootic transmission of the virus and pathogenesis in vertebrates. Finally, fitness studies and analysis of intrahost genetic diversity have contributed to our understanding of how WNV interacts with hosts, and helped clarify the population genetic determinants that facilitate perpetuation and adaptation of WNV in new environments.

Several important challenges remain. As mentioned earlier, the recent rapid increase in genetic data on WNV has led to the description of several proposed new WNV lineages. Antigenic studies using standardized immune reagents should be conducted to clarify the relationships of these newly proposed lineages to those that are widely accepted. Moreover, a reanalysis of the relationships among JEV serogroup viruses seems warranted. Second, the most obvious shortcoming of the extant literature on WNV genetic diversity and molecular epidemiology is an overarching reliance on partial genome sequences. While it does not appear that studies of sequence fragments are significantly biased in terms of ultimate conclusions, they may be significantly limited in their power to detect significant phylogenetic clusters, or any but the most obvious sequence motifs that may impact the phenotype. In addition, studies of selection are currently limited to examination of the most commonly sequenced genome region(s), usually the E glycoprotein. Studies of complete genomes should become the norm in the field, rather than the notable exception, with increasing availability of new technology for genome sequencing. Third, it is clear from all currently published molecular epidemiologic studies of WNV that there is abundant genetic diversity both within and between hosts. The significance of this diversity, however, is not at all clear. Nor is it clear how genetic diversity, replicative fitness and transmission phenotype are related. Studies addressing these issues, while rather complicated, are important given the success of WNV in the Americas. Finally, it is not clear whether results obtained from studies of WNV may be generalized to other arboviruses. It therefore seems important to broaden studies of WNV, which have been vigorous and productive, to include other agents that may emerge or re-emerge in the future such as Rift Valley fever and Chikungunya viruses. The complexity of the factors that impact evolution of mosquito-borne viruses needs to be addressed with carefully designed experiments in order to open new pathways to the eventual control of these pathogens.

References

Anderson RM, May RM (1982) Coevolution of hosts and parasites. Parasitology 85(Pt 2): 411–426

Anderson JF, Vossbrinck CR, Andreadis TG, Iton A, Beckwith WH, III, Mayo DR (2001) A phylogenetic approach to following West Nile virus in Connecticut. Proc Natl Acad Sci U S A 98:12885–12889

Asnis DS, Conetta R, Teixeira AA, Waldman G, Sampson BA (2000) The West Nile virus outbreak of 1999 in New York: the Flushing hospital experience. Clin Infect Dis 30:413–418

Bakonyi T, Hubalek Z, Rudolf I, Nowotny N (2005) Novel flavivirus or new lineage of West Nile virus, central Europe. Emerg Infect Dis 11:225–231

Bakonyi T, Ivanics E, Erdelyi K, Ursu K, Ferenczi E, Weissenbock H, Nowotny N (2006) Lineage 1 and 2 strains of encephalitic West Nile virus, central Europe. Emerg Infect Dis 12:618–623

Beasley DW, Davis CT, Guzman H, Vanlandingham DL, Travassos da Rosa APA, Parsons RE, Higgs S, Tesh RB, Barrett ADT (2003) Limited evolution of West Nile virus has occurred during its southwesterly spread in the United States. Virology 309:190–195

Beasley DW, Whiteman MC, Zhang S, Huang CY, Schneider BS, Smith DR, Gromowski GD, Higgs S, Kinney RM, Barrett AD (2005) Envelope protein glycosylation status influences mouse neuroinvasion phenotype of genetic lineage 1 West Nile virus strains. J Virol 79:8339–8347

Bertolotti L, Kitron U, Goldberg TL (2007) Diversity and evolution of West Nile virus in Illinois and The United States, 2002–2005. Virology 360(1):143–149

Bin H, Grossman Z, Pokamunski S, Malkinson M, Weiss L, Duvdevani P, Banet C, Weisman Y, Annis E, Gandaku D, Yahalom V, Hindyieh M, Shulman L, Mendelson E (2001) West Nile fever in Israel 1999–2000: from geese to humans. Ann N Y Acad Sci 951:127–142

Black WCIV, Moore CG (1996) Population biology as a tool for studying vector-borne diseases. In *The Biology of Disease Vectors*, pp. 393–416. Edited by Beaty BJ & Marquardt WC Niwot: University Press of Colorado.

Blackburn NK, Thompson DL, Jupp PG (1987) Antigenic relationship of West Nile strains by titre ratios calculated from cross-neutralization test results. Epidemiol Infect 99:551–557

Bondre VP, Jadi RS, Mishra AC, Yergolkar PN, Arankalle VA (2007) West Nile virus isolates from India: evidence for a distinct genetic lineage. J Gen Virol 88:875–884

Brault AC, Langevin SA, Bowen RA, Panella NA, Biggerstaff BJ, Miller BR, Nicholas K (2004) Differential virulence of West Nile strains for American crows. Emerg Infect Dis 10:2161–2168

Brault AC, Huang CY, Langevin SA, Kinney RM, Bowen RA, Ramey WN, Panella NA, Holmes EC, Powers AM, Miller BR (2007) A single positively selected West Nile viral mutation confers increased virogenesis in American crows. Nat Genet 39:1162–1166

Calisher CH, Karabatsos N, Dalrymple JM, Shope RE, Porterfield JS, Westaway EG, Brandt WE (1989) Antigenic relationships between flaviviruses as determined by cross-neutralization tests with polyclonal antisera. J Gen Virol 70(Pt 1):37–43

Charrel RN, Brault AC, Gallian P, Lemasson JJ, Murgue B, Murri S, Pastorino B, Zeller H, de Chesse R, de Micco P, de Lamballerie, X (2003) Evolutionary relationship between Old World West Nile virus strains. Evidence for viral gene flow between Africa, the Middle East, and Europe. Virology 315:381–388

Ciota AT, Lovelace AO, Jones SA, Payne A, Kramer LD (2007a) Adaptation of two flaviviruses results in differences in genetic heterogeneity and virus adaptability. J Gen Virol 88:2398–2406

Ciota AT, Lovelace AO, Ngo KA, Le AN, Maffei JG, Franke MA, Payne AF, Jones SA, Kauffman EB, Kramer LD (2007b) Cell-specific adaptation of two flaviviruses following serial passage in mosquito cell culture. Virology 357:165–174

Ciota AT, Ngo KA, Lovelace AO, Payne AF, Zhou Y, Shi P-Y, Kramer LD (2007c) Role of the mutant spectrum in adaptation and replication of West Nile virus. J Gen Virol 88:865–874

Davis CT, Beasley DW, Guzman H, Siirin M, Parsons RE, Tesh RB, Barrett AD (2004) Emergence of attenuated West Nile virus variants in Texas, 2003. Virology 330:342–350

Davis CT, Ebel GD, Lanciotti RS, Brault AC, Guzman H, Siirin M, Lambert A, Parsons RE, Beasley DW, Novak RJ, Elizondo-Quiroga D, Green EN, Young DS, Stark LM, Drebot MA, Artsob H, Tesh RB, Kramer LD, Barrett AD (2005) Phylogenetic analysis of North American West Nile virus isolates, 2001–2004: Evidence for the emergence of a dominant genotype. Virology 342:252–265

DeMadrid AT, Porterfield JS (1974) The flaviviruses (group B arboviruses): a cross-neutralization study. J Gen Virol 23:91–96

Ebel GD, Dupuis AP II, Ngo KA, Nicholas DC, Kauffman EB, Jones SA, Young DM, Maffei JG, Shi P-Y, Bernard KA, Kramer LD (2001) Partial genetic characterization of West Nile virus strains, New York State. Emerg Infect Dis 7:650–653

Ebel GD, Carricaburu J, Young D, Bernard KA, Kramer LD (2004) Genetic and phenotypic variation of West Nile virus in New York, 2000–2003. Am J Trop Med Hyg 71:493–500

Ferenczi E, Racz G, Faludi G, Czegledi A, Mezey I, Berencsi G (2005) Natural foci of classical and emerging viral zoonoses. In: Berencsi G, Khan AS, Halouzka J (eds) Emerging Biological Threat. IOS, Amsterdam, pp 43–49

Glavits R, Ferenczi E, Ivanics E, Bakonyi T, Mato T, Zarka P, Palya V (2005) Co-occurrence of West Nile Fever and circovirus infection in a goose flock in Hungary. Avian Pathol 34:408–414

Goddard LB, Roth AE, Reisen WK, Scott TW (2003) Vertical transmission of West Nile virus by three California Culex (Diptera: Culicidae) species. J Med Entomol 40:743–746

Goto A, Yoshii K, Obara M, Ueki T, Mizutani T, Kariwa H, Takashima I (2005) Role of the N-linked glycans of the prM and E envelope proteins in tick-borne encephalitis virus particle secretion. Vaccine 23:3043–3052

Gould EA, Lamballerie X, Zanotto PM, Holmes EC (2003) Origins, evolution, and vector/host coadaptations within the genus *Flavivirus*. Adv Virus Res 59:277–314

Hammam HM, Clarke DH, Price WH (1965) Antigenic variation of West Nile virus in relation to geography. Am J Epidemiol 82:40–55

Hanna SL, Pierson TC, Sanchez MD, Ahmed AA, Murtadha MM, Doms RW (2005) N-linked glycosylation of west nile virus envelope proteins influences particle assembly and infectivity. J Virol 79:13262–13274

Hayes EB, Gubler DJ (2005) West Nile virus: epidemiology and clinical features of an emerging epidemic in the United States. Annu Rev Med 57:181–194

Herring BL, Bernardin F, Caglioti S, Stramer S, Tobler L, Andrews W, Cheng L, Rampersad S, Cameron C, Saldanha J, Busch MP, Delwart E (2007) Phylogenetic analysis of WNV in North American blood donors during the 2003–2004 epidemic seasons. Virology 363(1):220–228

Hubalek Z, Halouzka J (1999) West Nile fever – a reemerging mosquito-borne viral disease in Europe. [Review] [73 refs]. Emerg Infect Dis 5:643–650

Jerzak G, Bernard KA, Kramer LD, Ebel GD (2005) Genetic variation in West Nile virus from naturally infected mosquitoes and birds suggests quasispecies structure and strong purifying selection. J Gen Virol 86:2175–2183

Jerzak GV, Bernard K, Kramer LD, Shi PY, Ebel GD (2007) The West Nile virus mutant spectrum is host-dependant and a determinant of mortality in mice. Virology 360:469–476

Jia Y, Moudy RM, Dupuis AP, Ngo KA, Maffei JG, Jerzak GV, Franke MA, Kauffman EB, Kramer LD (2007) Characterization of a small plaque variant of West Nile virus isolated in New York in 2000. Virology 367:339–347

Kilpatrick AM, Meola MA, Moudy RM, Kramer L (2008) Temperature, viral genetics, and the transmission of West Nile virus by Culex mosquitoes. PLoS Pathog 27;4(6):e1000092

Komar N, Langevin S, Hinten S, Nemeth N, Edwards E, Hettler D, Davis B, Bowen R, Bunning M (2003) Experimental infection of North American birds with the New York 1999 strain of West Nile virus. Emerg Infect Dis 9:311–322

Kramer LD, Bernard KA (2001) West Nile virus infection in birds and mammals. Ann N Y Acad Sci 951:84–93

Kuno G, Chang GJ, Tsuchiya KR, Karabatsos N, Cropp CB (1998) Phylogeny of the genus Flavivirus. J Virol 72:73–83

Ladeau SL, Kilpatrick AM, Marra PP (2007) West Nile virus emergence and large-scale declines of North American bird populations. Nature 447:710–713

Lanciotti RS, Roehrig JT, Deubel V, Smith J, Parker M, Steele K, Crise B, Volpe KE, Crabtree MB, Scherret JH, Hall RA, MacKenzie JS, Cropp CB, Panigrahy B, Ostlund E, Schmitt B, Malkinson M, Banet C, Weissman J, Komar N, Savage HM, Stone W, McNamara T, Gubler DJ (1999) Origin of the West Nile virus responsible for an outbreak of encephalitis in the northeastern United States. Science 286:2333–2337

Lanciotti RS, Ebel GD, Deubel V, Kerst AJ, Murri S, Meyer R, Bowen M, McKinney N, Morrill WE, Crabtree MB, Kramer LD, Roehrig JT (2002) Complete genome sequences and phylogenetic analysis of West Nile virus strains isolated from the United States, europe, and the middle East. Virology 298:96–105

Langevin SA, Brault AC, Panella NA, Bowen RA, Komar N (2005) Variation in virulence of West Nile virus strains for house sparrows (Passer domesticus). Am J Trop Med Hyg 72:99–102

Lukacik G, Anand M, Shusas EJ, Howard JJ, Oliver J, Chen H, Backenson PB, Kauffman EB, Bernard KA, Kramer LD, White DJ (2006) West Nile virus surveillance in mosquitoes in New York State, 2000–2004. J Am Mosq Control Assoc 22:264–271

Lvov DK, Butenko AM, Gromashevsky VL, Kovtunov AI, Prilipov AG, Kinney R, Aristova VA, Dzharkenov AF, Samokhvalov EI, Savage HM, Shchelkanov MY, Galkina IV, Deryabin PG, Gubler DJ, Kulikova LN, Alkhovsky SK, Moskvina TM, Zlobina LV, Sadykova GK, Shatalov AG, Lvov DN, Usachev VE, Voronina AG (2004) West Nile virus and other zoonotic viruses in Russia: examples of emerging-reemerging situations. Arch Virol Suppl 18:85–96

MacKenzie JS, Lindsay MD, Coelen RJ, Broom AK, Hall RA, Smith DW (1994) Arboviruses causing human disease in the Australasian zoogeographic region. Arch Virol 136:447–467

Macdonald G (1961) Epidemiologic models in studies of vector-borne diseases. Public Health Rep 76:753–764

Malkinson M, Banet C (2002) The role of birds in the ecology of West Nile virus in Europe and Africa. Current Topics in Microbiology and Immunology (267), 309–322

Moudy RM, Meola MA, Morin LL, Ebel GD, Kramer LD (2007) A newly emergent genotype of west nile virus is transmitted earlier and more efficiently by culex mosquitoes. Am J Trop Med Hyg 77:365–370

Murgue B, Zeller H, Deubel V (2002) The ecology and epidemiology of West Nile virus in Africa, Europe and Asia. Japanese Encephalitis and West Nile Viruses 267:195–221

Nasci RS, Savage HM, White D, Miller JR, Cropp BC, Godsey MS, Kerst AJ, Bennett P, Gottfried K, Lanciotti RS (2001) West Nile virus in overwintering *Culex* mosquitoes, New York City, 2000. Emerg Infect Dis 7:742–744

Parreira R, Severino P, Freitas F, Piedade J, Almeida AP, Esteves A (2007) Two distinct introductions of the West Nile virus in Portugal disclosed by phylogenetic analysis of genomic sequences. Vector Borne Zoonotic Dis 7:344–352

Petersen LR, Roehrig JT (2001) West Nile virus: a reemerging global pathogen. Emerg Infect Dis 7:611–614

Pisano MR, Tolou H (1996) The topotype notion and the quasispecies concept: The yellow fever virus as example. Travaux Scientifiques des Chercheurs du Service de Sante des Armees 69–70

Rosen L (1987) Overwintering mechanisms of mosquito-borne arboviruses in temperate climates. Am J Trop Med Hyg 37:69S–76S

Ryman KD, Ledger TN, Weir RC, Schlesinger JJ, Barrett AD (1997) Yellow fever virus envelope protein has two discrete type-specific neutralizing epitopes. J Gen Virol 78(Pt 6):1353–1356

Shirato K, Miyoshi H, Goto A, Ako Y, Ueki T, Kariwa H, Takashima I (2004) Viral envelope protein glycosylation is a molecular determinant of the neuroinvasiveness of the New York strain of West Nile virus. J Gen Virol 85:3637–3645

Smithburn KC, Hughes TP, Burke AW, Paul JH 1940 A neurotropic virus isolated from the blood of a native of Uganda. Am J Trop Med Hyg 20:471–473

Snappin KW, Holmes EC, Young DS, Bernard KA, Kramer LD, Ebel GD (2007) Declining Growth Rate of West Nile Virus in North America. J Virol 81:2531–2534

Varelas-Wesley I, Calisher CH (1982) Antigenic relationships of flaviviruses with undetermined arthropod-borne status. Am J Trop Med Hyg 31:1273–1284

Weaver SC (2006) Evolutionary influences in arboviral disease. Curr Top Microbiol Immunol 299:285–314

Zeller HG, Schuffenecker I (2004) West Nile virus: an overview of its spread in Europe and the Mediterranean basin in contrast to its spread in the Americas. Eur J Clin Microbiol Infect Dis 23:147–156

3. Vector Biology and West Nile Virus

Carol D. Blair

Abstract

West Nile virus (WNV) infection of competent mosquito vectors is essential for maintenance and amplification of the virus in natural mosquito–bird cycles. Bridge vectors are necessary to link the enzootic cycle to humans, and a complex set of vectorial, host and environmental factors determine the magnitude and dynamics of the epidemics. Identifying WNV vectors and controlling their numbers and/or reducing their rate of infection are the best current strategies for preventing human disease. Availability of new tools such as mosquito genome sequences presents opportunities for development of novel control strategies.

Keywords

arthropod-borne virus, mosquito, vector, transmission

1 Introduction: Importance of Vector Biology in West Nile Virus Ecology

West Nile virus (WNV) is an *ar*thropod-*bor*ne *virus* (arbovirus); its most important means of transmission between vertebrates is by arthropods, principally mosquitoes. Arboviruses infect and replicate in the cells of not only their vertebrate host, but also their arthropod vector, but whereas the virus infection is capable of causing serious cell damage and acute disease in the vertebrate, the accepted dogma is that it more rarely causes cytopathology or disease in arthropod cells or organisms. Identification of vectors and knowledge of virus–vector interactions is important because reducing vector-borne transmission is the only current option for protecting public health (Reisen and Brault, 2007). The recent incursion of WNV into North America has provided unprecedented opportunities to study the interactions of this virus with novel vectors as well as a naive vertebrate population, and a large number of important research articles on this subject have been published. It is beyond the scope of this chapter to review all these studies, rather it will give a selected overview of current knowledge of vector–virus interactions in North America. Several comprehensive reviews of arbovirus–vector interactions are available for more detailed information (Gubler, 2007; Hayes and Gubler, 2006; Hayes et al., 2005; Kramer and Ebel, 2003; Kuno and Chang, 2005).

2 Transmission of West Nile Virus

2.1 Natural Transmission Cycle

Prior to its appearance in North America, WNV occurred in parts of Africa, Asia, and Europe, and a number of different mosquito species were identified as its vectors in natural cycles involving birds as amplifying hosts. The natural enzootic cycle of WNV in North America involves principally transmission between passerine birds by *Culex* spp. mosquitoes (Hayes et al., 2005; Komar et al., 2003). A female mosquito is estimated to imbibe 0.5–1 μl of blood from a host, so an infectious host that perpetuates the transmission cycle must have a minimum viremia titer of 1,000–2,000 mosquito infectious units per ml, which is usually estimated at $10^{5.5}$–10^{7} PFU ml^{-1}, depending on mosquito species and population (Bowen and Nemeth, 2007; Reisen et al., 2005). North American birds have been shown to develop viremia titers as high as $10^{12.6}$ PFU ml^{-1} (Komar et al., 2003). Higher blood titers in birds are necessary to infect less efficient vectors such as *Cx. pipiens pipiens* (Reisen and Brault, 2007). Conversely, highly competent mosquitoes are capable of maintaining enzootic WNV transmission among less susceptible but widely distributed avian hosts (Reisen et al., 2006a).

Although WNV infects a number of non-avian vertebrates, most mammals, including humans, are tangential or "dead-end" hosts because they do not develop a viremia of sufficient titer to infect biting mosquitoes; however, some rodents, rabbits, and squirrels have been shown experimentally to develop viremia titers capable of infecting biting mosquitoes (Bowen and Nemeth, 2007). In addition, WNV-infected alligators can develop sustained blood titers sufficient to infect mosquitoes and thus could play a role in maintenance and transmission of the virus (Klenk et al., 2004).

Female mosquitoes become infected while seeking a blood meal for egg development. After identifying a suitable host, a female mosquito probes the dermis with her mouthparts until she locates a capillary from which to draw blood. During the probing process, the mosquito deposits saliva, which contains a number of bioactive substances to facilitate blood collection such as vasodilators, coagulation inhibitors, and analgesics, as well as immunomodulators, which may potentiate infection by mosquito-borne agents (Ribeiro et al., 2004; Ribeiro and Francischetti, 2003; Titus et al., 2006).

During her lifetime, a female mosquito may go through several egg development (gonotrophic) cycles, and some species are known to take more than one blood meal per cycle (Kramer and Ebel, 2003). After taking an infectious blood-meal, a mosquito can acquire a persistent, life-long infection and have the potential to transmit WNV to multiple vertebrate hosts during the next and all subsequent times she seeks blood. Multiple blood-meals contribute to increased virus transmission not only due to increased host contact, but also because of enhanced fecundity and longevity of the blood-fed mosquito (Kramer and Ebel, 2003).

2.2 Vector Competence and Vectorial Capacity

Vector competence is defined as the intrinsic ability of an arthropod (mosquito) to be infected by, support replication of, and transmit a virus (Hardy, 1988; Woodring et al., 1996). A number of physiologic/anatomic and genetic as well as environmental factors are involved in determining the competence of a particular mosquito species or population for transmission of WNV. After a mosquito has taken a blood-meal containing an infectious dose of WNV, the virus travels with the blood through the alimentary canal to the posterior midgut, where digestion of the blood is accomplished and virus infection is initiated ((Pennington and Wells, 2004) (Fig. 1). In order to infect the epithelial cells lining the midgut, WNV must presumably bind to an appropriate receptor, introduce its genome into the cytoplasm by fusion of its envelope with a cellular membrane, then carry out translation and RNA replication. Blockage at any of these steps constitutes a midgut infection barrier (MIB). Virion assembly and release, passage through the basal lamina surrounding the midgut, and dissemination into the hemocele, or body cavity, in order to infect secondary tissues are the next steps, and failure at any of these junctures constitutes a midgut escape barrier (MEB). Midgut infection and escape barriers are major determinants of WNV vector competence (Turell et al., 2005), but all the factors that contribute to each have not been precisely defined. Previous studies indicated that mosquito susceptibility to WNV infection is at least partially genetically controlled (Hayes et al., 1984).

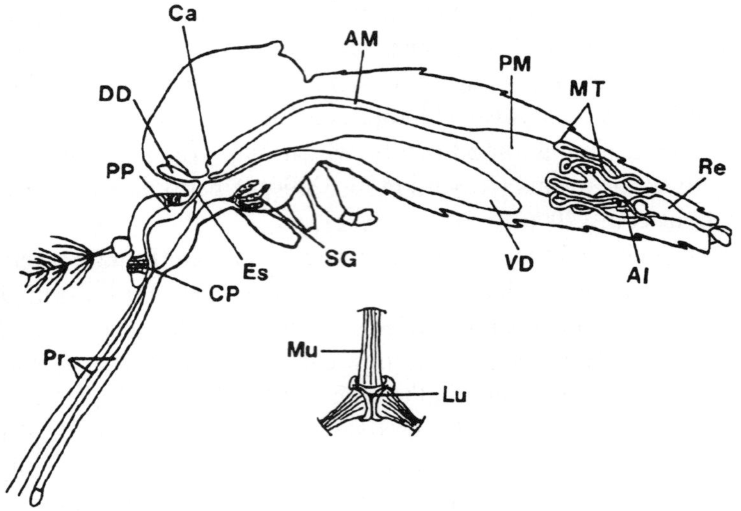

Figure 1. The mosquito alimentary canal. Arboviruses ingested in a blood-meal pass through the proboscis (Pr), esophagus (Es), and anterior midgut (AM); infect the posterior midgut (PM), and disseminate into the body cavity (hemocele) to infect the salivary gland (SG) (Pennington and Wells, 2004).

Virus attachment to and entry of midgut cells must occur within the short time before an acellular layer called the peritrophic matrix (Lehane, 1997) forms around the blood-meal if successful infection is to take place. Appropriate viral receptors on apical surfaces of midgut epithelial cells are required for attachment and entry. Several recent studies have shown that antibodies directed against particular membrane proteins of cultured mosquito cells inhibited flavivirus infection (Mercado-Curiel et al., 2006; Salas-Benito et al., 2007; Yazi Mendoza et al., 2002), including a paper by Chu et al. (Chu et al., 2005) in which antibodies to 95-kD and 70-kD proteins blocked mosquito cell infection by WNV. None of these proteins has been identified and it is not known if the same receptors are used by WNV and other flaviviruses in mosquito midguts as in cultured mosquito cells. Indeed, it is unknown whether WNV uses similar or different receptors and entry mechanisms in the tissues of its mosquito and vertebrate hosts. Similarly, most studies of the cellular components required for permissive replication and assembly of WNV have been conducted in cultured mammalian cells, and mosquito cell molecular factors that support infectious virus production are largely unknown.

After successful dissemination from the midgut, WNV infects and replicates in a number of tissues including the hemocytes, fat body, nervous tissue, and salivary glands. Salivary gland infection and escape barriers might also exist, but have not been characterized

(Turell et al., 2005). After further amplification in salivary gland tissues`, virus is secreted along with salivary proteins into the lumen for transmission during probing for the next blood-meal. The time from imbibing an infectious blood-meal until virus transmission in the saliva is the extrinsic incubation period (EIP), which is influenced by initial virus dose, virus and mosquito genetics, and environmental factors, most notably ambient temperature (Reisen et al., 2006c).

Estimates of WNV dose inoculated by each mosquito that were based on collection of saliva by placing the mouth-parts of individual mosquitoes in capillary tubes were $10^{1.2}$–$10^{4.3}$ PFU (Reisen et al., 2005; Styer et al., 2006; Vanlandingham et al., 2004). However, a recent study in which Styer et al. (2007a) made direct measurements of viral dose inoculated into tissues of chicks and mice as mosquitoes probed and fed demonstrated much higher mean doses of $10^{4.3}$ PFU by *Cx. tarsalis* and $10^{5.9}$ PFU by *Cx. pipiens*. In addition, Styer et al. (2007a) and Reisen et al. (2007b) demonstrated that probing mosquitoes inoculate mean doses of 10^{2}–$10^{4.2}$ PFU directly into the blood, suggesting that apparent "non-viremic" infection of mosquitoes (Higgs et al., 2005) actually could be due to uptake of virus in blood from this initial intravenous inoculum by co-feeding mosquitoes.

Risk of transmission of arbovirus disease is complex and has been modeled as vectorial capacity, which includes a combination of intrinsic (mosquito) factors such as vector competence, mosquito life-span, and host blood-meal preferences and extrinsic (environmental) factors such as larval nutrition, temperature, rainfall, host availability, and host immunity (Kramer and Ebel, 2003). A mathematical equation for vectorial capacity discussed by Black and Moore (1996) includes factors for vector competence, vector density, daily probability that a vector feeds on a host, daily probability of vector survival, duration of EIP, and vector lifespan after surviving EIP.

2.3 Mosquitoes Involved in Transmission of WNV

WNV has been detected in >60 species of mosquitoes in the US (http://www.cdc.gov/ncidod/dvbid/westnile/mosquitoSpecies.htm); however, detection of virus or viral RNA in a mosquito does not necessarily mean that the species is a competent vector, and a number of studies have examined the ability of various North American mosquitoes to transmit WNV under laboratory conditions (Goddard et al., 2002; Sardelis et al., 2001; Turell et al., 2005). In addition to isolation of WNV from a mosquito species in nature and demonstration that the species can be infected and transmit the virus in the laboratory, factors such as host feeding preference and mosquito density in the vertebrate hosts' environment are determinants of the importance of a particular species as a WNV vector (Turell et al., 2005). Also, geographically distinct mosquito populations within a species have been shown to vary widely in their competence to transmit flaviviruses (Bennett et al., 2002; Reisen et al., 2005; Vaidyanathan and Scott, 2007).

Turell and colleagues (2005) showed that several *Culex* species that feed preferentially on birds are efficient WNV vectors. These include *Cx. stigmatosoma, nigripalpus*, *pipiens*, *quincefasciatus*, *restuans,* and *tarsalis*, which are likely to be important in maintenance and amplification of WNV in enzootic cycles. In addition, they observed that *Oclerotatus canadensis* and *O. cantator* and *Aedes vexans* were likely to be competent WNV bridge vectors. Human epidemics necessitate spillover from the avian cycle by bridge vectors that feed on both birds and mammals. Although *Aedes aegypti*, the principal vector of yellow fever and dengue viruses, is a highly competent vector of WNV, its strict preference for human blood-feeding makes it unlikely to serve as a WNV bridge vector; however, the catholic feeding behavior and efficient transmission of WNV by *Ae. albopictus*, an invasive species in North America, make it a more serious candidate as a WNV bridge vector (Turell et al., 2005). Kilpatrick et al. (2005) developed a model for risk-assessment of human infection that incorporated vector abundance, infection prevalence, and biting behavior as well as vector competence. Kilpatrick's model suggested that *Cx. pipiens* and *restuans* in the northeastern and north central US are likely to be the most important bridge vectors from birds to humans.

Two forms of *Cx. pipiens*, *Cx. p. pipiens*, which feeds mainly on birds and *Cx. p. molestus*, which feeds mainly on mammals, are genetically isolated in northern Europe (Fonseca et al., 2004; Spielman, 2001). Fonseca et al. (2004) conducted microsatellite marker analysis of over 600 mosquitoes from the *Cx. pipiens* complex that were collected from populations distributed worldwide. They concluded that *Cx. p. pipiens* and *Cx. p. molestus* were separately introduced into the US and that the mixed mammalian and avian feeding preferences of North American *Cx. pipiens* result from their hybridization after introduction into the Western Hemisphere. Furthermore, they proposed that this mixed feeding behavior of American *Cx. pipiens* partially accounts for the explosive and sustained WNV epidemic in the US as compared to more confined recent outbreaks in Europe. However, the strict division of feeding behavior attributed to European *Cx. pipiens* populations by Fonseca et al. is controversial, and other factors could be responsible for the unique nature of the US epidemic (Spielman et al., 2004). Kilpatrick et al. (2007) also used microsatellite markers to derive genetic evidence of pipiens-molestus hybridization in North America, with increasing molestus ancestry leading to higher preference for mammalian blood. Their study showed that feeding preferences were influenced not only by genetics, but also by the availability of hosts. *Cx. pipiens* exhibited a feeding preference shift from birds to mammals late in the season, probably due to dispersal of birds after nesting and late summer migration of many bird populations. *Cx. quinquefasciatus*, which is also part of the *Cx. pipiens* complex, populates southern North America, is known to hybridize with both *Cx. pipiens* and *Cx. molestus*, and feeds on both mammals and birds (Urbanelli et al., 1997).

Cx. tarsalis, which are widely distributed west of the Mississippi (Bolling et al., 2007; Hayes et al., 2005) are more competent enzootic vectors than *Cx. pipiens* (Turell et al., 2005). *Cx. tarsalis*, like *Cx pipiens,* shift their feeding preference to mammals late in the transmission season (Kilpatrick et al., 2006) and thus are also better candidates for bridge vectors and human transmission than *Culex* spp. that feed exclusively on birds.

2.4 Other Potential Arthropod Vectors

WNV has been detected repeatedly in field-caught argasid (soft) and ixodid (hard) ticks, but, as with mosquitoes, acquisition of virus in a blood-meal does not necessarily indicate competence of ticks for transmission. Several laboratory studies have addressed the ability of various tick species to become infected by blood-feeding, maintain or amplify WNV, and transmit to new vertebrate hosts (Anderson et al., 2003; Hutcheson et al., 2005; Lawrie et al., 2004; Mumcuoglu et al., 2005; Reisen et al., 2007a). In biologically relevant studies involving ornithophilic tick species, Reisen et al. (2007a) concluded that juvenile ixodid ticks (*Ixodes pacificus*) transstadially maintained WNV acquired by feeding on viremic song sparrows but were unable to experimentally transmit infectious virus to naïve sparrows or western fence lizards. Hutcheson et al. (2005) showed that field collected adult seabird soft ticks (*Carios capensis*) could transmit WNV from infected to naïve ducklings after an EIP of 35 days, and concluded that the potential role of these ticks in WNV maintenance merits further investigation.

2.5 Role of Vectors in Overwintering

The North American outbreak of WNV in 1999 occurred in a temperate region where *Culex* spp. mosquitoes enter a dormant condition known as diapause in response to shorter day length and lower temperatures as winter approaches (Eldridge, 2004), and there was hope that the virus would not survive the winter. Most *Culex* spp. mosquitoes respond to day-length cues received in immature developmental stages (late-stage larvae or pupae) and enter diapause shortly after emergence as adults. Ovarian development ceases in females and if a blood-meal is taken before diapause it is used to produce stored fat rather than eggs (Eldridge, 2004). However, mosquitoes could serve as overwintering hosts if diapausing females were infected vertically (Goddard et al., 2003; Miller et al., 2000; Reisen et al., 2002; Turell et al., 2001) or took a pre-diapause infectious blood-meal that did not result in egg development (Eldridge, 2004) and survived the winter. To determine if adult female mosquitoes infected during the initial season of WNV transmission in the US would survive through the winter to sustain transmission in the following season, Nasci et al. (2001) collected >2,000 overwintering mosquitoes from various sites in New York City in January–February 2000, and assayed them for infectious WNV and viral RNA. Three pools of *Culex* pipiens mosquitoes collected from unheated, unoccupied stone structures near the area of the previous summer's outbreak were positive for infectious WNV, for an infection rate of 0.04%, indicating a possibility that virus could persist until spring to reestablish an enzootic transmission cycle. Indeed, WNV survived the winter and during summer of 2000, was transmitted over an expanded range in the northeastern US (http://www.cdc.gov/ncidod/dvbid/westnile/surv&controlCaseCount00_detailed.htm). However, Bolling et al. (2007) used similar techniques to test >9,000 mosquitoes collected in northern Colorado during the winters of 2003–2004 and 2004–2005 and detected no WNV RNA or infectious virus in diapausing *Cx. pipiens* or *Cx. tarsalis*. The roles of vertical transmission and infected diapausing female mosquitoes in maintaining WNV in temperate zones remain to be determined.

Reisen et al. (2006b) postulated three mechanisms that could enable trans-seasonal persistence of WNV in milder climates such as southern California (1) continued enzootic transmission throughout the year; (2) vertical transmission by *Culex* spp. mosquitoes; and (3) chronic infection in birds. Although birds that had died of WNV infection were found throughout the year, WNV RNA was not detected in concurrently trapped, reproductively active female mosquitoes. Vertical transmission to immature and male *Cx. p. quinquefasciatus* mosquitoes collected during summer and persistence of WNV RNA in experimentally infected birds of some species were demonstrated, leading to the conclusion that all three modes of WNV maintenance may operate at low levels.

3 Genetics and Molecular Biology of Virus–Mosquito Interactions

3.1 Influence of Mosquito Genetics on WNV Transmission

Variation in infectivity by flaviviruses among geographic strains (populations) of mosquitoes has suggested that susceptibility is at least partially under genetic control (Aitken et al., 1977; Bennett et al., 2002; Gubler and Rosen, 1976; Hayes et al., 1984). A number of studies aimed at selecting virus-resistant mosquito lines through genetic crosses have suggested that a major determinant of resistance in several species is infection of the midgut; however, the relative infectious titer in the midgut did not correlate with ability to disseminate and be transmitted (Turell et al., 2005), that is, midgut titer was not a determinant of transmission. Miller and Mitchell (1991) found that resistance to several flaviviruses in *Ae. aegypti* was likely to be controlled by multiple genetic loci. Relatively less variation in susceptibility has been demonstrated in *Culex* vectors of WNV than in *Aedes* vectors of other flaviviruses (Hayes et al., 2005); however, studies of WNV–vector interactions have tended to use more genetically homogeneous colonized mosquitoes rather than field-collected strains (Kramer and Ebel, 2003).

Until recently, identification of specific genes involved in mosquito–virus interactions was difficult; however, recent publication of the complete genome sequences of *Anopheles gambiae* (Holt et al., 2002) and *Ae. aegypti* (Nene et al., 2007) and progress in sequencing the *Culex p. quinquefasciatus* genome provide opportunities for identification of a number of genes that might influence WNV infection of and transmission by mosquitoes. Although annotation of mosquito genomes is far from complete, identification of homologies to the better-characterized *Drosophila melanogaster* genome has yielded productive results.

3.2 Key Interaction: Infection of and Dissemination from the Midgut

Earlier genetics studies indicated that flavivirus infection of and transmission by mosquitoes are quantitative variables influenced by multiple genes at more than one locus and subject to extrinsic environmental effects. Recent advances in molecular genetics and statistics have permitted mapping of quantitative trait loci (QTL) for midgut infection and dissemination by flaviviruses, two key steps in transmission by mosquitoes (Black et al., 2002). Bosio et al. (2000) mapped two QTL for midgut infection barriers of dengue-2 virus in *Ae. aegypti*, and Bennett et al. (2005) mapped three QTL for *Ae. aegypti* midgut escape barriers to this virus. Specific genes that condition susceptibility were not identified. Due to a relative paucity of genetic markers in *Culex* spp., mapping studies for WNV susceptibility have not been conducted.

3.3 Transmission by Bite: Mosquito Salivary Proteins

Adult female mosquitoes inject a number of antihemostatic substances in their saliva as they probe the skin of the vertebrate host to facilitate blood collection (Ribeiro and Francischetti, 2003). In addition, there are potent immunomodulators in saliva of the blood-feeding arthropods that may benefit the arthropod by preventing the vertebrate host from becoming sensitized to the saliva. As a consequence of these pharmacological activities, saliva can enhance transmission of arboviruses by arthropods (Gillespie et al., 2000; Titus et al., 2006). Ribeiro and colleagues have characterized the salivary transcriptomes and proteomes of a number of disease vectors, including *Cx. p. quinquefasciatus* (Ribeiro et al., 2004). They isolated RNA from salivary gland homogenates and sequenced transcripts that are translated into full-length proteins, including at least 13 that are secreted in saliva. Among the salivary proteins with potential immunomodulatory activity were a C-type lectin, a gambicin, and several members of the D7 family, which have been shown to sequester bioamines such as histamine and serotonin (Calvo et al., 2006).

Several studies have presented evidence for the effect of mosquito delivery of WNV on pathogenesis in the vertebrate host. Mice inoculated intradermally with WNV in an area where ~10 uninfected *Ae. aegypti* mosquitoes recently had fed had lower survival rates, longer-lasting and higher-titered viremia, and accelerated neuroinvasion (Schneider et al., 2004). Chickens infected with WNV by bite of *Cx. pipiens* exhibited viremia levels and viral shedding up to 1,000 times higher at early stages of infection than those inoculated intradermally by needle. Chickens infected by multiple mosquitoes had viremia titers 25–50 times higher than chickens infected by a single mosquito (Styer et al., 2006). In this study, enhanced early infection after mosquito inoculation could have been due to direct introduction of virus into the bloodstream (Reisen et al., 2007b; Styer et al., 2007a) as well as to potentiation by salivary proteins.

Growing evidence that mosquito salivary proteins enhance infectivity of pathogens has led to proposals that immunization with components of saliva might protect against or modulate infection (Titus et al., 2006).

Machain-Williams (2007) investigated the potential for protection by immunizing mice with proteins concentrated from *Cx. tarsalis* saliva before they were exposed to bites by WNV-infected mosquitoes. In unvaccinated controls, WNV RNA was detected in blood 24 h earlier and in brains 4 days earlier than in immunized mice. At 8 days postchallenge, infectious virus titers in brains were significantly lower in the vaccinated group. Anti-WNV antibodies were detected earlier and at higher titers in the immunized mice and they also had increased levels of the Th1 cytokines IL-2, IFN-γ and TNF-α; in contrast, IL-4 was up-regulated in the control group (Machain-Williams, 2007). Conversely, Schneider et al. (2007) concluded that sensitization to mosquito saliva exacerbates WNV infection. Mice were exposed to 2–4 feedings by uninfected *Ae. aegypti* mosquitoes, then challenged with WNV by the bite of a mosquito. Prior exposure of mice to mosquito feeding resulted in increased mortality following WNV infection. The disease in pre-exposed mice was associated with enhanced early viral replication, increased IL-10 expression, and elevated influx of WNV-susceptible cell types to the inoculation site. Differences in mosquito species, inbred mouse strains, and method of exposure to and dose of mosquito salivary proteins need to be further investigated.

3.4 Potential Barriers to Infection: Mosquito Defense Mechanisms

WNV infection is pathogenic in mammalian cells and organisms, but much less so in mosquitoes. The difference may be attributable to host defense systems for pathogens. Insects rely on innate immune mechanisms to combat pathogens. In *Drosophila*, defenses against cellular pathogens such as bacteria, fungi, and protozoa include induction of antimicrobial peptides, phagocytosis by hemocytes or encapsulation with melanin. Pathogen pattern recognition triggers the Toll pathway in the case of Gram-positive bacteria and fungi and the immune deficiency (Imd) pathway in the case of Gram-negative bacteria to induce immune pathways (Cherry and Silverman, 2006). Although these immune responses may be activated in some virus infections (Mizutani et al., 2003; Sanders et al., 2005; Zambon et al., 2005), they do not appear to have major roles in the defense against arboviruses. Instead, RNA interference seems to have evolved as the primary defense against pathogenic viruses in mosquitoes as in *Drosophila* (Cherry and Silverman, 2006; Wang et al., 2006; Zambon et al., 2006). Because vector competence to transmit arboviruses is at least partially genetically determined, it is possible that variability in genes that encode components of the RNAi pathway may be related to virus susceptibility/refractoriness.

RNAi is an important antiviral response in insects and other invertebrates. It is a post-transcriptional, sequence-specific phenomenon that is triggered by double-stranded RNA and results in degradation of any single-stranded RNA such as mRNA with sequence complementarity to either strand of the dsRNA (Fire et al., 1998; Hammond et al., 2001; Hannon, 2002). The dsRNA that is formed in infected cells as a replicative intermediate of WNV and other arboviruses with positive-sense RNA genomes is likely to be an effective inducer (Weber et al., 2006). In addition, the secondary structure of the viral genome may trigger RNAi, as shown for some plant viruses (Molnar et al., 2005). In *Drosophila*, the protein Dicer-2 recognizes and cleaves long dsRNA to produce 21–25 bp small interfering RNA (siRNA) (Bernstein et al., 2001; Lee et al., 2004). The siRNA duplex is denatured and one "guide" strand is loaded by Dicer in combination with R2D2 protein into a multicomponent RNA-induced silencing complex (RISC) (Elbashir et al., 2001; Liu et al., 2003; Pham and Sontheimer, 2005). The endonuclease component of RISC, Argonaute-2, uses annealing of the siRNA guide strand to its target RNA to identify the sequence destined for cleavage and inactivation (Hammond et al., 2001; Rand et al., 2004). When the target is a virus genome or messenger RNA, viral translation and replication are halted.

Upon publication of their genome sequences, comparison of *An. gambiae* (Christophides et al., 2002) and A*e. aegypti* (Waterhouse et al., 2007) with that of *Drosophila melanogaster* revealed the same immune-related gene families for pathogen recognition, signal modulation and effector systems, although several families were greatly expanded, possibly reflecting the different pathogens encountered by hematophagous insects. Significantly, both *Anopheles* and *Aedes* have homologues of the genes *dcr-2*, *r2d2*, and *ago-2*. It has been demonstrated that induction of RNAi by introduction of exogenous dsRNA can inhibit arbovirus replication in both *An. gambiae* (Keene et al., 2004) and *Ae. aegypti* (Adelman et al., 2001, 2002; Blair et al., 2006; Sanchez-Vargas et al., 2004) (Fig. 2). Importantly, RNAi is a natural anti-arboviral defense mechanism in mosquitoes (Keene et al., 2004; Sánchez-Vargas, unpublished). Although the role of RNAi as a defense against WNV in *Culex* mosquitoes has not been directly demonstrated, *Culex pipiens, tarsalis, tritaeniorhynchus* and *Aedes albopictus* and *triseriatus* have orthologous genes encoding Dcr-2, Ago-2, and other components of the RNAi pathway (Foy and Campbell, unpublished). WNV replication can be modulated in mammalian cells by transfection with virus-specific siRNA early in infection (Geiss et al., 2005).

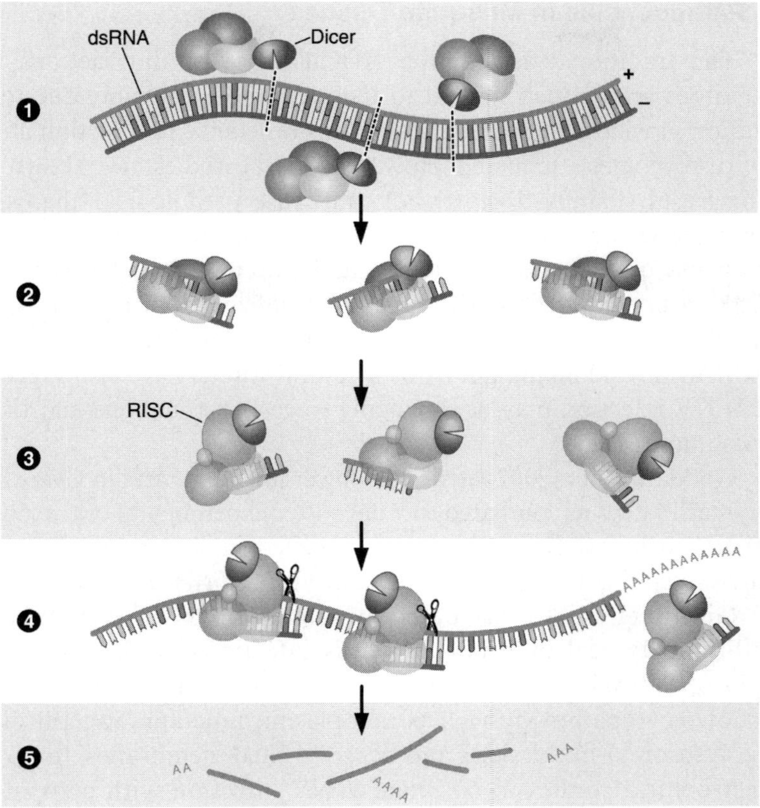

Figure 2. RNA interference in mosquitors. (1) Exogenously-introduced double-stranded (ds) RNA is recognized by the non-sequence specific endonuclease Dicer-2 and cleaved into ~21 bp small interfering (si) RNA. (2) Dicer-2 and R2D2 in combination load the siRNA into the RNA-induced silencing complex (RISC; 3), which retains the "guide" RNA strand. (4) Fully–assembled RISC, including Argonaute-2 as well as Dicer-2 and R2D2, uses the RNA guide sequence to identify and base-pair to complementary sequences on mRNA and Argonaute-2 cleaves the mRNA at the center of the complementary sequence. (5) The degraded mRNA is no longer functional for translation (Blair et al., 2006). (*See Color Plates*)

Apoptosis is used by mammalian cells as an antiviral mechanism (Shen and Shenk, 1995) but this function has been little-explored in insects. Goddard et al. (2002) noted that mosquitoes in a colony of *Cx. p. pipiens* from northern California often were refractory to transmission. They reasoned that a defense mechanism such as apoptosis might eliminate infected midgut cells before virus could replicate and be released into the hemocele. Vaidyanathan and Scott (2006) used acridine orange staining to demonstrate increased apoptosis in infected midguts. Examination by transmission electron microscopy showed ultrastructural changes consistent with apoptosis and, importantly, virions were present in the cytoplasm of apoptotic cells.

3.5 Pathogenesis in Mosquito Tissues

The traditional assumption is that arbovirus interactions with mosquitoes are nondetrimental to the arthropod, probably due to co-evolution in which the mosquito develops resistance to infection and/or the virus becomes attenuated. However, several studies have shown that Alphaviruses (Family *Togaviridae*) can cause pathological changes in midgut tissues or reduced rates of survival and reproduction in their natural mosquito hosts, e.g., (Scott and Lorenz, 1998; Weaver et al., 1992; Weaver et al., 1988). Although WNV infections of mosquitoes are generally considered persistent and nonpathogenic by comparison to those in avian and mammalian hosts, several studies have also suggested that WNV infection may be damaging to certain functions and tissues in mosquitoes.

Goddard et al. (2002) attributed lower infection rates in *Culex* mosquitoes at 14 days as compared to 7 days after experimental oral exposure to WNV to increased death rates in infected insects, particularly in a highly susceptible *Cx. tarsalis* population from northern California.

Girard et al. (2005) noted proliferation of intracytoplasmic membranes typical of other flavivirus infections in electron micrographs of WNV-infected *Cx. p. quinquefasciatus* salivary glands. They also noted cytopathology such as endoplasmic reticulum vacuolization, phagolysosomal-like vesicles and dense striated membranes. In further ultrastructural studies to correlate WNV infection with cytopathology, they observed morphologic features typical of apoptotic cells such as abnormally shaped, electron-dense nuclei with chromatin margination in the salivary glands of mosquitoes that had received an infectious blood-meal 21–28 days previously, although they did not perform specific assays for apoptosis. Ultrastructural changes attributable to apoptosis did not occur in mosquitoes that received blood-meals containing no virus (Girard et al., 2007). Although the mean titer of WNV in saliva did not change significantly, they observed that the percentage of saliva samples containing WNV decreased as cytopathology increased.

To determine reproductive fitness of WNV-infected mosquitoes, Styer et al. (2007b) conducted a life table study of *Cx. tarsalis* females that had fed on WNV-infected or uninfected chicks. They found that fecundity of infected mosquitoes was significantly lower, especially during the first oviposition. WNV-infected mosquitoes produced smaller egg rafts (fewer eggs) that had lower hatch rates. However, there were no significant differences in survival rates and blood-feeding rates were higher in infected mosquitoes; thus infection could serve to enhance vectorial capacity even though it caused reduced reproduction by individual mosquitoes.

These observations raise the question whether the recent introduction of WNV into new vector populations in the US has allowed insufficient time for adaptive evolution to occur, or if observed apoptosis in certain mosquito strains and tissues occurs as a defense mechanism.

3.6 Role of Virus Genetics in Vector Interactions

RNA virus genomes mutate rapidly (Steinhauer and Holland, 1987) and are generally thought to exist in nature as a genetically diverse population of genotypes around a consensus sequence (quasispecies), promoting rapid evolution during adaptation to new environments. Mosquito-borne viruses undergo lower rates of evolution than many other animal RNA viruses that replicate only in vertebrate hosts. One possible reason is potential genetic bottleneck effects that could decrease fitness due to low infectious dose in vertebrate blood and the small volume of virus inoculated by mosquitoes. However, the results published by Weaver et al. (1999) suggested that alternating host transmission cycles constrain the evolutionary rates of arboviruses but not their fitness for either host.

The recent introduction and rapid expansion of WNV across North America have presented unique opportunities to observe its evolution and adaptation to new hosts. Partial nucleotide sequences of virus isolates from restricted areas in the northeastern US in 2000 (Ebel et al., 2001) and southeast Texas in 2002 (Beasley et al., 2003) showed limited variability from the NY99 virus. Jerzak et al. (2005) prepared 20 cDNA clones from each of ten birds (corvids) and 10 mosquito pools (*Culex*) collected in a confined area of Long Island, NY, during the 2003 transmission season. The low proportion of nonsynonymous mutations in both intra- and inter-host populations was evidence of purifying selection; however, the sequences were significantly more diverse in mosquitoes than birds. They suggested that, although the rate of mutation in both hosts may be equal, persistence in mosquitoes as compared to rapid clearance or death in corvids may allow accumulation of more mutations in mosquitoes. Davis et al. (2003) examined partial genome sequences from 22 bird and mosquito isolates made during 2001 and 2002 in five different states of the US. They observed geographic clustering of genetically similar WNV isolates, but the predominance of certain mutations in most isolates suggested the possible emergence in 2002 of a dominant variant in much of the US.

The suggestion of emergence of a new dominant genotype prompted a large consortium of authors to compare partial genome sequences of 74 isolates and complete sequences of 25 isolates from birds and mosquitoes across the US between 2001 and 2004 (Davis et al., 2005). Although genetic variants continued to group temporally and geographically, they confirmed the emergence of a dominant genotype that replaced earlier genotypes. The emergence of this new phylogenetic clade corresponded with increased intensity of transmission and western progression of WNV across the US in 2002 (Davis et al., 2005). Ebel et al. (2004) suggested that the displacement and rapid spread may correlate with increased transmission efficiency in *Culex* spp. mosquitoes. Displacement by the new genotype appears to be due to an EIP 2–4 days shorter in both *Cx. pipiens* and *Cx. tarsalis* after oral infection but not parenteral infection, suggesting more efficient midgut interactions (Moudy et al., 2007). Selection of the new genotype thus could be due to more rapid dissemination, infection of the salivary glands, and transmission by mosquitoes, thereby increasing vectorial capacity.

4 Control of WNV Disease: Interruption of the Transmission Cycle

Strategies for protection of humans from WNV infection have relied on mosquito control. Various public health agencies in the US have employed either proactive integrated pest management programs or reactive emergency responses when human cases are reported. Proactive programs focus on preventing human infection by suppressing mosquito populations below levels where there is risk of spillover transmission from the enzootic cycle in birds, and frequently employ larval control. Reactive control usually relies on preventing additional infections by adult mosquito spraying to interrupt established transmission (Reisen and Brault, 2007). A major problem in reactive programs is that usually WNV has already amplified to epidemic levels and many human cases have occurred before they are initiated. Reisen and Brault (2007) pointed out that comparison of incidence of human infection in states where each approach is used provides clear evidence that proactive programs have been more effective in protecting public health.

Although traditional mosquito control programs have been shown to be effective in controlling WNV transmission, use of pesticides is subject to problems such as development of resistance and concerns about environmental consequences, so novel strategies are being developed as adjuncts. Among these is replacement of vector populations competent to transmit pathogens with pathogen incompetent vectors (Beaty, 2000; Knols et al., 2007; Olson et al., 2002). Requirements of such a strategy are to identify genes that express antiviral molecules in the mosquito and link this gene/these genes to a genetic drive system that will introgress the antiviral genes into natural mosquito populations (James, 2005). The first steps in this control strategy are to identify effector genes that will inhibit virus replication and insert them into the genome so that they will be appropriately expressed in the mosquito. An obvious mechanism for inhibition of arbovirus replication in mosquitoes is induction of RNA interference (RNAi). Franz et al. (2006) demonstrated proof of principle for this strategy by transforming *Ae. aegypti* using a nonautonomous transposable element with an inserted effector gene that expressed an inverted repeat RNA derived from the dengue virus genome. The RNA transcript was expressed under the control of the midgut-specific carboxypeptidase promoter, and formed a dsRNA in mosquito midgut cell cytoplasm. The dsRNA induced anti-dengue virus RNAi, rendering the mosquitoes resistant to virus infection after exposure to dengue virus in an infectious artificial blood-meal. To implement an RNAi-based strategy for control of an agent such as WNV, with multiple vectors and amplifying hosts, would be more complex than for dengue virus, which has a simple natural transmission cycle. Additionally, moving this strategy from the laboratory to the field would necessarily be a staged process to maximize the likelihood of public health benefits while identifying and minimizing potential risks (Blair et al., 2000; James, 2005; Knols et al., 2007).

5 Conclusion

The interactions of WNV with novel populations of mosquito vectors undoubtedly played an important role in the explosive epidemic of this virus that had previously been considered mildly pathogenic and confined to tropical and subtropical regions. The rapid spread of WNV across North America has presented unique opportunities to observe the co-evolution of the virus and its vectors, and many noteworthy studies have been published. However, a number of questions remain to be answered. Availability of new tools such as mosquito genome sequences and development of new technologies for their use will be critical additions to traditional methods of vector control in dealing with WNV.

References

Adelman, Z. N., Blair, C. D., Carlson, J. O., Beaty, B. J., and Olson,K. E. (2001). Sindbis virus-induced silencing of dengue viruses in mosquitoes. Insect Mol Biol *10*,265–273

Adelman,Z. N.,Sanchez-Vargas,I.,Travanty,E. A.,Carlson,J. O.,Beaty,B. J.,Blair,C. D., and Olson,K. E. (2002). RNA silencing of dengue virus type 2 replication in transformed

C6/36 mosquito cells transcribing an inverted-repeat RNA derived from the virus genome. J Virol *76*,12925–12933

Aitken,T. H. G.,Downs,W. G., and Shope,R. E. (1977). *Aedes aegypti* strain fitness for yellow fever virus transmission. Am J Trop Med Hyg *26*,985–989

Anderson,J. F.,Main,A. J.,Andreadis,T. G.,Wikel,S. K., and Vossbrinck,C. R. (2003). Transstadial transfer of West Nile virus by three species of ixodid ticks (Acari: Ixodidae). J Med Entomol *40*,528–533

Beasley,D. W. C.,Davis,C. T.,Guzman,H.,Vanlandingham,D. L.,Travassos da Rosa,A. P. A. ,Parsons,R. E.,Higgs,S.,Tesh,R. B., and Barrett,A. D. T. (2003). Limited evolution of West Nile virus has occurred during its southwesterly spread in the United States. Virology *309*,190–195

Beaty,B. J. (2000). Genetic manipulation of vectors: A potential novel approach for control of vector-borne diseases. Proc Nat Acad Sci *97*,10295–10297

Bennett,K. E.,Olson,K. E.,Munoz,M.,de,L.,Fernandez-Salas,I.,Farfan-Ale,J. A.,Higgs,S., Black,W. C. I., and Beaty,B. J. (2002). Variation in vector competence for dengue 2 virus among 24 collections of *Aedes aegypti* from Mexico and the United States. Am J Trop Med Hyg *67*,85–92

Bennett,K. E.,Flick,D.,Fleming,K. H.,Jochim,R.,Beaty,B. J., and Black,W. C.,IV (2005). Quantitative trait loci that control Dengue-2 virus dissemination in the mosquito *Aedes aegypti*. Genetics *170*,185–194

Bernstein,E.,Caudy,A. A.,Hammond,S. M., and Hannon,G. J. (2001). Role for a bidentate ribonuclease in the initiation step of RNA interference. Nature *409*,363–366

Black,W. C., and Moore,C. G. (1996). Population biology as a tool for studying vector-borne diseases, in The Biology of Disease Vectors,B. J. Beaty, and W. C. Marquardt, eds. (Niwot, CO: University Press of Colorado), pp. 393–416

Black,W. C.,Bennett,K. E.,Gorrochotegui-Escalante,N.,Barillas-Mury,C. V.,Fernandez-Salas,I.,Munoz,M. D. L.,Farfan-Ale,J. A.,Olson,K. E., and Beaty,B. J. (2002). Flavivirus susceptibility in *Aedes aegypti*. Arch Med Res *33*,379–388

Blair,C. D.,Adelman,Z. N., and Olson,K. E. (2000). Molecular strategies for interrupting arthropod-borne virus transmission by mosquitoes. Clin Microbiol Rev *13*,651–661

Blair,C. D.,Sanchez-Vargas,I.,Franz,A. W. E., and Olson,K. E. (2006). Rendering mosquitoes resistant to arboviruses through RNA interference. Microbe *1*,466–470

Bolling,B. G.,Moore,C. G.,Anderson,S. L.,Blair,C. D., and Beaty,B. J. (2007). Entomological studies along the Colorado front range during a period of intense West Nile virus activity. J Am Mosq Control Assoc *23*,37–46

Bosio,C. F.,Fulton,R. E.,Salasek,M. L.,Beaty,B. J., and Black,W. C.,IV (2000). Quantitative trait loci that control vector competence for Dengue-2 virus in the mosquito *Aedes aegypti*. Genetics *156*,687–698

Bowen,R. A., and Nemeth,N. M. (2007). Experimental infections with West Nile virus. Curr Opin Infect Dis *20*,293–297

Calvo,E.,Mans,B. J.,Andersen,J. F., and Ribeiro,J. M. C. (2006). Function and Evolution of a Mosquito Salivary Protein Family. J Biol Chem *281*,1935–1942

Cherry,S., and Silverman,N. (2006). Host–pathogen interactions in *Drosophila*: New tricks from an old friend. Nat Immunol *7*,911–917

Christophides,G. K.,Zdobnov,E.,Barillas-Mury,C.,Birney,E.,Blandin,S.,Blass,C.,Brey,P. T., Collins,F. H.,Danielli,A.,Dimopoulos,G., et al., .(2002). Immunity-related genes and gene families in *Anopheles gambiae*. Science *298*,159–165

Chu,J. J. H.,Leong,P. W. H., and Ng,M. L. (2005). Characterization of plasma membrane-associated proteins from *Aedes albopictus* mosquito (C6/36) cells that mediate West Nile virus binding and infection. Virology *339*,249–260

Davis,C. T.,Beasley,D. W. C.,Guzman,H.,Raj,P.,D'Anton,M.,Novak,R. J.,Unnasch,T. R., Tesh,R. B., and Barrett,A. D. T. (2003). Genetic variation among temporally and geographically distinct West Nile virus isolates, United States, 2001, 2002. Emerg Infect Dis *9*,1423–1429

Davis,C. T.,Ebel,G. D.,Lanciotti,R. S.,Brault,A. C.,Guzman,H.,Siirin,M.,Lambert,A.,Parsons, R. E.,Beasley,D. W. C.,Novak,R., J.et al. (2005). Phylogenetic analysis of North American West Nile virus isolates, 2001–2004: Evidence for the emergence of a dominant genotype. Virology *342*,252–265

Ebel,G. D.,Dupuis Ii,A. P.,Ngo,K.,Nicholas,D.,Kauffman,E.,Jones,S. A.,Young,D.,Maffei,J., Shi,P.-Y., Bernard,K., and Kramer,L. D. (2001). Partial genetic characterization of West Nile virus strains, New York State, 2000. Emerg Infect Dis *7*,650

Ebel,G. D.,Carricaburu,J.,Young,D.,Bernard,K. A., and Kramer,L. D. (2004). Genetic and phenotypic variation of West Nile virus in New York, 2000–2003. Am J Trop Med Hyg *71*,493–500

Elbashir,S. M.,Lendeckel,W., and Tuschl,T. (2001). RNA interference is mediated by 21- and 22-nucleotide RNAs. Genes Dev *15*,188–200

Eldridge,B. F. (2004). Mosquitoes, the Culicidae, in Biology of Disease Vectors,W. C. Marquardt, ed. (Burlington, MA: Elsevier Academic), pp. 95–111

Fire,A.,Xu,S.,Montgomery,M. K.,Kostas,S. A.,Driver,S. E., and Mello,C. C. (1998). Potent and specific genetic interference by double-stranded RNA in *Caenorhabditis elegans*. Nature *391*,806–811

Fonseca,D. M.,Keyghobadi,N.,Malcolm,C. A.,Mehmet,C.,Schaffner,F.,Mogi,M.,Fleischer,R. C., and Wilkerson,R. C. (2004). Emerging vectors in the culex pipiens complex. Science *303*,1535–1538

Franz,A. W. E.,Sanchez-Vargas,I.,Adelman,Z. N.,Blair,C. D.,Beaty,B. J.,James,A. A., and Olson,K. E. (2006). Engineering RNA interference-based resistance to dengue virus type 2 in genetically modified. *Aedes aegypt* PNAS *103*,4198–4203

Geiss,B. J.,Pierson,T. C., and Diamond,M. S. (2005). Actively replicating West Nile virus is resistant to cytoplasmic delivery of siRNA. Virol J *2*,53

Gillespie,R. D.,Mbow,M. L., and Titus,R. G. (2000). The immunomodulatory factors of blood-feeding arthropod saliva. Parasite Immunol *22*,319–331

Girard,Y. A.,Popov,V.,Wen,J.,Han,V., and Higgs,S. (2005). Ultrastructural study of West Nile virus pathogenesis in *Culex pipiens quinquefasciatus* (Diptera: Culicidae). J Med Entomol *42*,429–444

Girard,Y. A.,Schneider,B. S.,McGee,C. E.,Wen,J.,Han,V. C.,Popov,V.,Mason,P. W., and Higgs,S. (2007). Salivary gland morphology and virus transmission during long-term cytopathologic West Nile virus infection in *Culex* mosquitoes. Am J Trop Med Hyg *76*,118–128

Goddard,L. B.,Roth,A. E.,Reisen,W. K., and Scott,T. W. (2002). Vector competence of California mosquitoes for West Nile virus. Emerg Infect Dis *8*,1385–1391

Goddard,L. B.,Roth,A. E.,Reisen,W. K., and Scott,T. W. (2003). Vertical transmission of West Nile Virus by three California Culex (Diptera: Culicidae) species. J Med Entomol *40*,743–746

Gubler,D. (2007). Emerging infections: The continuing spread of West Nile Virus in the Western Hemisphere. Clin Infect Dis *45*,1039–1046

Gubler,D. J., and Rosen,L. (1976). Variation among geographic strains of *Aedes albopictus* in susceptibility to infection with dengue viruses. Am J Trop Med Hyg *25*,318–325

Hammond,S. M.,Boettcher,S.,Caudy,A. A.,Kobayashi,R., and Hannon,G. J. (2001). Argonaute2, a link between genetic and biochemical analyses of RNAi. Science *293*,1146–1150

Hannon,G. J. (2002). RNA interference. Nature *418*,244–251

Hardy,J. L. (1988). Susceptibility and resistance of vector mosquitoes, in The Arboviruses: Epidemiology and Ecology,T. P. Monath, ed. (Boca Raton, FL: CRC), pp. 87–126

Hayes,E. B., and Gubler,D. J. (2006). West Nile virus: Epidemiology and clinical features of an emerging epidemic in the United States. Annu Rev Med *57*,181–194

Hayes,C. G.,Baker,R. H.,Baqar,S., and Ahmed,T. (1984). Genetic variation for West Nile virus susceptibility in *Culex tritaeniorhynchus*. Am J Trop Med Hyg *33*,715–724

Hayes,E. B.,Komar,N.,Nasci,R. S.,Montgomery,S. P.,O'Leary,D. R., and Campbell,G. L. (2005). Epidemiology and transmission dynamics of West Nile virus disease. Emerg Infect Dis *11*,1167–1173

Higgs,S.,Schneider,B. S.,Vanlandingham,D. L.,Klingler,K. A., and Gould,E. A. (2005). Nonviremic transmission of West Nile virus. Proc Natl Acad Sci USA *102*,8871–8874

Holt,R. A.,Subramanian,G. M.,Halpern,A.,Sutton,G. G.,Charlab,R.,Nusskern,D. R., Wincker,P.,Clark,A. G.,Ribeiro,J. M. C.,Wides,R.,et-al. (2002). The genome sequence of the malaria mosquito *Anopheles gambiae*. Science *298*,129–149

Hutcheson,H. J.,Gorham,C. H.,Machain-Williams,C.,Lorono-Pino,M. A.,James,A. M., Marlenee,N. L.,Winn,B.,Beaty,B. J., and Blair,C. D. (2005). Experimental transmission of West Nile virus (Flaviviridae: Flavivirus) by *Carios capensis* ticks from North America. Vector Borne Zoonotic Dis *5*,293–295

James,A. A. (2005). Gene drive systems in mosquitoes: Rules of the road. Trends Parasitol *21*,64–67

Jerzak,G.,Bernard,K. A.,Kramer,L. D., and Ebel,G. D. (2005). Genetic variation in West Nile virus from naturally infected mosquitoes and birds suggests quasispecies structure and strong purifying selection. J Gen Virol *86*,2175–2183

Keene,K. M.,Foy,B. D.,Sanchez-Vargas,I.,Beaty,B. J.,Blair,C. D., and Olson,K. E. (2004). RNA interference acts as a natural antiviral response to O'nyong-nyong virus (Alphavirus; Togaviridae) infection of *Anopheles gambiae*. Proc Natl Acad Sci USA *101*,17240–17245

Kilpatrick,A. M.,Kramer,L. D.,Campbell,S. R.,Alleyne,E. O.,Dobson,A. P., and Daszak,P. (2005). West Nile virus risk assessment and the bridge vector paradigm. Emerg Infect Dis *11*,425–429

Kilpatrick,A. M.,Kramer,L. D.,Jones,M. J.,Marra,P. P., and Daszak,P. (2006). West Nile virus epidemics in North America are driven by shifts in mosquito feeding behavior. PLoS Biol *4*,e82

Kilpatrick,A. M.,Kramer,L. D.,Jones,M. J.,Marra,P. P.,Daszak,P., and Fonseca,D. M. (2007). Genetic influences on mosquito feeding behavior and the emergence of zoonotic pathogens. Am J Trop Med Hyg *77*,667–671

Klenk,K.,Snow,J.,Morgan,K.,Bowen,R.,Stephens,M.,Foster,F.,Gordy,P.,Beckett,S., Komar,N.,Gubler,D., and Bunning,M. (2004). Alligators as West Nile virus amplifiers. Emerg Infect Dis *10*,2150–2155

Knols,B. G. J.,Bossin,H. C.,Mukabana,W. R., and Robinson,A. S. (2007). Transgenic mosquitoes and the fight against malaria: Managing technology push in a turbulent GMO world. Am J Trop Med Hyg *77*,232–242

Komar,N.,Langevin,S.,Hinten,S.,Nemeth,N.,Edwards,E.,Hettler,D.,Davis,B.,Bowen,R., and Bunning,M. (2003). Experimental infection of North American birds with the New York 1999 strain of West Nile virus. Emerg Infect Dis *9*,311–322

Kramer,L. D., and Ebel,G. D. (2003). Dynamics of flavivirus infection in mosquitoes. Adv Virus Res *60*,187–232

Kuno,G., and Chang,G.-J. J. (2005). Biological transmission of arboviruses: Reexamination of and new insights into components, mechanisms, and unique traits as well as their evolutionary trends. Clin Microbiol Rev *18*,608–637

Lawrie,C. H.,Uzcategui,N. Y.,Gould,E. A., and Nuttall,P. A. (2004). Ixodid and argasid tick species and West Nile virus. Emerg Infect Dis *10*,653–657

Lee,Y. S.,Nakahara,K.,Pham,J. W.,Kim,K.,He,Z.,Sontheimer,E. J., and Carthew,R. W. (2004). Distinct roles for Drosophila Dicer-1 and Dicer-2 in the siRNA/miRNA silencing pathways. Cell *117*,69–81

Lehane,M. J. (1997). Peritrophic matrix structure and function. Annu Rev Entomol *42*,525–550

Liu,Q.,Rand,T. A.,Kalidas,S.,Du,F.,Kim,H.-E.,Smith,D. P., and Wang,X. (2003). R2D2, a bridge between the initiation and effector steps of the Drosophila RNAi pathway. Science *301*,1921–1925

Machain-Williams,C. I. (2007) The role of mosquito salivary proteins in the pathogenesis of flaviviruses (Flavivviridae: Flavivirus),PhD (Fort Collins,CO: Colorado State University)

Mercado-Curiel,R.,Esquinca-Aviles,H.,Tovar,R.,Diaz-Badillo,A.,Camacho-Nuez,M., and Munoz,M. (2006). The four serotypes of dengue recognize the same putative receptors in *Aedes aegypti* midgut and *Ae. albopictus* cells. BMC Microbiol *6*,85

Miller,B. R., and Mitchell,C. J. (1991). Genetic selection of a flavivirus-refractory strain of the yellow fever mosquito *Aedes aegypti*. Am J Trop Med Hyg *45*,399–407

Miller,B. R.,Nasci,R. S.,Godsey,M. S.,Savage,H. M.,Lutwama,J. J.,Lanciotti,R. S., and Peters,C. J. (2000). First field evidence for natural vertical transmission of West Nile virus in *Culex univittatus* complex mosquitoes from Rift Valley province, Kenya. Am J Trop Med Hyg *62*,240–246

Mizutani,T.,Kobayashi,M.,Eshita,V.,Shirato,K.,Kimura,T.,Ako,V.,Miyoshi,H.,Takasak,T., Kurane,I., Kadwa,H., et al. (2003). Involvement of the JNK-like protein of the Aedes albopictus mosquito cell line, C6/36, in phagocytosis, endocytosis and infection of West Nile virus. Insect Mol Biol *12*,491

Molnar,A., Csorba,T., Lakatos,L.,Varallyay,E., Lacomme,C., and Burgyan,J. (2005). Plant virus-derived small interfering RNAs originate predominantly from highly structured single-stranded viral RNAs. J Virol *79*,7812–7818

Moudy,R. M.,Meola,M. A.,Morin,L.-L. L.,Ebel,G. D., and Kramer,L. D. (2007). A newly emergent genotype of West Nile virus is transmitted earlier and more efficiently by culex mosquitoes. Am J Trop Med Hyg *77*,365–370

Mumcuoglu,K. Y., Banet-Noach,C., Malkinson,M.,Shalom,U., and Galun,R. (2005). Argasid ticks as possible vectors of West Nile virus in Israel. Vector Borne Zoonotic Dis *5*,65–71

Nasci,R. S.,Savage,H. M.,White,D. J.,Miller,J. R.,Cropp,B. C.,Godsey,M. S.,Kerst,A. J.,Bennett,P.,Gottfried,K., and Lanciotti,R. S. (2001). West Nile virus in overwintering culex mosquitoes, New York City, 2000. Emerg Infect Dis *7*,742

Nene,V.,Wortman,J. R.,Lawson,D.,Haas,B.,Kodira,C.,Tu,Z.,Loftus,B.,Xi,Z.,Megy,K.,Grabhe rr,M., et al. (2007). Genome sequence of *Aedes aegypti*, a major arbovirus vector. Science *316*,1718–1723

Olson,K. E.,Adelman,Z. N.,Travanty,E. A.,Sanchez-Vargas,I.,Beaty,B. J., and Blair,C. D. (2002). Developing arbovirus resistance in mosquitoes. Insect Biochem Mol Biol *32*,1333–1343

Pennington,J. E., and Wells,M. A. (2004). The adult midgut, structure and function, in Biology of Disease Vectors,W. C. Marquardt, ed. (Burlington, MA: Elsevier Academic), pp. 289–295

Pham,J. W., and Sontheimer,E. J. (2005). Molecular requirements for RNA-induced silencing complex assembly in the Drosophila RNA interference pathway. J Biol Chem *280*,39278–39283

Rand,T. A., Ginalski,K., Grishin,N. V., and Wang,X. (2004). Biochemical identification of Argonaute 2 as the sole protein required for RNA-induced silencing complex activity. Proc Natl Acad Sci USA *101*,14385–14389

Reisen,W., and Brault,A. C. (2007). West Nile virus in North America: Perspectives on epidemiology and intervention. Pest Manage Sci *63*,641–646

Reisen,W. K.,Kramer,L. D.,Chiles,R. E.,Wolfe,T. M., and Green,E.-G. N. (2002). Simulated overwintering of encephalitis viruses in diapausing female culex tarsalis (Diptera: Culicidae). J Med Entomol *39*,226233

Reisen,W. K.,Fang,Y., and Martinez,V. M. (2005). Avian host and mosquito (Diptera: Culicidae) vector competence determine the efficiency of West Nile and St. Louis encephalitis virus transmission. J Med Entomol *42*,367–375

Reisen,W. K., Barker,C. M., Carney,R., Lothrop,H. D.,Wheeler,S. S.,Wilson,J. L.,Madon,M. B., Takahashi,R.,Carroll,B.,Garcia,S., et al. (2006a). Role of corvids in epidemiology of West Nile virus in Southern California. J Med Entomol *43*,356–367

Reisen,W. K., Fang,Y., Lothrop,H. D., Martinez,V. M., Wilson,J.,O'Connor,P.,Carney,R., Cahoon-Young,B.,Shafii,M., and Brault,A. C. (2006b). Overwintering of West Nile virus in Southern California. J Med Entomol *43*,344–355

Reisen,W. K., Fang,Y., and Martinez,V. M. (2006c). Effects of temperature on the transmission of West Nile virus by Culex tarsalis (Diptera: Culicidae). J Med Entomol *43*,309–317

Reisen,W. K., Brault,A. C., Martinez,V. M., Fang,Y., Simmons,K., Garcia,S., Omi-Olsen,E., and Lane,R. S. (2007a). Ability of transstadially infected Ixodes pacificus (Acari: Ixodidae) to transmit West Nile virus to Song Sparrows or Western Fence Lizards. J Med Entomol *44*,320–327

Reisen,W. K., Fang,Y., and Martinez,V. (2007b). Is nonviremic transmission of West Nile virus by culex mosquitoes (Diptera: Culicidae) Nonviremic? J Med Entomol *44*,299–302

Ribeiro,J. M. C., and Francischetti,I. M. B. (2003). Role of arthropod saliva in blood feeding: Sialome and post-Sialome perspectives. Annu Rev Entomol *48*,73–88

Ribeiro,J.M.C., Charlab,R., Pham,V.M., Garfield,M.K., and Valenzuela,J.G. (2004). An insight into the salivary transcriptome and proteome of the adult female mosquito *Culex pipiens quinquefasciatus*. Insect Biochem Mol Biol *34*,543–563

Salas-Benito,J.,Valle,J. R.-D.,Salas-Benito,M., Ceballos-Olvera,I., Mosso,C., and del Angel,R. M. (2007). Evidence that the 45-kD glycoprotein, part of a putative dengue virus receptor complex in the mosquito cell line C6/36, is a heat-shock related protein. Am J Trop Med Hyg *77*,283–290

Sanchez-Vargas,I.,Travanty,E.A.,Keene,K.M.,Franz,A.W.,Beaty,B.J.,Blair,C.D., and Olson,K. E. (2004). RNA interference, arthropod-borne viruses, and mosquitoes. Virus Res *102*,65–74

Sanders,H. R.,Foy,B. D.,Evans,A. M.,Ross,L. S.,Beaty,B. J.,Olson,K. E., and Gill,S. S. (2005). Sindbis virus induces transport processes and alters expression of innate immunity pathway genes in the midgut of the disease vector, Aedes aegypti. Insect Biochem Mol Biol *35*,1293–1307

Sardelis,M. R.,Turell,M. J.,Dohm,D. J., and O'Guinn,M. L. (2001). Vector competence of selected North American culex and Coquillettidia mosquitoes for West Nile virus. Emerg Infect Dis *7*,1018–1022

Schneider,B. S.,Soong,L.,Zeidner,N. S., and Higgs,S. (2004). Aedes aegypti salivary gland extracts modulate anti-viral and TH1/TH2 cytokine responses to Sindbis virus infection. Viral Immunol *17*,565–573

Schneider,B. S.,McGee,C. E.,Jordan,J. M.,Stevenson,H. L.,Soong,L., and Higgs,S. (2007). Prior exposure to uninfected mosquitoes enhances mortality in naturally-transmitted West Nile virus infection. PLoS ONE *2*,e1171

Scott,T. W., and Lorenz,L. H. (1998). Reduction of Culiseta melanura fitness by eastern equine encephalomyelitis virus. Am J Trop Med Hyg *59*,341–346

Shen,Y., and Shenk,T. E. (1995). Viruses and apoptosis. Curr Opin Genet Dev *5*,105–111

Spielman,A. (2001). Structure and seasonality of nearctic Culex pipiens populations. Ann N Y Acad Sci *951*,220–234

Spielman,A.,Andreadis,T. G.,Apperson,C. S.,Cornel,A. J.,Day,J. F.,Edman,J. D., Fish,D., Harrington,L. C., Kiszewski,A. E.,Lampman,R., et al. (2004). Outbreak of West Nile virus in North America. Science *306*,1473e–1475e

Steinhauer,D. A., and Holland,J. J. (1987). Rapid evolution of RNA viruses. Annu Rev Microbiol *41*,409–431

Styer,L. M., Bernard,K. A., and Kramer,L. D. (2006). Enhanced early West Nile virus infection in young chickens infected by mosquito bite: Effect of viral dose. Am J Trop Med Hyg *75*,337–345

Styer,L. M., Kent,K.A., Albright,R.G., Bennett,C.J., Kramer,L.D., and Bernard,K.A. (2007a). Mosquitoes inoculate high doses of West Nile virus as they probe and feed on live hosts. PLoS Pathog *3*,e132

Styer,L.M., Meola,M.A., and Kramer,L.D. (2007b). West Nile virus infection decreases fecundity of Culex tarsalis females. J Med Entomol *44*,1074–1085

Titus,R. G.,Bishop,J. V., and Mejia,J. S. (2006). The immunomodulatory factors of arthropod saliva and the potential for these factors to serve as vaccine targets to prevent pathogen transmission. Parasite Immunol *28*,131–141

Turell,M. J.,O'Guinn,M. L.,Dohm,D. J., and Jones,J. W. (2001). Vector competence of North American mosquitoes (Diptera: Culicidae) for West Nile virus. J Med Entomol *38*,130–134

Turell,M. J., Dohm,D. J., Sardelis,M. R., O'Guinn,M. L., Andreadis,T. G., and Blow,J. A. (2005). An update on the potential of North American mosquitoes (Diptera: Culicidae) to transmit West Nile virus. J Med Entomol *42*,57–62

Urbanelli,S.,Silvestrini,F.,Reisen,W. K.,De Vito,E., and Bullini,L. (1997). Californian hybrid zone between Culex pipiens pipiens and Cx. p. quinquefasciatus revisited (Diptera: Culicidae). J Med Entomol *34*,116–127

Vaidyanathan,R., and Scott,T. (2006). Apoptosis in mosquito midgut epithelia associated with West Nile virus infection. Apoptosis *11*,1643–1651

Vaidyanathan,R., and Scott,T. W. (2007). Geographic variation in vector competence for West Nile virus in the Culex pipiens (Diptera: Culicidae) complex in California. Vector Borne Zoonotic Dis *7*,193–198

Vanlandingham,D. L.,Schneider,B. S.,Klingler,K.,Fair,J.,Beasley,D.,Huang,J.,Hamilton,P., and Higgs,S. (2004). Real-time reverse transcriptase-polymerase chain reaction quantification of West Nile virus transmitted by *Culex pipiens quinquefasciatus*. Am J Trop Med Hyg *71*,120–123

Wang,X.-H.,Aliyari,R.,Li,W.-X.,Li,H.-W.,Kim,K.,Carthew,R.,Atkinson,P., and Ding,S.-W. (2006). RNA interference directs innate immunity against viruses in adult Drosophila. Science *312*,452–454

Waterhouse,R. M.,Kriventseva,E. V.,Meister,S.,Xi,Z.,Alvarez,K. S.,Bartholomay,L. C., Barillas-Mury,C.,Bian,G.,Blandin,S.,Christensen,B. M.,et-al. (2007). Evolutionary dynamics of immune-related genes and pathways in disease-vector mosquitoes. Science *316*,1738–1743

Weaver,S. C.,Scott,T. W.,Lorenz,L. H.,Lerdthusnee,K., and Romoser,W. S. (1988). Togavirus-associated pathologic changes in the midgut of a natural mosquito vector. J Virol *62*,2083–2090

Weaver,S. C.,Lorenz,L. H., and Scott,T. W. (1992). Pathologic changes in the midgut of *Culex tarsalis* following infection with Western equine Encephalomyelitis virus. Am J Trop Med Hyg *47*,691–701

Weaver,S. C.,Brault,A. C.,Kang,W., and Holland,J. J. (1999). Genetic and Fitness changes accompanying adaptation of an arbovirus to vertebrate and invertebrate cells. J Virol *73*,4316–4326

Weber,F.,Wagner,V.,Rasmussen,S. B.,Hartmann,R., and Paludan,S. R. (2006). Double-stranded RNA is produced by positive-strand RNA viruses and DNA viruses but not in detectable amounts by negative-strand RNA viruses. J Virol *80*,5059–5064

Woodring,J. L.,Higgs,S., and Beaty,B. J. (1996). Natural cycles of vector-borne pathogens, in Biology of Disease Vectors,W. C. Marquardt, and B. Beaty, eds. (Boulder,CO: University Press of Colorado), pp. 51–72

Yazi Mendoza,M.,Salas-Benito,J. S.,Lanz-Mendoza,H.,Hernandez-Martinez,S., and del Angel,R. M. (2002). A putative receptor for dengue virus in mosquito tissues: Localization of a 45-kDa glycoprotein. Am J Trop Med Hyg *67*,76–84

Zambon,R. A.,Nandakumar,M.,Vakharia,V. N., andWu,L. P. (2005).The toll pathway is important for an antiviral response in *Drosophila*.Proc Natl Acad Sci *102*,7257–7262

Zambon,R. A.,Vakharia,V. N., and Wu,L. P. (2006). RNAi is an antiviral immune response against a dsRNA virus in Drosophila melanogaster. Cell Microbiol *8*,880–889

4. Clinical Manifestations of Neurological Disease

J. DAVID BECKHAM and KENNETH L. TYLER

Abstract

Since its introduction into North America in 1999, there have been over 10,000 confirmed cases of West Nile virus (WNV) neuroinvasive disease, and WNV has become by a large margin, the most common cause of arboviral encephalitis in the U.S. Since the ratio of asymptomatic WNV infections exceeds the number of neuroinvasive cases by ~150:1 it can be estimated that more than 1.5 million individuals have been infected by WNV in the U.S. Although the overall mortality in confirmed cases of WNV is only ~4%, almost all the deaths occur in patients with encephalitis (Bode et al., 2006) and in this group mortality approaches 20%. In this chapter, we review the epidemiology, risk factors, and clinical manifestations of WNV neurologic disease as well as the prognosis and potential therapies.

Keywords

neuroinvasive, encephalitis, meningitis, meningoencephalitis, flaccid paralysis, interferon.

1 Introduction

WNV is thought to reach the CNS by first replicating in regional lymphoid tissue and the spleen, followed by viremia that results in dissemination to the CNS (Diamond et al., 2003). While the exact mechanism of CNS invasion is not known, evidence from mouse models suggests that peripheral production of TNF-α increases the blood–brain barrier permeability and facilitates CNS invasion by blood borne virus (Wang et al., 2004). A recent study has also shown that the virus can spread through nerve axons as well as through the bloodstream, providing an additional potential mechanism of CNS entry (Samuel et al., 2007). Following penetration of the blood–brain-barrier the virus is able to directly infect neurons, with a propensity for those in the basal ganglia, thalamus, cerebellum, brainstem, and anterior horn of the spinal cord. The resulting clinical syndromes of WNV neuroinvasive disease are generally categorized as meningitis, encephalitis or poliomyelitis/acute flaccid paralysis, although these syndromes frequently occur in combination (Debiasi and Tyler, 2006; Petersen et al., 2002; Sejvar et al., 2003a,2003b, 2005; Kramer and Shi, 2007; Sayao et al., 2004). Less common presentations of WNV neuroinvasive disease include brainstem encephalitis, cerebellitis, cranial neuropathies, opsoclonus-myoclonus syndrome, polyneuropathy/radiculopathy, and ophthalmological disorders including chorioretinitis, and optic neuritis. The clinical and laboratory features of these syndromes are discussed below.

2 Epidemiology and Risk Factors for WNV Neuroinvasive Disease

Since its introduction in North America in 1999, approximately 27,000 cases of WNV infection have been reported to the centers for disease control (CDC), and 10,830 of these cases met diagnostic criteria for neuroinvasive disease (Fig.1; http://www.cdc.gov/ncidod/dvbid/westnile/index.htm). The relatively high percentage of WNV-infected patients reported to the CDC as having neuroinvasive disease (40%), almost certainly reflects a reporting bias, as many asymptomatic or mildly symptomatic patients are never tested for WNV infection, and the total number of confirmed WNV cases reported to the CDC certainly under-represents the total number of infected individuals. It is estimated that up to 80% of patients infected with WNV remain asymptomatic. Symptomatic disease occurs in approximately 20% of patients, usually in the form of West Nile fever (see below) and less than 1% of the WNV-infected population develops neuroinvasive disease.

Primary risk factors for neuroinvasive disease include older age and immunocompromized states. In a retrospective review of cases during the 2003 epidemic, patients with WNV encephalitis were significantly older (mean age of 60 years) compared to patients with WNV fever (mean age of 46 years) (Patnaik et al., 2006). Several other studies have confirmed the increased risk of encephalitis as a function of age and the associated increased death rate in the elderly (Tsai et al., 1998; Chowers et al., 2001; Nash et al., 2001; Weiss et al., 2001). The mechanisms that underlie age-related susceptibility to WNV encephalitis are not completely understood. Recent studies in animal models suggest that cell-mediated immunity is important for the control of WNV infection (Brien et al., 2007), and age-related decline in WNV-specific cell mediated immunity may be one factor contributing to age-related susceptibility.

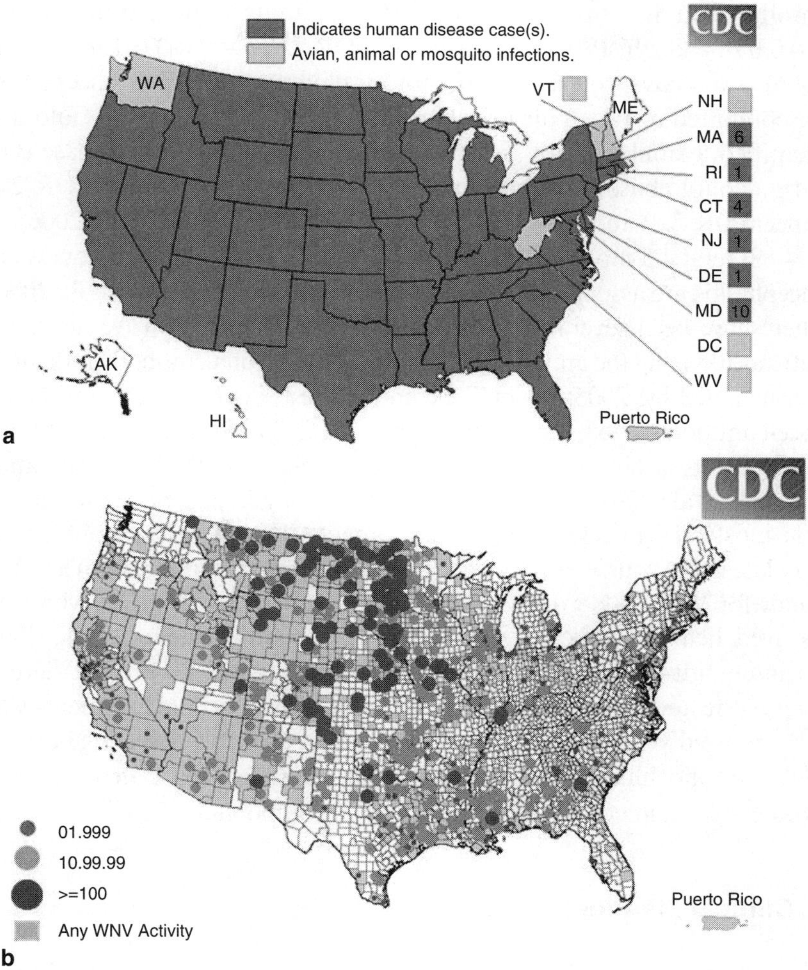

Figure 1. West Nile virus cases in 2007. (**a**) Number of WNV cases by state reported to the CDC for the 2007 season. (**b**) The incidence of neuroinvasive WNV infection during the year 2007. Values are updated as of November 7, 2007 (modified from CDC at http://www.cdc.gov/ncidod/dvbid/westnile/). (*See Color Plates*)

In contrast to the increased risk of WNV encephalitis in the elderly, is a decreased risk for development of neurological disease in children (Hayes and O'Leary, 2004). Compared to adults, children infected with WNV have shorter hospitalizations, fewer neurological symptoms, better neurological outcomes, and lower mortality (LaBeaud et al., 2006).

Risk factors in addition to age that are associated with risk of developing WNV encephalitis include cancer chemotherapy, diabetes, and renal disease (Patnaik et al., 2006). In a nested case control study, risk factors for WNV encephalitis included (odds ratio and p values shown): hepatitis C infection (OR 23.1, $p = 0.001$), black race (OR 12, $p < 0.001$)

chronic renal disease (OR 10.6, $p < 0.001$), immunosuppression (OR 3.9, $p = 0.03$), and a history of hypertension (OR 2.9, $p = 0.01$) (Murray et al., 2006). An analysis of risk factors for development of WNV encephalitis (as compared to West Nile fever) in 228 hospitalized patients in Colorado identified a similar set of risk factors including (OR): liver disease (OR 3.4), alcohol abuse (OR 3.4) diabetes (OR 3.1), age >50 years (OR 2.7) cancer (OR 2.2) and immunosuppression (OR 2.2) (Bode et al., 2006).

Several co-morbid conditions identified as risk factors for WNV encephalitis are associated with immune suppression (Murray et al., 2006). There have been several recent reports of WNV neuroinvasive disease in patients receiving the anti-TNF humanized monoclonal antibody infliximab (Batsis and Phy, 2005; Chan-Tack and Forrest, 2006), or the anti-CD20 B-cell antibody rituximab (Mawhorter et al., 2005). Surprisingly, only sporadic reports of neuroinvasive disease have occurred in the HIV population (Jamison et al., 2007; Torno et al., 2007). This may, in part be due to the fact that most HIV infected patients have intact CD8+ T cell responses until very late in the course of infection. The role of immunosuppression as a risk factor for WNV CNS disease is particularly obvious in patients who have received hematopoietic or solid organ transplants (Kumar et al., 2004; Kleinschmidt-Demasters et al., 2004). The highest risk (>40%) of developing severe and/or fatal WNV neuroinvasive disease occurs in patients who have received WNV-infected organ transplants, probably reflecting both the high viral inoculum and the concomitant transplant-associated immunosuppressive regimens (Iwamoto et al., 2003; Kumar et al., 2004).

3 Clinical Manifestations

Patients with WNV neuroinvasive disease typically have prodromal and non-neurological signs and symptoms preceding the onset of CNS disease. Patients typically develop a flu-like syndrome following an incubation period of 2–14 days. Common nonspecific symptoms include fever (75–100%), headache (50–100%), and altered mental status (50–100%). Gastrointestinal symptoms including nausea, vomiting, and diarrhea occur in 30–75% of cases (Debiasi and Tyler, 2006; Hayes et al., 2005; Petersen et al., 2002). The incidence of rash has been quite variable in different series generally ranging from 30–63% (Tilley et al., 2007).

Patients who develop neuroinvasive disease following WNV infection are in general, broadly categorized into three groups: West Nile meningitis (WNM), encephalitis (WNE), and acute flaccid paralysis (WNP). The CDC has established diagnostic criteria for these entities (Table 1), and more detailed clinical and laboratory criteria have been

Table 1. Diagnostic criteria for neuroinvasive West Nile virus infection

	West Nile meningitis	West Nile encephalitis	Acute flaccid paralysis
Clinical signs	Evidence of meningeal inflammation or meningismus (Nuchal rigidity, Photophobia, phonophobia, Kernig sign, or Brudzinski's sign)	Evidence of encephalopathy (altered level of consciousness, lethargy, or personality change lasting >24 h	Evidence of Acute onset Limb weakness with marked progression over 48 h
Additional evidence	One or more of the following:	Two or more of the following:	Two or more of the following:
	– Fever (temperature >38°C) or hypothermia (temperature <35°C)	– Fever (temperature >38°C) or hypothermia (temperature <35°C)	– Asymmetric weakness
	– CSF pleocytosis (>5 leukocytes mm^{-3})	– CSF pleocytosis (>5 leukocytes mm^{-3})	– Areflexia/hyporeflexia of affected limb(s)
	– Peripheral leukocyte count >10,000 mm^{-3}	– Peripheral leukocyte count > 10,000 mm^{-3}	– Absence of sensory changes in affected limb
	– Neuroimaging results consistent with acute meningeal inflammation	– Neuroimaging results consistent with acute inflammation or acute demyelination (with or without meningeal inflammation)	– CSF pleocytosis (>5 leukocytes mm^{-3})
		– Presence of focal neurological deficit	– Elevated CSF protein (>450 mg l^{-1})
		– Meningismus	– Electrodiagnostic studies consistent with an anterior horn cell process or poliomyelitis
		– Electroencephalography consistent with encephalitis	– Spinal cord MRI demonstrating abnormal signal in the anterior horn gray matter
		– Seizures (either new-onset or exacerbation of previously controlled seizures)	

Adapted from Debiasi and Tyler (2006)

proposed (Davis et al., 2006). The relative frequency of these subtypes is uncertain. In a series of hospitalized patients from Colorado, meningitis accounted for 61% of the neuroinvasive cases and encephalitis for 39% (Bode et al., 2006). Similarly, in a large review of CSF findings in 250 serologically documented cases of WNV neuroinvasive disease, 70% were classified as meningitis and 30% as encephalitis (Tyler et al., 2006). By contrast, one recent review estimated the relative frequencies of neuroinvasive disease as 55–60% encephalitis, 35–40% meningitis, and ~5–10% for acute flaccid paralysis (Sejvar, 2007). It is important to recognize that overlap syndromes are common and that the precise boundaries between syndromes (e.g. meningitis and encephalitis) may be difficult to define.

3.1 West Nile Meningitis

West Nile meningitis (WNM) presents as aseptic meningitis characterized by fever, nuchal rigidity, and phono- and photophobia (Bode et al., 2006; Nash et al., 2001; Sejvar et al., 2003a). Kernig and Brudzinski's signs are variably present (Klein et al., 2002). Cranial nerve palsies, notably unilateral or bilateral facial palsy, are present in ~20%, but may be delayed rather than present at presentation (Jeha et al., 2003). The key distinction between meningitis and encephalitis is the presence of signs, symptoms, or laboratory findings indicative of injury to the brain parenchyma in patients with encephalitis. As a result, patients with prominent alteration in mental status or with focal neurological signs including weakness, ataxia, or movement disorders should be classified as having encephalitis rather than meningitis. Patients with parenchymal CNS findings and evidence of inflammation in the subarachnoid space as demonstrated by CSF pleocytosis and/or meningeal enhancement are most accurately classified as having meningoencephalitis. Cranial nerve palsies may be present in patients with either meningitis or encephalitis. In one recent survey of hospitalized patients with WNV infection, the incidence of signs and symptoms of ataxia, visual disturbances, limb weakness, and myoclonus was actually higher in the group classified as "meningitis" compared to those classified as "encephalitis" (Bode et al., 2006). Similarly in a small series of cases from Louisiana, a significant percentage of patients classified as "meningitis" had tremor (80%), parkinsonism (40%), or myoclonus (20%) (Sejvar et al., 2003), suggesting that classification of patients into these subgroups is often arbitrary and probably inaccurate.

Patients with meningitis often have elevated peripheral leukocyte counts (>10,000 mm^{-3}). Neuroimaging studies are performed in virtually all cases. CT is almost invariably normal. MRI is also frequently normal but may show meningeal enhancement. Patients with more substantial abnormalities on neuroimaging studies should be classified as having meningoencephalitis rather than meningitis (see below). EEG abnormalities are rare in patients with meningitis and when present are typically limited to mild generalized or anteriorly predominant slowing (see below). In a large series of 250 patients with serologically confirmed WNV meningitis and encephalitis, the CSF in patients with meningitis exhibited a mean of 226 ± 50 white blood cells, and a median of 100 cells mm^{-3}. Almost 90% of patients had a cell count between 5–500 mm^{-3}. 45% of patients had a polymorphonuclear (PMN) predominance on their initial CSF exam and nearly 75% had >25% PMNs. Greater than 70% had an elevated CSF protein (mean 76 ± 2.5 mg $dL,^{-1}$ median 71 mg dL^{-1}). The CSF glucose level was >40 mg dL^{-1} in >99% of cases (Tyler et. al., 2006). In a series of 32 patients with meningoencephalitis the mean cell count was 156, mean protein concentration 79 mg dL^{-1} and all patients had a normal glucose (Rawal et al., 2006). In another series of 23 cases of meningoencephalitis from Ohio, the CSF cell counts had a mean of 223 and median of 171 during the first week of illness, declined to a mean of 79 and median of 47 during the second week, and declined further to a mean of 16 and median of 8 when obtained at >14 days post onset (Jeha et al., 2003). Similarly, the mean and median percentage of PMNs declined from a mean of 50% and median of 52% in week one, to a mean of 18% and median of 13% in week two, to a mean of 6% and median of 0% after 14 days (Jeha et al., 2003). The presence of Mollaret-like cells (large monocyte-like cells with cerebriform nuclei) (Procop et al., 2004) or of a plasma cell pleocytosis (Carson et al., 2003) have also been suggested as possible clues to the diagnosis of WNV meningoencephalitis, although their sensitivity and specificity are unknown.

Diagnosis of WNV meningitis is typically made by demonstration of WNV specific IgM in CSF by ELISA. IgM antibodies cross the blood brain barrier poorly, and their presence in CSF is therefore generally indicative of intrathecal synthesis. The percentage of patients with WNV meningitis or encephalitis with CSF WNV IgM increases by ~10%/day after the onset of illness and is >80% by the end of the first week (Fig. 2; Tardei et al., 2000). The duration for which CSF IgM antibodies persist following an acute infection has not been definitively established. However, in some cases serum IgM WNV antibodies have been shown to persist for >500 days (John et al., 2003; Kapoor et al., 2004), and those in CSF for up to 199 days (Kapoor et al., 2004). In one study of serum WNV IgM in patients with encephalitis and meningitis,

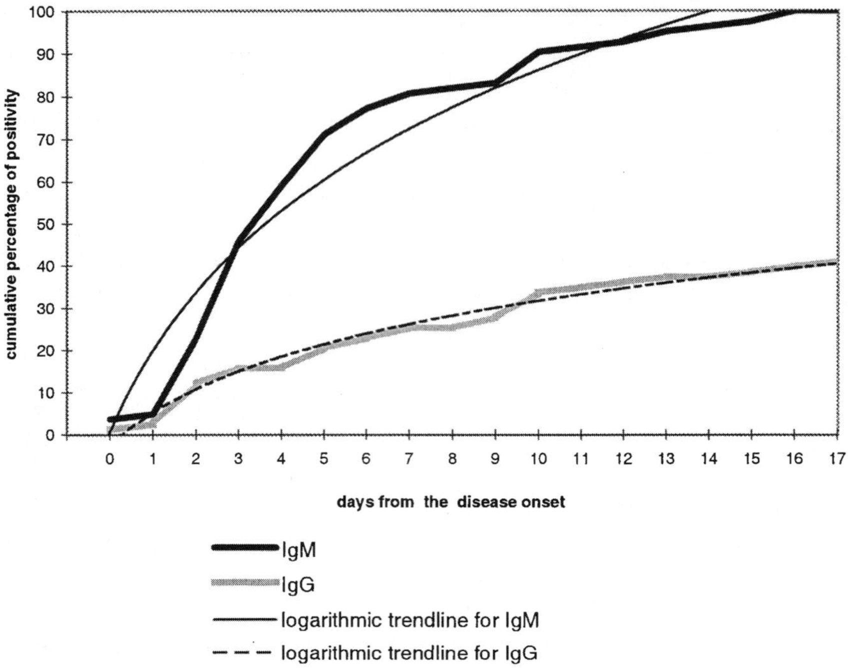

Figure 2. Cumulative percent positivity of anti-WNV IgM and IgG in the CSF of patients with confirmed and presumptive recent WNV infection (adapted from Tardei et al., 2000; Permission granted from American Society of Microbiology).

100% of the patients were still positive at 3 months post-infection, by 6 months this had dropped to 62% (encephalitis)-70% (meningitis), and by 9 months to 39% (encephalitis)-40% (meningitis). At eighteen months post-infection none of the seven tested meningitis cases remained positive, but 18% (2/11) of the encephalitis cases remained positive (Roehrig et al., 2003). The potential persistence of IgM antibodies may make it difficult to establish whether infection was acute or remote, and it is often useful to obtain both CSF IgG and IgM antibody titers to assist in this evaluation. In patients with positive CSF serologies in CSF specimens obtained during the first week of illness, 58% had only IgM and 42% had both IgG and IgM antibodies. CSF specimens obtained during the second and third week of illness are typically positive for both IgG and IgM antibodies (Tardei et al., 2000). Therefore, the presence of CSF WNV IgM in the absence of IgG is generally diagnostic of acute infection. Sequential analysis of CSF and/or serum antibody titers can also assist in diagnosis in problematic cases, as IgM titers in both serum and CSF generally decline over time.

It is important to recognize that infection with flaviviruses other than WNV (e.g. St. Louis encephalitis virus, Japanese encephalitis virus) or immunization against flaviviruses (e.g. yellow fever virus, Japanese encephalitis virus) can induce heterologous antibodies that cross-react with WNV antigens in ELISA assays. It is therefore important to obtain a complete travel and immunization history to enable proper interpretation of ELISA results. In difficult cases it may be necessary to obtain neutralizing antibody titers (available through the CDC and some state public health laboratories) against a panel of flaviviruses to establish a definitive diagnosis. Despite the potentially confounding issues in serodiagnosis, under most circumstances the presence of WNV specific IgM in the CSF of a patient is sufficient to establish a presumptive diagnosis both of WNV infection and of WNV CNS disease

RT-PCR of the serum or CSF is not as sensitive for diagnosis of WNV disease generally or neuroinvasive disease specifically as serology. Peak viremia occurs 3–4 days before symptom onset, and at the time of clinical presentation virus is often non-detectable in serum. The sensitivity of WNV RT-PCR compared to IgM serology in the CSF is 57% in neuroinvasive disease (Lanciotti and Kerst, 2001). CSF WNV PCR is useful in immunocompromised patients who may have delayed or absent antibody responses against WNV. CSF PCR also has extremely high specificity, and detection of WNV RNA in CSF establishes a definitive diagnosis of WNV CNS disease. RT-PCR and related nucleic acid amplification tests (NAAT) play an important role in screening donated blood and blood products for WNV infection (Pealer et al., 2003), and may also have a role in screening donated organs for WNV infection (Kiberd and Forward, 2004).

3.2 West Nile Encephalitis

Encephalitis is distinguished from meningitis by the presence of clinical signs or symptoms or evidence on tests such as neuroimaging or EEG indicative of involvement of the brain parenchyma. The frequency of specific signs and symptoms has varied widely across clinical studies (Bode et al., 2006; Jeha et al., 2003; Nash et al., 2001; Sejvar et al., 2003a). Patients with WNV encephalitis typically have fever (85–100%) (Nash et al., 2001; Weiss et al., 2001; Jeha et al., 2003; Brilla et al., 2004; Kramer and Shi, 2007; Shi et al., 2004). Meningeal symptoms occur in about 1/3rd of cases, although virtually all patients have a CSF pleocytosis and as such are most accurately considered as having "meningoencephalitis". Headache occurs in 47–59% of cases, but is likely to be under-reported and under-recognized due to the presence of associated mental status abnormalities (Nash et al., 2001; Weiss et al., 2001; Jeha et al., 2003; Brilla et al., 2004).

Movement disorders are often prominent and include tremors, myoclonus, and signs of Parkinsonism (bradykinesia, rigidity, postural instability). Tremor may be postural, kinetic or less commonly at rest. In one series of 65 patients with encephalitis from Colorado (Bode et al., 2006), 28% had tremor, and 9% had myoclonus. Tremor was noted in 26% of a series of 22 cases hospitalized at the Cleveland Clinic during 2002 (Jeha et al. 2003). Detailed descriptions of tremor in WNV are rare, although reported cases include examples of postural (static) and kinetic tremor. When present the tremor is often asymmetric and has a predilection for the upper extremities (Sejvar et al., 2003a). Rest tremor is rare, although patients frequently have other signs of Parkinsonism including bradykinesia, rigidity, and postural instability. In a small series from Louisiana that included eight cases of encephalitis, signs of Parkinsonism were noted in 75% (Sejvar et al., 2003a).

Weakness is an extremely prominent symptom and sign in patients with encephalitis. Generalized weakness and/or fatigue occur(s) in 40–80%, although objective evidence of focal limb weakness is less common. In a large series of encephalitis patients from Colorado, only 8% had signs or symptoms of limb weakness at hospital admission, although 42% developed limb weakness during the course of illness (Bode et al., 2006). Similarly, in a series of 23 cases hospitalized at the Cleveland Clinic in 2002, objective weakness was only present in 22% at presentation, but ultimately developed in 48% (Jeha et al., 2003).

Additional common signs and symptoms in patients with encephalitis include ataxia and limb dysmetria (~32%). Bulbar dysfunction occurred in 20%, typically manifesting as dysarthria or dysphagia. Cranial neuropathies occur in ~20% of cases, most commonly unilateral or bilateral peripheral facial nerve palsy. In some patients cranial neuropathies have a delayed onset into the second or third week of illness (Brilla et al., 2004; Hayes et al., 2005; Petersen et al., 2002; Sejvar et al., 2005).

Although EEG abnormalities are frequent in patients with encephalitis (see below), overt seizures occur in only 3–6%, probably reflecting the relative sparing of cortical neurons as opposed to those in the thalamus and basal ganglia (Doron et al., 2003).

General laboratory studies are usually not particularly useful in the diagnosis of WNV encephalitis. Patients may have a peripheral leukocytosis. In rare cases, lymphopenia (Cunha et al., 2004) and thrombocytopenia have been reported. It has been suggested that elevations in serum lipase (Batuello et al., 2005) or ferritin (Cunha et al., 2004) may provide a clue to the diagnosis of WNV infection (Batuello et al., 2005), although the sensitivity and specificity of these findings are uncertain. Hyponatremia occurs in 33–50%, and possibly reflects inappropriate anti-diuretic hormone secretion (SIADH). Some patients have abnormalities in liver function tests or elevations in muscle enzymes including creatine kinase (see Davis et al., 2006).

CSF studies in patients with encephalitis closely parallel those with meningitis (see above). In one large study the mean CSF white blood cell count in patients with encephalitis was 227 ± 47 (median 90). 37% of patients had a PMN predominance and >70% had greater than 25% PMNs. The mean CSF protein concentration was 101 ± 6 mg dL^{-1} (median 100 mg dL^{-1}) and none of the 76 patients analyzed had a CSF glucose <40 mg dL^{-1} (Tyler et al., 2006).

MRI abnormalities have been reported in between 20–70% of patients with neuroinvasive disease but may not develop until several weeks after onset of symptoms (Fig. 3; Ali et al., 2005; Petropoulou

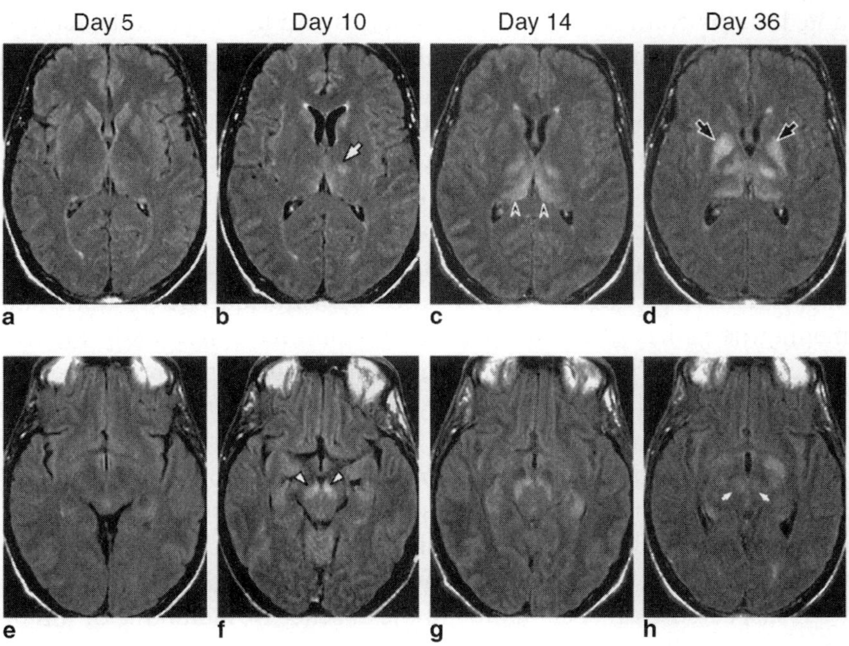

Figure 3. Sequential MRI images of a patient with WNV encephalitis showing evolution of the FLAIR sequence MRI abnormalities over time. Increased signal in the left thalamus is noted on day 10 (*arrow*, **b**) and in both thalami and the basal ganglia including the globus pallidi on day 36 (*arrows*, **d**). Increased signal in the substantia nigra of the midbrain is also initially noted on day 10 (*arrows*, **f**) and in the red nuclei on day 36 (*arrows*, **h**) (from Gea-Banacloche et al., 2004; permission granted from the American College of Physicians).

et al., 2005). In one study, 70% (12/17) patients with neuroinvasive WNV had an abnormal MRI. However, in four of the 12 patients with abnormalities these were only apparent on diffusion-weighted imaging (DWI) sequences. Three patients had increased signal lesions on FLAIR and/or T2-weighted sequences in the corona radiata and internal capsule or involving the pons, midbrain, basal ganglia, and thalamus. None of these lesions enhanced with contrast. The remaining five cases had either isolated meningeal increased signal on gadolinium enhanced T1 and/or FLAIR images or had abnormalities involving the spinal cord, cauda equina or lumbosacral roots (Ali et al., 2005). A second study involving 17 patients also found abnormal MRIs in 70% (12). No patient had parenchymal enhancement but meningeal enhancement was noted in one case. The most common abnormality consisted of increased signal on T2, FLAIR, Fast spin echo (FSE) T2 or diffusion weighted images that involved thalami, basal ganglia, and midbrain (Petropoulou et al., 2005). Surprisingly, five of 17 patients studied had signal abnormalities that included the mesial temporal lobes resembling those seen in herpes simplex encephalitis (Petropoulou et al., 2005). However, it has been noted that in contrast to HSE the lesions caused by WNV in the temporal lobes do not show mass effect, fail to enhance, and are not associated with hemorrhage (Zak et al., 2005).

Electroencephalography (EEG) may be helpful in diagnosis of WNV encephalitis. In one series of 13 hospitalized patients with encephalitis, 85% of them had at least one abnormal EEG, with the most common abnormality consisting of generalized slowing, often with an anterior or temporal prominence (Gandelman-Marton et al., 2003). The two patients in this series with normal EEGs both had meningitis rather than encephalitis. In another small study all five patients with WNV encephalitis had moderate to severe degrees of generalized slowing. Interestingly, three of the five patients had triphasic waves in the absence of renal, hepatic or other metabolic abnormalities with which these are typically associated (Rodriguez and Westmoreland, 2007). Triphasic waves have also been described in case reports of WNV encephalitis (Tyler et al., 2007), suggesting that they may provide a clue to diagnosis when they occur in patients with suspected viral encephalitis in the absence of metabolic or major organ system dysfunction.

The diagnosis of WNV encephalitis is similar to that of WNV meningitis (see above) and is typically made by detection of WNV-specific IgM in CSF.

3.3 West Nile Acute Flaccid Paralysis

West Nile acute flaccid paralysis (WNP) is defined as a syndrome of weakness in a single or multiple extremities associated with hypo-or areflexia and decreased tone in the absence of significant objective sensory abnormalities. Unlike WNV meningoencephalitis, WNP does not predominantly affect the elderly population but can occur in any age group. In a series of 32 cases from Colorado the mean age of those affected was 56 years; however the range was between 18 and 84 years and 75% cases were between 35–65 years (Sejvar et al., 2005).

WNP can occur in isolation, or in combination with encephalitis and/or meningitis (Sejvar et al., 2005; Sejvar et al., 2003b). The frequency of associated encephalitis is probably ~50–66% based on the reported frequency of concomitant findings of parkinsonism (25%), myoclonus (47%) and tremor (66%) (Sejvar et al., 2005). Most patients (90%) have an acute febrile illness characterized by headache (88%), malaise, and gastrointestinal disturbances (41%) (Sejvar et al., 2005). Rash is often absent. Patients may also have symptoms of lower back pain and bowel or bladder dysfunction (Leis et al., 2002; Sejvar et al., 2003b).

Weakness typically occurs abruptly and reaches its nadir in ~24 h (Sejvar et al., 2005). The pattern of weakness varies considerably among patients. In the large series reported by Sejvar et al. (2005) 64% had tetra-or quadriplegia, 24% had bilateral arm or leg weakness, and 20% had monoplegia. Although it was initially suspected that weakness associated with WNV infection was the result of a Guillain–Barré-like syndrome (GBS), subsequent studies have suggested that WNV-associated GBS is rare, and the majority of cases of WNP are due to a true poliomyelitis. In a study of 32 patients presenting with WNP in Colorado in 2003, 84% were considered to have poliomyelitis syndrome and only 13% Guillain–Barré syndrome, with one additional patient (3%) having a brachial plexus neuropathy (Sejvar et al., 2005). Another case with unilateral brachial plexopathy had been previously reported (Almhanna et al., 2003).

Eleven of the 25 (44%) patients with poliomyelitis had acute respiratory weakness requiring endotracheal intubation, as did one of the four (25%) GBS cases. Patients with respiratory failure were more likely than those without respiratory failure to have signs of associated bulbar dysfunction (dysarthria, dysphagia) (92% vs. 15%), to be immunocompromised (33% vs. 0%), and to have encephalitis (100% vs. 25%).

Cranial nerve involvement occurred in 70% and included dysphagia (52%), unilateral or bilateral facial weakness (40%), dysarthria (24%), paralysis of extraocular movements (16%), and vocal cord paralysis (8%).

Fourteen of the 27 patients with poliomyelitis had electromyography (EMG) and nerve conduction studies (NCS) and all had findings consistent with a motor axonopathy or an anterior horn cell process. Of the four patients classified as GBS, three had EMG and/or NCS findings consistent with demyelinating sensorimotor neuropathy, although these were not described in detail (Sejvar et al, 2005)

CSF findings in patients with WNP are generally similar to those seen in meningitis and encephalitis. In the series of Sejvar et al. (2005), the median CSF cell count was 108, median protein concentration 98 mg dL^{-1}, and lowest reported glucose concentration 36 mg dL^{-1}. In patients whose CSF was examined at or before the onset of weakness, approximately half had a PMN predominance, however in those whose CSF was examined after the onset of weakness, lymphocytes predominated in 77%. CSF exams were only performed in three GBS patients, and two had the classic albuminocytological dissociation, and one had both a pleocytosis and elevated protein.

A comprehensive study of neuroimaging findings in patients with WNP has not yet been reported. Imaging is often unremarkable, but MRI may show evidence of focal areas of increased T2 signal intensity in the anterior horns in spinal cord segments corresponding to the distribution of paralysis (Fig. 4; Li et al., 2003, Ali et al., 2005; Petropoulou et al., 2005; Sejvar et al., 2005).

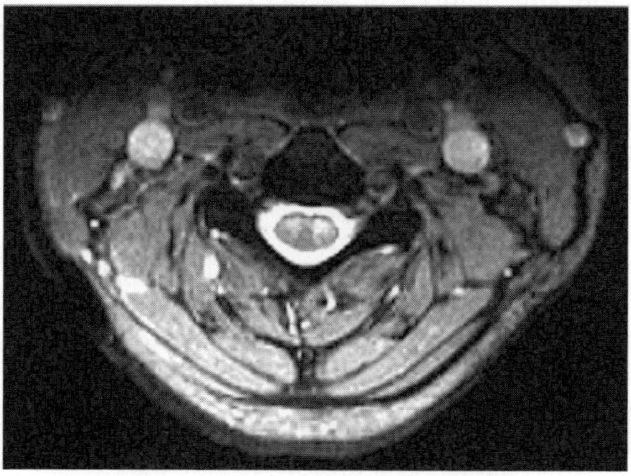

Figure 4. MRI image of a patient with WNV-induced AFP. Axial T2-weighted MR image of the spinal cord demonstrates bilateral hyperintense signal in the anterior horns.

Diagnosis is similar to other forms of neuroinvasive disease and is typically based on positive CSF IgM serology. As mentioned above, electrophysiological studies can be helpful in the diagnosis of WNP. NCS studies of nerves in symptomatic limbs will show decreased compound muscle action potential (CMAP) amplitudes with preserved amplitudes of sensory nerve action potentials (SNAPs). Demyelination does not occur except in cases with GBS, although conduction velocities may be decreased in proportion to axonal injury. EMG performed three or more weeks after illness shows evidence of denervation including fibrillations and positive sharp wave complexes (PSWC) (Jeha et al., 2003; Leis et al., 2003; Li et al., 2003).

Data regarding long term prognosis in patients with WNP is limited (Sejvar et al., 2006, Sejvar, 2007). In a one-year follow-up study of 22 surviving patients from Colorado, all showed at least some improvement in the strength of the affected muscles but the degree was quite variable. Patients with less initial weakness and fewer involved limbs showed the greatest improvement. When improvement occurred it tended to be maximal in the first 4 or 6 months after illness (Sejvar et al., 2006, Sejvar, 2007).

3.4 Other Clinical Manifestations

The extent of possible clinical presentations of WNV neuroinvasive disease is still being defined. Recent studies have emphasized the frequency of a variety of ocular and neuro-ophthalmologic presentations. Patients with WNV ocular involvement often present with fever, blurred vision, headache, nausea and vomiting. The visual symptoms can result from chorioretinitis with or without associated uveitis (Chan et al., 2006). Chorioretinitis is frequently multifocal and characterized by the presence of multiple, bilateral small round chorioretinal lesions scattered throughout the posterior as well as the peripheral fundus (Chan et al., 2006). In a small study of seven patients ("14 eyes") the most common ocular findings included multifocal chorioretinal target lesions (86%), retinal hemorrhages (50%), vitritis (43%), perivascular sheathing and vasculitis (29%), disc edema (29%), and optic atrophy (14%) (Chan et al., 2006). Chorioretinitis, occlusive vasculitis, and optic neuritis have all now been repeatedly described in patients with WNV neuroinvasive disease (Khairallah et al., 2004; Hershberger et al., 2003; Kaiser et al., 2003).

WNV neuroinvasive disease can also produce abnormalities of ocular motility including unilateral or bilateral VIth nerve palsy (Ross and Worthington, 2004; Cunha et al., 2006), and opsoclonus, the latter occurring with or without myoclonus (Sayao et al., 2004; Alshekhlee et al., 2006). Electroencephalography (EEG) of one reported patient showed mild slowing of the posterior dominant rhythm, with frequent prolonged runs of eye movements corresponding to opsoclonus. Over a period of 8 months visual symptoms resolved and the patient returned to baseline functional status. In a limited description of another case (Case #2) (Sayao et al., 2004) opsoclonus was associated with myoclonus, facial palsy, tremor and dysdiadocokinesis. The patient complained of gait unsteadiness and oscillopsia. There was a mild CSF pleocytosis with lymphocytic predominance. Brain MRI was normal. Symptoms resolved spontaneously over several weeks. In addition to opsoclonus, patients have also been reported with downbeat nystagmus and ocular flutter (Prasad et al., 2006). Prominent nystagmus may also occur as a manifestation of cerebellitis (Moon et al., 2005; Natarajan and Varman, 2007).

In addition to involvement of the IInd, VIth and VIIth cranial nerves (see above and earlier sections), at least one case report has described bilateral vocal cord paralysis resulting from injury to the recurrent laryngeal nerve. Interestingly, in the patient reported the vocal cord paralysis resolved spontaneously on one side while persisting on the other (Steele and Myssiorek, 2006).

Although WNV can produce persistent infections both in animals (Tesh et al., 2005) and in primary astrocyte cultures (Diniz et al., 2006), reports of chronic or persistent WNV neuroinvasive disease in humans have been limited to rare examples in immunocompromised individuals (Penn et al., 2006). To date there are no reported cases of acute disseminated encephalomyelitis (ADEM) associated with WNV. However, the report of a possible case associated with a related virus (St. Louis encephalitis virus) (Sejvar et al., 2004) and the presence in WNV infected patients of delayed cranial nerve palsies and of syndromes such as acute polyneuropathy (Guillain–Barre) and opsoclonus-myoclonus (see above) suggests that post-viral immune-mediated pathology may occur.

4 Outcomes and Prognoses

The outcomes associated with WNV neuroinvasive disease have been the subject of multiple investigations. A recent evaluation of 98 patients with West Nile fever (WNF) found that 96% of the patients report fatigue continuing after the acute infection for a median of 36 weeks (Watson et al., 2004). Other prolonged symptoms in these patients included headache (median of 10 weeks), muscle pain (median of 14 weeks), muscle weakness (median of 28 weeks), difficulty in concentrating (median of 14 weeks), and sensitivity to light (median of 14 weeks). Recent studies show that approximately 33% of patients infected with WNV suffer from prolonged depression as long as 1 year following acute infection (Murray et al., 2007).

A recent investigation of neurological sequelae of WNV infections indicated that symptoms can persist up to 8 months or longer following neuroinvasive disease (Sejvar et al., 2003a). One study of WNV infected patients 9 months after acute infection found that patients with neuroinvasive infection often had subtle but significantly worse neurocognitive function, concentration difficulty, and confusion than patients with WNF (Haaland et al., 2006).

In contrast to patients with encephalitis, outcomes in patients with meningitis tend to be indistinguishable from those found in WNF. Serious sequelae are rare but a high percentage of patients report persistent subjective complaints of fatigue, headache, muscle aches, memory, and concentration problems (Klee et al., 2004).

WNE has been associated with significant morbidity and mortality. Acute mortality associated with WNE is approximately 20% (Bode et al., 2006; Pepperell et al., 2003). Some studies suggest that in patients with WNE, a poor prognosis is associated with a prolonged or severe relative lymphopenia (Cunha, 2004; Cunha et al., 2000) and a high serum ferritin (>500 ng ml^{-1}) (Cunha et al., 2004). Of the patients that are hospitalized, those with a diagnosis of encephalitis are more likely to have prolonged hospitalization (Patnaik et al., 2006); discharge placement outside their home (approximately 75%), present at the 6th month follow-up evaluation in a wheelchair, and are more likely to be >65 years of age (Klee et al., 2004; Bode et al., 2006). In fact, age >65 years is associated with an increased likelihood of a prolonged recovery of physical, cognitive, or functional sequelae. Following acute WNE, approximately 47–65% of patients require some form of rehabilitation (Klee et al., 2004; Patnaik et al., 2006) and as few as 21% return to the pre-hospitalization level of function (Bode et al., 2006). Once discharged from the hospital, patients diagnosed with WNE still have high rates of mortality 1 year post-infection when compared to the general population (Green et al., 2005). Following acute WNE, patients also suffer from prolonged neurologic deficits such as movement disorders that may persist for months or years after the acute infection has resolved. New or persistent tremor was found in 20–38% of recovering WNE patients for 8 months to 1 year following acute infection (Carson et al., 2006; Sejvar et al., 2003a). In one cohort analysis of 228 WNV infected patients in Colorado, 18% of patients with neuroinvasive disease had persistent tremors, Parkinsonism, and ataxia for 1.5 years after the acute illness.

Recent data studying the long-term outcomes of patients with West Nile flaccid paralysis (WNP), or poliomyelitis, suggest that this disease is associated with significant morbidity and mortality. More than 50% of the mortality associated with WNP occurs in patients with acute neuromuscular respiratory failure (Sejvar et al., 2005, 2006) and of the patients that survive respiratory failure associated with WNP, a substantial number require long term supplemental oxygen. In one study, five patients were found to have quadriplegia and respiratory failure; only one patient survived and had a complete recovery. Of the patients that survive, persistent and prolonged neurological signs and symptoms are seen or reported in all patients with WNP. At one year of follow-up, symptoms can include tremor, myoclonus, Parkinsonism, cerebellar ataxia, and most commonly limb atrophy (Sejvar et al., 2006). Similar to WNE patients, the severity of the initial neurological deficit is not a good predictor of ultimate outcome. In patients that survive acute WNP, most strength recovery occurs in the first 6–8 months after which improvement seems to plateau (Cao et al., 2005, 2006).

5 Therapy of West Nile Virus Infection

A detailed review of the current status of experimental approaches to the treatment of WNV infection is presented in Chap. 21. There are currently no clinically proven specific therapies available for either treatment or prevention of WNV infection. Small studies and case reports have evaluated several therapies including ribavirin, interferon-α, and intravenous immunoglobulin therapy (IVIG) in WNV-infected humans, but few blinded, randomized clinical trials have been completed. New data from animal models of WNV infection suggest that new immunotherapies such as humanized WNV neutralizing monoclonal antibodies (Morrey et al., 2007), human IVIG (Planitzer et al., 2007), and single chain Fv-Fc fusion proteins (Gould et al., 2005) may be beneficial in the treatment of WNV infection (see Chap. 20).

There is evidence in animal models of WNV infection that passive transfer of antibodies may decrease morbidity and mortality associated with WNV CNS disease (see Chap. 20). Several case reports suggest that an Israeli IVIG preparation (Omr-IgG-am, Omrix Biopharmaceuticals) containing a naturally high titer of WNV neutralizing antibody (due to the high WNV seroprevalence rate among Israeli blood donors) may improve outcome for patients with WNV neuroinvasive disease. In one reported case a 70 year-old woman with chronic lymphocytic leukemia developed a viral prodrome that quickly progressed to coma in three days (Shimoni et al., 2001). WNV infection was diagnosed on the basis of the presence of WNV-specific IgM in both the serum and CSF. She was treated with Omr-IgG-am (0.4 mg kg^{-1} d^{-1} × 5 days) and recovered normal mental status after therapy. A second case involved a 42 year-old lung transplant recipient who developed WNV encephalitis with rapidly deteriorating neurologic status (Hamdan et al., 2002). He was also given Omr IgG-am and began to improve within 24 h and recovered fully 48 h following treatment. Despite these two encouraging reports, it is important to recognize that other case reports have shown no benefit from therapy (Haley et al., 2003).

Based on data from animal models and human case reports, a phase I/II multicenter, randomized, placebo-controlled trial of Omr-IgG-am was initiated by the Collaborative Antiviral Study Group (CASG)(CASG Trial #210; ClinicalTrials.gov identifier NCT00068055). This phase I/II study was designed to evaluate the safety and efficacy of Omr-IgG-am compared to a placebo group and to standard U.S. IVIG (which lacks antibody to WNV). Patients were enrolled if they were high risk for progression to WNV neuroinvasive disease or if they had advanced disease. The study was recently completed and the data is currently being analyzed.

In experimental models of WNV infection, monoclonal antibodies specific for domains on the WNV envelope (E) protein can protect mice against WNV CNS disease. Based on these studies, the safety, tolerability and efficacy of a humanized neutralizing monoclonal antibody (MGAWN1, MacroGenics) directed against an epitope on the WNV envelope protein has also been evaluated in Phase I human clinical trials (ClinicalTrials.gov Identifier NCT00515385).

Interferon-α (IFN-α) is an immunomodulatory protein used for therapy in several viral infections. IFN-α can inhibit WNV-induced cell death in tissue culture in vitro (Anderson and Rahal, 2002), and mice lacking IFN-α and IFN-β receptors have more severe disease when compared to wild-type controls (Samuel and Diamond, 2005). Several case reports have described the use of IFN-α treatment in human WNV disease. In one report, a 43 year-old male with lymphoblastic lymphoma developed headache, fever, confusion, and aphasia progressing to coma (Kalil et al., 2005). WNV-infection was diagnosed by detection of WNV IgM in the CSF and serum. Interferon-α was started 4 days after onset of CNS symptoms at a dose of 3 million units by intravenous injection on day one followed by 3 million units by subcutaneous injection daily for a total of 14 days. The patient fully recovered over a period of 9 months and returned to work. A second patient in the same case series was a 54 year-old woman receiving methotrexate and hydroxychloroquine for rheumatoid arthritis (Kalil et al., 2005). This patient developed fever, headache, progressive mental status changes, and weakness. She was treated with IFN-α 6 days after the onset of CNS symptoms and improved (Kalil et al., 2005). Three additional patients successfully treated with IFN-α are included in a report of seven cases from Calgary (Sayao et al., 2004). All received two weeks of treatment with subcutaneous IFN-α–2b (3 million units d^{-1}). The first patient (Case #4) (Sayao et al., 2004) was a 73 year-old woman with untreated CLL who developed fever, confusion and leg weakness that progressed to obtundation. She received IFN-α-2b about 5 days after the onset of neurological symptoms and began improving a few days later. Two weeks after discharge she was ambulating with a walker and had mild residual cognitive slowing and fatigue. The second patient (Case #5) (Sayao et al., 2005) was a 71 year-old man who developed headache, disabling ataxia and generalized weakness with intention tremor. He received IFN-α starting about 9 days after the onset of CNS symptoms and gradually improved, although he continued to have residual leg weakness. The third patient (Case #7) (Sayao et al., 2005) was a 50 year-old man who developed fever, headache, neck stiffness, ataxia and blurry vision that was followed by prominent brainstem abnormalities including bifacial weakness, dysarthria, palatal weakness, and tongue weakness. He received IFN-α beginning about 8 days after the onset of CNS symptoms and had only mild bifacial weakness and fatigue at discharge. It is important to emphasize that these cases were all treated in non-controlled trials and the investigators reporting on their progress were obviously unblinded. At least one patient who failed to respond to IFN-α therapy has also been reported (Chan-Tack and Forrest, 2005) and acute WNV infection has developed in patients receiving combined ribavirin and interferon treatment for hepatitis C infection (Hrnicek and Mailliard, 2004). The patient who died despite interferon treatment was a 76 year-old man who developed fever, headache, and confusion followed by lateral gaze deficits, decreased gag reflex, generalized weakness and respiratory insufficiency. Interferon therapy was not started until 17 days after the onset of CNS disease, and it is possible that this delay adversely affected efficacy.

No controlled clinical trials of IFN-α therapy in human WNV disease have been completed, however a major trial of its use in treatment of Japanese encephalitis was completed in Vietnam (Solomon et al., 2003). This was a placebo-controlled, double-blinded study evaluating

87 children with Japanese encephalitis. This study found that mortality did not differ between the interferon-α 2a treatment group (18% mortality at 3 month follow-up) when compared to the placebo treatment group (22% mortality at 3 month follow-up). Other outcome measures did not significantly differ between treatment groups including severe sequelae and the ability to walk. It is important to recognize that negative results in a treatment trial for Japanese encephalitis may not necessarily be predictive of results for treatment of WNV infection. A placebo controlled double-blind trial of IFN-α in human WNV disease was developed by Dr. James Rahal, an infectious disease specialist at New York Hospital, (http://www.nyhq.org/posting/rahal.html) but no results from this study have been reported.

Several in vitro studies have suggested that the antiviral drug ribavirin may inhibit WNV replication. No controlled clinical trials of ribavirin treatment in human WNV infection have been performed. However, during an outbreak of WNV in Israel in 2000, 37 patients received ribavirin in a non-blinded and non-controlled trial. Mortality in this group (41%) was higher than in untreated patients, although this may have reflected a bias towards treatment of sicker individuals with a poorer prognosis (Chowers et al., 2001). Nonetheless, unless subsequent trials provide evidence of efficacy, ribavirin is not recommended for therapy in WNV-infected patients as it may increase mortality.

The severity of WNV infection is increased in immunocompromised patients, and the presence of immunosuppression is a risk factor for the development of West Nile encephalitis in infected patients (see above). By contrast, there is at least one report of treatment of a patient with acute flaccid paralysis with high-dose steroids with apparently beneficial results (Pyrgos and Younus, 2004). The patient was a 68 year-old man with fever who developed progressive quadriparesis associated with rapid deterioration in his mental status progressing to coma. He received intravenous methylprednisone succinate (500 mg $d^{-1} \times 4$ days) and within 24 h his mental status improved and he became communicative followed by improvement in his weakness. He ultimately regained full upper extremity strength and was cognitively unimpaired but had residual paraparesis. Although this patient improved in response to steroid therapy, concerns about the potential deleterious effects of steroids on the ability of the host's immune system to control WNV infection preclude their routine use unless benefits can be confirmed in controlled clinical trials.

Several vaccines designed to prevent human WNV infection are currently under development and in early stage human clinical trials. A live chimeric vaccine (ChimeriVax-WN02, Acambis) in which genes encoding the WNV envelope (E) and pre-Membrane (PrM) proteins replace the analogous genes in the yellow fever virus 17D vaccine strain has been shown to be safe and induces neutralizing antibodies as well as WNV-specific T cell responses in phase I human clinical trials (Monath et al., 2006). A phase II trial of this vaccine is currently underway (ClinicalTrials.gov Identifier NCT00442169) to further examine the safety, tolerability and immunogenicity of this vaccine. A human phase I trial designed to evaluate safety and immunogenicity of another chimeric vaccine (WNV/DEN4delta30) in which WNV PrM and E genes are inserted into an attenuated dengue virus strain has also been initiated (ClinicalTrials.gov Identifier NCT00537147 and NCT00537147). Three WNV vaccines have been licensed for veterinary (equine) use (Gubler, 2007).

References

Ali, M., Safriel, Y., Sohi, J., Llave, A., and Weathers, S. (2005). West Nile virus infection: MR imaging findings in the nervous system. AJNR Am J Neuroradiol 26, 289–297.

Almhanna, K., Palanichamy, N., Sharma, M., Hobbs, R., Sil, A. (2003). Clin Infect Dis 36:1629–1630.

Alshekhlee, A., Sultan, B., and Chandar, K. (2006). Opsoclonus persisting during sleep in West Nile encephalitis. Arch Neurol 63, 1324–1326.

Anderson, J.F. and Rahal, J.J. (2002). Efficacy of interferon alpha-2b and ribavirin against West Nile virus in vitro. Emerg Infect Dis 8, 107–108.

Batsis, J.A., and Phy, M.P. (2005). West Nile virus meningitis in a chronic immunosuppressed patient with rheumatoid arthritis. Clin Rheumatol 24:548–550.

Batuello, J.T., Youngwerth, J., and Gabel, R. (2005). Increased serum lipase in West Nile virus infection. N Engl J Med 352, 420–421.

Bode, A.V., Sejvar, J.J., Pape, W.J., Campbell, G.L., and Marfin, A.A. (2006). West Nile virus disease: a descriptive study of 228 patients hospitalized in a 4-county region of Colorado in 2003. Clin Infect Dis 42, 1234–1240.

Brien, J.D., Uhrlaub, J.L., and Nikolich-Zugich, J. (2007). Protective capacity and epitope specificity of CD8(+) T cells responding to lethal West Nile virus infection. Eur J Immunol 37, 1855–1863.

Brilla, R., Block, M., Geremia, G., and Wichter, M. (2004). Clinical and neuroradiologic features of 39 consecutive cases of West Nile Virus meningoencephalitis. J Neurol Sci 220, 37–40.

Cao, N.J., Ranganathan, C., Kupsky, W.J., and Li, J. (2005). Recovery and prognosticators of paralysis in West Nile virus infection. J Neurol Sci 236, 73–80.

Carson, P.J., Steidler, T., Patron, R., Tate, J.M., Tight, R., SmegoJr., R.A. (2003). Plasma cell pleocytosis in cerebrospinal fluid in patients with West Nile virus encephalitis. Clin Infect Dis 37:e12–e15.

Carson, P.J., Konweko, P., Wold, K.S., Mariani, P., Goli, S., Bergloff, P., and Crosby, R.D. (2006). Long-term clinical and neuropsychological outcomes of West Nile virus infection. Clin Infect Dis 43, 723–730.

Chan, C.K., Limstrom, S.A., Tarasewicz, D.G., and Lin, S.G. (2006). Ocular features of West Nile virus infection in North America: a study of 14 eyes. Ophthalmology 113, 1539–1546.

Chan-Tack, K.M. and Forrest, G. (2005). Failure of interferon-alpha 2b in patient with West Nile virus meningoencephalitis and acute flaccid paralysis. Scand J Infect Dis 37:944–946.

Chan-Tack, K.M. and Forrest, G. (2006). West Nile virus meningoencephalitis and acute flaccid paralysis after infliximab treatment. J Rheumatol 33:191–192.

Chowers, M.Y., Lang, R., Nassar, F., Ben David, D., Giladi, M., Rubinshtein, E., Itzhaki, A., Mishal, J., Siegman-Igra, Y., Kitzes, R., Pick, N., Landau, Z., Wolf, D., Bin, H., Mendelson, E., Pitlik, S.D., and Weinberger, M. (2001). Clinical characteristics of the West Nile fever outbreak, Israel, 2000. Emerg Infect Dis 7, 675–678.

Cunha, B.A. (2004). Differential diagnosis of West Nile encephalitis. Curr Opin Infect Dis 17, 413–420.

Cunha, B.A., Minnaganti, V., Johnson, D.H., and Klein, N.C. (2000). Profound and prolonged lymphocytopenia with West Nile encephalitis. Clin Infect Dis 31, 1116–1117.

Cunha, B.A., Sachdev, B., and Canario, D. (2004). Serum ferritin levels in West Nile encephalitis. Clin Microbiol Infect 10, 184–186.

Cunha, B.A., Eisenstein, L.E., Wirkowski, E., Klein, N.H., Johnson, D.H. (2006). West Nile encephalitis relapse presenting with abducens and facial nerve palsies. Am J Med 119:e1–2.

Davis, L.E., Debiasi, R., Goade, D.E., Haaland, K.Y., Harrington, J.A., Harnar, J.B., Pergam, S.A., King, M.K., Demasters, B.K., and Tyler, K.L. (2006). West Nile virus neuroinvasive disease. Ann Neurol 60, 286–300.

Debiasi, R.L. and Tyler, K.L. (2006). West Nile virus meningoencephalitis. Nat Clin Pract Neurol 2, 264–275.

Diamond, M.S., Shrestha, B., Marri, A., Mahan, D., and Engle, M. (2003). B cells and antibody play critical roles in the immediate defense of disseminated infection by West Nile encephalitis virus. J Virol 77, 2578–2586.

Diniz, J.A., Da Rosa, A.P., Guzman, H., Xu, F., Xiao, S.Y., Popov, V.L., Vasconcelos, P.F., Tesh, R.B. (2006). West Nile virus infection of primary mouse neuronal and neuroglial lines: the role of astrocytes in chronic infection. Am J Trop Med Hyg 75:691–696.

Doron, S.I., Dashe, J.F., Adelman, L.S., Brown, W.F., Werner, B.G., and Hadley, S. (2003). Histopathologically proven poliomyelitis with quadriplegia and loss of brainstem function due to West Nile virus infection. Clin Infect Dis 37, e74–e77.

Gandelman-Marton, R., Kimiagar, I., Itzhaki, A., Klein, C., Theitler, J., and Rabey, J.M. (2003). Electroencephalography findings in adult patients with West Nile virus-associated meningitis and meningoencephalitis. Clin Infect Dis 37, 1573–1578.

Gea-Banacloche, J., Johnson, R.T., Bagic, A., Butman, J.A., Murray, P.R., Agrawal, A.G. (2004). West Nile virus: pathogenesis and therapeutic options. Ann Intern Med 140: 545–554.

Gould, L.H., Sui, J., Foellmer, H., Oliphant, T., Wang, T., Ledizet, M., Murakami, A., Noonan, K., Lambeth, C., Kar, K., Anderson, J.F., de Silva, A.M., Diamond, M.S., Koski, R.A., Marasco, W.A., and Fikrig, E. (2005). Protective and therapeutic capacity of human single-chain Fv-Fc fusion proteins against West Nile virus. J Virol 79, 14606–14613.

Green, M.S., Weinberger, M., Ben Ezer, J., Bin, H., Mendelson, E., Gandacu, D., Kaufman, Z., Dichtiar, R., Sobel, A., Cohen, D., and Chowers, M.Y. (2005). Long-term Death Rates, West Nile virus epidemic, Israel, 2000. Emerg Infect Dis 11, 1754–1757.

Gubler, D.J. (2007). The continuing spread of West Nile virus in the western hemisphere. Clin Infect Dis 45, 1039–1046.

Haaland, K.Y., Sadek, J., Pergam, S., Echevarria, L.A., Davis, L.E., Goade, D., Harnar, J., Nofchissey, R.A., Sewel, C.M., and Ettestad, P. (2006). Mental status after West Nile virus infection. Emerg Infect Dis 12, 1260–1262.

Haley, M., Retter, A.S., Fowler, D., Gea-Banacloche, J., and O'Grady, N.P. (2003). The role for intravenous immunoglobulin in the treatment of West Nile virus encephalitis. Clin Infect Dis 37, e88–e90.

Hamdan, A., Green, P., Mendelson, E., Kramer, M.R., Pitlik, S., and Weinberger, M. (2002). Possible benefit of intravenous immunoglobulin therapy in a lung transplant recipient with West Nile virus encephalitis. Transpl Infect Dis 4, 160–162.

Hayes, E.B. and O'Leary, D.R. (2004). West Nile virus infection: a pediatric perspective. Pediatrics 113, 1375–1381.

Hayes, E.B., Sejvar, J.J., Zaki, S.R., Lanciotti, R.S., Bode, A.V., and Campbell, G.L. (2005). Virology, pathology, and clinical manifestations of West Nile virus disease. Emerg Infect Dis 11, 1174–1179.

Hershberger, V.S., Augsburger, J.J., Hutchins, R.K., Miller, S.A., Horwitz, J.A., and Bergmann, M. (2003). Chorioretinal lesions in nonfatal cases of West Nile virus infection. Ophthalmology 110, 1732–1736.

Hrnicek, M.J., Mailliard, M.E. (2004). Acute west nile virus in two patients receiving interferon and ribavirin for hepatitis C. Am J Gastroenterol 99, 957.

Iwamoto, M., Jernigan, D.B., Guasch, A., Trepka, M.J., Blackmore, C.G., Hellinger, W.C., Pham, S.M., Zaki, S., Lanciotti, R.S., Lance-Parker, S.E., DiazGranados, C.A., Winquist, A.G., Perlino, C.A., Wiersma, S., Hillyer, K.L., Goodman, J.L., Marfin, A.A., Chamberland, M.E., and Petersen, L.R. (2003). Transmission of West Nile virus from an organ donor to four transplant recipients. N Engl J Med 348, 2196–2203.

Jamison, S.C., Michaels, S.R., Ratard, R., Sweet, J.M., and Deboisblanc, B.P. (2007). A 41-year-old HIV-positive man with acute onset of quadriplegia after West Nile virus infection. South Med J 100, 1051–1053.

Jeha, L.E., Sila, C.A., Lederman, R.J., Prayson, R.A., Isada, C.M., and Gordon, S.M. (2003). West Nile virus infection: a new acute paralytic illness. Neurology 61, 55–59.

John, T.R., Nash, D., Maldin, B., et al. (2003). Persistence of virus-reactive serum immunoglobulin m antibody in confirmed West Nile virus encephalitis cases. Emerg Infect Dis 9:376–379.

Kaiser, P.K., Lee, M.S., and Martin, D.A. (2003). Occlusive vasculitis in a patient with concomitant West Nile virus infection. Am J Ophthalmol 136, 928–930.

Kalil, A.C., Devetten, M.P., Singh, S., Lesiak, B., Poage, D.P., Bargenquast, K., Fayad, P., and Freifeld, A.G. (2005). Use of interferon-alpha in patients with West Nile encephalitis: report of 2 cases. Clin Infect Dis 40, 764–766.

Kapoor, H., Signs, K., Somel, P., Downes, F.P., Clark, P.A., Massey, J.P. (2004). Persistence of West Nile virus (WNV) IgM antibodies in cerebrospinal fluid from patients with CNS disease. J Clin Virol 31:289–291.

Khairallah, M., Ben Yahia, S., Ladjimi, A., Zeghidi, H., Ben Romdhane, F., Besbes, L., Zaouali, S., and Messaoud, R. (2004). Chorioretinal involvement in patients with West Nile virus infection. Ophthalmology 111, 2065–2070.

Kiberd, B.A., and Forward, K. (2004). Screening for West Nile virus in organ transplantation: a medical decision analysis. Am J Transplant 4, 1296–1301.

Klee, A.L., Maidin, B., Edwin, B., Poshni, I., Mostashari, F., Fine, A., Layton, M., and Nash, D. (2004). Long-term prognosis for clinical West Nile virus infection. Emerg Infect Dis 10, 1405–1411.

Klein, C., Kimiagar, I., Pollak, L., Gandelman-Marton, R., Itzhaki, A., Milo, R., and Rabey, J.M. (2002). Neurological features of West Nile virus infection during the 2000 outbreak in a regional hospital in Israel. J Neurol Sci 200, 63–66.

Kleinschmidt-Demasters, B.K., Marder, B.A., Levi, M.E., Laird, S.P., McNutt, J.T., Escott, E.J., Everson, G.T., and Tyler, K.L. (2004). Naturally acquired West Nile virus encephalomyelitis in transplant recipients: clinical, laboratory, diagnostic, and neuropathological features. Arch Neurol 61, 1210–1220.

Kramer, L.D., Li, J., and Shi, P.Y. (2007). West Nile virus. Lancet Neurol. 6, 171–181.

Kumar, D., Prasad, G.V., Zaltzman, J., Levy, G.A., and Humar, A. (2004). Community-acquired West Nile virus infection in solid-organ transplant recipients. Transplantation 77, 399–402.

LaBeaud, A.D., Lisgaris, M.V., King, C.H., and Mandalakas, A.M. (2006). Pediatric West Nile virus infection: neurologic disease presentations during the 2002 epidemic in Cuyahoga County, Ohio. Pediatr Infect Dis J 25, 751–753.

Lanciotti, R.S., and Kerst, A.J. (2001). Nucleic acid sequence-based amplification assays for rapid detection of West Nile and St. Louis encephalitis viruses. J Clin Microbiol 39, 4506–4513.

Leis, A.A., Stokic, D.S., Polk, J.L., Dostrow, V., and Winkelmann, M. (2002). A poliomyelitis-like syndrome from West Nile virus infection. N Engl J Med 347, 1279–1280.

Leis, A.A., Stokic, D.S., Webb, R.M., Slavinski, S.A., and Fratkin, J. (2003). Clinical spectrum of muscle weakness in human West Nile virus infection. Muscle Nerve 28, 302–308.

Li, J., Loeb, J.A., Shy, M.E., Shah, A.K., Tselis, A.C., Kupski, W.J., and Lewis, R.A. (2003). Asymmetric flaccid paralysis: a neuromuscular presentation of West Nile virus infection. Ann Neurol 53, 703–710.

Mawhorter, S.D., Sierk, A., Staugaitis, S.M., Avery, R.K., Sobecks, R., Prayson, R.A., Procop, G.W., Yen-Lieberman, B. (2005). Fatal West Nile virus infection after rituximab/fludarabine-induced remission for non-Hodgkin's lymphoma. Clin Lymphoma Myeloma 6:248–250.

Monath, T.P., Liu, J., Kanesa-Thasan, N., Myers, G.A., Nichols, R., et al. (2006). A live attenuated recombinant West Nile virus vaccine. Proc Natl Acad Sci USA 103:6694–6699.

Moon, T.D., Nadimpalli, A., Martin, E.B., Ortiz, M.A., and Van Dyke, R.B. (2005). Balance and gait abnormalities of a child with West Nile virus infection. Pediatr Infect Dis J 24, 568–570.

Morrey, J.D., Siddharthan, V., Olsen, A.L., Wang, H., Julander, J.G., Hall, J.O., Li, H., Nordstrom, J.L., Koenig, S., Johnson, S., and Diamond, M.S. (2007). Defining limits of treatment with humanized neutralizing monoclonal antibody for West Nile virus neurological infection in a hamster model. Antimicrob Agents Chemother 51, 2396–2402.

Murray, K., Baraniuk, S., Resnick, M., Arafat, R., Kilborn, C., Cain, K., Shallenberger, R., York, T.L., Martinez, D., Hellums, J.S., Hellums, D., Malkoff, M., Elgawley, N., McNeely, W., Khuwaja, S.A., and Tesh, R.B. (2006). Risk factors for encephalitis and death from West Nile virus infection. Epidemiol Infect 134, 1325–1332.

Murray, K.O., Resnick, M., and Miller, V. (2007). Depression after infection with West Nile virus. Emerg Infect Dis 13, 479–481.

Nash, D., Mostashari, F., Fine, A., Miller, J., O'Leary, D., Murray, K., Huang, A., Rosenberg, A., Greenberg, A., Sherman, M., Wong, S., and Layton, M. (2001). The outbreak of West Nile virus infection in the New York City area in 1999. N Engl J Med 344, 1807–1814.

Natarajan, N., and Varman, M. (2007). West Nile virus cerebellitis in a healthy 10-year-old child. Pediatr Infect Dis J 26, 767.

Patnaik, J.L., Harmon, H., and Vogt, R.L. (2006). Follow-up of 2003 human West Nile virus infections, Denver, Colorado. Emerg Infect Dis 12, 1129–1131.

Pealer, L.N., Marfin, A.A., Petersen, L.R., Lanciotti, R.S., Page, P.L., Stramer, S.L., Stobierski, M.G., Signs, K., Newman, B., Kapoor, H., Goodman, J.L., and Chamberland, M.E. (2003). Transmission of West Nile virus through blood transfusion in the United States in 2002. N Engl J Med 349, 1236–1245.

Penn, R.G., Guarner, J., Sejvar, J.J., Hartman, H., McComb, R.D., Nevins, D.L., Bhatnagar, J., Zaki, S.R. (2006). Persistent neuroinvasive West Nile virus infection in an immunocompromised patient. Clin Infect Dis 42:680–683.

Pepperell, C., Rau, N., Krajden, S., Kern, R., Humar, A., Mederski, B., Simor, A., Low, D.E., McGeer, A., Mazzulli, T., Burton, J., Jaigobin, C., Fearon, M., Artsob, H., Drebot, M.A., Halliday, W., and Brunton, J. (2003). West Nile virus infection in 2002: morbidity and mortality among patients admitted to hospital in southcentral Ontario. CMAJ 168, 1399–1405.

Petersen, L.R., Roehrig, J.T., and Hughes, J.M. (2002). West Nile virus encephalitis. N Engl J Med 347, 1225–1226.

Petropoulou, K.A., Gordon, S.M., Prayson, R.A., and Ruggierri, P.M. (2005). West Nile virus meningoencephalitis: MR imaging findings. AJNR Am J Neuroradiol 26, 1986–1995.

Planitzer, C.B., Modrof, J., and Kreil, T.R. (2007). West Nile virus neutralization by US plasma-derived immunoglobulin products. J Infect Dis 196, 435–440.

Prasad, S., Brown, M.J., and Galetta, S.L. (2006). Transient downbeat nystagmus from West Nile virus encephalomyelitis. Neurology 66, 1599–1600.

Procop, G.W., Yen-Lieberman, B., Prayson, R.A., Gordon, S.M. (2004). Mollaret-like cells in patients with West Nile virus infection. Emerg Infect Dis 10:753–754.

Pyrgos, V., and Younus, F. (2004). High-dose steroids in the management of acute flaccid paralysis due to West Nile virus infection. Scand J Infect Dis 36:509–512.

Rawal, A., Gavin, P.J., and Sturgis, C.D. (2006). Cerebrospinal fluid cytology in seasonal epidemic West Nile virus meningo-encephalitis. Diagn Cytopathol 34, 127–129.

Rodriguez, A.J., and Westmoreland, B.F. (2007). Electroencephalographic characteristics of patients infected with West Nile virus. J Clin Neurophysiol 24, 386–389.

Roehrig, J.T., Nash, D., Maldin, B., Labowitz, A., Martin, D.A., Lanciotti, R.S., Campbell, G.L. (2003). Persistence of virus-reactive serum immunoglobulin M antibody in confirmed West Nile virus encephalitis cases. Emerg Infect Dis 9:376–379.

Ross, J.J., and Worthington, M.G. (2004). Bilateral sixth nerve palsy in West Nile meningoencephalitis. J Neuro-ophthalmol 24:97–98.

Samuel, M.A., and Diamond, M.S. (2005). Alpha/beta interferon protects against lethal West Nile virus infection by restricting cellular tropism and enhancing neuronal survival. J Virol 79:13350–13361.

Samuel, M.A., Wang, H., Siddharthan, V., Morrey, J.D., and Diamond, M.S. (2007). Axonal transport mediates West Nile virus entry into the central nervous system and induces acute flaccid paralysis. Proc Natl Acad Sci USA 104, 17140–17145.

Sayao, A.L., Suchowersky, O., Al-Khathaami, A., Klassen, B., Katz, N.R. et al. (2004). Calgery experience with West Nile virus neurological syndrome during the late summer of 2003. Can J Neurol Sci 31:194–203.

Sejvar, J.J. (2007). The long-term outcomes of human West Nile virus infection. Clin Infect Dis 44, 1617–1624.

Sejvar, J.J., Haddad, M.B., Tierney, B.C., Campbell, G.L., Marfin, A.A., Van Gerpen, J.A., Fleischauer, A., Leis, A.A., Stokic, D.S., and Petersen, L.R. (2003a). Neurologic manifestations and outcome of West Nile virus infection. JAMA 290, 511–515.

Sejvar, J.J., Leis, A.A., Stokic, D.S., Van Gerpen, J.A., Marfin, A.A., Webb, R., Haddad, M.B., Tierney, B.C., Slavinski, S.A., Polk, J.L., Dostrow, V., Winkelmann, M., and Petersen, L.R. (2003b). Acute flaccid paralysis and West Nile virus infection. Emerg Infect Dis 9, 788–793.

Sejvar, J.J., Bode, A.V., Curiel, M., Marfin, A.A. (2004). Post-infectious encephalomyelitis associated with St. Louis encephalitis virus infection. Neurology 63:1719–21.

Sejvar, J.J., Bode, A.V., Marfin, A.A., Campbell, G.L., Ewing, D., Mazowiecki, M., Pavot, P.V., Schmitt, J., Pape, J., Biggerstaff, B.J., and Petersen, L.R. (2005). West Nile virus-associated flaccid paralysis. Emerg Infect Dis 11, 1021–1027.

Sejvar, J.J., Bode, A.V., Marfin, A.A., Campbell, G.L., Pape, J., Biggerstaff, B.J., and Petersen, L.R. (2006). West Nile Virus-associated flaccid paralysis outcome. Emerg Infect Dis 12, 514–516.

Shimoni, Z., Niven, M.J., Pitlick, S., and Bulvik, S. (2001). Treatment of West Nile virus encephalitis with intravenous immunoglobulin. Emerg Infect Dis 7, 759.

Solomon, T., Dung, N.M., Wills, B., Kneen, R., Gainsborough, M., Diet, T.V., Thuy, T.T., Loan, H.T., Khanh, V.C., Vaughn, D.W., White, N.J., and Farrar, J.J. (2003). Interferon alfa-2a in Japanese encephalitis: a randomised double-blind placebo-controlled trial. Lancet 361, 821–826.

Steele, N.P., and Myssiorek, D. (2006).West Nile virus induced vocal fold paralysis. Laryngoscope 116:494–496.

Tardei, G., Ruta, S., Chitu, V., Rossi, C., Tsai, T.F., Cernescu, C. (2000). Evaluation of immunoglobulin M (IgM) and IgG enzyme immunoassays in serologic diagnosis of West Nile virus infection. J Clin Microbiol 38:2232–2239.

Tesh, R.B., Siirin, M., Guzman, H., Travassos da Rosa, A.P., Wu, X., Duan, T., Lei, H., Nunes, M.R., Xiao, S.Y. (2005). Persistent west nile virus infection in the golden hamster: studies on its mechanism and possible implications for other flavivirus infections. J Infect Dis 192:287–295.

Tilley, P.A., Fox, J.D., Jayaraman, G.C., and Preiksaitis, J.K. (2007). Maculopapular rash and tremor are associated with West Nile fever and neurological syndromes. J Neurol Neurosurg Psychiatry 78, 529–531.

Torno, M., Vollmer, M., and Beck, C.K. (2007). West Nile virus infection presenting as acute flaccid paralysis in an HIV-infected patient: a case report and review of the literature. Neurology 68, E5–E7.

Tsai, T.F., Popovici, F., Cernescu, C., Campbell, G.L., and Nedelcu, N.I. (1998). West Nile encephalitis epidemic in southeastern Romania. Lancet 352, 767–771.

Tyler, K.L. (2004). West Nile virus infection in the United States. Arch Neurol 61, 1190–1195.

Tyler, K.L., Pape, J., Goody, R.J., Corkill, M., and Kleinschmidt-Demasters, B.K. (2006). CSF findings in 250 patients with serologically confirmed West Nile virus meningitis and encephalitis. Neurology 66, 361–365.

Tyler, K.L., AksamitJr., A.J., Keegan, B.M., Parisi, J.E. (2007). An 85-year-old man with chronic leukemia and altered mental status. Neurology 68:460–467.

Wang, T., Town, T., Alexopoulou, L., Anderson, J.F., Fikrig, E., and Flavell, R.A. (2004). Toll-like receptor 3 mediates West Nile virus entry into the brain causing lethal encephalitis. Nat Med 10, 1366–1373.

Watson, J.T., Pertel, P.E., Jones, R.C., Siston, A.M., Paul, W.S., Austin, C.C., and Gerber, S.I. (2004). Clinical characteristics and functional outcomes of West Nile Fever. Ann Intern Med 141, 360–365.

Weiss, D., Carr, D., Kellachan, J., Tan, C., Phillips, M., Bresnitz, E., and Layton, M. (2001). Clinical findings of West Nile virus infection in hospitalized patients, New York and New Jersey, 2000. Emerg Infect Dis 7, 654–658.

Zak, I.T., Altinok, D., Merline, J.R., Chander, S., Kish, K.K. (2005). West Nile virus infection. AJR 184:957–961.

5. Molecular Biology of West Nile Virus

Margo A. Brinton

Abstract

The single-stranded, positive-sense West Nile virus (WNV) RNA genome is about 11 kb in length and encodes a single polyprotein that is processed during and after translation into three structural (C, prM/M, and E) and seven nonstructural (NS1, NS2A, NS2B, NS3, NS4A, NS4B, and NS5) proteins. Recently obtained crystal structures of two structural and two nonstructural proteins of WNV provide new insights about the functions of these proteins. Although there are still many questions to be answered, a significant amount of data on the molecular biology of WNV and other flaviviruses has already been obtained. In this chapter, molecular aspects of virion and genome structure, the stages of the viral replication cycle in the cytoplasm of infected cells, viral protein function, conserved genomic elements, host factors involved in viral RNA replication, virus remodeling of cells, host genetic resistance, and virus virulence are discussed.

Keywords

Replication, Structural proteins, Nonstructural proteins, Protein–protein interactions, cis-Acting sequences, Conserved RNA structures, Host factors, RNA-protein interactions, Host cell remodeling

1 Introduction

West Nile virus (WNV) is a mosquito-borne flavivirus that primarily infects birds and occasionally infects humans and horses. In recent years, the frequency of WNV outbreaks in humans has increased and these outbreaks have been associated with a higher incidence of severe disease. The expansion of WNV into the Western Hemisphere in 1999 resulted in an increased effort to analyze the components of this virus and its interactions with the host at the molecular level. However, as data are not yet available for all aspects of WNV molecular biology, and since it is likely that in many cases the basic properties of other flaviviruses will be similar to those of WNV, information obtained for other flaviviruses is described when appropriate.

2 Virus Classification

WNV is a member of the family Flaviviridae, genus *Flavivirus*. The members of the genus *Flavivirus* are distantly related to the members of the two other genera in this family, *Pestivirus* and *Hepacivirus* (Heinz et al., 2000b). Although all Flaviviridae family members have a similar gene order, share conserved nonstructural protein motifs, and produce a single polyprotein, they differ in the *cis*-acting elements controlling viral RNA replication and translation. Translation of the pestivirus and hepacivirus polyproteins is controlled by an internal ribosome entry site (IRES), while the translation of the flavivirus polyprotein is dependent on a 7-methyl guanosine cap. The genomes of members of the genus *Flavivirus* contain 3′ and 5′ cyclization sequences that are not found in the genomes of members of the other two genera. Additional information on flaviviruses can be found in other recent reviews (Gubler et al., 2007; Lindenbach et al., 2007). The 73 members of the genus *Flavivirus* are subdivided into 12 antigenic serogroups (Poidinger et al., 1996). WNV is a member of the Japanese encephalitis virus (JEV) serogroup. Kunjin virus, endemic to Australia and Asia, is now considered a WNV subtype (Scherret et al., 2001). Although all WNV isolates constitute a single serotype, they have been grouped into two genetic lineages (1 and 2) on the basis of signature amino acid substitutions and deletions in the envelope protein sequence (Berthet et al., 1997). All of the WNV isolates so far associated with outbreaks of human disease are lineage 1 viruses (Jia et al., 1999; Lanciotti et al., 1999). Lineage 2 viruses are restricted to endemic enzootic infections in Africa. However, attenuated and virulent isolates from both lineages have been identified (Beasley et al., 2002).

3 Genome RNA

The WNV genome is a single-stranded RNA of ~11,000 nucleotides (Lanciotti et al., 1999). This positive-polarity genome functions as the only viral mRNA and is the template for synthesis of complementary negative strand RNA. The 3′ end of the genome RNA terminates with a conserved

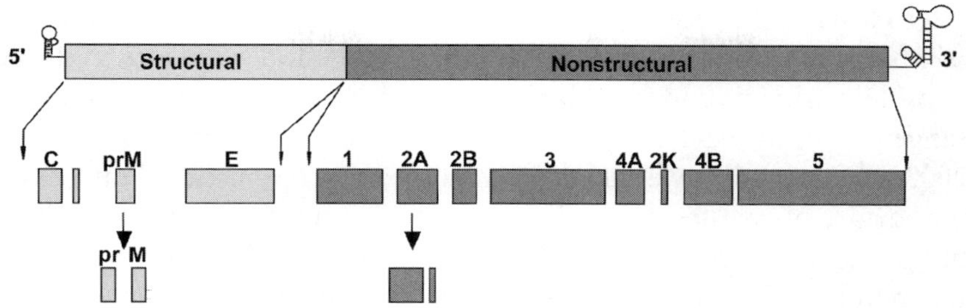

Figure 1. The genome of WNV. The noncoding regions are indicated by *black lines*, and the conserved terminal structures are indicated. The single open reading frame is indicated by a *box*. The structural gene region is *light gray*, whereas the nonstructural gene region is *dark gray*. The single polyprotein is co- and post-translationally processed by the viral NS2B–NS3 protease as well as cell proteases to the mature viral proteins, which are shown below the genome. The genome and products are not drawn to scale.

CU_{OH} (Rice et al., 1985; Brinton et al., 1986). Flaviviruses are the only known mammalian positive-strand RNA viruses that do not have a 3′ poly (A) tract. A type 1 cap structure (m^7GpppAmp) is added to the 5′ end of the genome in the cytoplasm of infected cells (Cleaves and Dubin, 1979). The 5′ noncoding region (NCR) of the WNV genome RNA is 96 nucleotides in length, while the 3′ NCR varies from 337 to 649 nucleotides. The variable region of the 3′ NCR is located just 3′ of the coding region stop codon (Shurtleff et al., 2001). A single open reading frame (ORF) of 10,301 nucleotides in most WNV isolates encodes a polyprotein that is co- and post-translationally processed by the viral serine protease complex (NS2b–NS3) and various cellular proteases into ten mature viral proteins (Fig. 1; Nowak et al., 1989; Lindenbach et al., 2007). The three viral structural proteins, capsid (C), membrane (prM/M), and envelope (E), are encoded within the 5′ portion of the genomic ORF, while the seven nonstructural proteins (NS1, NS2A, NS2B, NS3, NS4A, NS4B, and NS5) are encoded within the 3′ portion (Rice et al., 1985).

4 Virion Morphology and Proteins

Virions are small (~50 nm in diameter), spherical, and enveloped and have a buoyant density of ~1.2 g/cm^{-3}. In immature virions, prM, and E exist as trimers of heterodimers that form spikes. In this conformation, the fusion peptide of E is covered by the pr portion of prM and protected from premature fusion with cell membranes in the mildly acidic compartments of the cell secretory pathway during virion egress (Heinz et al., 1994; Zhang et al., 2003b). Following the release of pr after cleavage of prM by a cell furin-like protease in the *trans*-Golgi compartment, mature virions assume an icosahedral symmetry as the 60 E protein trimers undergo conformational change, rotation, and rearrangement to form 90 antiparallel dimers (Stadler et al., 1997; Kuhn et al., 2002; Mukhopadhyay et al., 2003; Zhang et al., 2007). The E dimers on the surface of mature virions lie relatively flat against the envelope lipid bilayer. The C-terminal portion (M) remains inserted in the envelope of the mature virion and contains a short ectodomain and two membrane spanning domains (Murray et al., 1993). The efficiency of prM cleavage varies, and some uncleaved prM is present on virions in most virus populations (Heinz et al., 1994).

The pr portion of WNV prM contains an N-linked glycosylation site at residue 15. The E proteins of most of the WNV strains are glycosylated at a single N-linked glycosylation site (residue 154). Other strains of WNV have a mutated N-linked glycosylation site and their E proteins are not glycosylated (Adams et al., 1995; Berthet et al., 1997). Glycosylation of both prM and E proteins affects the infectivity and release of virions but the effects vary with the strain of WNV and the cell type (Hanna et al., 2005). E and prM are type I integral membrane proteins with C-terminal membrane anchors. Cysteine residues in the E protein ectodomain are strictly conserved and form intramolecular disulfide bonds (Nowak and Wengler, 1987). The E protein is composed of three domains joined by flexible hinges. Domain III forms an immunoglobulin-like fold and has been postulated to interact with cell surface factors and facilitate virion entry. Domain II is a finger-like domain that contains the 13 conserved, hydrophobic residues of the internal fusion loop. Domain I is a β-barrel structure that links Domains II and III (reviewed by Mukhopadhyay et al., 2005).

The virion capsid (~30 nm in diameter) is composed solely of the C protein. No discernible symmetry for flavivirus nucleocapsids was detected in virion cryo-EM reconstructions (Zhang et al., 2003a). The precursor of C, designated anchored C, contains a hydrophobic region at its C-terminus, which is the signal sequence for translocation of prM in the polyprotein across the ER membrane. Mature C is generated by cleavage of the C-terminal anchor by the viral protease. The structure of C dimers determined by NMR and crystallography indicates that they are composed of monomers containing four α-helices (Jones et al., 2003; Dokland et al., 2004; Ma et al., 2004). The RNA binding domains are located at the C- and N-termini and are separated by a hydrophobic region. The C dimer, which is thought to be the basic component for assembling nucleocapsids, has a very high net positive charge, with half of the basic residues located on one face and the conserved hydrophobic region forming an apolar surface on the opposite side of the dimer (Ma et al., 2004). An encapsidation signal sequence in flavivirus genomic RNA that is recognized by the C protein has not been identified. RNA probes from either the 3′ or 5′ NCRs of Kunjin genomic RNA bind strongly to glutathione-*S*-transferase fusion proteins made from different regions of Kunjin C protein (Khromykh and Westaway, 1996). It is possible that the charged face of the C dimer binds to the genomic RNA in a nonspecific manner similar to the interactions between histones and cellular DNA, while the hydrophobic face binds to the inner side of the viral lipid membrane (Dokland et al., 2004; Ma et al., 2004). This idea is consistent with the observation that large deletions in the central hydrophobic region of the tick-borne encephalitis virus C protein were tolerated (Kofler et al., 2002). A recent study suggested that in addition to their role in nucleocapsid formation, flavivirus C proteins may also function as RNA chaperones, facilitating RNA structural rearrangements during the virus life cycle (Ivanyi-Nagy et al., 2007).

5 WNV Replication Cycle

WNV replicates in various types of primary and continuous cell cultures from a wide variety of avian, mammal, amphibian, and insect species. Cytopathology is observed in some cell cultures but not in others, even though virus replication is efficient (Brinton, 1986). The cell proteins that function as co-receptors for virion attachment and those that facilitate virion fusion/entry have not been definitively characterized for WNV or other flaviviruses. Various glycosaminoglycans including heparin have been reported to be low-affinity receptors for several flaviviruses (reviewed by Mukhopadhyay et al., 2005). Although the dendritic cell-specific lectins DC-SIGNR and DC-SIGN both bind mannose-rich glycans, only DC-SIGNR efficiently promotes WNV infection (Davis et al., 2006). Glycosylation of either prM or E is sufficient for WNV interaction with DC-SIGNR. However, WNV efficiently infects many types of cells that do not express DC-SIGNR, suggesting that other cell proteins can provide virus attachment and entry functions. WNV has been reported to adhere to red blood cells, but the cell proteins that facilitate this interaction have not been identified (Rios et al., 2007). WNV was reported to interact with $\alpha_v\beta_3$ integrin (Chu and Ng, 2004). Silencing the expression of this integrin increased the resistance of cells to infection and overexpression increased cell susceptibility. The E protein Domain III regions of several flaviviruses including WNV contain an RGD/RGE sequence, which is an integrin recognition motif.

However, RGD peptides did not inhibit WNV entry (Chu and Ng, 2004). Similarly, mutation of this motif in YFV did not affect the ability of the virus to infect cells (van der Most et al., 1999). Integrin interactions may be mediated by other regions of E or, alternatively, by the virion M protein. For reoviruses, β1 integrins mediate virion internalization via clathrin-dependent endocytosis but not virion attachment (Maginnis et al., 2006). However, a recent study showed that neither $\alpha_v\beta_3$ integrin nor focal adhesion kinase are utilized for WNV entry (Medigeshi et al., 2008). Rab 5 has been reported to be required for WNV and dengue entry (Krishnan et al., 2007). Other cellular proteins reported as putative flavivirus receptors include GRP78 (BiP), CD14-associated molecules, and HSP70/90 (Chen et al., 1999; Jindadamrongwech et al., 2004; Reyes-Del Valle et al., 2005).

After attachment, WNV enters both mammalian and mosquito cells via receptor-mediated endocytosis of clathrin-coated pits (Chu and Ng, 2004; Chu et al., 2006). The acidic environment of the endosome triggers the E protein to irreversibly associate into trimers. This leads to fusion of the viral membrane with the endosomal vesicle membrane and release of the nucleocapsid into the cytoplasm (Fig. 2a; Allison et al., 1995; Heinz and Allison, 2000). The genomic RNA is released from the capsid by an unknown mechanism and is translated (Fig. 2b). The polyprotein is co- and post-translationally cleaved at multiple sites to generate the mature viral proteins (Fig. 2c). Flavivirus infection does not result in the shut-off of host translation. Results from both differential centrifugation (Grun and Brinton, 1987; Chu and Westaway, 1992) and colocalization (Westaway et al., 1999; Mackenzie et al., 2007a) experiments indicate that flavivirus RNA replication complexes are associated with perinuclear, endoplasmic reticular (ER) membranes. The viral RNA-dependent RNA polymerase (RdRp), NS5, copies complementary negative polarity (–) strands from the positive polarity genomic (+) RNA template (Fig. 2d) and these (–) strand RNAs serve as templates for the synthesis of new genomic RNAs (Fig. 2e). The (–) strand RNA is synthesized throughout the replication cycle, but viral RNA synthesis is asymmetric; genomic RNA synthesis is more than ten times more efficient than (–) strand RNA synthesis (Cleaves et al., 1981). Data obtained with Kunjin virus suggested that only one nascent (–) RNA is copied from a (+) strand template at a time, while (–) strand templates are efficiently reinitiated and multiple nascent (+) strands are simultaneously copied from a single template in a semiconservative manner (Chu and Westaway, 1987). Once established, viral RNA synthesis can continue even in the absence of protein synthesis, indicating that transient viral polyprotein precursors are not required (Cleaves et al., 1981; Chu and

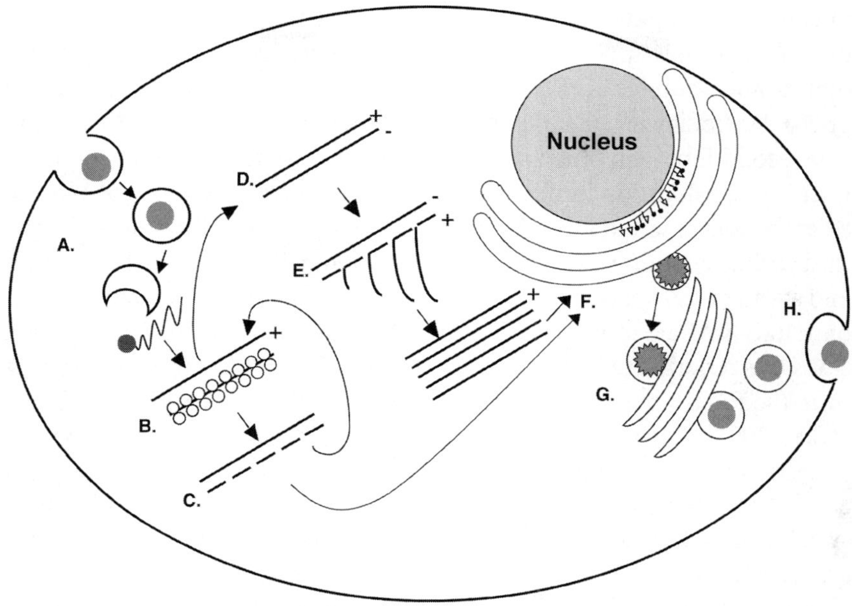

Figure 2. The replication cycle of WNV. (**a**) Attachment of a virion to a cell and entry via endocytosis followed by fusion of the viral and endocytic vesicle membranes and release of the nucleocapsid and genome RNA. (**b**) Translation of the genome RNA. (**c**). Co- and post-translational proteolytic processing of the polyprotein. (**d**) Synthesis of (–) strand RNA from the genomic (+) RNA. (**e**) Synthesis of nascent (+) RNA from (–) strand RNA. RNA synthesis occurs in perinuclear, ER membrane-associated replication complexes. (**f**) Association of viral structural proteins with cytoplamic vesicle membranes, encapsidation of nascent genome RNA, and assembly of immature virions (spiked particles). (**g**). Transport of immature virions through the cell secretory pathway to the cell surface. Mature virions form after prM is cleaved (smooth particles). (**h**) Release of nascent virions by fusion of vesicle and plasma membranes.

Westaway, 1987). Extensive reorganization and proliferation of cytoplasmic, perinuclear ER membranes redistribution of cellular cholesterol are observed in infected cells (Lindenbach et al., 2007; Mackenzie et al., 2007b). Data obtained with a Kunjin replicon suggest that translation is a prerequisite for replication of genomic RNAs (Khromykh et al., 1999b) and that RNA replication is a prerequisite of encapsidation (Khromykh et al., 2001b). However, translation must predominate over RNA replication to generate nascent virions, which contain 180 copies of each structural protein and one genome.

Virion assembly occurs in association with rough ER membranes (Fig. 2f). Although the viral envelope is obtained from cell membrane, little is known about this process because budding intermediates and free nucleocapsids in the cytoplasm have rarely been observed in flavivirus-infected cells. Although nucleocapsid-like particles have been successfully assembled in vitro from C dimers and RNA or DNA (Kiermayr et al., 2004), assembly of virions in cells appears to occur by coordinated interactions between C dimers and genomic RNA as well as C dimers and membrane-associated prM-E heterodimers rather than by the budding of preformed nucleocapsids. The second E protein transmembrane domain, which is the signal sequence for NS1 translocation in the polyprotein, was recently reported to play an essential role in virion assembly (Orlinger et al., 2006). Intracellular immature virions containing heterodimers of E and prM accumulate in vesicles and are then transported through the host secretory pathway where glycans on E and prM are modified (Fig. 2g). The activity of the Src family kinase c-Yes was reported to be required for transit of assembled immature virions from the ER through the secretory pathway in WNV-infected cells (Hirsch et al., 2005). The N-terminal portion of prM on immature virions is cleaved in the *trans*-Golgi network by a cellular furin-like protease (Stadler et al., 1997) and then virions are transported to the plasma membrane in vesicles and released by exocytosis (Fig. 2h; Mason, 1989). Progeny virions are released from infected cells starting at 10–12 h after infection and maximal extracellular virus titers are usually observed by 24 h. The Sarafend strain of WNV has been reported to assemble virions at the plasma membrane and studies with WNV chimeras mapped this characteristic to a 5′ region that included the C, prM, E, and NS1 genes (Li et al., 2005). Infected cells also produce smooth, noninfectious, subviral particles called slowly sedimenting hemagglutinin (SHA), which are composed only of cell membranes inserted with E and M proteins as well as some prM (Russell et al., 1980; Allison et al., 2003).

6 Viral Nonstructural Proteins

The functions of the WNV nonstructural proteins have not yet been completely characterized. All seven of the flavivirus nonstructural proteins appear to be directly or indirectly involved in viral RNA synthesis. Structural information for several of the enzymatic nonstructural proteins has recently been obtained. It is likely that each of the flavivirus nonstructural proteins has multiple functions. Little is known about the interactions between individual viral nonstructural proteins or between viral nonstructural proteins and cell proteins that are required for remodeling the cell environment and for the formation of active viral RNA replication complexes.

NS2A, NS2B, NS4A, and NS4B are small, hydrophobic proteins that do not have conserved motifs characteristic of known enzymes. The structures and functions of these four proteins have not yet been well characterized. Each of these proteins has membrane-spanning regions and some or all of these proteins may facilitate the assembly and/or anchoring of the viral replication complexes on perinuclear membranes. NS2A, NS2B, NS4A, and NS4B could not be complemented *in trans* in a Kunjin replicon system suggesting that production *in cis* is required for correct insertion into ER membranes and incorporation into replication complexes (Khromykh et al., 2000). NS2A is membrane associated and localizes with viral replication complexes in infected cells (Mackenzie et al., 1998). Data from mutational analyses suggest that NS2A also participates in virion assembly (Kummerer and Rice, 2002; Liu et al., 2003). WNV and Kunjin NS2A have been reported to block type 1 IFN signaling (Liu et al., 2004, 2006). NS2B complexes with the viral serine protease NS3 and functions as a protease cofactor. The central domain of NS2B is required for cofactor activity and intercalates into the NS3 serine protease domain (Erbel et al., 2006).

Studies of dengue NS4A suggest that this protein contains three central transmembrane regions with the N-terminus located in the cytoplasm and the highly hydrophobic C-terminus located in the ER lumen (Miller et al., 2007). The C-terminal region of NS4A contains the signal sequence for translocation of NS4B into the lumen of the ER. Cleavage of a site upstream of this signal sequence by the YFV NS2B–NS3 on the cytoplasmic side of the ER membrane is a prerequisite for cleavage of the NS4A–NS4B junction by cell signalase (Lindenbach et al., 2007). The cleaved C-terminal fragment is called the 2K peptide. In infected cells, NS4A is located primarily in ER membrane–associated viral replication complexes (Miller et al., 2007) and interacts with NS1 (Lindenbach and Rice, 1999). Overexpression of NS4A in mammalian cells resulted in membrane rearrangements similar to those observed in infected cells (Roosendaal et al., 2006; Miller et al., 2007). NS4B colocalizes with viral replication complexes and associates with membranes in the absence of 2K (Westaway et al., 2002; Miller et al., 2006). NS4B may be post-translationally modified (Chambers et al., 1990; Preugschat and Strauss, 1991). Expression of Kunjin NS4B in mammalian cells resulted in protein accumulation in the perinuclear region, induction of cytoplasmic membrane proliferation, and some nuclear localization (Westaway et al., 1997a). Both NS4A and NS4B have been reported to block type I IFN signaling (Munoz-Jordan et al., 2003).

NS1 is a glycoprotein with 2 or 3 conserved N-linked glycosylation sites and 12 conserved cysteines that form disulfide bonds essential for virus viability (Blitvich et al., 2001; Wallis et al., 2004). The N-terminus of NS1 is inserted into the lumen of ER membranes via an E protein signal sequence that is cleaved by cell signalase (Falgout et al., 1989). NS2A is cleaved from the C-terminus of NS1 by an unknown membrane-bound host protease associated with the ER (Falgout and Markoff, 1995). Although most of NS1 is located within infected cells, it has also been detected on the cell surface and in the supernatant of infected mammalian cells (Smith and Wright, 1985). Immunization with NS1 can provide protection against flavivirus-induced encephalitis (Schlesinger et al., 1985; Gould et al., 1986). NS1 monomers and heat-labile dimers as well as truncated and elongated forms of this protein are detected in infected cells (Mason, 1989; Blitvich et al., 1999). NS1 monomers are soluble and hydrophilic but homodimers associate with membranes (Winkler et al., 1988, 1989). Secretion of NS1 from WNV infected Vero cells begins between 16 and 24 h after infection, while the release of virus particles begins between 8 and 16 h (Macdonald et al., 2005). NS1 is secreted as a soluble hexamer consisting of three homodimers (Flamand et al., 1999). A proline to leucine substitution at position 250 in Kunjin NS1 completely inhibited NS1 dimer formation, but NS1 was still secreted efficiently from cells infected with this mutant (Hall et al., 1999). However, virus production in Vero cells and infectivity for weanling mice were reduced for this mutant. Complete processing of exposed complex sugar chains is essential for NS1 secretion; NS1 is not secreted by mammalian cells treated with glycosylation inhibitors or by mosquito cells, which produce glycoproteins with high-mannose chains rather than complex sugars. NS1 and virus circulate in the blood of infected individuals and the abundance of WNV NS1 has been linked to disease severity in a hamster model (Macdonald et al., 2005). Both recombinant and cell-surface-associated WNV NS1 bind to glycoprotein factor H, which is the main circulating factor regulating activation of the alternative complement pathway (Chung et al., 2006). This interaction accelerates the breakdown of C3b which protects cells from complement-mediated lysis.

Several reports suggest a role for NS1 in viral RNA replication. Colocalization of NS1 with viral replication complexes has been observed in infected cells (Mackenzie et al., 1996; Westaway et al., 1997b). Mutation of N-linked glycosylation sites on NS1 dramatically reduced viral RNA replication levels (Muylaert et al., 1996), and a yellow fever virus (YFV) with a temperature-sensitive mutation in NS1 did not accumulate viral RNA at the nonpermissive temperature (Muylaert et al., 1997). Successful *trans*-complementation of YFV RNA containing a large in-frame deletion in NS1 with wild-type NS1 expressed from a Sindbis replicon has been reported (Lindenbach and Rice, 1997). Neither (+) nor (–) strand viral RNA accumulated unless NS1 was supplied *in trans*, suggesting that NS1 functions prior to or early during (–)

strand RNA synthesis, the first type of viral RNA synthesized in infected cells (Fig. 2). The inability of a dengue NS1 to *trans*-complement the YFV NS1 deletion mutant suggested that NS1 interacts with another viral protein in a sequence-specific manner (Lindenbach and Rice, 1999). YFV variants that could utilize the dengue NS1 had a single point mutation in the NS4A gene, suggesting that an interaction between NS1 and NS4A is required for viral RNA synthesis.

A glycosyl–phosphatidylinositol linked form of NS1 was detected on the surface of infected cells as well as on cells expressing only recombinant NS1 (Jacobs et al., 2000). Incubation of infected cells with anti-NS1 antibody induced signal transduction as detected by tyrosine phosphorylation of cellular proteins. The role of an NS1-mediated cell activation mechanism during the flavivirus replication cycle is not yet known. Recombinant dengue 2 NS1 has been reported to interact with activated STAT3b in cytoplasmic vesicles (Chua et al., 2005).

NS3 is a highly conserved, multifunctional protein. The N-terminal 175 residues of NS3 comprise a serine protease that is a member of the trypsin superfamily (Bazan and Fletterick, 1989; Gorbalenya et al., 1989a; Wengler et al., 1991). However, this region of NS3 is not an active protease until it forms a stable complex with NS2B, and membrane association of the NS3–NS2B complex is required for efficient polyprotein processing. The NS3–NS2B protease complex cleaves the viral polyprotein at multiple sites within the nonstructural region consisting of two basic amino acids followed by an amino acid with a short side chain (Yusof et al., 2000). Crystal structures of WNV NS2B–NS3pro with and without a substrate mimetic supported an "induced fit" mechanism of catalysis with the oxyanion hole forming correctly only in the presence of an authentic substrate or a close mimic with an appropriate P1′ residue (Aleshin et al., 2007). The residue interactions provided a rationale for the preference of the WNV protease for glycine at the P1′ position of a cleavage site as well as for the preferences of other flavivirus proteases for serine or threonine at this position. NS2B–NS3pro can exist in two conformations that differ in the position of NS2B in relation to NS3 (Aleshin et al., 2007). In one, NS2B wraps around NS3 and completes the active site by contributing a β-fold to the chymotrypsin-like fold, and also provides a negatively charged surface at the protease S2 specificity site. In the other, the NS2B chain changes direction shortly after entering the C-terminal lobe of NS3 and this conformation is not expected to have protease activity. The second conformation may provide a mechanism for inactivating the protease when the C-terminal region of the protein becomes involved in RNA replication.

The remainder of NS3 contains motifs with homology to supergroup 2 RNA helicases (Gorbalenya et al., 1989a,b), to an RNA-stimulated nucleoside triphosphatase (NTPase) (Wengler and Wengler, 1991), and to an RNA triphosphatase (RTPase) (Wengler and Wengler, 1993); the activities of each of these enzymes have been confirmed. Crystal structures of this region of NS3 show that the NTPase and helicase domains overlap within two domains and that there is a third helicase domain that interacts with viral RNA and proteins (Wu et al., 2005; Xu et al., 2005). Although the RTPase is located in the C-terminal region of NS3, it utilizes the Walker B motif of the helicase-NTPase catalytic center for phosphodiester bond hydrolysis (Benarroch et al., 2004b). The RTPase is thought to dephosphorylate the 5′ end of nascent viral RNAs prior to cap addition. RNA helicases are motor proteins that can travel along RNA in a 3′ to 5′ direction fueled by ATP hydrolysis and this movement opens secondary structures and displaces proteins bound to RNA (Frick et al., 2007). Helicase activity is required for flavivirus replication, but the exact functions of the helicase during the virus replication cycle have not yet been defined. Dengue infectious clones with helicase inactivating mutations produced no virus (Matusan et al., 2001) and infectious clones of a pestivirus (bovine viral diarrhea virus) containing deletions or point mutations that inactivated either the helicase or the helicase and NTPase of NS3 produced no detectable (–) strand RNA and no infectious virus (Gu et al., 2000). Deletions that inactivated the NS3 protease domain could not be complemented *in trans* (Liu et al., 2002).

Interaction between NS3 and NS5 has been demonstrated with recombinant proteins as well as in infected cells by co-immunoprecipitation and immunoblotting (Kapoor et al., 1995). This interaction can occur in the absence of other viral proteins, but may depend on the phosphorylation state of NS5. The interacting regions were mapped to the C-terminus of dengue NS3 (residues 303–618) and the N-terminus of dengue NS5 (residues 320–368). Both the in vitro NTPase and RTPase activities of NS3 were enhanced by addition of recombinant NS5 (Cui et al., 1998; Yon et al., 2005), suggesting that these two NS3 activities, and possibly also the helicase activity, might be regulated by the interaction between these two proteins. The interaction between NS3 and NS5 may also coordinate helicase, NTPase, polymerase, and capping activities during viral RNA synthesis. An interaction between a recombinant dengue NS3 protein and regions of the viral 3′ NCR RNA has also been reported (Chen et al., 1997; Cui et al., 1998). However, although apparently specific, this RNA–protein interaction is relatively weak, which may explain why an interaction between a WNV 3′ RNA probe and NS3 in an infected cytoplasmic extract was not detected under more stringent assay conditions (Blackwell and Brinton, 1995). Although RNA replication by a Kunjin replicon containing deletions in NS3 could be complemented *in trans* by a helper replicon, this mutant genome was not packaged (Liu et al., 2002). Further studies showed that translation of NS3 *in cis* was required for genome packaging (Pijlman et al., 2006). However, the mechanism by which NS3 associates with the genome RNA template from which it is translated is not known. Mutations in NS2A were shown to block virion assembly and the effect of a mutation in NS2A on virus assembly was suppressed by a mutation on the surface of NS3 (Kummerer and Rice, 2002; Liu et al., 2003). These results suggest that NS2A interacts with replication complexes and is involved in coordinating the assembly of nacent genomes into virions. Overexpression of the NS3 proteins of dengue 2, Langat, or WNV was reported to induce apoptosis, possibly due to activation of caspase-8 (Prikhod'ko et al., 2002; Shafee and AbuBakar, 2003; Ramanathan et al., 2006). However, whether or not NS3 is involved in inducing cytopathology in infected cells is not known.

NS5 is the C-terminal protein of the viral polyprotein (Fig. 1) and is the largest and most conserved of the mature flavivirus proteins. The N-terminal region of NS5 contains an *S*-adenosyl methionine methyltransferase (MTase) domain, which is part of the virus RNA capping machinery (Koonin, 1991). Both N7 and 2′ *O* MTase activities have been demonstrated for the C-terminal regions of WNV, YFV, and dengue NS5 proteins (Ray et al., 2006; Zhou et al., 2007). The N7 methyl transfer precedes the 2′*O* methyl transfer. These data suggested that the flavivirus MTases have an active center that can sequentially transfer a methyl group onto two acceptor positions that differ from each other chemically and conformationally. A hypothetical scheme for the flavivirus capping pathway was proposed based on the crystal structures of a dengue MTase complexed with either a cap analog, a capped RNA, or an uncapped RNA (Egloff et al., 2002, 2007). According to this scheme, GTP first binds to a GTP-binding site in the viral MTase and is then transferred by the MTase to the 5′ end of a nascent viral RNA that has been dephosphorylated by the NS3 RTPase. Repositioning of the capped RNA within the MTase allows methylation at the N7 position. The cofactor *S*-adensyl-l-methionine is then reloaded, and the capped RNA is repositioned so that the cap on the RNA is in the GTP binding site and 2′*O*-methylation occurs. This is followed by dissociation of the capped RNA and MTase. The proposed guanylyltranferase activity of the flavivirus MTases has not yet been demonstrated biochemically. The first two nucleotides of the type 1 cap structure ${}^{7me}GpppA_{2'OMe}G$ are completely conserved among flaviviruses (Cleaves and Dubin, 1979) and an A as the first nucleotide appears to be required for a unique stacked intermediate conformation in the proposed capping pathway (Egloff et al., 2007). The MTase interaction sites on the WNV 5′ RNA for the two methyltransferase reactions have been mapped (Dong et al., 2007). N7-methylation requires specific nucleotides at the second and third positions from the 5′ end of the genome RNA followed by a 5′ stem loop structure, while 2′*O*-methylation requires specific nucleotides at the first and third 5′ positions followed by a minimum of 20 nucleotides. Ribavirin 5′ triphosphate and acyclovir inhibit the 2′*O*-methyltransferase activity of flavivirus MTases and compete with GTP for binding to the MTase (Benarroch et al., 2004a). Mutation of the MTase GTP binding site in a dengue infectious clone was lethal (Hanley et al., 2002). Although small deletions in the MTase region could be complemented *in trans*, larger deletions in this part of NS5 could not (Khromykh et al., 1999b).

The C-terminal portion of NS5 contains motifs characteristic of all RdRps, including the highly conserved GDD motif. Mutation of this motif is lethal for virus replication (Kamer and Argos, 1984; Koonin, 1991). Both large deletions and inactivating mutations in the polymerase region of NS5 in a Kunjin replicon were complemented *in trans* by a helper virus (Khromykh et al., 1998, 1999a,b, 2000). However, complementation of NS5-deficient Kunjin replicons was much more efficient when all seven of the viral nonstructural proteins were expressed than when NS5 was expressed alone (Khromykh et al., 1999b). WNV NS5 colocalizes with viral replication complexes in infected cells (Mackenzie et al., 2007a) and recombinant NS5 has polymerase activity (Ackermann and Padmanabhan, 2001; Guyatt et al., 2001; Selisko et al., 2006). Crystal structures have been obtained for the polymerase regions of WNV and dengue NS5 (Malet et al., 2007; Yap et al., 2007). These polymerases form canonical right-hand RdRp structures consisting of finger, palm, and thumb domains but have some unique characteristics such as a priming loop with a different fold than the hepatitis C (HCV) RdRp. The tryptophan-800 is conserved among the members of the genus *Flavivirus* and is the critical residue in the priming loop required for de novo initiation. This residue is positioned in the WNV RdRp structure with its aromatic base stacked against the priming nucleotide and provides the initiation platform (Malet et al., 2007). Exactly how the MTase and RdRp regions interact is not yet known. However, *in silico* docking of a dengue MTase structure on the WNV RdRp structure suggested that the RNA substrate binding groove of the MTase domain is positioned near the RNA exit tunnel. A region containing nuclear localization sequences (NLS) in dengue NS5 was previously reported to interact with the 300 C-terminal amino acids of NS3 (Johansson et al., 2001).

This region is predicted to be located on the top of the back side of both the dengue and WNV RdRp structures (Malet et al., 2007). Interaction between this region and the C-terminus of NS3 would position the helicase close to the entrance of the RNA template tunnel. Part of the NLS region, which was originally proposed to be an interdomain region between the MTase and RdRp domains (Brooks et al., 2002), actually forms the two most N-terminal structural elements of the RdRp (Selisko et al., 2006).

Dengue NS5 has been reported to localize to both the nucleus and cytoplasm (Kapoor et al., 1995), and the NLS region was shown to facilitate NS5 nuclear localization. Two adjacent NLSs were defined in this region. The bNLS (aa 320–368) bound to NLS-binding importin α/β heterodimeric nuclear import receptor, while the a/bNLS (aa 369–389) bound to importin β (Johansson et al., 2001; Brooks et al., 2002). However, the NS5 of two strains of WNV did not localize to the nucleus (Mackenzie et al., 2007a). Cytoplasmic NS5 is associated with convoluted membranes and concentrated in replication complexes. This distribution pattern of WNV NS5 was consistent with the results of previous WNV-infected cell fractionation studies (Grun and Brinton, 1987; Chu and Westaway, 1992). Only about 10% of the total amount of NS5 was associated with the "heavy" membrane fractions that were enriched for NS3, NS2A, and NS2B/NS4A and contained the WNV polymerase activity, suggesting that NS5 may cycle in and out of the membrane-associated replication complexes formed by the other viral nonstructural proteins. Whether the nuclear localization of the NS5 of some flaviviruses plays a role in the viral replication cycle or in remodeling the cell is currently not known. Other functions have been reported for the NS5 proteins of some flaviviruses. Dengue 2 NS5 has been reported to upregulate the transcription and secretion of interleukin 8 via activation of various transcription factors (Medin et al., 2005). The NS5 proteins of both dengue and WNV are phosphorylated on serine and threonine residues by cellular serine/threonine kinases and the phosphorylation state of NS5 may regulate the association between NS5 and NS3 (Kapoor et al., 1995; Reed et al., 1998; Mackenzie et al., 2007a). A number of cellular phosphoproteins as well as a cellular kinase(s) coimmunoprecipitated with NS5, suggesting that cellular proteins may also complex with NS5 (Reed et al., 1998). The NS5 of a tick-borne encephalitis virus has been reported to bind to the receptors for IFNα/β and IFNγ and block phosphorylation of the Tyk2 and Jak1 kinases which prevents IFN signaling (Best et al., 2005).

7 In Vitro Polymerase Assays

It has not yet been possible to purify intact active flavivirus replication complexes from infected cells (Grun and Brinton, 1988) or to reconstruct them from recombinant proteins. In vitro polymerase activity was detected in cytoplasmic extracts from WNV and Kunjin infected cells (Grun and Brinton, 1986; Chu and Westaway, 1987). Enzyme activity was inhibited by high concentrations of NaCl and was dependent on the presence of Mg^{2+}, but Mn^{2+} could replace Mg^{2+} over a limited concentration range (Grun and Brinton, 1986). Elongation and release of nascent (+) strands from already initiated, endogenous (–) strand templates constituted most of the in vitro polymerase activity and only a limited amount of reinitiation was observed (Grun and Brinton, 1986; Chu and Westaway, 1987). Both uridylyl and adenylyl terminal transferase activities were associated with the cytoplasmic fractions that contained WNV polymerase activity (Grun and Brinton, 1986). Exogenous template-dependent synthesis was reported for nuclease-treated, dengue-infected cell extract. After addition of a 3′ viral RNA template, only × products twice the length of the template (2×) were generated by self-priming from the 3′ terminal hairpin on the template RNA. Short chimeric templates that contained both the 5′ and 3′ terminal genomic sequences produced mostly 2× product but also some template length (1×) product (You and Padmanabhan, 1999; You et al., 2001). However, the possibility that the 1× RNA represented end labeling of the template by RdRp terminal transferase activity was not excluded. Similar products were obtained in assays with recombinant flavivirus RdRps (Tan et al., 1996; Steffens et al., 1999; Ackermann and Padmanabhan, 2001). A low level of de novo (primer independent) synthesis of 1× RNA was reported when both the 5′ and 3′ cyclization sequences were included in a chimeric template RNA (Ackermann and Padmanabhan, 2001). However, sequences flanking the cyclization sequences were also required for initiation of the 1× product. A recent study compared the in vitro activities of several recombinant flavivirus RdRps (Selisko et al., 2006) and confirmed that most (80%) of the product copied from a chimeric viral template containing both the 3′ and 5′ cyclization sequences was self-primed 2× RNA. The labeled, template-sized 1× products observed were attributed to end labeling of the template by the terminal transferase activity of the dengue and WNV RdRps (Selisko et al., 2006). Both recombinant hepatitis C and bovine viral diarrhea virus RdRps were also reported to have an intrinsic terminal transferase activity (Ranjith-Kumar et al., 2001). The nucleotides added by these viral terminal transferases differed depending on the 3′ terminal sequence of the template RNA. The function of the RdRp terminal transferase activity may be to repair the 3′ termini of viral RNA templates. Consistent with the results obtained with WNV-infected cell extracts (Grun and Brinton, 1986), both Mg^{2+} and Mn^{2+} could act as catalytic ions for recombinant flavivirus RdRps (Selisko et al., 2006). However, de novo synthesis of short products by dengue, WNV, and Kunjin RdRps from homopolymeric poly(rC) templates was observed only when Mn^{2+} was included in the reactions. The observations that poly(rC) was the only homopolymeric template that was efficiently copied by recombinant flavivirus RdRps (Selisko et al., 2006) and that de novo synthesis was detected with chimeric viral templates only when a very high concentration of GTP was used (Nomaguchi et al., 2003) suggest that either initiation occurs at the conserved C in the second position of both (+) and (–) strand templates, or, alternatively, that a GTP binding site in the RdRp forms the initiation platform as has been proposed for HCV NS5B (Ferron et al., 2005; Selisko et al., 2006). The GTP-binding site located in the MTase does not interfere with RdRp activity (Egloff et al., 2002), and similar RdRp activities were observed for full-length NS5 and the C-terminal RdRp (Selisko et al., 2006). The WNV and dengue RdRps show less rate limitation at the early steps of synthesizing short products in vitro than does the HCV RdRp, suggesting that the conformations of the WNV and dengue RdRps are more flexible during the transition from initiation to elongation (Selisko et al., 2006). Template specificity is a property of viral polymerase complexes found in infected cells. These complexes are membrane bound and associated with cell proteins as well as other viral proteins. The in vitro polymerase activities of recombinant flavivirus NS5s are not template specific. Template specificity was not restored by the addition of Vero cell extracts to in vitro polymerase assays done with recombinant Kunjin NS5 (Tan et al 1996; Steffens, Thiel, and Behrens, 1999; Ackermann and Padmanabhan, 2001; Guyatt, Westaway, and Khromykh, 2001).

8 Conserved Viral RNA Terminal Structures and Sequences

8.1 Conserved Sequences

Short conserved sequences located within the 3′ terminal stem loop (SL) structure of flavivirus genomic RNA include the 3′ terminal 5′CU 3′ and a 5′CACAGA3′ sequence near the 3′ terminus (Fig. 3; Brinton et al., 1986). These conserved terminal sequences are also found in the same relative positions within the 3′ terminal region of the (–) strand RNA (complementary to the 5′ end of the genome), although they are in different structural contexts. Mutation of the top loop sequence of the 3′ terminal SL in a WNV infectious clone showed that

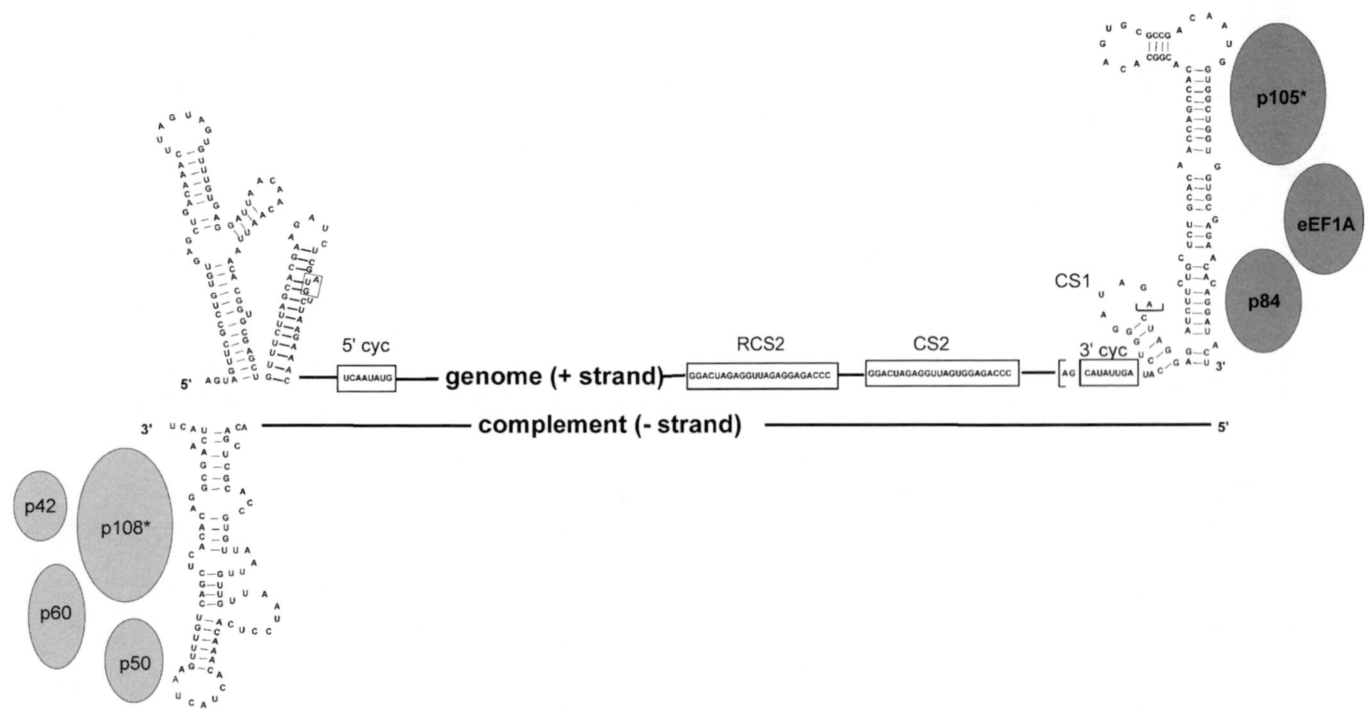

Figure 3. Conserved RNA structures and sequences of the WNV genome and complementary (–) strand RNA. Conserved sequences and the translation start codon are indicated by a *box* or *bracket*. Cell proteins that bind to the 3′ terminal regions of the (+) and (–) RNAs are indicated by *gray ovals*. The *asterisk* indicates that the 105 and 108 kDa proteins may be identical. CS – conserved sequence. RCS – Repeated conserved sequence. cyc – cyclization sequence.

most of these nucleotides were *cis*-acting and that three (5′ACAGUGC3′) were critical for virus viability (Elghonemy et al., 2005). A single copy of a sequence designated conserved sequence 1 (CS1) is located upstream of the 3′ terminal SL and partially overlaps the small 3′ SL in mosquito-borne flavivirus genomes (Fig. 3). An eight-nucleotide sequence, designated 3′ cyclization sequence (3′Cyc), is located within the 5′ portion of CS1 and is an exact complement of the eight-nucleotide 5′ Cyc sequence that is located in the capsid coding region near the 5′ end of the genome (Hahn et al., 1987). Base pairing between the 3′ and 5′ Cyc sequences, even though this interaction spans more than 10,000 nucleotides, was calculated as thermodynamically feasible (Hahn et al., 1987). Data from experiments using a chimeric dengue RNA (Alvarez et al., 2005), a dengue replicon in an in vitro system (You et al., 2001), and a Kunjin replicon in an in vivo system (Khromykh et al., 2001a), suggested that base pairing between the 5′ and 3′ Cyc sequences is required for RNA replication but not translation. Base pairing between the 3′ and 5′ Cyc sequences, not the sequences themselves, is required for this function. However, the mechanism by which the Cyc sequences facilitate virus RNA replication is not known. Much longer regions of complementarity between 3′ and 5′ sequences flanking the Cyc sequences have been predicted in genomes of the members of the genus *Flavivirus* (Gritsun and Gould, 2007a). Two copies of a second conserved sequence designated CS2 and RCS2 are also located in the 3′ NCR of the genomic RNA (Fig. 3; Lindenbach et al., 2007).

8.2 Secondary Structures

Conserved RNA secondary structures have been identified at both the 3′ and 5′ termini of the genome RNA (Fig. 3; Brinton et al., 1986; Brinton and Dispoto, 1988) The sizes and shapes of these structures are conserved among divergent flaviviruses even though most of the sequences composing them are not conserved (Rauscher et al., 1997). The predicted structures of the 3′ terminal (+) and (–) strand RNAs were demonstrated by structure probing (Brinton et al., 1986; Shi et al., 1996b). The small SL adjacent to the 3′ terminal SL has been predicted to form in both partial and full-length genomic sequences (Rauscher et al., 1997; Sgro and Palmenberg, unpublished data). Deletion of this small SL in a WNV infectious clone was lethal and mutation of the nucleotides in the loop of this structure reduced the efficiency of virus replication indicating that they are *cis*-acting (Davis and Brinton, unpublished data). Although composed of complementary sequences, the 5′(+) and 3′(–) SL structures differ because of the formation of G–U base pairs (Fig. 3). Deletion of either the 3′ or 5′ terminal genomic SL RNA sequences was lethal for flavivirus infectious clones (Lai et al., 1992; Cahour et al., 1995; Yu and Markoff, 2005). Replacement of the lower part of the 3′(+) SL in a dengue infectious clone with WNV sequences was also lethal, suggesting that this part of the 3′(+) SL was not exchangeable between these two viruses (Zeng et al., 1998). However, the effect of this sequence exchange on the predicted RNA structures was not assessed. Although dengue infectious clones with large deletions in other regions of the 3′ NCR produced virus that was attenuated and/or had altered phenotypes, the functions of these regions are not currently known (Men et al., 1996; Mandl et al., 1998). The regions containing the direct repeats (CS2 and RCS2) were predicted to form conserved "dumbbell-like" secondary structures and a recent analysis suggested that these repeats are evolutionary remnants of an ancient sequence consisting of six repeats (Olsthoorn and Bol, 2001; Gritsun and Gould, 2007b).

8.3 Tertiary Structures

Several tertiary interactions were previously predicted between sequences within the 3′ NCR of the flavivirus genome RNA. A prior study suggested the possibility that the 5′ side of the 3′ terminal SL could form a tertiary interaction with the loop of the adjacent small SL (Shi et al., 1996a). However, recent data do not support the existence of this tertiary interaction (Davis, Germann, and Brinton, unpublished data). Two additional pseudoknots were predicted in the regions containing the CS2 and RCS2 sequences in various flavivirus genomic RNAs but these interactions have not been experimentally confirmed (Olsthoorn and Bol, 2001).

9 Host Cell Proteins Interact with the WNV 3′ Terminal SLs and Facilitate RNA Synthesis

Data from several laboratories indicate that the 3′ terminal regions of both (+) and (–) strand RNA viruses have evolved specificity for interaction with different sets of cellular RNA binding proteins (Lai, 1998). For several viruses, including flaviviruses, different sets of cellular proteins bind to the 3′ terminal regions of the (+) and (–) strand RNAs (Fig. 3), suggesting that host proteins could be involved in differential regulation of RNA synthesis from (+) and (–) strand templates. Although general aspects of the regulatory mechanisms may be similar, the molecular details likely vary for the different families of RNA viruses. A number of cell proteins have been reported to bind to the 3′ terminal regions of flavivirus RNAs (Shi et al., 1996b; Blackwell and Brinton, 1997; Ta and Vrati, 2000; De Nova-Ocampo et al., 2002; Kim and Jeong, 2006; Paranjape and Harris, 2007; Yocupicio-Monroy et al., 2007). Evidence supporting the functional relevance in RNA replication for two of these cell proteins has recently been reported (Davis et al., 2007a; Emara and Brinton, 2008).

9.1 Cellular Proteins Bind to the 3′(+) SL RNA

Cellular proteins with molecular masses of 52, 84, and 105 kDa have been reported to bind specifically to the 3′ terminal SL of the genome RNAs of WNV (Fig. 3; Blackwell and Brinton, 1995). The 52 kDa protein was identified as eukaryotic elongation factor 1 alpha (eEF1A), and one major (60% of the binding activity) and two minor eEF1A binding sites (each 20% of the binding activity) on the WNV 3′(+) SL RNA were mapped by RNase footprinting and filter-binding assays (Fig. 4a; Blackwell and Brinton, 1997). The primary cellular function of eEF1A is to ferry aminoacylated tRNAs to the ribosomes (Riis et al., 1990). Anti-eEF1A antibody supershifts RNA–protein complexes formed between proteins in S100 cell extracts and the WNV 3′ SL RNA probe, and also co-immunoprecipitates WNV genomic RNA from infected cells (Blackwell and Brinton, 1997). The dissociation constant (K_d) for the eEF1A-WNV 3′(+) SL RNA interaction was calculated as 2.1×10^{-9} M, which is similar to that between eEF1A and aminoacylated tRNA (10^{-9} –10^{-10} M), suggesting that the interaction between the viral 3′(+) RNA and eEF1A could occur in infected cells (Blackwell and Brinton, 1997). eEF1A was also shown to bind with similar efficiency to the 3′(+) SL RNAs of three other flaviviruses (Davis et al., 2007). Since eEF1A is the second most abundant protein in cells after actin, the interaction between the viral 3′(+) SL RNA and eEF1A in infected cells is not expected to reduce the efficiency of cell translation. Base substitutions in the major eEF1A binding site or in adjacent areas of the 3′(+) SL were engineered into a WNV infectious clone. Mutations that decreased, as well as ones that increased, eEF1A binding to the 3′ SL RNA in in vitro interaction assays had a negative affect on virus growth (Davis et al., 2007a). Although the efficiency of polyprotein translation from the mutant RNAs was similar to that of the parental RNA, all of the mutations that decreased in vitro eEF1A binding also decreased (–) strand and (+)

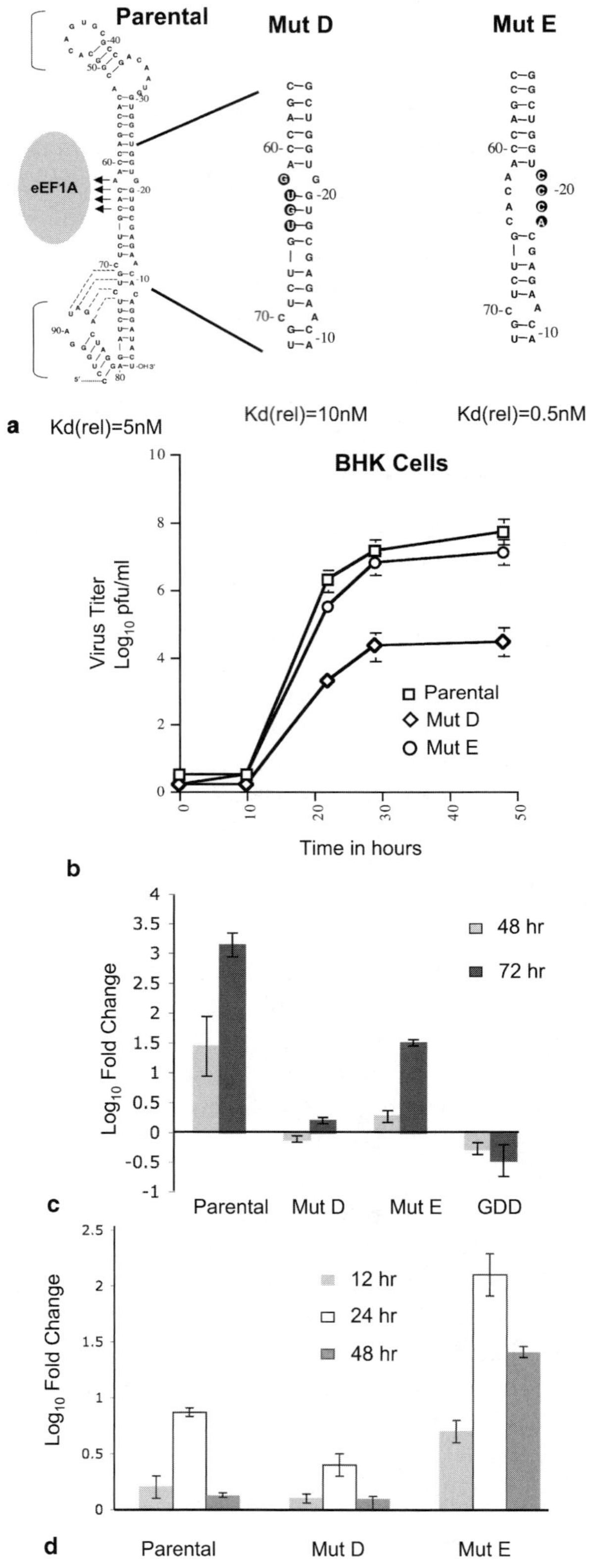
Parental
Mut D
Mut E
eEF1A
a
Kd(rel)=5nM
Kd(rel)=10nM
Kd(rel)=0.5nM
BHK Cells
Virus Titer Log10 pfu/ml
Parental
Mut D
Mut E
Time in hours
b
Log10 Fold Change
48 hr
72 hr
Parental
Mut D
Mut E
GDD
c
Log10 Fold Change
12 hr
24 hr
48 hr
d
Parental
Mut D
Mut E

strand viral RNA synthesis (Fig. 4). Of special note, a mutation that increased the efficiency of eEF1A binding to the 3′(+) SL RNA increased (–) strand RNA synthesis (Fig. 4d) but decreased genomic RNA synthesis (Fig. 4c). These results suggest that overproduction of (–) strand RNA is not beneficial for the virus and that the interaction between eEF1A and the WNV 3′(+) SL facilitates viral (–) strand synthesis. Focal concentrations of eEF1A colocalize with WNV replication complexes in infected cells and antibody to eEF1A co-immunoprecipitates viral nonstructural proteins from infected cell lysates, suggesting that eEF1A facilitates the interaction between the 3′ end of the genomic RNA and viral replication complexes in infected cells. In addition, eEF1A also colocalizes with viral replication complexes in dengue virus–infected cells and eEF1A bound in in vitro interaction assays with similar efficiency to the 3′(+) SL RNAs of four divergent flaviviruses (Davis et al., 2007).

9.2 Cellular Proteins Bind to the 3′(–) SL

Four cell proteins, p108, p60, p50, and p42, that bind specifically to the 3′ terminal SL (75 nucleotides) of the WNV complementary (–) strand RNA were detected in RNA–protein interaction assays (Fig. 3; Shi et al., 1996b). The 42 kDa protein was identified as T cell-restricted intracellular antigen-1 (TIA-1)-related protein (TIAR). TIA-1 was subsequently shown to also bind to the WNV 3′(–) SL RNA but with a ten times lower efficiency (Li et al., 2002). These two multifunctional proteins are members of the RRM family of RNA-binding proteins and are expressed in most tissues. They are normally present in both the cytoplasm and nucleus of cells and

Figure 4. Effects of mutations in the major eEF1A binding site of the 3′(+) SL of a WNV infectious clone. (**a**) The parental 3′ SL RNA is shown on the *left*. The minor eEF1A binding sites are indicated by *brackets*. The major eEF1A binding site is indicated by *arrows*. Substituted nucleotides in mutant RNAs are indicated by *black circles*. Relative binding dissociation constants ($K_{d\ rel}$) for interactions between recombinant eEF1A and various 3′ SL RNAs are shown. (**b**) Virus growth in BHK cell monolayers infected with mutant or parental WNV (MOI of 0.1). Error bars represent the standard error of the mean ($n = 4$). (**c**). Relative quantification (RQ) of intracellular WNV (+) genomic RNA at 48 and 72 h after RNA transfection compared to the level of viral RNA (input) present at 6 h by quantitative real-time reverse transcriptase polymerase chain reaction (qRT-PCR). (**d**) RQ of intracellular WNV (–) strand RNA at 12, 24, and 48 h after transfection compared to the level at 2 h by strand-specific real time qRT-PCR. Values in (**c**) and (**d**) are expressed as the log fold change in RQ units. Each RNA sample was normalized to cellular GAPDH mRNA. Error bars represent the standard error of the mean ($n = 3$).

shuttle between these compartments (Taupin et al., 1995; Dember et al., 1996). The K_d for the interaction between recombinant TIAR and the WNV 3′(–) SL RNA was estimated as greater than 10^{-8} M (Li et al., 2002). The replication of WNV, but not that of viruses from other families, was less efficient in TIAR-knockout cell lines, suggesting a functional role for this protein during WNV replication (Li et al., 2002). TIA-1 and TIAR were previously reported to bind to relatively long AU-rich elements (ARE) in the 3′ UTRs of some cellular mRNAs and silence translation (Dixon et al., 2003; Cok et al., 2004). The binding sites for TIAR and TIA-1 on the WNV 3′(–) SL RNA were mapped to two short AU sequences located in the two side loops, L1 and L2, of this RNA structure (Fig. 3; Emara and Brinton, 2008). The AU sequences in both loops were required for efficient binding of either protein in vitro. Deletion or substitution of the AU sequences in either loop in a WNV infectious clone produced a lethal phenotype and partial deletion or substitution of these sequences reduced the efficiency of virus replication. The translation efficiencies of mutant viral RNAs that produced virus with large, small plaque, or pinpoint plaque, or that had a lethal phenotype were similar to that of the parental RNA. The 3′(–) SL functions as the promoter for genomic RNA synthesis. The relative efficiencies of (+) strand RNA synthesis and virus production for mutant RNAs decreased with decreasing efficiency of in vitro protein binding efficiency. The detection of rescuing reversions in viral RNA after transfection of some mutant RNAs that bound inefficiently to TIAR/TIA-1 in vitro indicated that initial low level (+) and (–) strand RNA synthesis could occur in the absence of an efficient TIA-1/TIAR- 3′(–) SL RNA interaction and suggested that this RNA-protein interaction facilitated the asymetric amplification of genome RNA.

Colocalization of TIAR and TIA-1 with flavivirus replication complexes was detected in both WNV and dengue 2 virus infected cells (Emara and Brinton, 2007). The kinetics of TIAR accumulation in the perinuclear region of infected cells was similar to that of genomic RNA amplification. In contrast, relocation of TIA-1 to the perinuclear region began only after maximal levels of viral RNA synthesis were achieved, except when TIAR was absent in TIAR(–/–) cells. Neither WNV nor dengue 2 virus induced stress granule formation in infected cells. The results suggest that flaviviruses derive two benefits from the interaction between TIAR/TIA-1 and the 3′(–) SL RNA: enhancement of genomic RNA synthesis and prevention of cell stress granule formation which causes translational shutoff.

9.3 Virus Interactions with the Host Cell

At early times after infection, flaviviruses can suppress the activation of cell responses that lead to death. In some WNV infected cells, the expression of a subset of interferon-stimulated genes involved in apoptotic pathways is suppressed (Scherbik et al., 2007b). Also, stress granule formation is not induced in flavivirus infected cells (Emara and Brinton, 2007). However, accumulation of viral products at later times after infection can result in cytopathology. Bax-dependent apoptosis observed in WNV-infected K562 and Neuro-2a cells was shown to depend on virus replication since UV-inactivated virus failed to induce apoptosis (Parquet et al., 2001). Analysis of WNV-induced apoptosis in the brain tumor cell line T98G indicated that caspases 3, 8, and 9 were activated, poly(ADP-ribose) polymerase (PARP) was cleaved, and cytochrome C was released from the mitochondria, suggesting that both extrinsic and intrinsic apoptosis pathway components were involved (Kleinschmidt et al., 2007). Pan-caspase inhibition prevented PARP cleavage and WNV-induced apoptosis without decreasing the efficiency of virus replication and caspase 3 (–/–) mice were more resistant to lethal WNV infection even though virus tissue titers and the kinetics of viral spread were similar to those in control mice (Samuel et al., 2007). WNV infection in human neuroblastoma cells and primary rat hippocampal neurons was observed to activate multiple unfolded protein response (UPR) and apoptosis pathways (Medigeshi et al., 2007). The activation of the cellular UPR by other flaviviruses has also been reported (Yu et al., 2006). Induction of the proapoptotic cyclic AMP response element-binding transcription factor homologous protein (CHOP) was observed in cells infected with WNV and in cells transfected with a WNV replicon. The observation that the titers of WNV produced were significantly higher in CHOP knockout cells than in wild-type cells suggested that more rapid cell death limits virus production (Medigeshi et al., 2007).

Overexpression of the WNV C protein in HeLa cells induced apoptosis via a mitochondrial pathway that resulted in caspase-3 and -9 activation (Yang et al., 2002). The WNV C protein was subsequently reported to bind to the human homolog of murine double minute 2 (HDM2) oncogene and sequester it in nucleoli(Yang et al., 2008). This prevented the formation of the HDM2–p53 complex and led to induction of Bax (Yang et al., 2007). Flavivirus C proteins have been detected in both the cytoplasm and nucleoli of infected cells, and residues essential for nuclear localization were identified in the C-terminal region (Wang et al., 2002). Mutation of these residues in a Japanese encephalitis virus infectious clone prevented nuclear localization of C and also reduced the efficiency of viral replication suggesting that the C protein may have additional roles in viral replication. Jab-1 was identified as a putative binding partner of the WNV capsid protein, and additional studies suggested that the interaction of C with Jab-1 facilitates nuclear export via the CRM1 complex followed by proteasome degradation of the C protein (Oh et al., 2006). Other putative cell protein binding partners of the WNV C protein are Hsp70 (Oh and Song, 2006), B23 (Tsuda et al., 2006), and I2PP2A (Hunt et al., 2007).

9.4 Host Genetic Resistance to Flavivirus-Induced Disease

Variation in disease outcome among individuals after a flavivirus infection has been observed in humans as well as in other host species. In mice, the alleles of a single, autosomal gene, *Flv*, can determine whether or not a lethal infection develops (Brinton and Perelygin, 2003). Resistant strains include BRVR, BSVS, C3H/RV, Det, PRI, and most of the outbred, wild *Mus musculus domesticus* populations are resistant. The majority of inbred laboratory mouse strains are susceptible to flavivirus-induced disease. The resistance conferred by the Flv^r allele is flavivirus specific. Resistant animals and cell cultures can be infected by flaviviruses, but produce lower yields of virus than susceptible animals and cells. The *Flv* gene was first mapped to a region of murine chromosome 5 (Urosevic et al., 1995, 1997) and then identified as the 2′–5′ oligoadenylate synthetase gene 1b (*Oas1b*) (Mashimo et al., 2002; Perelygin et al., 2002). The Oas1b transcript in susceptible mice contains a premature stop codon and encodes a truncated protein. Although there is only a single *OAS1* gene in the human genome, there are eight *Oas1* genes in mice. Six of the mouse *Oas1* genes, including *Oas1b*, are inactive synthetases (Elbahesh and Brinton, unpublished data; Kakuta et al., 2002). After activation by double-stranded RNA (dsRNA), enzymatically active Oas proteins polymerize ATP into 2′–5′ linked oligoadenylates, which activate the single-strand specific, cytoplasmic endoribonuclease, RNase L. Activated RNase L degrades both viral and cellular single-stranded RNA. Although RNase L is activated in WNV-infected cells and this contributes to the cellular antiviral response to WNV infection, RNase L activity is not required for the resis-tance phenotype (Scherbik et al., 2006). Knock-in of the *Oas1b* resistance allele into a susceptible mouse strain produced mice that were resistant to flaviviruses and confirmed that this phenotype is mediated by a single gene (Scherbik et al., 2007a). The mechanism(s) by which the products of the different *Oas1b* alleles confer differential susceptibility to flavivirus-induced disease in mice and affect the level of viral RNA and virus produced by infected cells is not currently known. Even though there is no *Oas1b* homolog in humans, analysis of Oas1b-mediated resistance in mice may still provide new information about cell pathways which can reduce the efficiency of flavivirus replication. One study suggested that a polymorphism in the human *OASL* gene correlated with increased susceptibility to WNV-induced disease. However, only a small number of samples were analyzed and the frequency of the associated allele in the control population was lower than in the general population (Yakub et al., 2005). Although a deletion in the human CCR5 gene did not correlate with increased disease severity in WNV-infected individuals, a putative association with a fatal outcome was reported in one of two sample sets but only a few fatal cases were genotyped (Glass et al., 2006).

9.5 Virulence Determinants

For some strains of WNV, glycosylation of E at Asn 154 increases virion stability in mildly acidic conditions and facilitated neuroinvasion (Shirato et al., 2004; Beasley et al., 2005; Hanna et al., 2005). A T249P amino acid substitution in WNV NS3 increases the level of viremia and virulence in American crows but not in mammals (Brault et al., 2007). Mutations in the NS4B proteins of a number of flaviviruses correlate with attenuation and variation in plaque phenotype (Wicker et al., 2006). Site-directed mutation of three of the four highly conserved cysteines in NS4B in a WNV infectious clone had no effect but mutation of C102, which is predicted to be located near the junction of the central ectodomain and the third transmembrane region, attenuated both neuroinvasion and neurovirulence and produced a temperature-sensitive phenotype (Wicker et al., 2006). An adaptive mutation in persistently replicating WNV replicons in mouse cells mapped to E249G in the C-terminal tail of NS4B (Puig-Basagoiti et al., 2007). This mutation decreased the RNA replication efficiency of a WNV replicon and an infectious clone in rodent cells but had little effect on the efficiency of replication in mosquito cells, suggesting an interaction between NS4B and a mammalian cell protein. The presence of this mutation did not alter the ability of the virus to block type I IFN signaling.

In some cases, multiple mutations are responsible for the altered phenotypes observed. A small plaque phenotype, temperature sensitivity, and attenuation in mice were observed for virus with an E249G substitution in NS4B combined with either an A804V substitution in NS5 or three 3′ NCR mutations (A10596G, C10774U, and A10799G) (Davis et al., 2007b). A small plaque, temperature-sensitive WNV variant isolated from the kidney of an American crow, which replicated less efficiently in mammalian, mosquito, and especially in avian cells, had mutations in prM (P54S) and NS2A (V61A) (Jia et al., 2007). In a study of chimeras containing regions of an attenuated linage 2 WNV (strain W956) and the virulent lineage 1 WNV strain NY99, nonstructural gene variations as well as E glycosylation were reported to contribute to tissue culture cytopathology and neuroinvasion/virulence phenotypes of NY99 (Borisevich et al., 2006).

10 Conclusions

Although much has already been learned about the molecular biology of flaviviruses, there are still many unanswered questions. A growing body of literature suggests that viruses utilize cell proteins during each step of their replication cycles. Because WNV alternates between insect vectors and vertebrates in nature, the cellular proteins that this virus uses during its replication may be evolutionarily conserved, or, alternatively, more than one cell protein may be able to provide a particular function required by the virus. Little is currently known about the molecular mechanisms utilized for the initiation or regulation of flavivirus RNA synthesis or for the regulation of genome switching between translation and replication. Although a number of interesting insights have been obtained, our knowledge of how flaviviruses remodel the cells that they infect remains, is in many ways, rudimentary.

Acknowledgements The author thanks W. Davis for producing the graphics, W. Davis and M. Emara for assistance with manuscript production, and G. Radu and S. Scherbik for proofreading.

References

Ackermann, M., and Padmanabhan, R. 2001. De novo synthesis of RNA by the dengue virus RNA-dependent RNA polymerase exhibits temperature dependence at the initiation but not elongation phase. J Biol Chem 276:39926–39937.

Adams, S. C., Broom, A. K., Sammels, L. M., Hartnett, A. C., Howard, M. J., Coelen, R. J., Mackenzie, J. S., and Hall, R. A. 1995. Glycosylation and antigenic variation among Kunjin virus isolates. Virology 206:49–56.

Aleshin, A. E., Shiryaev, S. A., Strongin, A. Y., and Liddington, R. C. 2007. Structural evidence for regulation and specificity of flaviviral proteases and evolution of the Flaviviridae fold. Protein Sci 16:795–806.

Allison, S. L., Schalich, J., Stiasny, K., Mandl, C. W., Kunz, C., and Heinz, F. X. 1995. Oligomeric rearrangement of tick-borne encephalitis virus envelope proteins induced by an acidic pH. J Virol 69:695–700.

Allison, S. L., Tao, Y. J., O'Riordain, G., Mandl, C. W., Harrison, S. C., and Heinz, F. X. 2003. Two distinct size classes of immature and mature subviral particles from tick-borne encephalitis virus. J Virol 77:11357–11366.

Alvarez, D. E., Lodeiro, M. F., Luduena, S. J., Pietrasanta, L. I., and Gamarnik, A. V. 2005. Long-range RNA-RNA interactions circularize the dengue virus genome. J Virol 79:6631–6643.
Bazan, J. F., and Fletterick, R. J. 1989. Detection of a trypsin-like serine protease domain in flaviviruses and pestiviruses. Virology 171:637–639.
Beasley, D. W., Li, L., Suderman, M. T., and Barrett, A. D. 2001. West Nile virus strains differ in mouse neurovirulence and binding to mouse or human brain membrane receptor preparations. Ann N Y Acad Sci 951:332–335.
Beasley, D. W., Li, L., Suderman, M. T., and Barrett, A. D. 2002. Mouse neuroinvasive phenotype of West Nile virus strains varies depending upon virus genotype. Virology 296:17–23.
Beasley, D. W., Whiteman, M. C., Zhang, S., Huang, C. Y., Schneider, B. S., Smith, D. R., Gromowski, G. D., Higgs, S., Kinney, R. M., and Barrett, A. D. 2005. Envelope protein glycosylation status influences mouse neuroinvasion phenotype of genetic lineage 1 West Nile virus strains. J Virol 79:8339–8347.
Benarroch, D., Egloff, M. P., Mulard, L., Guerreiro, C., Romette, J. L., and Canard, B. 2004a. A structural basis for the inhibition of the NS5 dengue virus mRNA 2′-*O*-methyltransferase domain by ribavirin 5′-triphosphate. J Biol Chem 279:35638–35643.
Benarroch, D., Selisko, B., Locatelli, G. A., Maga, G., Romette, J. L., and Canard, B. 2004b. The RNA helicase, nucleotide 5′-triphosphatase, and RNA 5 -triphosphatase activities of Dengue virus protein NS3 are Mg2+-dependent and require a functional Walker B motif in the helicase catalytic core. Virology 328:208–218.
Berthet, F. X., Zeller, H. G., Drouet, M. T., Rauzier, J., Digoutte, J. P., and Deubel, V. 1997. Extensive nucleotide changes and deletions within the envelope glycoprotein gene of Euro-African West Nile viruses. J Gen Virol 78 (Pt 9):2293–2297.
Best, S. M., Morris, K. L., Shannon, J. G., Robertson, S. J., Mitzel, D. N., Park, G. S., Boer, E., Wolfinbarger, J. B., and Bloom, M. E. 2005. Inhibition of interferon-stimulated JAK-STAT signaling by a tick-borne flavivirus and identification of NS5 as an interferon antagonist. J Virol 79:12828–12839.
Blackwell, J. L., and Brinton, M. A. 1995. BHK cell proteins that bind to the 3′ stem-loop structure of the West Nile virus genome RNA. J Virol 69:5650–5658.
Blackwell, J. L., and Brinton, M. A. 1997. Translation elongation factor-1 alpha interacts with the 3 stem-loop region of West Nile virus genomic RNA. J Virol 71:6433–6444.
Blitvich, B. J., Scanlon, D., Shiell, B. J., Mackenzie, J. S., and Hall, R. A. 1999. Identification and analysis of truncated and elongated species of the flavivirus NS1 protein. Virus Res 60:67–79.
Blitvich, B. J., Scanlon, D., Shiell, B. J., Mackenzie, J. S., Pham, K., and Hall, R. A. 2001. Determination of the intramolecular disulfide bond arrangement and biochemical identification of the glycosylation sites of the nonstructural protein NS1 of Murray Valley encephalitis virus. J Gen Virol 82:2251–2256.
Borisevich, V., Seregin, A., Nistler, R., Mutabazi, D., and Yamshchikov, V. 2006. Biological properties of chimeric West Nile viruses. Virology 349:371–381.
Brault, A. C., Huang, C. Y., Langevin, S. A., Kinney, R. M., Bowen, R. A., Ramey, W. N., Panella, N. A., Holmes, E. C., Powers, A. M., and Miller, B. R. 2007. A single positively selected West Nile viral mutation confers increased virogenesis in American crows. Nat Genet 39:1162–1166.
Brinton, M. A. 1986. Replication of flaviviruses. *In* Togaviridae and Flaviviridae, The viruses. (Schlesinger, S. and Schlesinger, M., eds.), Plenum, New York. 329–376.
Brinton, M. A., and Dispoto, J. H. 1988. Sequence and secondary structure analysis of the 5′-terminal region of flavivirus genome RNA. Virology 162:290–299.
Brinton, M. A., and Perelygin, A. A. 2003. Genetic resistance to flaviviruses. Adv Virus Res 60:43–85.
Brinton, M. A., Fernandez, A. V., and Dispoto, J. H. 1986. The 3′-nucleotides of flavivirus genomic RNA form a conserved secondary structure. Virology 153:113–121.
Brooks, A. J., Johansson, M., John, A. V., Xu, Y., Jans, D. A., and Vasudevan, S. G. 2002. The interdomain region of dengue NS5 protein that binds to the viral helicase NS3 contains independently functional importin beta 1 and importin alpha/beta-recognized nuclear localization signals. J Biol Chem 277:36399–36407.

Cahour, A., Pletnev, A., Vazielle-Falcoz, M., Rosen, L., and Lai, C. J. 1995. Growth-restricted dengue virus mutants containing deletions in the 5′ noncoding region of the RNA genome. Virology 207:68–76.

Chambers, T. J., McCourt, D. W., and Rice, C. M. 1990. Production of yellow fever virus proteins in infected cells: identification of discrete polyprotein species and analysis of cleavage kinetics using region-specific polyclonal antisera. Virology 177:159–174.

Chen, C. J., Kuo, M. D., Chien, L. J., Hsu, S. L., Wang, Y. M., and Lin, J. H. 1997. RNA-protein interactions: involvement of NS3, NS5, and 3′ noncoding regions of Japanese encephalitis virus genomic RNA. J Virol 71:3466–3473.

Chen, Y. C., Wang, S. Y., and King, C. C. 1999. Bacterial lipopolysaccharide inhibits dengue virus infection of primary human monocytes/macrophages by blockade of virus entry via a CD14-dependent mechanism. J Virol 73:2650–2657.

Chu, J. J., and Ng, M. L. 2004. Infectious entry of West Nile virus occurs through a clathrin-mediated endocytic pathway. J Virol 78:10543–10555.

Chu, P. W., and Westaway, E. G. 1987. Characterization of Kunjin virus RNA-dependent RNA polymerase: reinitiation of synthesis in vitro. Virology 157:330–337.

Chu, P. W., and Westaway, E. G. 1992. Molecular and ultrastructural analysis of heavy membrane fractions associated with the replication of Kunjin virus RNA. Arch Virol 125:177–191.

Chu, J. J., Leong, P. W., and Ng, M. L. 2006. Analysis of the endocytic pathway mediating the infectious entry of mosquito-borne flavivirus West Nile into Aedes albopictus mosquito (C6/36) cells. Virology 349:463–475.

Chua, J. J., Bhuvanakantham, R., Chow, V. T., and Ng, M. L. 2005. Recombinant non-structural 1 (NS1) protein of dengue-2 virus interacts with human STAT3beta protein. Virus Res 112:85–94.

Chung, K. M., Liszewski, M. K., Nybakken, G., Davis, A. E., Townsend, R. R., Fremont, D. H., Atkinson, J. P., and Diamond, M. S. 2006. West Nile virus nonstructural protein NS1 inhibits complement activation by binding the regulatory protein factor H. Proc Natl Acad Sci USA 103:19111–19116.

Cleaves, G. R., and Dubin, D. T. 1979. Methylation status of intracellular dengue type 2 40 S RNA. Virology 96:159–165.

Cleaves, G. R., Ryan, T. E., and Schlesinger, R. W. 1981. Identification and characterization of type 2 dengue virus replicative intermediate and replicative form RNAs. Virology 111:73–83.

Cok, S. J., Acton, S. J., Sexton, A. E., and Morrison, A. R. 2004. Identification of RNA-binding proteins in RAW 264.7 cells that recognize a lipopolysaccharide-responsive element in the 3-untranslated region of the murine cyclooxygenase-2 mRNA. J Biol Chem 279: 8196–8205.

Cui, T., Sugrue, R. J., Xu, Q., Lee, A. K., Chan, Y. C., and Fu, J. 1998. Recombinant dengue virus type 1 NS3 protein exhibits specific viral RNA binding and NTPase activity regulated by the NS5 protein. Virology 246:409–417.

Davis, C. W., Nguyen, H. Y., Hanna, S. L., Sanchez, M. D., Doms, R. W., and Pierson, T. C. 2006. West Nile virus discriminates between DC-SIGN and DC-SIGNR for cellular attachment and infection. J Virol 80:1290–1301.

Davis, W. G., Blackwell, J. L., Shi, P. Y., and Brinton, M. A. 2007a. Interaction between the cellular protein eEF1A and the 3′-terminal stem-loop of West Nile virus genomic RNA facilitates viral minus-strand RNA synthesis. J Virol 81:10172–10187.

Davis, C. T., Galbraith, S. E., Zhang, S., Whiteman, M. C., Li, L., Kinney, R. M., and Barrett, A.D. 2007b. A combination of naturally occurring mutations in North American West Nile virus nonstructural protein genes and in the 3′ untranslated region alters virus phenotype. J. Virol. 81:6111–6116.

Dember, L. M., Kim, N. D., Liu, K. Q., and Anderson, P. 1996. Individual RNA recognition motifs of TIA-1 and TIAR have different RNA binding specificities. J Biol Chem 271:2783–2788.

De Nova-Ocampo, M., Villegas-Sepulveda, N., and del Angel, R. M. 2002. Translation Elongation Factor-1[alpha], La, and PTB Interact with the 3′ Untranslated Region of Dengue 4 Virus RNA. Virology 295:337–347.

Dixon, D. A., Balch, G. C., Kedersha, N., Anderson, P., Zimmerman, G. A., Beauchamp, R. D., and Prescott, S. M. 2003. Regulation of cyclooxygenase-2 expression by the translational silencer TIA-1. J Exp Med 198:475–481.

Dokland, T., Walsh, M., Mackenzie, J. M., Khromykh, A. A., Ee, K. H., and Wang, S. 2004. West Nile virus core protein; tetramer structure and ribbon formation. Structure 12:1157–1163.

Dong, H., Ray, D., Ren, S., Zhang, B., Puig-Basagoiti, F., Takagi, Y., Ho, C. K., Li, H., and Shi, P. Y. 2007. Distinct RNA elements confer specificity to flavivirus RNA cap methylation events. J Virol 81:4412–4421.

Egloff, M. P., Benarroch, D., Selisko, B., Romette, J. L., and Canard, B. 2002. An RNA cap (nucleoside-2′-*O*-)-methyltransferase in the flavivirus RNA polymerase NS5: crystal structure and functional characterization. EMBO J 21:2757–2768.

Egloff, M. P., Decroly, E., Malet, H., Selisko, B., Benarroch, D., Ferron, F., and Canard, B. 2007. Structural and functional analysis of methylation and 5 -RNA sequence requirements of short capped RNAs by the methyltransferase domain of dengue virus NS5. J Mol Biol 372:723–736.

Elghonemy, S., Davis, W. G., and Brinton, M. A. 2005. The majority of the nucleotides in the top loop of the genomic 3′ terminal stem loop structure are cis-acting in a West Nile virus infectious clone. Virology 331:238–246.

Emara, M. M., and Brinton, M. A. 2007. Interaction of TIA-1/TIAR with West Nile and dengue virus products in infected cells interferes with stress granule formation and processing body assembly. Proc Natl Acad Sci USA 104:9041–9046.

Emara, M. M., and Brinton, M. A. 2008. Mutation of mapped TIA-1/TIAR binding sites within the West Nile virus 3′ terminal minus-strand RNA sequence in an infectious clone negatively affects viral plus-strand synthesis. J Virol, in press.

Erbel, P., Schiering, N., D'Arcy, A., Renatus, M., Kroemer, M., Lim, S. P., Yin, Z., Keller, T. H., Vasudevan, S. G., and Hommel, U. 2006. Structural basis for the activation of flaviviral NS3 proteases from dengue and West Nile virus. Nat Struct Mol Biol 13:372–373.

Falgout, B., and Markoff, L. 1995. Evidence that flavivirus NS1-NS2A cleavage is mediated by a membrane-bound host protease in the endoplasmic reticulum. J Virol 69:7232–7243.

Falgout, B., Chanock, R., and Lai, C. J. 1989. Proper processing of dengue virus nonstructural glycoprotein NS1 requires the N-terminal hydrophobic signal sequence and the downstream nonstructural protein NS2a. J Virol 63:1852–1860.

Ferron, F., Bussetta, C., Dutartre, H., and Canard, B. 2005. The modeled structure of the RNA dependent RNA polymerase of GBV-C virus suggests a role for motif E in Flaviviridae RNA polymerases. BMC Bioinformatics 6:255.

Flamand, M., Megret, F., Mathieu, M., Lepault, J., Rey, F. A., and Deubel, V. 1999. Dengue virus type 1 nonstructural glycoprotein NS1 is secreted from mammalian cells as a soluble hexamer in a glycosylation-dependent fashion. J Virol 73:6104–6110.

Frick, D. N., Banik, S., and Rypma, R. S. 2007. Role of divalent metal cations in ATP hydrolysis catalyzed by the hepatitis C virus NS3 helicase: magnesium provides a bridge for ATP to fuel unwinding. J Mol Biol 365:1017–1032.

Glass, W. G., McDermott, D. H., Lim, J. K., Lekhong, S., Yu, S. F., Frank, W. A., Pape, J., Cheshier, R. C., and Murphy, P. M. 2006. CCR5 deficiency increases risk of symptomatic West Nile virus infection. J Exp Med 203:35–40.

Gorbalenya, A. E., Donchenko, A. P., Koonin, E. V., and Blinov, V. M. 1989a. N-terminal domains of putative helicases of flavi- and pestiviruses may be serine proteases. Nucleic Acids Res 17:3889–3897.

Gorbalenya, A. E., Koonin, E. V., Donchenko, A. P., and Blinov, V. M. 1989b. Two related superfamilies of putative helicases involved in replication, recombination, repair and expression of DNA and RNA genomes. Nucleic Acids Res 17:4713–4730.

Gould, E. A., Buckley, A., Barrett, A. D., and Cammack, N. 1986. Neutralizing (54 K) and non-neutralizing (54 K and 48 K) monoclonal antibodies against structural and non-structural yellow fever virus proteins confer immunity in mice. J Gen Virol 67 (Pt 3):591–595.

Gritsun, T. S., and Gould, E. A. 2007a. Origin and evolution of flavivirus 5′UTRσ and panhandles: trans-terminal duplications? Virology 366:8–15.

Gritsun, T. S., and Gould, E. A. 2007b. Direct repeats in the flavivirus 3′ untranslated region; a strategy for survival in the environment? Virology 358:258–265.

Grun, J. B., and Brinton, M. A. 1986. Characterization of West Nile virus RNA-dependent RNA polymerase and cellular terminal adenylyl and uridylyl transferases in cell-free extracts. J Virol 60:1113–1124.

Grun, J. B., and Brinton, M. A. 1987. Dissociation of NS5 from cell fractions containing West Nile virus-specific polymerase activity. J Virol 61:3641–3644.

Grun, J. B., and Brinton, M. A. 1988. Separation of functional West Nile virus replication complexes from intracellular membrane fragments. J Gen Virol 69 (Pt 12):3121–3127.

Gu, B., Liu, C., Lin-Goerke, J., Maley, D. R., Gutshall, L. L., Feltenberger, C. A., and Del Vecchio, A. M. 2000. The RNA helicase and nucleotide triphosphatase activities of the bovine viral diarrhea virus NS3 protein are essential for viral replication. J Virol 74:1794–1800.

Gubler, D. J., Kuno, G., and Markoff, J. 2007. Flaviviruses. Pp. 1153–1252 *in*P. M. H. David M Knipe, ed. Fields Virology. Lippincott Williams and Wilkins, Philadelphia.

Guyatt, K. J., Westaway, E. G., and Khromykh, A. A. 2001. Expression and purification of enzymatically active recombinant RNA-dependent RNA polymerase (NS5) of the flavivirus Kunjin. J Virol Methods 92:37–44.

Hahn, C. S., Hahn, Y. S., Rice, C. M., Lee, E., Dalgarno, L., Strauss, E. G., and Strauss, J. H. 1987. Conserved elements in the 3 untranslated region of flavivirus RNAs and potential cyclization sequences. J Mol Biol 198:33–41.

Hall, R. A., Khromykh, A. A., Mackenzie, J. M., Scherret, J. H., and Khromykh, T. I. 1999. Loss of dimerisation of the nonstructural protein NS1 of Kunjin virus delays viral replication and reduces virulence in mice, but still allows secretion of NS1. Virology 264:66.

Hanley, K. A., Lee, J. J., Blaney, J. E., Jr., Murphy, B. R., and Whitehead, S. S. 2002. Paired charge-to-alanine mutagenesis of dengue virus type 4 NS5 generates mutants with temperature-sensitive, host range, and mouse attenuation phenotypes. J Virol 76:525–531.

Hanna, S. L., Pierson, T. C., Sanchez, M. D., Ahmed, A. A., Murtadha, M. M., and Doms, R. W. 2005. N-linked glycosylation of West Nile virus envelope proteins influences particle assembly and infectivity. J Virol 79:13262–13274.

Heinz, F. X., and Allison, S. L. 2000. Structures and mechanisms in flavivirus fusion. Adv Virus Res 55:231–269.

Heinz, F. X., Auer, G., Stiasny, K., Holzmann, H., Mandl, C., Guirakhoo, F., and Kunz, C. 1994. The interactions of the flavivirus envelope proteins: implications for virus entry and release. Arch Virol Suppl 9:339–348.

Heinz, F. X., Purcell, M.S., Gould, E.A., Howard, C.R., Houghton, M. et-al. 2000. Family Flaviviridae. Pp. 860–878 *in*C. F. MHV Regenmortel, DHL Bishop, E.B Carstens, M K Estes, ed. Virus Taxonomy. Academic Press, San Diego.

Hirsch, A. J., Medigeshi, G. R., Meyers, H. L., DeFilippis, V., Fruh, K., Briese, T., Lipkin, W. I., and Nelson, J. A. 2005. The Src family kinase c-Yes is required for maturation of West Nile virus particles. J Virol 79:11943–11951.

Hunt, T. A., Urbanowski, M. D., Kalani, K., Law, L.-M. J., Brinton, M. A., and Hobman, T. C. 2007. Interactions between the West Nile virus capsid protein and the host cell-encoded phosphatase inhibitor, I2PP2A. Cell Microbiol 9(11):2756–2766.

Ivanyi-Nagy, R., Lavergne, J. P., Gabus, C., Ficheux, D., and Darlix, J. L. 2008. RNA chaperoning and intrinsic disorder in the core proteins of Flaviviridae. Nucleic Acids Res 36(3):712–725.

Jacobs, M. G., Robinson, P. J., Bletchly, C., Mackenzie, J. M., and Young, P. R. 2000. Dengue virus nonstructural protein 1 is expressed in a glycosyl-phosphatidylinositol-linked form that is capable of signal transduction. FASEB J 14:1603–1610.

Jia, X. Y., Briese, T., Jordan, I., Rambaut, A., Chi, H. C., Mackenzie, J. S., Hall, R. A., Scherret, J., and Lipkin, W. I. 1999. Genetic analysis of West Nile New York 1999 encephalitis virus. Lancet 354:1971–1972.

Jia, Y., Moudy, R. M., Dupuis, A. P., 2nd, Ngo, K. A., Maffei, J. G., Jerzak, G. V., Franke, M. A., Kauffman, E. B., and Kramer, L. D. 2007. Characterization of a small plaque variant of West Nile virus isolated in New York in 2000. Virology 367:339–347.

Jindadamrongwech, S., Thepparit, C., and Smith, D. R. 2004. Identification of GRP 78 (BiP) as a liver cell expressed receptor element for dengue virus serotype 2. Arch Virol 149:915–927.

Johansson, M., Brooks, A. J., Jans, D. A., and Vasudevan, S. G. 2001. A small region of the dengue virus-encoded RNA-dependent RNA polymerase, NS5, confers interaction with both the nuclear transport receptor importin-beta and the viral helicase, NS3. J Gen Virol 82:735–745.

Jones, C. T., Ma, L., Burgner, J. W., Groesch, T. D., Post, C. B., and Kuhn, R. J. 2003. Flavivirus capsid is a dimeric alpha-helical protein. J Virol 77:7143–7149.

Kakuta, S., Shibata, S., and Iwakura, Y. 2002. Genomic structure of the mouse 2′,5′-oligoadenylate synthetase gene family. J Interferon Cytokine Res 22:981– 993.

Kamer, G., and Argos, P. 1984. Primary structural comparison of RNA-dependent polymerases from plant, animal and bacterial viruses. Nucleic Acids Res 12:7269–7282.

Kapoor, M., Zhang, L., Ramachandra, M., Kusukawa, J., Ebner, K. E., and Padmanabhan, R. 1995. Association between NS3 and NS5 proteins of dengue virus type 2 in the putative RNA replicase is linked to differential phosphorylation of NS5. J Biol Chem 270:19100–19106.

Khromykh, A. A., and Westaway, E. G. 1996. RNA binding properties of core protein of the flavivirus Kunjin. Arch Virol 141:685–699.

Khromykh, A. A., Kenney, M. T., and Westaway, E. G. 1998. *trans*-Complementation of flavivirus RNA polymerase gene NS5 by using Kunjin virus replicon-expressing BHK cells. J Virol 72:7270–7279.

Khromykh, A. A., Sedlak, P. L., Guyatt, K. J., Hall, R. A., and Westaway, E. G. 1999a. Efficient trans-complementation of the flavivirus kunjin NS5 protein but not of the NS1 protein requires its coexpression with other components of the viral replicase. J Virol 73:10272–10280.

Khromykh, A. A., Sedlak, P. L., and Westaway, E. G. 1999b. *trans*-Complementation analysis of the flavivirus Kunjin ns5 gene reveals an essential role for translation of its N-terminal half in RNA replication. J Virol 73:9247–9255.

Khromykh, A. A., Sedlak, P. L., and Westaway, E. G. 2000. *cis*- and *trans*-acting elements in flavivirus RNA replication. J Virol 74:3253–3263.

Khromykh, A. A., Meka, H., Guyatt, K. J., and Westaway, E. G. 2001a. Essential role of cyclization sequences in flavivirus RNA replication. J Virol 75:6719–6728.

Khromykh, A. A., Varnavski, A. N., Sedlak, P. L., and Westaway, E. G. 2001b. Coupling between replication and packaging of flavivirus RNA: evidence derived from the use of DNA-based full-length cDNA clones of Kunjin virus. J Virol 75:4633–4640.

Kiermayr, S., Kofler, R. M., Mandl, C. W., Messner, P., and Heinz, F. X. 2004. Isolation of capsid protein dimers from the tick-borne encephalitis flavivirus and in vitro assembly of capsid-like particles. J Virol 78:8078–8084.

Kim, S. M., and Jeong, Y. S. 2006. Polypyrimidine tract-binding protein interacts with the 3′ stem-loop region of Japanese encephalitis virus negative-strand RNA. Virus Res 115:131–140.

Kleinschmidt, M. C., Michaelis, M., Ogbomo, H., Doerr, H. W., and Cinatl, J., Jr. 2007. Inhibition of apoptosis prevents West Nile virus induced cell death. BMC Microbiol 7:49.

Kofler, R. M., Heinz, F. X., and Mandl, C. W. 2002. Capsid protein C of tick-borne encephalitis virus tolerates large internal deletions and is a favorable target for attenuation of virulence. J Virol 76:3534–3543.

Koonin, E. V. 1991. The phylogeny of RNA-dependent RNA polymerases of positive-strand RNA viruses. J Gen Virol 72 (Pt 9):2197–2206.

Krishnan, M. N., Sukumaran, B., Pal, U., Agaisse, H., Murray, J. L., Hodge, T. W., and Fikrig, E. 2007. Rab 5 is required for the cellular entry of dengue and West Nile viruses. J Virol 81:4881–4885.

Kuhn, R. J., Zhang, W., Rossmann, M. G., Pletnev, S. V., Corver, J., Lenches, E., Jones, C. T., Mukhopadhyay, S., Chipman, P. R., Strauss, E. G., Baker, T. S., and Strauss, J. H. 2002. Structure of dengue virus: implications for flavivirus organization, maturation, and fusion. Cell 108:717–725.

Kummerer, B. M., and Rice, C. M. 2002. Mutations in the yellow fever virus nonstructural protein NS2A selectively block production of infectious particles. J Virol 76:4773–4784.

Lai, M. M. 1998. Cellular factors in the transcription and replication of viral RNA genomes: a parallel to DNA-dependent RNA transcription. Virology 244:1–12.

Lai, C. J., Men, R., Pethel, M., and Bray, M. 1992. Infectious RNA transcribed from stably cloned full-length cDNA: Construction of growth-restricted dengue mutants. Pp. 265–270 *in* R. M. C. F Brown, H. S. Ginsberg, and R. A. Lerner, ed. Vaccines 92. Cold Spring Harbor Laboratory Press, Cold Spring Harbor.

Lanciotti, R. S., Roehrig, J. T., Deubel, V., Smith, J., Parker, M., Steele, K., Crise, B., Volpe, K. E., Crabtree, M. B., Scherret, J. H., Hall, R. A., MacKenzie, J. S., Cropp, C. B., Panigrahy, B., Ostlund, E., Schmitt, B., Malkinson, M., Banet, C., Weissman, J., Komar, N., Savage, H. M., Stone, W., McNamara, T., and Gubler, D. J. 1999. Origin of the West Nile virus responsible for an outbreak of encephalitis in the northeastern United States. Science 286:2333–2337.

Li, W., Li, Y., Kedersha, N., Anderson, P., Emara, M., Swiderek, K. M., Moreno, G. T., and Brinton, M. A. 2002. Cell proteins TIA-1 and TIAR interact with the 3′ stem-loop of the West Nile virus complementary minus-strand RNA and facilitate virus replication. J Virol 76:11989–12000.

Li, J., Bhuvanakantham, R., Howe, J., and Ng, M. L. 2005. Identifying the region influencing the cis-mode of maturation of West Nile (Sarafend) virus using chimeric infectious clones. Biochem Biophys Res Commun 334:714–720.

Lindenbach, B. D., and Rice, C. M. 1997. Trans-complementation of yellow fever virus NS1 reveals a role in early RNA replication. J Virol 71:9608–9617.

Lindenbach, B. D., and Rice, C. M. 1999. Genetic interaction of flavivirus nonstructural proteins NS1 and NS4A as a determinant of replicase function. J Virol 73:4611–4621.

Lindenbach, B. D., Thiel, H. J., and Rice, C. M. 2007. Flaviviridae: The Viruses and Their Replication. Pp. 1101–1152 *in* D. M. Knipe, and Howley, P. M., ed. Fields Virology. Lippincott Williams and Wilkins, Philadelphia.

Liu, W. J., Sedlak, P. L., Kondratieva, N., and Khromykh, A. A. 2002. Complementation analysis of the flavivirus Kunjin NS3 and NS5 proteins defines the minimal regions essential for formation of a replication complex and shows a requirement of NS3 in *cis* for virus assembly. J Virol 76:10766–10775.

Liu, W. J., Chen, H. B., and Khromykh, A. A. 2003. Molecular and functional analyses of Kunjin virus infectious cDNA clones demonstrate the essential roles for NS2A in virus assembly and for a nonconservative residue in NS3 in RNA replication. J Virol 77:7804–7813.

Liu, W. J., Chen, H. B., Wang, X. J., Huang, H., and Khromykh, A. A. 2004. Analysis of adaptive mutations in Kunjin virus replicon RNA reveals a novel role for the flavivirus nonstructural protein NS2A in inhibition of beta interferon promoter-driven transcription. J Virol 78:12225–12235.

Liu, W. J., Wang, X. J., Clark, D. C., Lobigs, M., Hall, R. A., and Khromykh, A. A. 2006. A single amino acid substitution in the West Nile virus nonstructural protein NS2A disables its ability to inhibit alpha/beta interferon induction and attenuates virus virulence in mice. J Virol 80:2396–2404.

Ma, L., Jones, C. T., Groesch, T. D., Kuhn, R. J., and Post, C. B. 2004. Solution structure of dengue virus capsid protein reveals another fold. Proc Natl Acad Sci USA 101:3414–3419.

Macdonald, J., Tonry, J., Hall, R. A., Williams, B., Palacios, G., Ashok, M. S., Jabado, O., Clark, D., Tesh, R. B., Briese, T., and Lipkin, W. I. 2005. NS1 protein secretion during the acute phase of West Nile virus infection. J Virol 79:13924–13933.

Mackenzie, J. M., Jones, M. K., and Young, P. R. 1996. Immunolocalization of the dengue virus nonstructural glycoprotein NS1 suggests a role in viral RNA replication. Virology 220:232–240.

Mackenzie, J. M., Khromykh, A. A., Jones, M. K., and Westaway, E. G. 1998. Subcellular localization and some biochemical properties of the flavivirus Kunjin nonstructural proteins NS2A and NS4A. Virology 245:203–215.

Mackenzie, J. M., Kenney, M. T., and Westaway, E. G. 2007a. West Nile virus strain Kunjin NS5 polymerase is a phosphoprotein localized at the cytoplasmic site of viral RNA synthesis. J Gen Virol 88:1163–1168.

Mackenzie, J. M., Khromykh, A. A., and Parton, R. G. 2007b. Cholesterol manipulation by West Nile virus perturbs the cellular immune response. Cell Host Microbe 2:229–239.

Maginnis, M. S., Forrest, J. C., Kopecky-Bromberg, S. A., Dickeson, S. K., Santoro, S. A., Zutter, M. M., Nemerow, G. R., Bergelson, J. M., and Dermody, T. S. 2006. Beta1 integrin mediates internalization of mammalian reovirus. J Virol 80:2760–2770.

Malet, H., Egloff, M. P., Selisko, B., Butcher, R. E., Wright, P. J., Roberts, M., Gruez, A., Sulzenbacher, G., Vonrhein, C., Bricogne, G., Mackenzie, J. M., Khromykh, A. A., Davidson, A. D., and Canard, B. 2007. Crystal structure of the RNA polymerase domain of the West Nile virus non-structural protein 5. J Biol Chem 282:10678–10689.

Mandl, C. W., Holzmann, H., Meixner, T., Rauscher, S., Stadler, P. F., Allison, S. L., and Heinz, F. X. 1998. Spontaneous and engineered deletions in the 3′ noncoding region of tick-borne encephalitis virus: construction of highly attenuated mutants of a flavivirus. J Virol 72:2132–2140.

Mashimo, T., Lucas, M., Simon-Chazottes, D., Frenkiel, M. P., Montagutelli, X., Ceccaldi, P. E., Deubel, V., Guenet, J. L., and Despres, P. 2002. A nonsense mutation in the gene encoding 2′-5′-oligoadenylate synthetase/L1 isoform is associated with West Nile virus susceptibility in laboratory mice. Proc Natl Acad Sci USA 99:11311–11316.

Mason, P. W. 1989. Maturation of Japanese encephalitis virus glycoproteins produced by infected mammalian and mosquito cells. Virology 169:354–364.

Matusan, A. E., Pryor, M. J., Davidson, A. D., and Wright, P. J. 2001. Mutagenesis of the Dengue virus type 2 NS3 protein within and outside helicase motifs: effects on enzyme activity and virus replication. J Virol 75:9633–9643.

Medigeshi, G. R., Hirsh, A. J., Streblow, D. N., Nikolich-Zugich, J., and Nelson, J. A. 2008. West Nile virus entry requires cholesterol-rich membrane microdomains and is independent of alphavbeta3 integrin. J Virol 82:5212–5219.

Medigeshi, G. R., Lancaster, A. M., Hirsch, A. J., Briese, T., Lipkin, W. I., Defilippis, V., Fruh, K., Mason, P. W., Nikolich-Zugich, J., and Nelson, J. A. 2007. West Nile virus infection activates the unfolded protein response, leading to CHOP induction and apoptosis. J Virol 81:10849–10860.

Medin, C. L., Fitzgerald, K. A., and Rothman, A. L. 2005. Dengue virus nonstructural protein NS5 induces interleukin-8 transcription and secretion. J Virol 79:11053–11061.

Mehlhop, E., and Diamond, M. S. 2006. Protective immune responses against West Nile virus are primed by distinct complement activation pathways. Pp. 1371–1381.

Men, R., Bray, M., Clark, D., Chanock, R. M., and Lai, C. J. 1996. Dengue type 4 virus mutants containing deletions in the 3′ noncoding region of the RNA genome: analysis of growth restriction in cell culture and altered viremia pattern and immunogenicity in rhesus monkeys. J Virol 70:3930–3937.

Miller, S., Sparacio, S., and Bartenschlager, R. 2006. Subcellular localization and membrane topology of the Dengue virus type 2 Non-structural protein 4B. J Biol Chem 281:8854–8863.

Miller, S., Kastner, S., Krijnse-Locker, J., Buhler, S., and Bartenschlager, R. 2007. The non-structural protein 4A of dengue virus is an integral membrane protein inducing membrane alterations in a 2 K-regulated manner. J Biol Chem 282:8873–8882.

van der Most, R. G., Corver, J., and Strauss, J. H. 1999. Mutagenesis of the RGD motif in the yellow fever virus 17D envelope protein. Virology 265:83–95.

Mukhopadhyay, S., Kim, B. S., Chipman, P. R., Rossmann, M. G., and Kuhn, R. J. 2003. Structure of West Nile virus. Science 302:248.

Mukhopadhyay, S., Kuhn, R. J., and Rossmann, M. G. 2005. A structural perspective of the flavivirus life cycle. Nat Rev Microbiol 3:13–22.

Munoz-Jordan, J. L., Sanchez-Burgos, G. G., Laurent-Rolle, M., and Garcia-Sastre, A. 2003. Inhibition of interferon signaling by dengue virus. Proc Natl Acad Sci USA 100:14333–14338.

Murray, J. M., Aaskov, J. G., and Wright, P. J. 1993. Processing of the dengue virus type 2 proteins prM and C-prM. J Gen Virol 74 (Pt 2):175–182.

Muylaert, I. R., Chambers, T. J., Galler, R., and Rice, C. M. 1996. Mutagenesis of the N-linked glycosylation sites of the yellow fever virus NS1 protein: effects on virus replication and mouse neurovirulence. Virology 222:159–168.

Muylaert, I. R., Galler, R., and Rice, C. M. 1997. Genetic analysis of the yellow fever virus NS1 protein: identification of a temperature-sensitive mutation which blocks RNA accumulation. J Virol 71:291–298.

Nomaguchi, M., Ackermann, M., Yon, C., You, S., and Padmanabhan, R. 2003. De novo synthesis of negative-strand RNA by Dengue virus RNA-dependent RNA polymerase in vitro: nucleotide, primer, and template parameters. J Virol 77:8831–8842.

Nowak, T., and Wengler, G. 1987. Analysis of disulfides present in the membrane proteins of the West Nile flavivirus. Virology 156:127–137.

Nowak, T., Farber, P. M., Wengler, G., and Wengler, G. 1989. Analyses of the terminal sequences of West Nile virus structural proteins and of the in vitro translation of these proteins allow the proposal of a complete scheme of the proteolytic cleavages involved in their synthesis. Virology 169:365–376.

Oh, W. K., and Song, J. 2006. Hsp70 functions as a negative regulator of West Nile virus capsid protein through direct interaction. Biochem Biophys Res Commun 347:994–1000.

Oh, W., Yang, M. R., Lee, E. W., Park, K. M., Pyo, S., Yang, J. S., Lee, H. W., and Song, J. 2006. Jab1 mediates cytoplasmic localization and degradation of West Nile virus capsid protein. J Biol Chem 281:30166–30174.

Olsthoorn, R. C., and Bol, J. F. 2001. Sequence comparison and secondary structure analysis of the 3′ noncoding region of flavivirus genomes reveals multiple pseudoknots. RNA 7:1370–1377.

Orlinger, K. K., Hoenninger, V. M., Kofler, R. M., and Mandl, C. W. 2006. Construction and mutagenesis of an artificial bicistronic tick-borne encephalitis virus genome reveals an essential function of the second transmembrane region of protein e in flavivirus assembly. J Virol 80:12197–12208.

Paranjape, S. M., and Harris, E. 2007. Y box-binding protein-1 binds to the dengue virus 3′-untranslated region and mediates antiviral effects. J Biol Chem 282:30497–30508.

Parquet, M. C., Kumatori, A., Hasebe, F., Morita, K., and Igarashi, A. 2001. West Nile virus-induced bax-dependent apoptosis. FEBS Lett 500:17–24.

Perelygin, A. A., Scherbik, S. V., Zhulin, I. B., Stockman, B. M., Li, Y., and Brinton, M. A. 2002. Positional cloning of the murine flavivirus resistance gene. Proc Natl Acad Sci USA 99:9322–9327.

Pijlman, G. P., Kondratieva, N., and Khromykh, A. A. 2006. Translation of the flavivirus kunjin NS3 gene in cis but not its RNA sequence or secondary structure is essential for efficient RNA packaging. J Virol 80:11255–11264.

Poidinger, M., Hall, R. A., and Mackenzie, J. S. 1996. Molecular characterization of the Japanese encephalitis serocomplex of the flavivirus genus. Virology 218:417–421.

Preugschat, F., and Strauss, J. H. 1991. Processing of nonstructural proteins NS4A and NS4B of dengue 2 virus in vitro and in vivo. Virology 185:689–697.

Prikhod'ko, G. G., Prikhod'ko, E. A., Pletnev, A. G., and Cohen, J. I. 2002. Langat flavivirus protease NS3 binds caspase-8 and induces apoptosis. J Virol 76:5701–5710.

Puig-Basagoiti, F., Tilgner, M., Bennett, C. J., Zhou, Y., Munoz-Jordan, J. L., Garcia-Sastre, A., Bernard, K. A., and Shi, P. Y. 2007. A mouse cell-adapted NS4B mutation attenuates West Nile virus RNA synthesis. Virology 361:229–241.

Ramanathan, M. P., Chambers, J. A., Pankhong, P., Chattergoon, M., Attatippaholkun, W., Dang, K., Shah, N., and Weiner, D. B. 2006. Host cell killing by the West Nile Virus

NS2B-NS3 proteolytic complex: NS3 alone is sufficient to recruit caspase-8-based apoptotic pathway. Virology 345:56–72.

Ranjith-Kumar, C. T., Gajewski, J., Gutshall, L., Maley, D., Sarisky, R. T., and Kao, C. C. 2001. Terminal nucleotidyl transferase activity of recombinant Flaviviridae RNA-dependent RNA polymerases: implication for viral RNA synthesis. J Virol 75:8615–8623.

Rauscher, S., Flamm, C., Mandl, C. W., Heinz, F. X., and Stadler, P. F. 1997. Secondary structure of the 3′-noncoding region of flavivirus genomes: comparative analysis of base pairing probabilities. RNA 3:779–791.

Ray, D., Shah, A., Tilgner, M., Guo, Y., Zhao, Y., Dong, H., Deas, T. S., Zhou, Y., Li, H., and Shi, P. Y. 2006. West Nile virus 5′-cap structure is formed by sequential guanine N-7 and ribose 2′-*O* methylations by nonstructural protein 5. J Virol 80:8362–8370.

Reed, K. E., Gorbalenya, A. E., and Rice, C. M. 1998. The NS5A/NS5 proteins of viruses from three genera of the family flaviviridae are phosphorylated by associated serine/threonine kinases. J Virol 72:6199–6206.

Reyes-Del Valle, J., Chavez-Salinas, S., Medina, F., and Del Angel, R. M. 2005. Heat shock protein 90 and heat shock protein 70 are components of dengue virus receptor complex in human cells. J Virol 79:4557–4567.

Rice, C. M., Lenches, E. M., Eddy, S. R., Shin, S. J., Sheets, R. L., and Strauss, J. H. 1985. Nucleotide sequence of yellow fever virus: implications for flavivirus gene expression and evolution. Science 229:726–733.

Riis, B., Rattan, S. I., Clark, B. F., and Merrick, W. C. 1990. Eukaryotic protein elongation factors. Trends Biochem Sci 15:420–424.

Rios, M., Daniel, S., Chancey, C., Hewlett, I. K., and Stramer, S. L. 2007. West Nile virus adheres to human red blood cells in whole blood. Clin Infect Dis 45:181–186.

Roosendaal, J., Westaway, E. G., Khromykh, A., and Mackenzie, J. M. 2006. Regulated cleavages at the West Nile virus NS4A-2 K-NS4B junctions play a major role in rearranging cytoplasmic membranes and Golgi trafficking of the NS4A protein. J Virol 80:4623–4632.

Russell, P. K., Brandt, W. E., and Dalrymple, J. M. 1980. Chemical and antigenic structure of flaviviruses. Pp. 503–529 *in* R. W. Schlesinger, ed. The Togaviruses. Academic, New York.

Samuel, M. A., Morrey, J. D., and Diamond, M. S. 2007. Caspase 3-dependent cell death of neurons contributes to the pathogenesis of West Nile virus encephalitis. J Virol 81:2614–2623.

Scherbik, S. V., Paranjape, J. M., Stockman, B. M., Silverman, R. H., and Brinton, M. A. 2006. RNase L plays a role in the antiviral response to West Nile virus. J Virol 80:2987–2999.

Scherbik, S. V., Kluetzman, K., Perelygin, A. A., and Brinton, M. A. 2007a. Knock-in of the Oas1b(r) allele into a flavivirus-induced disease susceptible mouse generates the resistant phenotype. Virology 368:232–237.

Scherbik, S. V., Stockman, B. M., and Brinton, M. A. 2007b. Differential activation of interferon regulatory factors (IRFs) and interferon-stimulated genes (ISGs) at early times after West Nile virus (WNV) infection of mouse embryo fibroblasts (MEFs). J Virol 81:12005–12018.

Scherret, J. H., Poidinger, M., Mackenzie, J. S., Broom, A. K., Deubel, V., Lipkin, W. I., Briese, T., Gould, E. A., and Hall, R. A. 2001. The relationships between West Nile and Kunjin viruses. Emerg Infect Dis 7:697–705.

Schlesinger, J. J., Brandriss, M. W., and Walsh, E. E. 1985. Protection against 17D yellow fever encephalitis in mice by passive transfer of monoclonal antibodies to the nonstructural glycoprotein gp48 and by active immunization with gp48. J Immunol 135:2805–2809.

Selisko, B., Dutartre, H., Guillemot, J. C., Debarnot, C., Benarroch, D., Khromykh, A., Despres, P., Egloff, M. P., and Canard, B. 2006. Comparative mechanistic studies of de novo RNA synthesis by flavivirus RNA-dependent RNA polymerases. Virology 351:145–158.

Shafee, N., and AbuBakar, S. 2003. Dengue virus type 2 NS3 protease and NS2B-NS3 protease precursor induce apoptosis. J Gen Virol 84:2191–2195.

Shi, P. Y., Brinton, M. A., Veal, J. M., Zhong, Y. Y., and Wilson, W. D. 1996a. Evidence for the existence of a pseudoknot structure at the 3′ terminus of the flavivirus genomic RNA. Biochemistry 35:4222–4230.

Shi, P. Y., Li, W., and Brinton, M. A. 1996b. Cell proteins bind specifically to West Nile virus minus-strand 3′ stem-loop RNA. J Virol 70:6278–6287.

Shirato, K., Miyoshi, H., Goto, A., Ako, Y., Ueki, T., Kariwa, H., and Takashima, I. 2004. Viral envelope protein glycosylation is a molecular determinant of the neuroinvasiveness of the New York strain of West Nile virus. J Gen Virol 85:3637–3645.

Shurtleff, A. C., Beasley, D. W., Chen, J. J., Ni, H., Suderman, M. T., Wang, H., Xu, R., Wang, E., Weaver, S. C., Watts, D. M., Russell, K. L., and Barrett, A. D. 2001. Genetic variation in the 3' non-coding region of dengue viruses. Virology 281:75–87.

Smith, G. W., and Wright, P. J. 1985. Synthesis of proteins and glycoproteins in dengue type 2 virus-infected vero and Aedes albopictus cells. J Gen Virol 66 (Pt 3):559–571.

Stadler, K., Allison, S. L., Schalich, J., and Heinz, F. X. 1997. Proteolytic activation of tick-borne encephalitis virus by furin. J Virol 71:8475–8481.

Steffens, S., Thiel, H. J., and Behrens, S. E. 1999. The RNA-dependent RNA polymerases of different members of the family Flaviviridae exhibit similar properties in vitro. J Gen Virol 80 (Pt 10):2583–2590.

Ta, M., and Vrati, S. 2000. Mov34 protein from mouse brain interacts with the 3′ noncoding region of Japanese encephalitis virus. J Virol 74:5108–5115.

Tan, B. H., Fu, J., Sugrue, R. J., Yap, E. H., Chan, Y. C., and Tan, Y. H. 1996. Recombinant dengue type 1 virus NS5 protein expressed in Escherichia coli exhibits RNA-dependent RNA polymerase activity. Virology 216:317–325.

Taupin, J. L., Tian, Q., Kedersha, N., Robertson, M., and Anderson, P. 1995. The RNA-binding protein TIAR is translocated from the nucleus to the cytoplasm during Fas-mediated apoptotic cell death. Proc Natl Acad Sci USA 92:1629–1633.

Tsuda, Y., Mori, Y., Abe, T., Yamashita, T., Okamoto, T., Ichimura, T., Moriishi, K., and Matsuura, Y. 2006. Nucleolar protein B23 interacts with Japanese encephalitis virus core protein and participates in viral replication. Microbiol Immunol 50:225–234.

Urosevic, N., Mansfield, J. P., Mackenzie, J. S., and Shellam, G. R. 1995. Low resolution mapping around the flavivirus resistance locus (Flv) on mouse chromosome 5. Mamm Genome 6:454–458.

Urosevic, N., Mann, K., Hodgetts, S. I., and Shellam, G. R. 1997. High-resolution genetic mapping of the chromosomal region around the mouse flavivirus resistance locus (Flv). Arbovir Res Aust 7:296–299.

Wallis, T. P., Huang, C. Y., Nimkar, S. B., Young, P. R., and Gorman, J. J. 2004. Determination of the disulfide bond arrangement of dengue virus NS1 protein. J Biol Chem 279:20729–20741.

Wang, S. H., Syu, W. J., Huang, K. J., Lei, H. Y., Yao, C. W., King, C. C., and Hu, S. T. 2002. Intracellular localization and determination of a nuclear localization signal of the core protein of dengue virus. J Gen Virol 83:3093–3102.

Wengler, G., and Wengler, G. 1991. The carboxy-terminal part of the NS 3 protein of the West Nile flavivirus can be isolated as a soluble protein after proteolytic cleavage and represents an RNA-stimulated NTPase. Virology 184:707–715.

Wengler, G., and Wengler, G. 1993. The NS 3 nonstructural protein of flaviviruses contains an RNA triphosphatase activity. Virology 197:265–273.

Wengler, G., Czaya, G., Farber, P. M., and Hegemann, J. H. 1991. In vitro synthesis of West Nile virus proteins indicates that the amino-terminal segment of the NS3 protein contains the active centre of the protease which cleaves the viral polyprotein after multiple basic amino acids. J Gen Virol 72 (Pt 4):851–858.

Westaway, E. G., Khromykh, A. A., Kenney, M. T., Mackenzie, J. M., and Jones, M. K. 1997a. Proteins C and NS4B of the flavivirus Kunjin translocate independently into the nucleus. Virology 234:31–41.

Westaway, E. G., Mackenzie, J. M., Kenney, M. T., Jones, M. K., and Khromykh, A. A. 1997b. Ultrastructure of Kunjin virus-infected cells: colocalization of NS1 and NS3 with double-stranded RNA, and of NS2B with NS3, in virus-induced membrane structures. J Virol 71:6650–6661.

Westaway, E. G., Khromykh, A. A., and Mackenzie, J. M. 1999. Nascent flavivirus RNA colocalized in situ with double-stranded RNA in stable replication complexes. Virology 258:108–117.

Westaway, E. G., Mackenzie, J. M., and Khromykh, A. A. 2002. Replication and gene function in Kunjin virus. Curr Top Microbiol Immunol 267:323–351.

Wicker, J. A., Whiteman, M. C., Beasley, D. W., Davis, C. T., Zhang, S., Schneider, B. S., Higgs, S., Kinney, R. M., and Barrett, A. D. 2006. A single amino acid substitution in the central portion of the West Nile virus NS4B protein confers a highly attenuated phenotype in mice. Virology 349:245–253.

Winkler, G., Randolph, V. B., Cleaves, G. R., Ryan, T. E., and Stollar, V. 1988. Evidence that the mature form of the flavivirus nonstructural protein NS1 is a dimer. Virology 162:187–196.

Winkler, G., Maxwell, S. E., Ruemmler, C., and Stollar, V. 1989. Newly synthesized dengue-2 virus nonstructural protein NS1 is a soluble protein but becomes partially hydrophobic and membrane-associated after dimerization. Virology 171:302–305.

Wu, J., Bera, A. K., Kuhn, R. J., and Smith, J. L. 2005. Structure of the Flavivirus helicase: implications for catalytic activity, protein interactions, and proteolytic processing. J Virol 79:10268–10277.

Xu, T., Sampath, A., Chao, A., Wen, D., Nanao, M., Chene, P., Vasudevan, S. G., and Lescar, J. 2005. Structure of the Dengue virus helicase/nucleoside triphosphatase catalytic domain at a resolution of 2.4 A. J Virol 79:10278–10288.

Yakub, I., Lillibridge, K. M., Moran, A., Gonzalez, O. Y., Belmont, J., Gibbs, R. A., and Tweardy, D. J. 2005. Single nucleotide polymorphisms in genes for 2′–5′-oligoadenylate synthetase and RNase L inpatients hospitalized with West Nile virus infection. J Infect Dis 192:1741–1748.

Yang, J. S., Ramanathan, M. P., Muthumani, K., Choo, A. Y., Jin, S. H., Yu, Q. C., Hwang, D. S., Choo, D. K., Lee, M. D., Dang, K., Tang, W., Kim, J. J., and Weiner, D. B. 2002. Induction of inflammation by West Nile virus capsid through the caspase-9 apoptotic pathway. Emerg Infect Dis 8:1379–1384.

Yang, M. R., Lee, S. R., Oh, W., Lee, E. W., Yeh, J. Y., Nah, J. J., Joo, Y. S., Shin, J., Lee, H. W., Pyo, S., and Song, J. 2008. West Nile virus capsid protein induces p53-mediated apoptosis via the sequestration of HDM2 to the nucleolus. Cell Microbiol 10(1):165–176

Yap, T. L., Xu, T., Chen, Y. L., Malet, H., Egloff, M. P., Canard, B., Vasudevan, S. G., and Lescar, J. 2007. Crystal structure of the dengue virus RNA-dependent RNA polymerase catalytic domain at 1.85-angstrom resolution. J Virol 81:4753–4765.

Yocupicio-Monroy, M., Padmanabhan, R., Medina, F., and del Angel, R. M. 2007. Mosquito La protein binds to the 3′ untranslated region of the positive and negative polarity dengue virus RNAs and relocates to the cytoplasm of infected cells. Virology 357:29–40.

Yon, C., Teramoto, T., Mueller, N., Phelan, J., Ganesh, V. K., Murthy, K. H., and Padmanabhan, R. 2005. Modulation of the nucleoside triphosphatase/RNA helicase and 5′-RNA triphosphatase activities of Dengue virus type 2 nonstructural protein 3 (NS3) by interaction with NS5, the RNA-dependent RNA polymerase. J Biol Chem 280:27412–27419.

You, S., and Padmanabhan, R. 1999. A novel in vitro replication system for Dengue virus. Initiation of RNA synthesis at the 3′-end of exogenous viral RNA templates requires 5′- and 3′-terminal complementary sequence motifs of the viral RNA. J Biol Chem 274:33714–33722.

You, S., Falgout, B., Markoff, L., and Padmanabhan, R. 2001. In vitro RNA synthesis from exogenous dengue viral RNA templates requires long range interactions between 5′- and 3′-terminal regions that influence RNA structure. J Biol Chem 276:15581–15591.

Yu, L., and Markoff, L. 2005. The topology of bulges in the long stem of the flavivirus 3′ stem-loop is a major determinant of RNA replication competence. J Virol 79:2309–2324.

Yu, C. Y., Hsu, Y. W., Liao, C. L., and Lin, Y. L. 2006. Flavivirus infection activates the XBP1 pathway of the unfolded protein response to cope with endoplasmic reticulum stress. J Virol 80:11868–11880.

Yusof, R., Clum, S., Wetzel, M., Murthy, H. M., and Padmanabhan, R. 2000. Purified NS2B/NS3 serine protease of dengue virus type 2 exhibits cofactor NS2B dependence for cleavage of substrates with dibasic amino acids in vitro. J Biol Chem 275:9963–9969.

Zeng, L., Falgout, B., and Markoff, L. 1998. Identification of specific nucleotide sequences within the conserved 3′-SL in the dengue type 2 virus genome required for replication. J Virol 72:7510–7522.

Zhang, W., Chipman, P. R., Corver, J., Johnson, P. R., Zhang, Y., Mukhopadhyay, S., Baker, T. S., Strauss, J. H., Rossmann, M. G., and Kuhn, R. J. 2003a. Visualization of membrane protein domains by cryo-electron microscopy of dengue virus. Nat Struct Biol 10:907–912.

Zhang, Y., Corver, J., Chipman, P. R., Zhang, W., Pletnev, S. V., Sedlak, D., Baker, T. S., Strauss, J. H., Kuhn, R. J., and Rossmann, M. G. 2003b. Structures of immature flavivirus particles. EMBO J 22:2604–2613.

Zhang, Y., Kaufmann, B., Chipman, P. R., Kuhn, R. J., and Rossmann, M. G. 2007. Structure of immature West Nile virus. J Virol 81:6141–6145.

Zhou, Y., Ray, D., Zhao, Y., Dong, H., Ren, S., Li, Z., Guo, Y., Bernard, K. A., Shi, P. Y., and Li, H. 2007. Structure and function of flavivirus NS5 methyltransferase. J Virol 81:3891–3903.

6. Virulence of West Nile Virus in Different Animal Hosts

David W.C. Beasley and Alan D.T. Barrett

1 Introduction

West Nile virus (WNV) is one of the most widely distributed of all arboviruses, with an extensive distribution throughout Africa, the Middle East, parts of Europe and the former Soviet Union, South Asia, and Australia. Until the mid-1990s, human infections with WNV were mostly associated with a mild undifferentiated fever. However, outbreaks in Europe, Israel, and North America involving humans and animals have been associated with significant rates of neurological disease (see Hayes and Gubler, 2006 for a review). The most important event was the introduction of WNV into the Western Hemisphere in the summer of 1999 when the virus was first isolated in New York City before spreading along the eastern seaboard of the US. The initial outbreak in 1999 involved meningitis and encephalitis with 62 human cases, including seven fatalities (a case: fatality rate of 12%); 25 clinical cases in equines, including nine deaths (a case: fatality rate of 36%); and an accompanying epizootic in birds. The virus subsequently spread along the Eastern seaboard of the US in 2000 and 2001, before moving westwards across the US. In 2002 the virus was found in 44 states and the District of Columbia in the US, and five provinces in Canada. At least 4,161 human cases, including 277 fatalities, and more than 14,000 equine cases were reported in 2002, representing the largest recorded epidemic of arboviral meningoencephalitis in the Western Hemisphere. WNV has continued to be a major public health problem in the US and this is expected to continue for the foreseeable future.

2 Classification of West Nile Virus Strains: Antigenic and Nucleotide Sequence Diversity

Serological studies and nucleotide sequencing and phylogenetic analyses of WNV and other flaviviruses have resulted in antigenic and phylogenetic classification into major subgroups such as the Japanese encephalitis complex of the genus *Flavivirus*, to which WNV belongs (Billoir et al., 2000; Gaunt et al., 2001). Serological comparisons of WNV strains have distinguished four major antigenic subtypes: a group of strains from Africa; strains from Europe and some African strains; strains from India; and strains of Kunjin virus (KUNV), a variant of WNV found in Australia. Prior to the identification of WNV in North America, molecular sequencing studies of WNV strains had defined two major genetic lineages of WNV (lineages 1 and 2), of which lineage 1 strains had a wide global distribution, and lineage 2 strains were primarily of African origin (Berthet et al., 1997) (Fig. 1). Subsequent analyses determined that the 1999 introduction of WNV into North America involved a virus most closely related to a lineage 1 isolate from Israel associated with an outbreak that occurred in 1998 (Jia et al., 1999; Lanciotti et al., 1999), suggesting that WNV was introduced into the United States from the Middle East by unknown mechanism(s).

Detailed phylogenetic analyses of representative lineage 1 WNV strains, based on whole or partial genomic sequences, have identified three major clades corresponding to African/Middle Eastern/European isolates (clade 1a), KUNV strains (1b), or Indian (1c) WNV strains (Beasley et al., 2002; Charrel et al., 2003; Lanciotti et al., 2002). Subclades within these groups have also been proposed (Briese et al., 2002), and a recent phylogenetic study employing a relatively large number of Indian isolates has offered support for establishing these isolates as members of a separate lineage, rather than a subclade of lineage 1 (Bondre et al., 2007). Some other individual virus isolates have been identified that are antigenically and molecularly related to lineage 1 and 2 WNV strains and have been proposed to represent distinct WNV lineages, or possibly even distinct types within the JEV complex. These include a KUNV isolate from Malaysia (Scherret et al., 2001b) and recent isolates from Central Europe (Bakonyi et al., 2005) and Russia (Lvov et al., 2004). One of these isolates, the Rabensberg virus from the Czech Republic, has been found to be attenuated for adult mice even when given by the intracranial route but was as virulent as lineage 1 strains for suckling mice (Bakonyi et al., 2005).

Very close genetic relationships between some lineage 1a strains from widely dispersed geographical areas suggested that significant movement

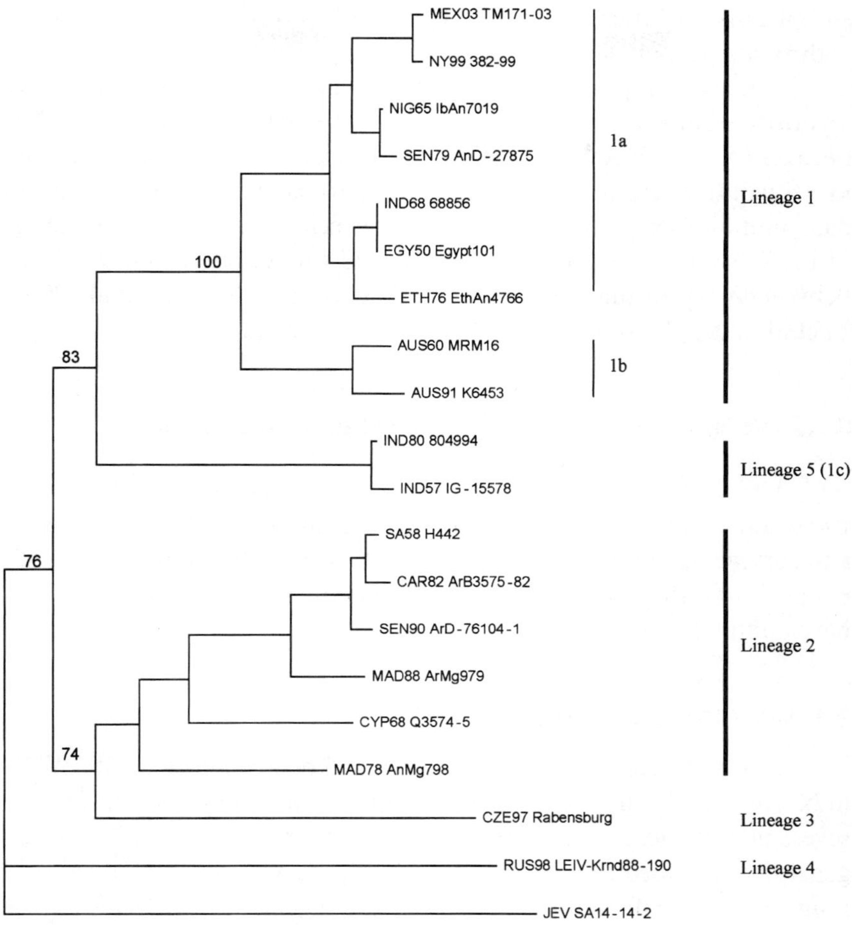

Figure 1. Neighbor-joining phylogenetic tree based on 3′ noncoding region sequences of representative West Nile virus strains showing major genetic lineages. Japanese encephalitis virus strain SA14–14–2 was used as the outgroup. For details on individual strains and definition of major lineages, see Bakonyi et al. (2005), Beasley et al. (2002), Bondre et al. (2007), Charrel et al. (2003), Lanciotti et al. (2002), Lvov et al. (2004).

of WNV occurs, perhaps via migrating birds (Charrel et al., 2003; Rappole and Hubalek, 2003). Similar findings have also been reported for lineage 2 strains, which are primarily of African origin (Burt et al., 2002). Lineage 1 WNV strains, particularly in clade 1a, have been linked to the increased incidence of severe neurological WNV disease observed since the mid-1990s, suggesting that they may be inherently more pathogenic (Lanciotti et al., 2002). However, it may be incorrect to assume that strains of lineage 2 are not associated with severe disease, owing to the limited potential for accurate clinical diagnosis of WNV and other arboviral encephalitis in Africa and to the epidemiological differences associated with WNV transmission in the African setting (Burt et al., 2002).

Studies with polyclonal antisera and monoclonal antibodies have identified significant antigenic overlap between strains in different WNV lineages. Although several WNV subtype-specific antibodies have been identified, their epitopes are not restricted to particular lineages and, to date, mutations associated with subtype-specific epitopes in domain III of the WNV E protein have not been linked with differences in virulence between WNV strains (Besselaar and Blackburn, 1988; Burt et al., 2002; Li et al., 2005; Razumov et al., 2005; Scherret et al., 2001b).

3 WNV: Natural Hosts, Animal Models and Disease

The virus is maintained in an enzootic cycle between birds and *Culex* mosquitoes with birds developing prolonged high levels of viremia and serving as amplifying hosts (Fig. 2). Although mammals are incidental hosts, they can show clinical signs of infection and sometimes have a fatal disease.

3.1 WNV Disease in Humans

It is estimated that approximately 1 in 5 people infected with WNV in North America develops clinical symptoms, and only 1 in 150 develop severe neurological disease (Weaver and Barrett, 2004). West Nile fever, generally described as the mildest form of disease, is characterized by a range of nonspecific symptoms including fatigue, fever, headache, muscle pain and weakness, rash, and neck pain or stiffness (Watson et al., 2004). In many cases, these symptoms can be very debilitating, resulting in hospitalization and a lengthy recovery time, and in older individuals can be associated with fatal outcome, perhaps due to exacerbation of underlying conditions (Sejvar, 2007).

Severe WNV disease is associated with neurological involvement that can range from meningitis and/or encephalitis to a poliomyelitis-like condition with acute flaccid paralysis, often with respiratory failure. A small number of cases involving significant viscerotropic disease, including hepatic or hemorrhagic manifestations, have also been reported (Georges et al., 1988; Mathiot et al., 1990; Paddock et al., 2006). Although a wide range of symptoms and the pathology associated with these conditions have been defined since the introduction of WNV into North America in 1999, it is only recently that investigators have been able to consider the long-term outcomes of these severe infections. Although it

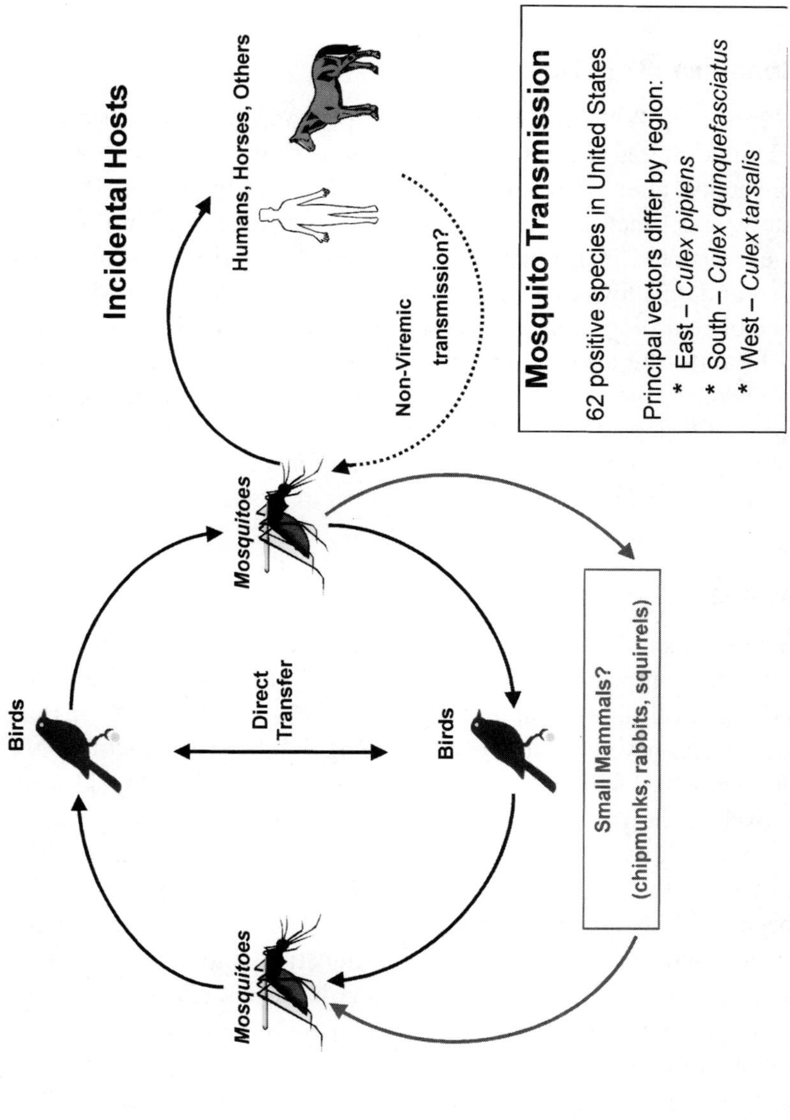

Figure 2. The West Nile virus transmission cycle. (*See Color Plates*)

is estimated that approximately 10% of neuroinvasive WNV disease cases have a fatal outcome, survivors have significant long-term deficits, with more than half of a group of patients from the 1999 New York City outbreak reporting physical or cognitive deficits 18 months or more after infection (Klee et al., 2004).

3.2 Nonhuman Primates

Reports of WNV infections among primates in outdoor colonies in the United States suggest clinical disease is rare, despite high annual seroconversion rates (Hukkanen et al., 2006; Ratterree et al., 2003). In one study, experimental infection of rhesus macaques via an intradermal route resulted in no clinical illness directly attributable to WNV infection and only low level transient viremia (<100 pfu mL^{-1}), although evidence of neuroinvasion was evident for one animal (Ratterree et al., 2004). Despite the absence of overt clinical disease in these reports, earlier primate studies with WNV have reported neurological disease and death following intranasal inoculation in a high proportion of animals, suggesting that route of exposure may be important in facilitating entry to the central nervous system (Goverdhan et al., 1992).

3.3 Avians

The high susceptibility of many North American avian species to WNV infection appears to have been a significant factor in the intensity of WNV epizootic activity in that region. In the United States more than 300 bird species have been identified as being fatally infected with WNV through dead bird surveillance (http://www.cdc.gov/ncidod/dvbid/westnile/birdspecies.htm), with the major impact observed among corvid species, particularly crows and bluejays. Major declines in certain bird populations in the US, including American crows and the sage grouse, have been attributed to WNV (LaDeau et al., 2007). Postmortem analysis of dead avians as well as experimental infection studies in susceptible species using the NY99 or similar strains have described widespread distribution of virus throughout peripheral organs and the nervous system, accompanied by very high viremias (>10^{10} pfu mL^{-1} in American crows and bluejays) (Clark et al., 2006; Komar et al., 2003; Nemeth et al., 2006; Weingartl et al., 2004; Wunschmann et al., 2004). These infected birds can secrete significant quantities of virus in their feces (up to > 10^8 pfu g^{-1} reported for American crows), which contribute to direct bird-to-bird and environmental transmission of the virus (Kipp et al., 2006; Komar et al., 2003). The duration of infection and illness is generally short, with many susceptible birds in experimental studies dying within one week of infection, often 24–48 h after onset of symptoms.

In addition to crows and bluejays, which are highly susceptible to NY99 infection, several other bird species including house finches and house sparrows, which have significantly lower mortality rates, are capable of efficiently amplifying WNV to high titers and are considered to be significant reservoirs for WNV amplification (Komar et al., 2003).

3.4 Equines

Following infection by the bite of a virus-infected mosquito, clinical studies indicate a 20–40% mortality among clinically ill horses. As with other species, a number of factors play a role in the progression of infection, including age, vaccination status, inability to rise, and female gender, which are all associated with the risk of death (Salazar et al., 2004). Viremia in horses is low and of short duration; therefore, horses are unlikely to serve as important amplifying hosts for WNV in nature. During the early years of WNV activity in the US, disease and mortality among horses was a significant finding, and cases of encephalomyelitis had been previously reported in horses in Europe and elsewhere. An apparently high mortality rate among horses in the areas of the initial WNV epidemic in New York State (up to 36% in some areas) and elsewhere during subsequent epidemic years led to the rapid development and approval of an inactivated WNV vaccine for veterinary applications.

Approximately 10% of experimentally infected horses develop clinical illness. An experimental infection of horses via biting of infected *Aedes albopictus* mosquitoes resulted in only one case of clinical disease among 12 animals, and no evidence for sufficient virus amplification to allow re-infection of feeding naive mosquitoes, suggesting that horses did not contribute significantly to amplification of WNV in nature (Bunning et al., 2002). However, there is a report of a possible role for horses and other large mammals in the WNV transmission cycle via nonviremic transmission during co-feeding of infected and uninfected mosquitoes that has been suggested (Higgs et al., 2005). It is also not clear whether feeding by *Culex spp.* mosquitoes, which are the major vectors of WNV in North America, might result in a different outcome to that reported in this study.

3.5 Chipmunks, Rabbits, and Tree Squirrels

Since the introduction of WNV into the US in 1999, over 30 mammal and reptile species have been shown to at least seroconvert and many show clinical signs of infection but all, with three exceptions, are considered to be "dead-end" hosts (i.e., the viremia is too low to enable infection of a mosquito during feeding). The exceptions are tree squirrels, chipmunks, and rabbits. The fox squirrel (*Sciurus niger*), eastern gray squirrel (*S. griseus*), and western gray squirrel (*S. carolinensis*) all succumb to WNV-mediated neurological disease and have viremias (10^5 pfu mL^{-1}) that are sufficient to infect a feeding mosquito (Padgett et al., 2007). Eastern chipmunks (*Tamias striatus*) (Platt et al., 2007) and eastern cottontail rabbits (*Sylvilagus floridanus*) also develop viremias sufficient to infect mosquitoes. The role these species play as amplifying hosts in the enzootic cycle of WNV remains to be established.

3.6 Small Animal Models: Mice and Hamsters

Mice and hamsters are the small animal models of choice for experimental studies of the pathogenesis of WNV neuroinvasive disease. The mouse has proven to be a useful small animal model for WNV neurological disease that in many ways mimics the pathogenesis of severe human WNV infections. Murine models have been used to investigate the tissue tropism of WNV (Kramer and Bernard, 2001) and the pathology associated with WNV neuroinvasion (e.g., Shrestha et al., 2003), to study the immune responses to and immune control of WNV infection (e.g., Diamond et al., 2003b; Shrestha and Diamond, 2004; Wang et al., 2003), and to evaluate candidate antiviral therapies and vaccines against WNV (e.g., Martina et al., 2008; Morrey et al., 2004a). Strains of both species can display a fatal neuroinvasive disease phenotype, although significant variation in susceptibility, particularly between strains of mice, has been reported (Beasley et al., 2002; Brown et al., 2007; Garcia-Tapia et al., 2007; Morrey et al., 2004b; Xiao et al., 2001). In susceptible mice and hamsters, the disease course following peripheral inoculation is relatively short. Virus is usually detected in the blood by the first day post infection (dpi), peaking by 3–4 dpi. Neuroinvasion generally occurs by around 5 dpi, usually accompanied by neurological signs such as tremors or hind limb paralysis, and death occurs around 7 dpi; note that clinical signs may vary depending on the strain and age of the host. In mice that die following WNV infection, widespread distribution of virus throughout peripheral organs and the central nervous system has been reported (Garcia-Tapia et al., 2007; Kramer and Bernard, 2001). The disease outcome appears to depend on a combination of virus strain, route of inoculation, genetic background of the host, and quantity of virus in the inoculum. Studies with strains of outbred Swiss mice have shown them to be highly susceptible to infection with the NY99 strain, with little evidence of age-related resistance (Beasley et al., 2002; Davis et al., 2004). Inbred strains appear to be variable in their susceptibility: BALB/c and C3H mice are considered to be highly susceptible, whereas C57BL/6 mice are more resistant. Interestingly, a recent study of infection kinetics and virus distribution in C3H (susceptible) and C57BL/6 (resistant) mice revealed no significant differences in either viral loads or tissue distribution, suggesting that the relative morbidity and mortality of WNV infections are not solely due to differences in tropism or viral load (Brown et al., 2007).

Extensive use has been made of knockout mice on the C57BL/6 background to study the importance of various immune system components in pathogenesis of and recovery from WNV infection, with most facets of the innate, humoral, and cell-mediated immune responses shown to be important at some level (Diamond et al., 2003a; Samuel and Diamond, 2006). Significantly, some aspects of innate immune responses may also contribute to the pathogenesis of WNV neurological disease, particularly in facilitating movement of the virus into the central nervous system (Arjona et al., 2007; Wang et al., 2004).

4 Molecular Determinants of Natural Virulence Variations Between WNV Strains

The WNV genome is a single-stranded, positive-sense RNA approximately 11,000 nucleotide in length. The genome is composed of a short 5′ noncoding region, a single long open reading frame containing more than 10,000 nucleotides, and a 3′ noncoding region of variable length. The 3400-amino acid open reading frame encodes three structural proteins at the 5′ end, which are the capsid (C), pre-membrane (preM), and envelope (E) proteins, followed by seven nonstructural (NS) proteins (NS1, NS2A, NS2B, NS3, NS4A,NS4B, NS5) in the 3′ two-thirds of the genome.

4.1 Contribution of Individual Viral-Encoded Proteins to WNV Virulence

There have been a number of studies identifying the contribution of viral genes to the virulence phenotype of the virus on the basis of single or multiple mutations. A number of studies have shown that the

NS4B protein is involved in resistance to interferon due to inhibition of the interferon signal transduction pathway (Evans and Seeger, 2007; Puig-Basagoiti et al., 2007). Mutation of NS4B at residues E22A/K24A or E249G resulted in reduction of both viral RNA synthesis and virus multiplication. Mutation of NS4B C102S resulted in a virus that replicated similar to wild-type virus in cell culture but was temperature sensitive and attenuated for neurovirulence and neuroinvasiveness in the 4-week-old mouse model (Wicker et al., 2006). Mutations of prM P54S and NS2A V61A resulted in a virus that has reduced replication in cell culture and in *Culex pipiens* mosquitoes, and was temperature sensitive (Jia et al., 2007). An earlier study (Liu et al., 2006) demonstrated that NS2A is also involved in interference with transcription of interferon-β, and that mutation of NS2A A30P resulted in a virus that was attenuated for both neurovirulence and neuroinvasiveness in the 3-week-old mouse model. Not surprisingly, these studies show that virulence is multigenic and that the same phenotype can be derived by different mutations.

4.2 Comparative Studies of WNV Virulence in Mouse and Hamster Models

Several studies have examined the relative virulence of WNV strains in mice, hamsters, and birds in an effort to identify specific viral determinants of virulence. Although it had been hypothesized that enhanced virulence leading to higher rates of neuroinvasive disease in humans was primarily associated with lineage 1 WNV strains, studies in outbred mice and hamsters using strains representing major subtypes of lineages 1 and 2 identified both highly attenuated and highly neuroinvasive strains in both lineages (Beasley et al., 2001, 2002, 2004b). These and a number of other studies indicate that WNV is more virulent (in terms of lethality) in mice than hamsters such that mice may be considered a lethality model while the hamster is an infection model where a major criterion is level of viremia rather than lethality. The strain of WNV introduced into North America was among the most virulent in these studies, although one comparison in 9–11-week-old Syrian Golden hamsters reported higher neuroinvasive disease and mortality rates among animals infected with the prototype Uganda 37 strain (lineage 2) compared to a NY99 strain (Morrey et al., 2004b).

Two studies comparing cytokine and other gene expression profiles in brain and/or peripheral tissues of mice infected with neuroinvasive or attenuated WNV strains reported that many of the same genes were upregulated in either case, although the magnitude of responses did vary for some genes (Shirato et al., 2004a; Venter et al., 2005)

On the basis of the identification of virulence differences between WNV strains, attempts have been made to identify specific virus-encoded determinants of neuroinvasive WNV disease. Nucleotide sequence comparisons of an attenuated lineage 1 strain, ETH76a, with the virulent North American prototype strain NY99, and site-directed mutagenesis using an infectious clone derived from NY99, determined that absence of a glycosylation motif in the viral envelope (E) protein was associated with attenuation of virus multiplication in vitro and reduced neuroinvasiveness in mice (Beasley et al., 2005). Similar observations were also reported following characterization of plaque-purified variants of North American isolates lacking the E glycosylation site (Shirato et al., 2004b, 2006) and from studies of glycosylation variants of KUNV (Scherret et al., 2001a), although other authors had reported that loss of glycosylation was associated with enhanced neuroinvasivness of an Israeli WNV strain (Chambers et al., 1998).

Thus, E protein glycosylation of WNV may contribute to growth and virulence in several ways, and has been reported to affect assembly and infectivity of virions (Hanna et al., 2005), virion stability (Beasley et al., 2005), and interactions with cell surface ligands (Davis et al., 2006a,b). Comparisons of cytokine responses identified higher expression of TNFα from mouse peritoneal macrophages infected with glycosylated compared with non-glycosylated WNV variants, which may also contribute to enhanced neuroinvasiveness (Shirato et al., 2006).

There have been comparatively few studies on the virulence phenotype in the hamster model. Ding et al. (2005) developed an asymptomatic persistent infection in the kidneys of experimentally infected hamsters, including shedding of virus in the urine. Genomic sequencing revealed that the persistent viruses had five amino acid substitutions (E-V21M, E-L167F, NS1-I183T, NS2B-M99I, NS5-M167I, and NS5-E390G), of which only E-L167F was shared by all persisting viruses.

4.3 Emergence and Characterization of Attenuated WNV Variants in the Americas

Although the vast majority of WNV strains isolated in North America retain the highly neuroinvasive phenotype of the originally introduced strain, several strains have been identified with significant attenuation and other phenotypic changes. An isolate from Tabasco, Mexico, included a mixed population with one variant lacking E protein glycosylation (Estrada-Franco et al., 2003). This non-glycosylated variant was completely attenuated for neuroinvasiveness in outbred mice, suggesting that additional amino acid mutations identified in the prM, NS4B, and/or NS5 proteins had an additive attenuating effect (Beasley et al., 2004a). Several WNV isolates from Texas and New York with small plaque, temperature-sensitive and/or mouse attenuated phenotypes have also been described (Davis et al., 2004; Jia et al., 2007). Analysis of those variants and studies with infectious clones have determined that combinations of mutations in structural (prM) and nonstructural (NS2A, NS4B, NS5) genes, and the 3′ noncoding region are the likely determinants of attenuation (Davis et al., 2007; Jia et al., 2007). In one study, virulence was found to be multigenic involving at least NS4B, NS5, and the 3′ noncoding region (Davis et al., 2007).

4.4 Determinants of Virulence in Avians

Although smaller in scope than studies comparing virulence of WNV strains in mice, inoculation of American crows or house sparrows with three lineage 1 WNV strains (NY99, a Kenyan isolate KN-3829, and Kunjin strain K-6453) has also demonstrated clear differences in the ability of these strains to infect and cause disease in birds. Strains NY99 and KN-3829 are equally virulent in the mouse model (Beasley et al., 2004b), but NY99 caused 100% mortality in American crows and was associated with very high viremias (approx. 10^9 pfu mL^{-1}), whereas K-6453 caused no mortality and KN-3829 killed only 1 of 8 crows, with viremias $< 10^4$ pfu mL^{-1} (Brault et al., 2004). In sparrows, which are more resistant to NY99 infection (~50% mortality), virulence of KN-3829 and NY99 were more comparable (Langevin et al., 2005), suggesting that host as well as virus characteristics are significant for determining the ultimate outcome of infection.

Detailed comparison of the multiplication characteristics of NY99 and KN-3829 determined that NY99 was capable of more efficient replication at elevated temperatures (>40°C) suggesting that the relative attenuation of KN-3829 in crows was driven by less efficient replication in that host (Kinney et al., 2006). Subsequent studies with infectious clones of these two strains have identified a single mutation at residue 249 of the NS3 protein that is responsible for thermostable replication of NY99 and its increased virulence in American crows (Brault et al., 2007). This mutation is also present in other lineage 1 WNV strains associated with epidemics of encephalitis in Eastern Europe and Russia.

5 Summary

It is clear that there is no single virus-encoded determinant of increased virulence. Although individual mutations have been identified as contributing to the highly virulent characteristics of the NY99 strain in mouse/hamster models and/or in birds when compared with some other WNV strains, emergence of attenuated variants in the Americas has been associated with a spectrum of mutations throughout the viral genome. Furthermore, individual mutations have not necessarily been reported to have similar attenuating effects in mammalian versus avian hosts. However, further studies of these WNV variants and the effects of specific mutations on virus growth, replication, and pathogenesis in vitro and in avian and mammalian systems should provide useful insights into the specific viral characteristics that contribute to neuroinvasive WNV disease.

References

Arjona, A., Foellmer, H.G., Town, T., Leng, L., McDonald, C., Wang, T., Wong, S.J., Montgomery, R.R., Fikrig, E., and Bucala, R. (2007). Abrogation of macrophage migration inhibitory factor decreases West Nile virus lethality by limiting viral neuroinvasion. J Clin Invest *117*, 3059–3066.

Bakonyi, T., Hubalek, Z., Rudolf, I., and Nowotny, N. (2005). Novel flavivirus or new lineage of West Nile virus, central Europe. Emerg Infect Dis *11*, 225–231.

Beasley, D.W.C., Davis, C.T., Estrada-Franco, J., Navarro-Lopez, R., Campomanes-Cortes, A., Tesh, R.B., Weaver, S.C., and Barrett, A.D.T. (2004a). Genome sequence and attenuating mutations in West Nile virus isolate from Mexico. Emerg Infect Dis *10*, 2221–2224.

Beasley, D.W.C., Li, L., Suderman, M.T., and Barrett, A.D.T. (2001). West Nile virus strains differ in mouse neurovirulence and binding to mouse or human brain membrane receptor preparations. Ann N Y Acad Sci *951*, 332–335.

Beasley, D.W.C., Li, L., Suderman, M.T., and Barrett, A.D.T. (2002). Mouse neuroinvasive phenotype of West Nile virus strains varies depending upon virus genotype. Virology *296*, 17–23.

Beasley, D.W.C., Whiteman, M.C., Zhang, S., Huang, C.Y., Schneider, B.S., Smith, D.R., Gromowski, G.D., Higgs, S., Kinney, R.M., and Barrett, A.D.T. (2005). Envelope protein glycosylation status influences mouse neuroinvasion phenotype of genetic lineage 1 West Nile virus strains. J Virol *79*, 8339–8347.

Berthet, F.X., Zeller, H.G., Drouet, M.T., Rauzier, J., Digoutte, J.P., and Deubel, V. (1997). Extensive nucleotide changes and deletions within the envelope glycoprotein gene of Euro-African West Nile viruses. J Gen Virol *78* (Pt 9), 2293–2297.

Besselaar, T.G., and Blackburn, N.K. (1988). Antigenic analysis of West Nile virus strains using monoclonal antibodies. Arch Virol *99*, 75–88.

Billoir, F., de Chesse, R., Tolou, H., de Micco, P., Gould, E.A., and de Lamballerie, X. (2000). Phylogeny of the genus flavivirus using complete coding sequences of arthropod-borne viruses and viruses with no known vector. J Gen Virol *81(Pt 9)*, 2339.

Bondre, V.P., Jadi, R.S., Mishra, A.C., Yergolkar, P.N., and Arankalle, V.A. (2007). West Nile virus isolates from India: evidence for a distinct genetic lineage. J Gen Virol *88*, 875–884.

Brault, A.C., Huang, C.Y., Langevin, S.A., Kinney, R.M., Bowen, R.A., Ramey, W.N., Panella, N.A., Holmes, E.C., Powers, A.M., and Miller, B.R. (2007). A single positively selected West Nile viral mutation confers increased virogenesis in American crows. Nat Genet *39*, 1162–1166.

Brault, A.C., Langevin, S.A., Bowen, R.A., Panella, N.A., Biggerstaff, B.J., Miller, B.R., and Komar, N. (2004). Differential virulence of West Nile strains for American crows. Emerg Infect Dis *10*, 2161–2168.

Briese, T., Rambaut, A., Pathmajeyan, M., Bishara, J., Weinberger, M., Pitlik, S., and Lipkin, W.I. (2002). Phylogenetic analysis of a human isolate from the 2000 Israel West Nile virus epidemic. Emerg Infect Dis *8*, 528–531.

Brown, A.N., Kent, K.A., Bennett, C.J., and Bernard, K.A. (2007). Tissue tropism and neuroinvasion of West Nile virus do not differ for two mouse strains with different survival rates. Virology *368*, 422–430.

Bunning, M.L., Bowen, R.A., Cropp, C.B., Sullivan, K.G., Davis, B.S., Komar, N., Godsey, M.S., Baker, D., Hettler, D.L., Holmes, D.A., et al. (2002). Experimental infection of horses with West Nile virus. Emerg Infect Dis *8*, 380–386.

Burt, F.J., Grobbelaar, A.A., Leman, P.A., Anthony, F.S., Gibson, G.V., and Swanepoel, R. (2002). Phylogenetic relationships of southern African West Nile virus isolates. Emerg Infect Dis *8*, 820–826.

Chambers, T.J., Halevy, M., Nestorowicz, A., Rice, C.M., and Lustig, S. (1998). West Nile virus envelope proteins: nucleotide sequence analysis of strains differing in mouse neuroinvasiveness. J Gen Virol 79(Pt 10), 2375–2380.

Charrel, R.N., Brault, A.C., Gallian, P., Lemasson, J.J., Murgue, B., Murri, S., Pastorino, B., Zeller, H., de Chesse, R., de Micco, P., et al. (2003). Evolutionary relationship between Old World West Nile virus strains. Evidence for viral gene flow between Africa, the Middle East, and Europe. Virology *315*, 381–388.

Clark, L., Hall, J., McLean, R., Dunbar, M., Klenk, K., Bowen, R., and Smeraski, C.A. (2006). Susceptibility of greater sage-grouse to experimental infection with West Nile virus. J Wildl Dis *42*, 14–22.

Davis, C.T., Beasley, D.W.C., Guzman, H., Siirin, M., Parsons, R.E., Tesh, R.B., and Barrett, A.D.T. (2004). Emergence of attenuated West Nile virus variants in Texas, 2003. Virology *330*, 342–350.

Davis, C.T., Galbraith, S.E., Zhang, S., Whiteman, M.C., Li, L., Kinney, R.M., and Barrett, A.D.T. (2007). A combination of naturally occurring mutations in North American West Nile virus nonstructural protein genes and in the 3 untranslated region alters virus phenotype. J Virol *81*, 6111–6116.

Davis, C.W., Mattei, L.M., Nguyen, H.Y., Ansarah-Sobrinho, C., Doms, R.W., and Pierson, T.C. (2006a). The location of asparagine-linked glycans on West Nile virions controls their interactions with CD209 (dendritic cell-specific ICAM-3 grabbing nonintegrin). J Biol Chem *281*, 37183–37194.

Davis, C.W., Nguyen, H.Y., Hanna, S.L., Sanchez, M.D., Doms, R.W., and Pierson, T.C. (2006b). West Nile virus discriminates between DC-SIGN and DC-SIGNR for cellular attachment and infection. J Virol *80*, 1290–1301.

Diamond, M.S., Shrestha, B., Mehlhop, E., Sitati, E., and Engle, M. (2003a). Innate and adaptive immune responses determine protection against disseminated infection by West Nile encephalitis virus. Viral Immunol *16*, 259–278.

Diamond, M.S., Sitati, E.M., Friend, L.D., Higgs, S., Shrestha, B., and Engle, M. (2003b). A critical role for induced IgM in the protection against West Nile virus infection. J Exp Med *198*, 1853–1862.

Ding, X., Wu, X., Duan, T., Siirin, M., Guzman, H., Yang, Z., Tesh, R.B., and Xiao, S.Y. (2005). Nucleotide and amino acid changes in West Nile virus strains exhibiting renal tropism in hamsters. Am J Trop Med Hyg *73*, 803–807.

Estrada-Franco, J.G., Navarro-Lopez, R., Beasley, D.W.C., Coffey, L., Carrara, A.S., Travassos da Rosa, A., Clements, T., Wang, E., Ludwig, G.V., Cortes, A.C. et al., (2003). West Nile virus in Mexico: evidence of widespread circulation since July 2002. Emerg Infect Dis *9*, 1604–1607.

Evans, J.D., and Seeger, C. (2007). Differential effects of mutations in NS4B on West Nile virus replication and inhibition of interferon signaling. J Virol *81*, 11809–11816.

Garcia-Tapia, D., Hassett, D.E., Mitchell, W.J., Jr., Johnson, G.C., and Kleiboeker, S.B. (2007). West Nile virus encephalitis: sequential histopathological and immunological events in a murine model of infection. J Neurovirol *13*, 130–138.

Gaunt, M.W., Sall, A.A., de Lamballerie, X., Falconar, A.K., Dzhivanian, T.I., and Gould, E.A. (2001). Phylogenetic relationships of flaviviruses correlate with their epidemiology, disease association and biogeography. J Gen Virol *82*, 1867–1876.

Georges, A.J., Lesbordes, J.L., Georges-Courbot, M.C., Meunier, D.M.Y., and Gonzalez, J.P. (1988). Fatal hepatitis from West Nile virus. Ann Inst Pasteur Virol *138*, 237–244.

Goverdhan, M.K., Kulkarni, A.B., Gupta, A.K., Tupe, C.D., and Rodrigues, J.J. (1992). Two-way cross-protection between West Nile and Japanese encephalitis viruses in bonnet macaques. Acta Virol *36*, 277–283.

Hanna, S.L., Pierson, T.C., Sanchez, M.D., Ahmed, A.A., Murtadha, M.M., and Doms, R.W. (2005). N-linked glycosylation of west nile virus envelope proteins influences particle assembly and infectivity. J Virol *79*, 13262–13274.

Hayes, E.B., and Gubler, D.J. (2006). West Nile virus: epidemiology and clinical features of an emerging epidemic in the United States. Annu Rev Med *57*, 181–194.

Higgs, S., Schneider, B.S., Vanlandingham, D.L., Klingler, K.A., and Gould, E.A. (2005). Nonviremic transmission of West Nile virus. Proc Natl Acad Sci USA *102*, 8871–8874.

Hukkanen, R.R., Liggitt, H.D., Kelley, S.T., Grant, R., Anderson, D.M., Hall, R.A., Tesh, R.B., Travassos DaRosa, A.P., and Bielefeldt-Ohmann, H. (2006). West Nile and St. Louis encephalitis virus antibody seroconversion, prevalence, and persistence in naturally infected pig-tailed macaques (Macaca nemestrina). Clin Vaccine Immunol *13*, 711–714.

Jia, X.Y., Briese, T., Jordan, I., Rambaut, A., Chi, H.C., Mackenzie, J.S., Hall, R.A., Scherret, J., and Lipkin, W.I. (1999). Genetic analysis of West Nile New York 1999 encephalitis virus. Lancet *354*, 1971–1972.

Jia, Y., Moudy, R.M., Dupuis, A.P., III, Ngo, K.A., Maffei, J.G., Jerzak, G.V., Franke, M.A., Kauffman, E.B., and Kramer, L.D. (2007). Characterization of a small plaque variant of West Nile virus isolated in New York in 2000. Virology *367*, 339–347.

Kinney, R.M., Huang, C.Y., Whiteman, M.C., Bowen, R.A., Langevin, S.A., Miller, B.R., and Brault, A.C. (2006). Avian virulence and thermostable replication of the North American strain of West Nile virus. J Gen Virol *87*, 3611–3622.

Kipp, A.M., Lehman, J.A., Bowen, R.A., Fox, P.E., Stephens, M.R., Klenk, K., Komar, N., and Bunning, M.L. (2006). West Nile virus quantification in feces of experimentally infected American and fish crows. Am J Trop Med Hyg *75*, 688–690.

Klee, A.L., Maidin, B., Edwin, B., Poshni, I., Mostashari, F., Fine, A., Layton, M., and Nash, D. (2004). Long-term prognosis for clinical West Nile virus infection. Emerg Infect Dis *10*, 1405–1411.

Komar, N., Langevin, S., Hinten, S., Nemeth, N., Edwards, E., Hettler, D., Davis, B., Bowen, R., and Bunning, M. (2003). Experimental infection of North American birds with the New York 1999 strain of West Nile virus. Emerg Infect Dis *9*, 311–322.

Kramer, L.D., and Bernard, K.A. (2001). West Nile virus infection in birds and mammals. Ann N Y Acad Sci *951*, 84–93.

LaDeau, S.L., Kilpatrick, A.M., and Marra, P.P. (2007). West Nile virus emergence and large-scale declines of North American bird populations. Nature *447*, 710–713.

Lanciotti, R.S., Ebel, G.D., Deubel, V., Kerst, A.J., Murri, S., Meyer, R., Bowen, M., McKinney, N., Morrill, W.E., Crabtree, M.B.et al., (2002). Complete genome sequences and phylogenetic analysis of West Nile virus strains isolated from the United States, Europe, and the Middle East. Virology *298*, 96–105.

Lanciotti, R.S., Roehrig, J.T., Deubel, V., Smith, J., Parker, M., Steele, K., Crise, B., Volpe, K.E., Crabtree, M.B., Scherret, J.H.et al., (1999). Origin of the West Nile virus responsible for an outbreak of encephalitis in the northeastern United States. Science *286*, 2333–2337.

Langevin, S.A., Brault, A.C., Panella, N.A., Bowen, R.A., and Komar, N. (2005). Variation in virulence of West Nile virus strains for house sparrows (Passer domesticus). Am J Trop Med Hyg *72*, 99–102.

Li, L., Barrett, A.D.T., and Beasley, D.W.C. (2005). Differential expression of domain III neutralizing epitopes on the envelope proteins of West Nile virus strains. Virology *335*, 99–105.

Liu, W.J., Wang, X.J., Clark, D.C., Lobigs, M., Hall, R.A., and Khromykh, A.A. (2006). A single amino acid substitution in the West Nile virus nonstructural protein NS2A disables its ability to inhibit alpha/beta interferon induction and attenuates Kinney, R., virus virulence in mice. J Virol *80*, 2396–2404.

Lvov, D.K., Butenko, A.M., Gromashevsky, V.L., Kovtunov, A.I., Prilipov, A.G., Aristova, V.A., Dzharkenov, A.F., Samokhvalov, E.I., Savage, H.M., et al. (2004). West Nile virus and other zoonotic viruses in Russia: examples of emerging-reemerging situations. Arch Virol Suppl 85–96.

Martina, B.E., Koraka, P., van den Doel, P., van Amerongen, G., Rimmelzwaan, G.F., and Osterhaus, A.D. (2008). Immunization with West Nile virus envelope domain III protects mice against lethal infection with homologous and heterologous virus. Vaccine *26*, 153–157.

Mathiot, C.C., Georges, A.J., and Deubel, V. (1990). Comparative analysis of West Nile virus strains isolated from human and animal hosts using monoclonal antibodies and cDNA restriction digest profiles. Res Virol *141*, 533–543.

Morrey, J.D., Day, C.W., Julander, J.G., Blatt, L.M., Smee, D.F., and Sidwell, R.W. (2004a). Effect of interferon-alpha and interferon-inducers on West Nile virus in mouse and hamster animal models. Antivir Chem Chemother *15*, 101–109.

Morrey, J.D., Day, C.W., Julander, J.G., Olsen, A.L., Sidwell, R.W., Cheney, C.D., and Blatt, L.M. (2004b). Modeling hamsters for evaluating West Nile virus therapies. Antiviral Res *63*, 41–50.

Nemeth, N., Gould, D., Bowen, R., and Komar, N. (2006). Natural and experimental West Nile virus infection in five raptor species. J Wildl Dis *42*, 1–13.

Paddock, C.D., Nicholson, W.L., Bhatnagar, J., Goldsmith, C.S., Greer, P.W., Hayes, E.B., Risko, J.A., Henderson, C., Blackmore, C.G., Lanciotti, R.S., et-al. (2006). Fatal hemorrhagic fever caused by West Nile virus in the United States. Clin Infect Dis *42*, 1527–1535.

Padgett, K.A., Reisen, W.K., Kahl-Purcell, N., Fang, Y., Cahoon-Young, B., Carney, R., Anderson, N., Zucca, L., Woods, L., Husted, S., et al. (2007). West Nile virus infection in tree squirrels (Rodentia: Sciuridae) in California, 2004–2005. Am J Trop Med Hyg *76*, 810–813.

Platt, K.B., Tucker, B.J., Halbur, P.G., Tiawsirisup, S., Blitvich, B.J., Fabiosa, F.G., Bartholomay, L.C., and Rowley, W.A. (2007). West Nile virus viremia in eastern chipmunks (Tamias striatus) sufficient for infecting different mosquitoes. Emerg Infect Dis *13*, 831–837.

Puig-Basagoiti, F., Tilgner, M., Bennett, C.J., Zhou, Y., Munoz-Jordan, J.L., Garcia-Sastre, A., Bernard, K.A., and Shi, P.Y. (2007). A mouse cell-adapted NS4B mutation attenuates West Nile virus RNA synthesis. Virology *361*, 229–241.

Rappole, J.H., and Hubalek, Z. (2003). Migratory birds and West Nile virus. J Appl Microbiol 94(Suppl), 47S–58S.

Ratterree, M.S., da Rosa, A.P., Bohm, R.P. , CogswellJr., F.B., Phillippi, K.M., Caillouet, K., Schwanberger, S., Shope, R.E., and Tesh, R.B. (2003). West Nile virus infection in nonhuman primate breeding colony, concurrent with human epidemic, southern Louisiana. Emerg Infect Dis *9*, 1388–1394.

Ratterree, M.S., Gutierrez, R.A., Travassos da Rosa, A.P., Dille, B.J., Beasley, D.W.C., Bohm, R.P., Desai, S.M., Didier, P.J., Bikenmeyer, L.G., Dawson, G.J., et al. (2004). Experimental infection of rhesus macaques with West Nile virus: level and duration of viremia and kinetics of the antibody response after infection. J Infect Dis *189*, 669–676.

Razumov, I.A., Kazachinskaia, E.I., Ternovoi, V.A., Protopopova, E.V., Galkina, I.V., Gromashevskii, V.L., Prilipov, A.G., Kachko, A.V., Ivanova, A.V., L'Vov, D.K., et al.

(2005). Neutralizing monoclonal antibodies against Russian strain of the West Nile virus. Viral Immunol *18*, 558–568.

Salazar, P., Traub-Dargatz, J.L., Morley, P.S., Wilmot, D.D., Steffen, D.J., Cunningham, W.E., and Salman, M.D. (2004). Outcome of equids with clinical signs of West Nile virus infection and factors associated with death. J Am Vet Med Assoc *225*, 267–274.

Samuel, M.A., and Diamond, M.S. (2006). Pathogenesis of West Nile Virus infection: a balance between virulence, innate and adaptive immunity, and viral evasion. J Virol *80*, 9349–9360.

Scherret, J.H., Mackenzie, J.S., Khromykh, A.A., and Hall, R.A. (2001a). Biological significance of glycosylation of the envelope protein of Kunjin virus. Ann N Y Acad Sci *951*, 361–363.

Scherret, J.H., Poidinger, M., Mackenzie, J.S., Broom, A.K., Deubel, V., Lipkin, W.I., Briese, T., Gould, E.A., and Hall, R.A. (2001b). The relationships between West Nile and Kunjin viruses. Emerg Infect Dis *7*, 697–705.

Sejvar, J.J. (2007). The long-term outcomes of human West Nile virus infection. Clin Infect Dis *44*, 1617–1624.

Shirato, K., Kimura, T., Mizutani, T., Kariwa, H., and Takashima, I. (2004a). Different chemokine expression in lethal and non-lethal murine West Nile virus infection. J Med Virol *74*, 507–513.

Shirato, K., Miyoshi, H., Goto, A., Ako, Y., Ueki, T., Kariwa, H., and Takashima, I. (2004b). Viral envelope protein glycosylation is a molecular determinant of the neuroinvasiveness of the New York strain of West Nile virus. J Gen Virol *85*, 3637–3645.

Shirato, K., Miyoshi, H., Kariwa, H., and Takashima, I. (2006). The kinetics of proinflammatory cytokines in murine peritoneal macrophages infected with envelope protein-glycosylated or non-glycosylated West Nile virus. Virus Res *121*, 11–16.

Shrestha, B., and Diamond, M.S. (2004). Role of CD8± T cells in control of West Nile virus infection. J Virol *78*, 8312–8321.

Shrestha, B., Gottlieb, D., and Diamond, M.S. (2003). Infection and injury of neurons by West Nile encephalitis virus. J Virol *77*, 13203–13213.

Venter, M., Myers, T.G., Wilson, M.A., Kindt, T.J., Paweska, J.T., Burt, F.J., Leman, P.A., and Swanepoel, R. (2005). Gene expression in mice infected with West Nile virus strains of different neurovirulence. Virology *342*, 119–140.

Wang, T., Scully, E., Yin, Z., Kim, J.H., Wang, S., Yan, J., Mamula, M., Anderson, J.F., Craft, J., and Fikrig, E. (2003). IFN-gamma-producing gamma delta T cells help control murine West Nile virus infection. J Immunol *171*, 2524–2531.

Wang, T., Town, T., Alexopoulou, L., Anderson, J.F., Fikrig, E., and Flavell, R.A. (2004). Toll-like receptor 3 mediates West Nile virus entry into the brain causing lethal encephalitis. Nat Med *10*, 1366–1373.

Watson, J.T., Pertel, P.E., Jones, R.C., Siston, A.M., Paul, W.S., Austin, C.C., and Gerber, S.I. (2004). Clinical characteristics and functional outcomes of West Nile Fever. Ann Intern Med *141*, 360–365.

Weaver, S.C., and Barrett, A.D.T. (2004). Transmission cycles, host range, evolution and emergence of arboviral disease. Nat Rev Microbiol *2*, 789–801.

Weingartl, H.M., Neufeld, J.L., Copps, J., and Marszal, P. (2004). Experimental West Nile virus infection in blue jays (Cyanocitta cristata) and crows (Corvus brachyrhynchos). Vet Pathol *41*, 362–370.

Wicker, J.A., Whiteman, M.C., Beasley, D.W.C., Davis, C.T., Zhang, S., Schneider, B.S., Higgs, S., Kinney, R.M., and Barrett, A.D.T. (2006). A single amino acid substitution in the central portion of the West Nile virus NS4B protein confers a highly attenuated phenotype in mice. Virology *349*, 245–253.

Wunschmann, A., Shivers, J., Carroll, L., and Bender, J. (2004). Pathological and immunohistochemical findings in American crows (Corvus brachyrhynchos) naturally infected with West Nile virus. J Vet Diagn Invest *16*, 329–333.

Xiao, S.Y., Guzman, H., Zhang, H., Travassos da Rosa, A.P., and Tesh, R.B. (2001). West Nile virus infection in the golden hamster (Mesocricetus auratus): a model for West Nile encephalitis. Emerg Infect Dis *7*, 714–721.

7. Innate immune Response and Mechanisms of Interferon Antagonism Against West Nile Virus

Jared D. Evans and Christoph Seeger

1 Introduction

Like other arboviruses, West Nile virus (WNV) has the remarkable ability to replicate and assemble virus particles in both insect and mammalian cells and, hence, can complete its life cycle under very different environmental conditions. Both invertebrates and vertebrates rely on cellular antiviral programs, the innate immune response, to restrict amplification of viral genomes and to protect cells from infection. How do cells recognize invading WNV and how does the virus overcome cellular defense mechanisms? In this chapter, we summarize major advances to our knowledge about the nature of the complex interactions between host and pathogen.

Keywords

West Nile virus, innate immunity, interferon response, Toll-like receptors, retionic acid inducible gene

2 Recognition of WNV by Cellular Sensors: Activation of the First Wave of the Innate Immune Response

Infection of humans begins with a bite of a WNV-infected mosquito. On the basis of in vivo experiments with mice, the size of extravascular inoculum varies among different mosquito species and can reach 10^4–10^6 plaque forming units (PFUs) (Styer et al., 2007). As a consequence, bone marrow–derived dendritic cells residing in the epidermis, called Langerhans cells (LC), become infected and "activated," leading to their migration to the draining lymph node via the afferent lymphatic system (Johnston et al., 2000). Migration of LCs is interleukin-1β dependent (Byrne et al., 2001).

Cells use extracellular and cytoplasmic pathogen recognition receptors (PRRs) to sense pathogen associated molecular patterns (PAMPs) derived from viral proteins and nucleic acids (reviewed in Akira et al., 2006). PRRs known to play a role in WNV biology include the toll-like receptor 3 (TLR3), which is present primarily on endosomes, as well as the cytoplasmic RNA helicases retinoic acid inducible gene I (RIG-I) and melanoma differentiation associated protein 5 (MDA5) (Fig. 1). RIG-I contains two N-terminal caspase recruitment domains (CARD), known to form homotypic interactions with other CARD domains, and a C-terminal DExD/H box RNA helicase domain. The latter can recognize PAMPs leading to the binding to the adaptor, IPS-1, also known as VISA, Cardif, or MAVS. In turn, this interaction triggers activation of the IκB kinase related kinase IKK-ε, also known as IKK-i, and TANK-binding kinase 1 (TBK1) by unknown mechanisms. The kinases then activate the latent transcription factors IRF3 and IRF7. In parallel, IPS-1 can interact with FADD and RIP1 to activate IKK-β and, in turn, NF-κB (reviewed in Takeuchi and Akira, 2007). Thus, the IPS-1 axis results in the activation of two transcription factors that are required

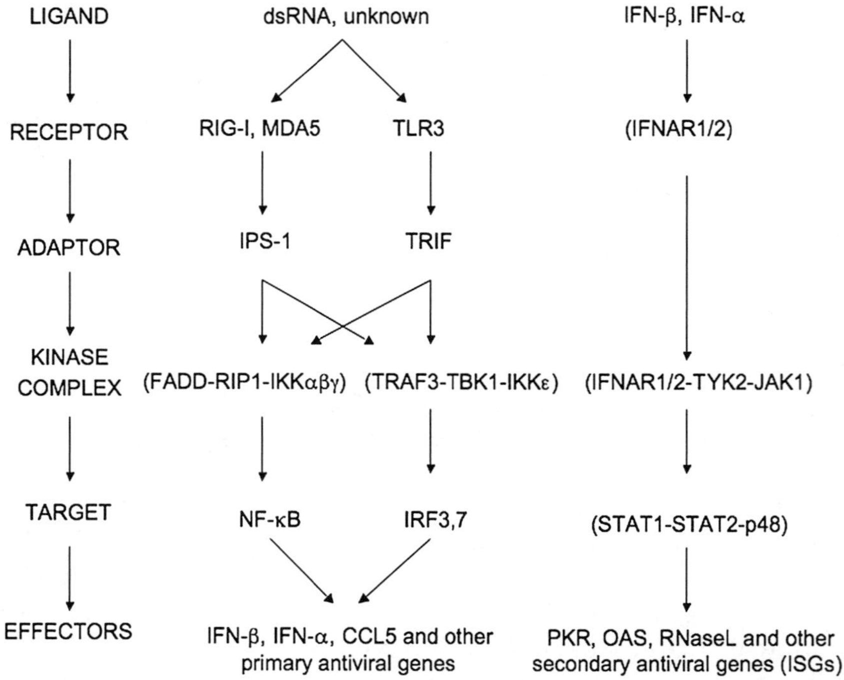

Figure 1. Model for innate signal transduction pathways. The figure depicts the components of two innate signal transduction pathways that are activated in WNV infections and lead to the expression of primary and secondary antiviral genes as described in detail in the text.

both for the induction of genes with direct antiviral and inflammatory properties, such as IFN-β and IFNα4.

The observation that WNV can stimulate infected LC to migrate to lymph nodes indicated that one or several viral components are recognized as PAMPs and trigger activation of an innate immune response (Johnston et al., 1996, 2000). Experiments with WNV-infected mouse embryo fibroblasts (MEFs) demonstrated that WNV could induce the expression of IFN-β and other genes via the RIG-I–IPS-1 signal transduction pathway (Fredericksen and Gale, 2006; Fredericksen et al., 2004). These experiments also revealed that RIG-I, but not IPS-1, deficiency could be partially complemented by MDA5 (Fredericksen et al., 2007). Infection of MEFs with Japanese encephalitis virus (JEV), a close relative of WNV, revealed a requirement for RIG-I, but not MDA5 (Kato et al., 2006). Importantly, JEV caused uniform lethality in RIG-I-deficient mice in comparison to wild-type or MDA5-deficient animals. Consistent with the role of RIG-I as an "IFN inducer," IFN levels in RIG-I-deficient mice were significantly reduced compared with control animals (Kato et al., 2006). In contrast, bone marrow–derived dendritic cells and macrophages obtained from MDA5-deficient mice produced normal levels of type I IFNs following WNV infections (Gitlin et al., 2006).

Consistent with the results described so far, mice lacking IRF3 displayed enhanced WNV replication and uniformly succumbed to infections in comparison to wild-type mice (Daffis et al., 2007). Interestingly, however, IRF3-deficient mice produce normal levels of IFNs indicating that, at least in some cell types, other transcription factors, most likely IRF7, also respond to the activation of the IPS-1 signal transduction pathway by WNV-specific PAMPs (Daffis et al., 2007) (Fig. 1). Induction of IFN-β has also been reported from secondary sites of viral infection including the central nervous system (CNS), indicating that multiple cell types including neurons can recognize WNV infections (Samuel and Diamond, 2005).

A comparison of the kinetics of activation of the innate immune response in response to WNV and vesicular stomatitis virus (VSV) infections revealed an unexplained delay of IRF3 phosphorylation in the case of WNV (Fredericksen and Gale, 2006). The reason for this delayed induction is not known. One possibility is that WNV expresses less effective PAMPs compared with vesicular stomatitis virus (VSV), which requires larger concentrations for biological activity that are achieved at later time points during an infection. Alternatively, it is conceivable that one of the WNV proteins acts as a weak inhibitor that can delay activation of IRF3.

In addition to RIG-I and MDA5, cells express an additional protein with PRR-like properties: the double-stranded RNA-dependent protein kinase PKR containing a dsRNA binding motif and a serine/threonine kinase domain (Sadler and Williams, 2007). Although the best characterized function of PKR is to inhibit translation through phosphorylation of the alpha subunit of eukaryotic translation initiation factor 2 (eIF2α), the kinase also activates NF-κB axis and, hence, can act in parallel with the IPS-1 pathway (Kumar et al., 1994). Infection of MEFs with virus-like particles (VLPs) secreted from cells replicating WNV subgenomic replicons and expressing viral structural proteins from cDNA can lead to an apparent activation of PKR, resulting in the production of IFN (Gilfoy and Mason, 2007). A role for PKR in controlling WNV infections has also been observed in vivo with mice deficient in PKR expression (Samuel et al., 2006).

The nature of the biologically relevant PAMPs of WNV is not yet known. One possibility is that double-stranded viral RNA present in intramolecular hairpins or derived from replication complexes containing plus strand genomes and minus strand RNAs provides a substrate for RIG-I and MDA5. Both proteins are known to recognize RNA derived from viral genomes. At least in tissue culture cells analyzed, TLR3 does not appear to be necessary for inducing an antiviral program against WNV (Fredericksen and Gale, 2006) and in vivo TLR3 deficiency was associated with a relatively small increase in viral load (Wang et al., 2004). Nevertheless, it cannot be excluded that TLR3 or the ss RNA-specific TLR7 and TLR8 as well as TLR4, which can recognize some viral envelope proteins, could act as PRRs in LC and trigger the initial stimulus required for migration of these cells into the proximal lymph node. Finally, we have to consider the possibility that LC activation is caused indirectly by inflammatory cytokines or IFN-α produced by infected dermal antigen presenting cells. However, this scenario appears less likely: a study with the distantly related Venezuelan equine encephalitis alphavirus demonstrated that infected LCs appear in draining popliteal lymph nodes within 30–120 min following footpad inoculation (MacDonald and Johnston, 2000).

Surprisingly, TLR3-deficient mice were more resistant to WNV-induced CNS infections than wild-type animals (Wang et al., 2004). Survival of mutant mice was associated with decreased leakiness of the blood–brain barrier (BBB) compared with wild-type mice infected with WNV. Despite higher levels of viremia, TLR3-deficient mice produced reduced amounts of tumor necrosis factor alpha (TNF-α), IL6, and IFN-β. Interestingly, TNF-α-receptor1-deficient mice exhibited higher survival rates than wild-type animals, suggesting that TNF-α could promote the leakiness of the BBB and in turn facilitate CNS infections

(Wang et al., 2004). Although TNF-α antibody therapy (infliximab) is available for humans against Crohn' disease and rheumatoid arthritis, it can also be associated with viral disease. In fact, one case report described a patient on infliximab therapy, who died from WNV-induced meningoencephalitis (Chan-Tack and Forrest, 2006). Hence, in humans infected with WNV, TNF-α might exert a protective effect from WNV-induced CNS infections.

An important consequence of viral infection and activation of the IPS-1 pathway is the expression of inflammatory cytokines and chemokines, which leads to the activation and recruitment of lymphocytes into the infected tissue. Expression of these factors represents a link between innate and adaptive immunity. Among the chemokines induced by activation of the IPS-1 pathway is CCL5 (RANTES), which acts as a mediator for leukocyte trafficking. About 1% of Caucasians carry a deletion mutant, CCR5δ32, which confers resistance to HIV infections (Lim et al., 2006). In contrast, this mutation appears to be associated with more severe WNV disease (Glass et al., 2006). Correspondingly, CCR5-deficient mice exhibit impaired leukocyte trafficking into the brain and higher viral loads in the brain, and uniformly succumb to WNV infections (Glass et al., 2005).

In summary, experiments in animals and cell culture demonstrated that cellular sensors, PRRs, recognize one or several WNV components and in turn activate the so-called first wave of the innate immune response leading to the expression of IFN-β, IFN-α subtypes, CCL5, and many other genes that induce an antiviral state and activate the adaptive immune response. These initial steps activate through a positive feedback loop known as the IFN signal transduction pathway, a second wave of gene expression prolonging and enhancing the innate antiviral response.

3 INF-α and IFN Stimulated Genes (ISGs) Are Essential for Survival of WNV Infections

It is generally assumed that the first site of WNV amplification is the draining lymph node. Here, virus released from the incoming LC presumably infects macrophages and dendritic cells, then spreads via the efferent lymphatics and the thoracic duct into the blood stream, causing a systemic infection.

There are several lines of evidence that replication of WNV is associated with the production of IFN-α and ISGs, which represents the second wave of the innate immune response. First, mice deficient in the expression of functional IFN-α receptor (IFNAR) succumb much more rapidly to WNV infections than wild-type animals, indicating that IFN-α is not only induced by WNV infections, but also exhibits an attenuating effect on virus production and neuroinvasion (Keller et al., 2006; Samuel and Diamond, 2005). Lack of functional IFNAR can lead to a 100- to 1,000-fold increase in virus titers in peripheral organs, brain, and spinal cord compared to wild-type animals. The differences in virus load in serum can be even more pronounced, reaching 4–5 $\log_{10}$ genomic equivalents/milliliter. Second, DNA microarray analyses with mRNA isolated from livers, spleens, and brains of WNV-infected mice revealed activation of many ISGs including transcription factors and genes with known antiviral activities (Venter et al., 2005).

In vivo, the primary role of IFN-α is to attenuate viral replication in peripheral organs with the exception of the pancreas (Samuel and Diamond, 2005). The spleen appeared to be particularly affected by IFNAR deficiency because WNV-infected mice lacking the receptor exhibited a significant loss in splenocytes and a marked increase in the frequency of infected macrophages. Moreover, IFN protected neurons from WNV-induced death, which in part explains the difference in frequency of lethal WNV infections between mutant and wild-type mice. In this context, it should be noted that type I IFNs alone are not sufficient for survival of WNV-infected animals, because mice deficient in IFN-γ function also succumbed more rapidly to WNV infections compared to wild-type animals and exhibited enhanced viremia and replication in spleen and lymph nodes (Shrestha et al., 2006b). Interestingly, mice lacking IFN-γ producing γδ T cells exhibited a similar phenotype, suggesting that γδ T cells rather than NK cells are the major source of IFN-γ production. Hence, besides its role in activating an acquired immune response to clear the infection, IFN-γ also plays a role early in infection to reduce viral burden and the chance for lethal CNS infections (Shrestha et al., 2006a).

4 Effectors of the IFN Response

The hallmark of the second wave of the IFN response is the activation of many ISGs. The first evidence that an ISG could play a major role in the outcome of WNV infections stems from the observation that only certain laboratory-inbred mouse strains are susceptible to disease induced by WNV and other flaviviruses (Sabin, 1952; and reviewed in Samuel, 2002). Oas1b, a member of the IFN inducible 2′-5′ oligoadenylate synthetase (OAS) family of enzymes, is the determinant for susceptibility

(Mashimo et al., 2002; Perelygin et al., 2002; Scherbik et al., 2007). In susceptible mouse strains, Oas1b bears a nonsense mutation, resulting in expression of truncated protein. In the presence of double-stranded RNA, OAS genes produce oligoadenylates, which in turn activate the enzymatic activity of the latent endoribonuclease RNAseL. However, Oas1b apparently does not exhibit synthetase activity and hence does not play a role in the antiviral activity attributed to RNAseL (Scherbik et al., 2007). Thus, the mechanism by which this gene controls susceptibility to WNV infections remains elusive.

Evidence from experiments in mice and tissue culture cells invoked a role for RNAseL in controlling WNV replication and pathogenesis. While RNAseL-deficient mice exhibited a modestly increased susceptibility to WNV infections compared to wild-type animals, experiments with MEFs demonstrated that the enzyme could reduce viral RNA levels up to tenfold (Samuel et al., 2006; Scherbik et al., 2006). A more dramatic effect was observed in mice deficient in both PKR and RNAseL expression. About 90% of infected mice succumbed to WNV disease under conditions where 70% of wild-type mice survived the infection (Samuel et al., 2006). The reason for the increased lethality of WNV observed in the mutant mice might have stemmed from an attenuated IFN response in macrophages and cortical neurons (Samuel et al., 2006). As mentioned above, the IFN-inducible PKR exhibited an antiviral effect in vivo by reducing viral burden, but whether this was due to inhibition of translation or induction of IFN is not known (Gilfoy and Mason, 2007; Samuel et al., 2006). Also, it should be noted that in vivo the antiviral effect of PKR was indirectly determined through a comparison of infections in mice deficient for both PKR and RNAseL, and RNAseL alone (Samuel et al., 2006).

Efforts to translate observations from mouse models to human disease led to the identification of single nucleotide polymorphisms (SNPs) in RNAseL and OASL, the human homolog of Oas1b (Yakub et al., 2005). The results from nucleotide sequence analyses of 33 patients with severe WNV disease revealed an SNP in OASL that was associated with higher frequency in symptomatic patients than in controls. On the basis of computer modeling, the mutation could enhance RNA splicing and possibly the expression of full-length mRNA, and thus increase the production of OASL. Because OASL is enzymatically inactive, it might act as dominant negative inhibitor of other OAS enzymes. While these observations are exciting and might provide a link with the model, they require experimental testing before firm conclusions can be drawn. Also, larger cohorts need to be analyzed to validate the observed correlation between the mutation and WNV disease.

5 Viral Antagonism of the IFN Response

While early studies showed that IFN was protective against WNV infections in tissue culture cells, subsequent work revealed that the cytokine exhibited a significantly diminished therapeutic effect in contrast to other RNA viruses, such as vesicular stomatitis virus (VSV) and Sindbis virus (Darnell et al., 1974; Isaacs and Westwood, 1959). Moreover, additional studies with WNV and dengue virus revealed that while IFN-α could prevent an infection within a cell, it did not inhibit viral replication following the establishment of an infection, suggesting that expression of one or more viral proteins could inhibit the IFN-α response (Diamond et al., 2000; Lucas et al., 2003; Morrey et al., 2004). These observations stand in marked contrast to results reported with hepatitis C virus, which demonstrated that this virus is very sensitive to the antiviral program induced by IFN-α in tissue culture cells (Blight et al., 2000; Guo et al., 2001). Interestingly, other members of the flavivirus genus, including JEV and Dengue virus also interfere with the activation of the IFN signal transduction pathways (Lin et al., 2004; Munoz-Jordan et al., 2003).

These observations raised the question about the mechanism used by WNV and other flaviviruses to inhibit the IFN response. Under normal conditions, type I IFNs activate IFN signaling by binding to IFNAR, which is a heterodimeric complex consisting of two transmembrane proteins IFNAR1 and IFNAR2c. The receptor subunits are associated with the Janus kinases Tyk2 and JAK1, respectively (Fig. 2). Moreover, IFNAR2c provides a constitutive binding site for the latent transcription factor Stat2 and possibly Stat1. IFN-α induces the dimerization of the receptor subunits and leads to the juxtaposition of the kinases, which can now activate each other by cross-phosphorylation on specific tyrosine residues (Gauzzi et al., 1996; Irie-Sasaki et al., 2001). The Janus kinases then phosphorylate tyrosine residues on IFNAR1 and IFNAR2c, leading to the binding and phosphorylation of the latent transcription factors Stat1 and Stat2 (for details, see Stark et al., 1998). Once activated, the Stat proteins form together with p48, the hetero-trimeric transcription factor, ISGF3, which binds to the ISRE present in the enhancers of ISGs.

Biochemical studies demonstrated that WNV replication interfered with a very early step of the signal transduction pathway required for the activation of IFN-induced genes (Guo et al., 2005). In WNV-infected cells, IFN does not induce efficient phosphorylation of JAK1 and Tyk2. As a consequence, the Stat transcription factors remain in a latent stage and the induction of ISGs is either absent or weak and therefore insufficient to establish an antiviral state.

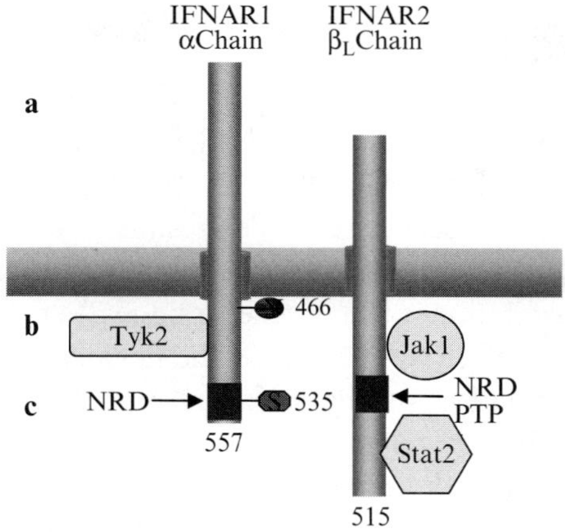

Figure 2. Inhibition of IFN signal transduction by WNV. The figure depicts the binding sites for the Janus kinases Tyk2 and Jak1 and the transcription factor Stat2 on cytoplasmic domains of the two chains of the IFNα/βR. Potential targets of WNV to inhibit IFN-induced phosphorylation of Tyk2 and Jak1 could include (**a**) Blocking ligand binding, inhibiting receptor dimerization, or destruction of a receptor subunit; (**b**) Inhibition of JAK kinase activation or activity, or activation or relocalization of a phosphatase to the receptor complex; (**c**) Activation of the negative regulatory domain (NRD), perhaps by phoshphorylation of Ser_{535}. PTP, protein tyrosine phosphatase. (*See Color Plates*)

These findings raised questions about the nature of the viral determinants that play a role in the observed inhibition of the IFN response. Although several reports addressing this question have been published, they have yielded conflicting conclusions. For example, one group reported that NS4B of WNV and dengue virus is the major determinant for inhibition (Munoz-Jordan et al., 2005, 2003). In contrast, another laboratory reported that all NS proteins, except NS1 and NS5, could inhibit the IFN response (Liu et al., 2005). Finally, investigations with JEV, the closest relative of WNV, pointed to a role for NS5, the viral polymerase in the inhibition of the IFN response (Lin et al., 2006). The latter result supported a previous study with Langat virus, a tick-borne encephalitis virus, which also demonstrated that NS5 could antagonize the IFN response through direct binding to the IFN-α and IFN-γ receptor complexes (Best et al., 2005). Finally, results from mutagenesis studies in our laboratory showed that NS4B from WNV is a determinant for blocking IFN signaling in cells harboring a replicating genome. However, homologous mutations in infectious virus do not phenocopy these results. Thus, there are additional viral factors that play a role in IFN antagonism during virus infection that are not observed with expression of individual viral proteins or subgenomic replicons (Evans and Seeger, 2007). Further investigations are needed to more completely dissect the complex interplay of virus proteins and IFN signaling components.

6 What Is the Mechanism Responsible for this Inhibition of the IFN Response?

One possibility is that WNV and other flaviviruses activate one or more cellular pathways that negatively regulate IFN signaling. (a) Ligand binding of the receptor complex results in targeted degradation of IFNAR1. A hitherto unknown kinase tags the IFNAR complex through phosphorylation of a serine residue (Ser_{535}) located in the C-terminal domain of IFNAR1, called the negative regulatory domain (NRD). Phosphorylation of the serine residue acts as a signal for a ubiqutin ligase complex, which induces the transport of the IFNAR into endosomes where the two subunits are separated. Subsequently, IFNAR1 is degraded via the proteasome and IFNAR2c is recycled back to the cell surface (Kumar et al., 2004; Marijanovic et al., 2006). (b) The protein tyrosine phosphatases HCP (SHP-1) and SHP-2 can act as inhibitors of IFN signaling (Yasukawa et al., 2000). (c) A family of proteins known as suppressors of cytokine signaling (SOCS) has been shown to influence stability of and signaling through IFN receptors (Alexander and Hilton, 2004). Hence, it is conceivable that NS4B and/or other viral protein modulate the phosphorylation-modifying enzymes or SOCS proteins, thereby interrupting the IFN signal transduction cascade. Alternatively, one or several viral proteins may directly or indirectly disrupt proper dimerization of the receptor subunits. Further, viral proteins could also prevent binding of either JAK1 or Tyk2 to their respective subunits. Taken together, one or all the pathways described above may contribute to abrogating signaling through the IFN receptor, thus blocking antiviral activities.

7 Implications for Natural Infections in Humans

During the past decade, we witnessed significant progress in our understanding of WNV biology, particularly in the area of host–virus interactions that play a role in defining the outcome of WNV infections. In particular, the availability of gene knockout mice helped in identifying the critical components of the IFN system. A major question concerns the relevance of these findings for infections and sequelae in humans. For example, can polymorphisms in genes belonging to the innate immune system indeed be linked to WNV induced CNS disease in humans? Moreover, gaps in knowledge remain about the molecular mechanisms regulating the innate immune response against WNV. For example, what are the viral ligands (PAMPs) that activate the cellular receptors (PRRs)? What are the mechanisms responsible for the observed delay of activation of primary antiviral genes and inhibition of the IFN response? Does the latter occur in vivo? The advances made in the last few years have provided the basis for exciting new discoveries that should yield answers to these and many other questions about the biology of WNV infections.

Acknowledgments We thank Siddharth Balachandran for a critical reading of the manuscript and helpful suggestions. This work was supported by grants from the NIH (CS) and an appropriation from the Commonwealth of Pennsylvania. JDE was supported by grant T32 CA-09035–30.

References

Akira, S., Uematsu, S., and Takeuchi, O. 2006. Pathogen recognition and innate immunity. Cell 124: 783–801.

Alexander, W. S., and Hilton, D. J. 2004. The role of suppressors of cytokine signaling (SOCS) proteins in regulation of the immune response. Annu Rev Immunol 22: 503–529.

Best, S. M., Morris, K. L., Shannon, J. G., Robertson, S. J., Mitzel, D. N., Park, G. S., Boer, E., Wolfinbarger, J. B., and Bloom, M. E. 2005. Inhibition of interferon-stimulated JAK-STAT signaling by a tick-borne flavivirus and identification of NS5 as an interferon antagonist. J Virol 79: 12828–12839.

Blight, K. J., Kolykhalov, A. A., and Rice, C. M. 2000. Efficient initiation of HCV RNA replication in cell culture. Science 290: 1972–1974.

Byrne, S. N., Halliday, G. M., Johnston, L. J., and King, N. J. 2001. Interleukin-1beta but not tumor necrosis factor is involved in West Nile virus-induced Langerhans cell migration from the skin in C57BL/6 mice. J Invest Dermatol 117: 702–709.

Chan-Tack, K. M., and Forrest, G. 2006. West Nile virus meningoencephalitis and acute flaccid paralysis after infliximab treatment. J Rheumatol 33: 191–192.

Daffis, S., Samuel, M. A., Keller, B. C., Gale, M., Jr., and Diamond, M. S. 2007. Cell-specific IRF-3 responses protect against West Nile virus infection by interferon-dependent and -independent mechanisms. PLoS Pathog 3: e106.

Darnell, M. B., Koprowski, H., and Lagerspetz, K. 1974. Genetically determined resistance to infection with group B arboviruses. I. Distribution of the resistance gene among various mouse populations and characteristics of gene expression in vivo. J Infect Dis 129: 240–247.

Diamond, M. S., Roberts, T. G., Edgil, D., Lu, B., Ernst, J., and Harris, E. 2000. Modulation of dengue virus infection in human cells by alpha, beta, and gamma interferons. J Virol 74: 4957–4966.

Evans, J. D., and Seeger, C. 2007. Differential effects of mutations in NS4B on West Nile virus replication and inhibition of interferon signaling. J Virol 81: 11809–11816.

Fredericksen, B. L., and Gale, M., Jr. 2006. West Nile virus evades activation of interferon regulatory factor 3 through RIG-I-dependent and -independent pathways without antagonizing host defense signaling. J Virol 80: 2913–2923.

Fredericksen, B. L., Smith, M., Katze, M. G., Shi, P. Y., and Gale, M., Jr. 2004. The Host response to West Nile virus infection limits viral spread through the activation of the interferon regulatory factor 3 pathway. J Virol 78: 7737–7747.

Fredericksen, B. L., Keller, B. C., Fornek, J., Katze, M. G., and Gale, M., Jr. 2007. Establishment and maintenance of the innate antiviral response to West Nile virus involves both RIG-I and MDA5 signaling through IPS-1. J Virol

Gauzzi, M. C., Velazquez, L., McKendry, R., Mogensen, K. E., Fellous, M., and Pellegrini, S. 1996. Interferon-alpha-dependent activation of Tyk2 requires phosphorylation of positive regulatory tyrosines by another kinase. J Biol Chem 271: 20494–20500.
Gilfoy, F. D., and Mason, P. W. 2007. West Nile virus-induced interferon production is mediated by the double-stranded RNA-dependent protein kinase PKR. J Virol 81: 11148–11158.
Gitlin, L., Barchet, W., Gilfillan, S., Cella, M., Beutler, B., Flavell, R. A., Diamond, M. S., and Colonna, M. 2006. Essential role of mda-5 in type I IFN responses to polyriboinosinic:p olyribocytidylic acid and encephalomyocarditis picornavirus. Proc Natl Acad Sci USA 103: 8459–8464.
Glass, W. G., Lim, J. K., Cholera, R., Pletnev, A. G., Gao, J. L., and Murphy, P. M. 2005. Chemokine receptor CCR5 promotes leukocyte trafficking to the brain and survival in West Nile virus infection. J Exp Med 202: 1087–1098.
Glass, W. G., McDermott, D. H., Lim, J. K., Lekhong, S., Yu, S. F., Frank, W. A., Pape, J., Cheshier, R. C., and Murphy, P. M. 2006. CCR5 deficiency increases risk of symptomatic West Nile virus infection. J Exp Med 203: 35–40.
Guo, J. T., Bichko, V. V., and Seeger, C. 2001. Effect of alpha interferon on the hepatitis C virus replicon. J Virol 75: 8516–8523.
Guo, J. T., Hayashi, J., and Seeger, C. 2005. West Nile virus inhibits the signal transduction pathway of alpha interferon. J Virol 79: 1343–1350.
Irie-Sasaki, J., Sasaki, T., Matsumoto, W., Opavsky, A., Cheng, M., Welstead, G., Griffiths, E., Krawczyk, C., Richardson, C. D., Aitken, K., et-al.2001. CD45 is a JAK phosphatase and negatively regulates cytokine receptor signalling. Nature 409: 349–354.
Isaacs, A., and Westwood, M. A. 1959. Duration of protective action of interferon against infection with West Nile virus. Nature 184(Suppl 16): 1232–1233.
Johnston, L. J., Halliday, G. M., and King, N. J. 1996. Phenotypic changes in Langerhans' cells after infection with arboviruses: a role in the immune response to epidermally acquired viral infection? J Virol 70: 4761–4766.
Johnston, L. J., Halliday, G. M., and King, N. J. 2000. Langerhans cells migrate to local lymph nodes following cutaneous infection with an arbovirus. J Invest Dermatol 114: 560–568.
Kato, H., Takeuchi, O., Sato, S., Yoneyama, M., Yamamoto, M., Matsui, K., Uematsu, S., Jung, A., Kawai, T., Ishii, K. J., et al.2006. Differential roles of MDA5 and RIG-I helicases in the recognition of RNA viruses. Nature 441: 101–105.
Keller, B. C., Fredericksen, B. L., Samuel, M. A., Mock, R. E., Mason, P. W., Diamond, M. S., and Gale, M., Jr. 2006. Resistance to alpha/beta interferon is a determinant of West Nile virus replication fitness and virulence. J Virol 80: 9424–9434.
Kumar, A., Haque, J., Lacoste, J., Hiscott, J., and Williams, B. R. 1994. Double-stranded RNA-dependent protein kinase activates transcription factor NF-kappa B by phosphorylating I kappa B. Proc Natl Acad Sci USA 91: 6288–6292.
Kumar, K. G., Krolewski, J. J., and Fuchs, S. Y. 2004. Phosphorylation and specific ubiquitin acceptor sites are required for ubiquitination and degradation of the IFNAR1 subunit of type I interferon receptor. J Biol Chem 279: 46614–46620.
Lim, J. K., Glass, W. G., McDermott, D. H., and Murphy, P. M. 2006. CCR5: no longer a "good for nothing" gene–chemokine control of West Nile virus infection. Trends Immunol 27: 308–312.
Lin, R. J., Liao, C. L., Lin, E., and Lin, Y. L. 2004. Blocking of the alpha interferon-induced Jak-Stat signaling pathway by Japanese encephalitis virus infection. J Virol 78: 9285–9294.
Lin, R. J., Chang, B. L., Yu, H. P., Liao, C. L., and Lin, Y. L. 2006. Blocking of interferon-induced Jak-Stat signaling by Japanese encephalitis virus NS5 through a protein tyrosine phosphatase-mediated mechanism. J Virol 80: 5908–5918.
Liu, W. J., Wang, X. J., Mokhonov, V. V., Shi, P. Y., Randall, R., and Khromykh, A. A. 2005. Inhibition of Interferon Signaling by the New York 99 Strain and Kunjin Subtype of West Nile Virus Involves Blockage of STAT1 and STAT2 Activation by Nonstructural Proteins. J Virol 79: 1934–1942.

Lucas, M., Mashimo, T., Frenkiel, M. P., Simon-Chazottes, D., Montagutelli, X., Ceccaldi, P. E., Guenet, J. L., and Despres, P. 2003. Infection of mouse neurones by West Nile virus is modulated by the interferon-inducible 2′-5′ oligoadenylate synthetase 1b protein. Immunol Cell Biol 81: 230–236.

MacDonald, G. H., and Johnston, R. E. 2000. Role of dendritic cell targeting in Venezuelan equine encephalitis virus pathogenesis. J Virol 74: 914–922.

Marijanovic, Z., Ragimbeau, J., Kumar, K. G., Fuchs, S. Y., and Pellegrini, S. 2006. TYK2 activity promotes ligand-induced IFNAR1 proteolysis. Biochem J 397: 31–38.

Mashimo, T., Lucas, M., Simon-Chazottes, D., Frenkiel, M. P., Montagutelli, X., Ceccaldi, P. E., Deubel, V., Guenet, J. L., and Despres, P. 2002. A nonsense mutation in the gene encoding 2′-5′-oligoadenylate synthetase/L1 isoform is associated with West Nile virus susceptibility in laboratory mice. Proc Natl Acad Sci USA 99: 11311–11316.

Morrey, J. D., Day, C. W., Julander, J. G., Blatt, L. M., Smee, D. F., and Sidwell, R. W. 2004. Effect of interferon-alpha and interferon-inducers on West Nile virus in mouse and hamster animal models. Antivir Chem Chemother 15: 101–109.

Munoz-Jordan, J. L., Sanchez-Burgos, G. G., Laurent-Rolle, M., and Garcia-Sastre, A. 2003. Inhibition of interferon signaling by dengue virus. Proc Natl Acad Sci USA 100: 14333–14338.

Munoz-Jordan, J. L., Laurent-Rolle, M., Ashour, J., Martinez-Sobrido, L., Ashok, M., Lipkin, W. I., and Garcia-Sastre, A. 2005. Inhibition of alpha/beta interferon signaling by the NS4B protein of flaviviruses. J Virol 79: 8004–8013.

Perelygin, A. A., Scherbik, S. V., Zhulin, I. B., Stockman, B. M., Li, Y., and Brinton, M. A. 2002. Positional cloning of the murine flavivirus resistance gene. Proc Natl Acad Sci USA 99: 9322–9327.

Sabin, A. B. 1952. Nature of inherited resistance to viruses affecting the nervous system. Proc Natl Acad Sci USA 38: 540–546.

Sadler, A. J., and Williams, B. R. 2007. Structure and function of the protein kinase R. Curr Top Microbiol Immunol 316: 253–292.

Samuel, C. E. 2002. Host genetic variability and West Nile virus susceptibility. Proc Natl Acad Sci USA 99: 11555–11557.

Samuel, M. A., and Diamond, M. S. 2005. Alpha/beta interferon protects against lethal West Nile virus infection by restricting cellular tropism and enhancing neuronal survival. J Virol 79: 13350–13361.

Samuel, M. A., Whitby, K., Keller, B. C., Marri, A., Barchet, W., Williams, B. R., Silverman, R. H., Gale, M., Jr., and Diamond, M. S. 2006. PKR and RNase L contribute to protection against lethal West Nile Virus infection by controlling early viral spread in the periphery and replication in neurons. J Virol 80: 7009–7019.

Scherbik, S. V., Paranjape, J. M., Stockman, B. M., Silverman, R. H., and Brinton, M. A. 2006. RNase L plays a role in the antiviral response to West Nile virus. J Virol 80: 2987–2999.

Scherbik, S. V., Kluetzman, K., Perelygin, A. A., and Brinton, M. A. 2007. Knock-in of the Oas1b(r) allele into a flavivirus-induced disease susceptible mouse generates the resistant phenotype. Virology 368: 232–237.

Shrestha, B., Samuel, M. A., and Diamond, M. S. 2006a. CD8 + T cells require perforin to clear West Nile virus from infected neurons. J Virol 80: 119–129.

Shrestha, B., Wang, T., Samuel, M. A., Whitby, K., Craft, J., Fikrig, E., and Diamond, M. S. 2006b. Gamma interferon plays a crucial early antiviral role in protection against West Nile virus infection. J Virol 80: 5338–5348.

Stark, G. R., Kerr, I. M., Williams, B. R., Silverman, R. H., and Schreiber, R. D. 1998. How cells respond to interferons. Annu Rev Biochem 67: 227–264.

Styer, L. M., Kent, K. A., Albright, R. G., Bennett, C. J., Kramer, L. D., and Bernard, K. A. 2007. Mosquitoes inoculate high doses of West Nile virus as they probe and feed on live hosts. PLoS Pathog 3: 1262–1270.

Takeuchi, O., and Akira, S. 2007. Recognition of viruses by innate immunity. Immunol Rev 220: 214–224.

Venter, M., Myers, T. G., Wilson, M. A., Kindt, T. J., Paweska, J. T., Burt, F. J., Leman, P. A., and Swanepoel, R. 2005. Gene expression in mice infected with West Nile virus strains of different neurovirulence. Virology 342: 119–140.

Wang, T., Town, T., Alexopoulou, L., Anderson, J. F., Fikrig, E., and Flavell, R. A. 2004. Toll-like receptor 3 mediates West Nile virus entry into the brain causing lethal encephalitis. Nat Med 10: 1366–1373.

Yakub, I., Lillibridge, K. M., Moran, A., Gonzalez, O. Y., Belmont, J., Gibbs, R. A., and Tweardy, D. J. 2005. Single nucleotide polymorphisms in genes for 2′-5′-oligoadenylate synthetase and RNase L inpatients hospitalized with West Nile virus infection. J Infect Dis 192: 1741–1748.

Yasukawa, H., Sasaki, A., and Yoshimura, A. 2000. Negative regulation of cytokine signaling pathways. Annu Rev Immunol 18: 143–164.

8. Innate Immune Responses to West Nile Virus Infection

Alvaro Arjona and Erol Fikrig

Abstract

Innate antiviral immunity is activated by the detection of conserved virus-associated molecular motifs by host-encoded pathogen-recognition receptors (PRRs). This phenomenon triggers the production of antiviral and proinflammatory cytokines as well as the expression of costimulatory molecules in immune cells, leading to the establishment of an antiviral state and the induction of adaptive immune responses. In this chapter we review our current understanding of the innate immune mechanisms that mediate the recognition of West Nile virus (WNV) infection. The role of innate immune cells and cytokines in WNV immunopathogenesis is also discussed. Paradoxically, although many of the innate responses induced by WNV infection are protective, others favor WNV neuroinvasion by their detrimental effect on blood–brain barrier permeability.

1 Introduction

Any antimicrobial immune response involves the initial recognition of the invading pathogen and the subsequent reaction to eliminate it. In vertebrates, immune responses have been traditionally divided into two categories: innate and adaptive (also denominated combinatorial). Innate immunity comprises a plethora of ancient defense mechanisms that have been highly preserved throughout evolution. In contrast to the adaptive immune response, innate immune mechanisms are relatively unspecific and do not improve with each successive encounter with the pathogen. However, owing to its immediacy, innate responses are posited to play a vital role during the early stages of infection, when adaptive mechanisms have not yet been mounted. Additionally, it has become increasingly evident during the past decade that the innate/adaptive dichotomy is somewhat artificial, and the overall immune response is currently considered a continuum in which both innate and adaptive branches of the immune system cross-talk constantly. Moreover, it is now recognized that the activation of certain innate immune pathways specialized in the detection of invading pathogens is necessary for triggering and sustaining the adaptive response (Iwasaki and Medzhitov, 2004).

The early stage of a viral infection can be envisioned as a competition between virulence pathogenic mechanisms and the host innate defense system. The net balance of these two opposing forces often determines the outcome of infection. This is true especially in the case of viruses presenting with acute and subacute pathogenic time courses, such as West Nile virus (WNV) (Petersen and Marfin, 2002). The skin and mucous membranes constitute the initial barriers against viral invasion. In the case of arboviruses these barriers are easily breached by the arthropod vector, which renders the host defense completely reliant on antiviral innate immune mechanisms. The antiviral innate immune response comprises both humoral and cellular phenomena, including the interferon (IFN) response, the complement system, natural antibodies, phagocytosis, cytotoxic mechanisms, and apoptosis, among others. Some of these relatively unspecific defenses are induced by infection (e.g., IFNs), whereas others exist constitutively prior to pathogen invasion (e.g., complement). The IFN response and the complement system are among the most important of the host innate defenses against viral infection. Their importance in the early control of WNV has been demonstrated clearly, and it is reviewed in depth in elsewhere in this book. Table 1 summarizes the main components of the innate immune system that are known to be involved directly in the protective and/or pathogenic response to WNV infection. In the present chapter, we will focus on innate immune sensors that mediate the initial detection of WNV infection. The role of specific innate immune cell types and certain cytokines involved in the early response to WNV is also discussed.

2 Sensing WNV Infection: Role of Pattern-Recognition Receptors

As mentioned earlier, the initial recognition of the invading pathogen is an integral part of the overall immune response. The question of how host cells first sense infection has been object of intense research efforts in the recent years, and it has become a fundamental theme in the innate immunity field. It is now clear that the initial detection of viruses

Table 1. Role of innate immunity components in WNV infection

Component		Experimental manipulation	Effect on WNV infection/pathogenesis
Complement	Multiple factors	Gene deletion	Increased infection and lethality
PRRs	TLR3	Gene deletion	Decreased CNS infection and lethality
	RIG-I	Gene deletion	Delayed IFN response – increased replication[a]
	MDA5	Gene deletion	Slight decrease in IFN production[a]
Leukocytes	Myeloid cells	Cell depletion	Increased infection and lethality
	γδT cells	Gene deletion	Increased and accelerated CNS invasion Increased lethality
	NK cells	Cell depletion	No change in lethality
IFN system	IFNα/βR	Gene deletion	Increased infection and lethality
	Type I IFNs	Treatment	Reduced viral replication[a]
	RNase L PKR	Gene deletion	Increased infection and lethality
	IRF3	Gene deletion	Increased infection and lethality
Cytokines	IFN-γ IFN-γR	Gene deletion	Increased infection and lethality
	IL-1β	Ab blockade	Decreased Langerhans cell migration
	IL-6	Gene deletion	No change in lethality
	TNF-α	Gene deletion	Decreased lethality
	MIF	Gene deletion	Decreased CNS infection and lethality
		Pharm. inhib.	Decreased lethality
		Ab blockade	Decreased lethality

[a]In vitro

by specialized germline-encoded host receptors is critical for triggering an effective antiviral immune response. These receptors are generally referred to as pattern-recognition receptors (PRRs). PRRs recognize conserved molecular motifs (pathogen-associated molecular patters, PAMPs) that distinguish foreign organisms (viruses, bacteria, fungi, and parasites) from host cells. Virus-associated PAMPs include viral nucleic acids such as single-stranded RNA (ssRNA) and CpG DNA as well as viral replication intermediates such as double-stranded RNA (dsRNA). There is evidence suggesting that viral glycoproteins may act as virus-associated PAMPs as well (Kurt-Jones et al., 2000; Boehme et al., 2004). After recognition of viral components, PRRs initiate rapid antiviral responses, which generally (but not exclusively) involve genomic effects resulting from the rapid activation of critical transcription factors such as nuclear factor (NF)-κB and interferon regulatory factors (IRFs). This transcriptional activation leads to the production of a variety of cytokines and the subsequent induction of antiviral, inflammatory, and, ultimately, adaptive immune responses. In particular, as pointed out above, type I IFNs (both IFN-β and the multiple IFN-α molecules) are key cytokines produced soon after viral infection that mediate the induction of the innate immune response and the subsequent development of adaptive immunity. Type I IFNs also induce the expression of multiple effector molecules (e.g., PKR, RNAse L) that influence protein synthesis, cell growth, and survival in the process of establishing an antiviral state (Theofilopoulos et al., 2005). These molecules have proven necessary for cellular restraint of WNV replication (Samuel et al., 2006; Scherbik et al., 2006) and for WNV-induced secretion of type I IFNs (Gilfoy and Mason, 2007). It has been demonstrated that type I IFNs are required to restrict WNV replication and spread in vivo (Samuel and Diamond, 2005). Not surprisingly, WNV has evolved mechanisms to antagonize IFN-induced responses during infection (see Chap. 7).

2.1 Toll-Like Receptors

The best characterized group of PRRs to date is the Toll-like receptor (TLR) family. Mammalian TLRs are homologs of the *Drosophila* Toll gene, which mediates the innate antifungal response in insects (Lemaitre et al., 1996). TLRs constitute a group of at least 11 type I transmembrane glycoproteins expressed either on the cell surface or intracellularly on endosomal membranes. Each receptor comprises a leucine-rich repeat (LRR) motif in the pathogen-binding ectodomain and a cytoplasmic Toll/IL-1R homology domain (known as the TIR domain) responsible for signal transduction. Following LRR domain binding to its respective ligand, the TIR domain signals through specific adaptor molecules such as myeloid differentiation factor 88 (MyD88) or IL-1R domain-containing adaptor-inducing IFN-β (TRIF). The structural and biochemical details of the interaction between PAMPs and the LRR domain in TLRs are now beginning to be ascertained (Bell et al., 2006a,b; Jin et al., 2007; Kim et al., 2007). TLR engagement initiates rapid signaling events that lead to activation of the transcription factors NF-κB and IRF-3, production of proinflamatory mediators including type I IFNs, and induction of the costimulatory molecules CD80, CD86, and CD40 (Kawai and Akira, 2006). Each TLR recognizes specific molecular motifs shared by a relatively wide group of pathogens. For example, TLR4 detects lipopolysaccharide of Gram-negative bacteria, while TLR5 senses flagellin. This feature confers TLR-mediated innate responses with certain degree of pathogen specificity.

Among the TLRs that have been implicated in viral recognition, TLR3 senses dsRNA (Alexopoulou et al., 2001), a replication derivative that is produced during the life cycle of many viruses, including WNV (Samuel, 2002). TLR3 localizes into intracellular vesicles of myeloid cells, whereas it is expressed on the surface of fibroblasts (Kawai and Akira, 2006). The crystal structure of the TLR3 ectodomain showed that the LRR domain of TLR3 includes a positively charged region predicted to bind nucleic acids (Choe et al., 2005), and the recent solving of the TLR3 ectodomain structure has revealed a horseshoe-shaped solenoid composed of 23 LRRs (Bell et al., 2006a). In contrast to most TLRs, which signal through MyD88, TLR3 uses TRIF as the adaptor molecule (Yamamoto et al., 2003). Mouse infection models have yielded substantial evidence demonstrating that TLR3 is critically involved in WNV immunopathogenesis: TLR3-deficient mice present with increased viral load and reduced production of antiviral and proinflammatory cytokines (IFN-β, TNF-α, and IL-6) in blood and spleen, which clearly indicates that TLR3 has a direct antiviral effect on WNV infection. Paradoxically, because of the detrimental effect of proinflammatory cytokines (chiefly TNF-α) on blood–brain barrier (BBB) permeability (Abbott, 2000; Wang et al., 2004), the reduced level of peripheral inflammatory responses in the absence of TLR3 renders mice more resistant to viral neuroinvasion and, accordingly, TLR3-deficient mice show an increased survival rate upon intraperitoneal WNV challenge (Wang et al., 2004).

It is well known that despite the crucial role that TLRs exert in the early host defense against invading pathogens, their excessive or inappropriate activation may lead to exacerbated inflammatory responses (Kobayashi and Flavell, 2004). The findings in TLR3-deficient mice constitute a perfect example of how PRR-induced innate immune system activation can be a double-edged sword for the host. This phenomenon may be particularly relevant during infection with encephalitic viruses, when overproduction of inflammatory cytokines may compromise the integrity of the BBB. On the other hand, it is yet to be investigated which cell type is the major contributor to TLR3-mediated TNF-α production in response to WNV peripheral infection. In regard to TLR3-mediated detection of WNV in the CNS, cultured murine microglia (the resident macrophage of the central nervous system (CNS)) respond to WNV infection by upregulating TLR3 expression. It was also observed that a reduced proinflammatory cytokine production by TLR3-deficient microglia (which indicates TLR3-mediated detection of WNV) is associated with the inability of these cells to promote neuronal injury upon

WNV challenge (Wang et al., 2004). In contrast, results obtained from in vitro studies indicate that TLR3 may be dispensable for WNV recognition in certain cell types (Fredericksen and Gale, 2006).

Other TLRs involved in viral recognition include TLR7 and the phylogenetically related TLR8, both of which have the ability to detect ssRNA and ribonucleic acid homologs (Kawai and Akira, 2006). TLR7 is highly expressed in plasmocytoid dendritic cells (pDCs), where it mediates the production of type I IFNs in response to viral infection (Diebold et al., 2004). IFN-α production by pDCs from TLR7-deficient mice is impaired after infection with ssRNA viruses such as influenza or vesicular stomatitis virus (Diebold et al., 2004; Lund et al., 2004). Whether TLR7 and/or TLR8 sense WNV infection and how this recognition influences WNV immunopathogenesis await investigation. On the other hand, it is unlikely that TLR9, which specifically recognizes unmethylated CpG DNA motifs that are present in DNA viruses (Lund et al., 2003), contributes to the detection of WNV.

2.2 RNA Helicases

Another important family of PRRs specialized in viral recognition are the RNA helicases retinoic acid-inducible gene I (RIG-I) and melanoma differentiation-associated gene 5 (MDA5, also known as Helicard) (Saito and Gale, 2007). These are cytoplasmic proteins that sense dsRNA inside the cell (not in intracellular vesicles or within endosomal spaces, as it occurs with TLRs). RIG-I and MDA5 have a C-terminal DExD/H box RNA helicase domain (required for the interaction with dsRNA) and two N-terminal caspase activation and recruitment domains (CARDs), which allow interaction with the adaptor protein IPS-1 and transmission of the signal (Saito and Gale, 2007). Stimulation of RIG-I and MDA5 by intracellular dsRNA results in the activation of IRF-3 and NF-κB and the subsequent upregulation of type I IFN and proinflammatory cytokines (Kawai and Akira, 2006). In vitro studies have revealed that IRF-3 activation is significantly delayed in WNV-infected mouse embryonic fibroblasts lacking RIG-I, and this phenomenon was correlated with an increase in viral titers and cytopathic effects (Fredericksen and Gale, 2006). These results indicate that RIG-I is a critical sensor during the initial immune response to WNV infection. MDA5 appears to be less important, as the WNV-induced type I IFN production detected in MDA5-deficient dendritic cells (DCs) and macrophages was only slightly reduced when compared to that seen in control cells (Gitlin et al., 2006). It is possible that MDA5 has a complementary role in sensing WNV dsRNA, particularly once productive infection has been established in nonimmune cell types, as MDA5 levels were significantly increased in neurons 48 h after WNV infection (Daffis et al., 2007). Nevertheless, no difference in survival was observed in MDA5-deficient mice after infection with Japanese encephalitis virus, a closely related flavivirus (Kato et al., 2006).

2.3 Nonconventional PRRs

Receptors other than the TLRs and the aforementioned RNA helicases may also bind viral molecules and, in some cases, can regulate the expression of innate response genes. For example, C-type lectin receptors (CLRs) such as the mannose-binding lectin, bind to various sugar moieties (e.g., *N*-acetyl-d-glucosamine, mannose, *N*-acetly-mannosamine, fucose, and glucose) that decorate many microbial products, including viral proteins (Takahashi et al., 2006). In the case of WNV, the C-type lectins DC-SIGN and DC-SIGNR, which bind mannose-rich glycans with high affinity, have been implicated in cellular attachment and infection via their recognition of the envelope glycoprotein (Davis et al., 2006a,b). As the engagement of DC-SIGN and other DC-SIGN-like molecules by several bacterial pathogens has been shown to modulate immune cell activation and cytokine production (Geijtenbeek et al., 2003; Caparros et al., 2006), it is tempting to speculate that detection of WNV envelope glycoprotein (WNV-E), the first WNV molecule to interact with the host, by DC-SIGN and possibly other CLRs may constitute the first WNV-derived signal for triggering host innate immune responses.

3 Role of Innate Immune Cells in WNV Infection

3.1 Macrophages

Macrophages are a critical component of the innate immune system. In addition to their role of phagocytosis of pathogens and infected cells, they are involved in the regulation of both humoral and cell-mediated immune responses (Morrissette et al., 1999). Upon phagocytosis, macrophages degrade proteins and process antigens for their presentation on major histocompatibility complex (MHC) molecules for T-cell recognition. Macrophages are readily infected by WNV (Garcia-Tapia et al., 2006; Rios et al., 2006; Shirato et al., 2006), and respond to this challenge by secreting antiviral and proinflammatory cytokines. Mouse studies have shown that in microglia this antiviral response is partially dependent on TLR3 (Wang et al., 2004; Town et al., 2006). The susceptibility of monocytes and macrophages to productive infection in vitro is compatible with a potential role in initial WNV replication and propagation after transmission by transfusion (Rios et al., 2006). In addition, macrophages constitute an important fraction of the inflammatory infiltrate seen in the CNS of patients with WNV infection (Kelley et al., 2003). Macrophages can function as effective WNV antigen-presenting cells (APCs), and they have been shown to promote WNV-specific T-cell proliferation (Kulkarni et al., 1991). Of note, macrophage depletion in WNV-infected mice results in increased viremia and mortality, which underscores the importance of this innate cell type in the immune response to WNV infection (Ben-Nathan et al., 1996). Furthermore, even a non-neurotropic WNV strain was able to cross the BBB in the absence of macrophages (Ben-Nathan et al., 1996). The clear protective role of macrophages in WNV infection suggests that this cell type is not the major contributor to the TLR3-mediated TNF-α production induced by WNV infection. Experimental evidences indicate that macrophages require constitutive expression of key host defense molecules (e.g., RIG-I, MDA5, PKR, and RNase L) to control WNV infection, in ways both dependent on and independent of type I IFN production (Samuel et al., 2006; Daffis et al., 2007). Interestingly, murine macrophages deficient in IRF-3 lacked basal expression of some of these host defense genes and supported increased WNV infection as well as increased production of IFN-α and IFN-β (Daffis et al., 2007). Although it is likely that macrophages restrict WNV infection by the secretion of antiviral cytokines and by presenting WNV antigens to lymphocytes, it has been proposed that they may also control WNV infection by direct effector mechanisms such as the production of nitric oxide intermediates (Kreil and Eibl, 1996; Lin et al., 1997; Garcia-Tapia et al., 2006; Samuel and Diamond, 2006).

3.2 Dendritic Cells

DCs constitute a heterogeneous group of immune cells that are functionally located in the interface of innate and adaptive responses. Although rare, they are ubiquitously distributed as sentinel cells in multiple tissues, where they recognize pathogens, modulate the inflammatory response, and function as highly efficient APCs (Steinman and Hemmi, 2006). DCs, like macrophages, can be infected by WNV (Pierson et al., 2005) and, by virtue of their expression of DC-SIGN-like attachment molecules (Davis et al., 2006b), are believed to be initial targets for WNV replication after skin inoculation. Langerhans cells, the resident DC type of the skin, have been shown to carry WNV from the skin to local lymph nodes in murine models of infection (Johnston et al., 2000), a phenomenon that is dependent on the cytokine IL-1β (Byrne et al., 2001). During an epidermally acquired viral infection, local Langerhans cells mature to a phenotype that resembles that of lymphoid DCs (Johnston et al., 1996). This change may be important in the activation of naïve T cells and the subsequent clearance of viral infection. CD4 + T cell proliferation studies have demonstrated that, as expected, DCs are able to present WNV antigens (Kulkarni et al., 1991), albeit some reports suggest that they do so to a lesser extent than macrophages (Pisarev et al., 2003). Unfortunately, there is not much information about the specific protective roles of DCs in WNV infection in vivo, although it is likely that they produce type I IFNs soon after infection (particularly pDCs) and function as APCs to prime the adaptive immune response. Indeed, in vitro studies have shown that WNV triggers type I IFN production in human pDCs and monocyte-derived DCs. Strikingly, pDCs produce IFN-α when stimulated with WNV grown in mammalian (host) cells but not when stimulated with WNV derived from mosquito (vector) cells (Silva et al., 2007). These observations concur with our recent studies demonstrating that WNV-E protein can block dsRNA-induced cytokine production in macrophages only when carrying a mosquito-derived glycosylation profile (Arjona et al., in press). Whether these and other putative DC responses to WNV are mediated by conventional PRRs needs to be confirmed experimentally.

3.3 γ/δT Cells

γ/δ T cells represent a small fraction of the T-cell population in lymphoid organs, but they are well represented in the peripheral blood and are abundant at epithelial and mucosal sites (Hayday, 2000). γ/δ T cells can be polarized to produce Th1- or Th2-type cytokines, although their TCR repertoire is significantly limited when compared to α/β T cells (Hayday, 2000; Yin et al., 2000). γ/δ T cells rapidly produce cytokines in response to microbial antigens (Wang et al., 2001). Moreover, they lack MHC restriction and have the capacity to react to antigens that have not undergone conventional processing (Carding and Egan, 2002). Thus, despite their T-cell ontogeny, these features suggest a role for γ/δ T cells in early viral control, and posit them as pivotal participants of the innate antiviral response. Along this line, recent studies in mice have shown that the γ/δ T-cell population expands significantly after WNV infection, and that mice deficient in γ/δ T cells develop higher viral load and have increased mortality (Wang et al., 2003). Soon after WNV infection, γ/δ T cells produce high amounts of IFN-γ, which correlates with an increase in perforin expression in splenic T cells. Bone marrow chimera reconstitution experiments have further demonstrated that IFN-γ derived from γ/δ T cells is required for the early control of WNV dissemination (Shrestha et al., 2006b). In addition, γ/δ T cells have been shown to promote protective adaptive immune responses to WNV infection. In a series of elegant experiments, Wang et al. (2006) have found that TCRδ-deficient mice were more susceptible to secondary challenge with WNV than the wild-type mice that survived the primary challenge. This finding was correlated with a numeric and functional reduction of CD8 memory T cells in these mice. Thus, γδ T cells constitute an important functional link between innate and adaptive immune responses to WNV infection.

3.4 NK Cells

NK cells participate in innate immune responses by secreting IFN-γ, which activates macrophages and DCs and shapes the adaptive immune response toward Th1-type immunity. They recognize and kill host somatic cells that are coated with antibody, or cells that fail to express an appropriate MHC repertoire. NK cells are also an important source of antiviral and inflammatory cytokines. It has been long recognized that NK cells have the ability to detect and eliminate virus-infected cells (Andoniou et al., 2006). All the evidences accumulated to date, however, point towards the notion that they are dispensable for immunity against WNV. Antibody-mediated depletion of NK cells in mice did not alter survival after WNV infection (Shrestha et al., 2006a). Similarly, transgenic mice lacking functional circulating NK cells did not show an increase in WNV-induced lethality (Samuel and Diamond, 2006). It is therefore plausible to speculate that WNV actively inhibits the NK cell response. Indeed, NK cell activity was transiently activated and then suppressed following flavivirus infection in mice (Vargin and Semenov, 1986). It is well known that some viruses evade NK cell activation by increasing the cell surface expression of MHC-I molecules (Herzer et al., 2003; King and Kesson, 2003), which function as negative regulators of NK cell activity thorough their interaction with surface inhibitory receptors in NK cells. Although there is currently no evidence supporting that WNV employs this immune evasion mechanisms in vivo, WNV infection in vitro has been shown to increase MHC-I expression in some cell types (Liu et al., 1988, 1989; King et al., 1989; Cheng et al., 2004).

4 Cytokines Involved in Innate Responses to WNV Infection

As discussed at the beginning of this chapter, type I IFNs are key cytokines produced during the early cellular response to viral infection and have important antiviral and immunomodulatory effects (Parmar and Platanias, 2005). Accordingly, they play a critical role in WNV infection, which has been comprehensively discussed elsewhere in this book. Obviously there are innate immune responses to WNV that are independent of type I IFNs (Daffis et al., 2007). We will conclude this chapter by discussing how other innate cytokines produced in the early stages of infection influence WNV immunopathogenesis.

4.1 Macrophage Migration Inhibitory Factor

Migration inhibitory factor (MIF) was originally identified as a T-cell cytokine (David, 1966), but is currently envisioned as a critical mediator of the inflammatory cascade and therefore the innate immune response (Calandra and Roger, 2003). MIF is produced by virtually all leukocytes as well as by several nonimmune cell types, and is rapidly released from preformed intracellular pools after bacterial and viral challenge (Arndt et al., 2002; Bacher et al., 2002; Calandra and Roger, 2003). Among other effects, MIF induces the production of inflammatory cytokines, which in turn further stimulate MIF production, forming a positive feedback loop during the initial stages of the inflammatory cascade (Leng and Bucala, 2005). A recent study has elucidated the role of MIF in WNV pathogenesis. WNV-infected patients presented with increased MIF levels in plasma and in cerebrospinal fluid (Arjona et al., 2007). These findings are in line with previous reports showing increased MIF levels in patients infected with dengue and hepatitis B viruses (Zhang et al., 2002; Chen et al., 2006). In the mouse model, WNV infection also induces a transient increase in MIF levels during the early stage of infection. Remarkably, abrogation of MIF action by genetic deletion, pharmacologic inhibition, or antibody blockade rendered mice more resistant to WNV lethality. Although wild-type and MIF-deficient mice presented with comparable viral titers in the periphery, WNV neuroinvasion was delayed and overall reduced in animals lacking MIF

(Arjona et al., 2007). Rather than having an effect on viral replication or on the migration of infected leukocytes into the CNS, MIF appears to promote WNV neuroinvasion partly by its adjuvant action on TNF-α secretion and subsequent increase in BBB permeability. It cannot be ruled out that MIF also increases BBB breaching during WNV infection by inducing the synthesis of matrix metalloproteinases (Pakozdi et al., 2006; Yu et al., 2007). These findings indicate that, as occurs with the TLR3-mediated production of inflammatory cytokines (Wang et al., 2004), MIF has a detrimental role in WNV infection (Fig. 1). The sensing mechanisms that trigger MIF release have yet to be ascertained, although the fact that MIF is promptly secreted from preformed intracellular stores upon challenge argues against the possibility of canonical PRRs (which generally act at the genomic level) mediating this response. In any case, the results obtained from MIF abrogation studies emphasize the notion that activation of certain innate/inflammatory pathways has deleterious rather than protective consequences in the case of infection with WNV, and perhaps other neurotropic viruses.

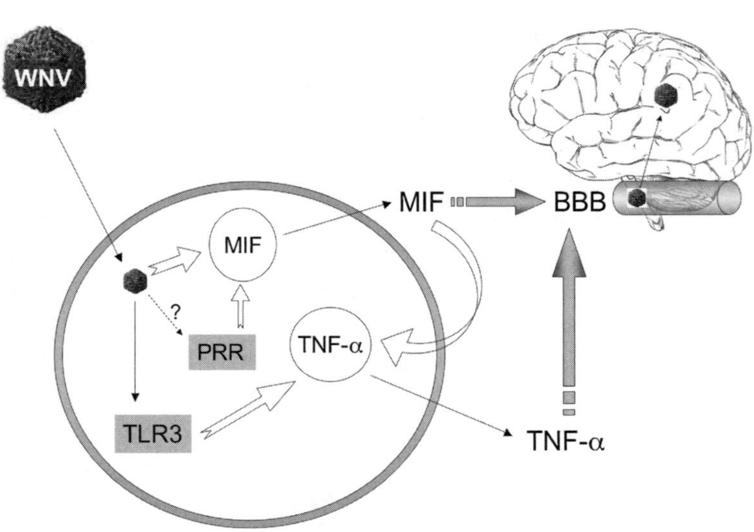

Figure 1. Detrimental effect of TLR3-mediated TNF-α production and MIF in WNV pathogenesis. WNV infection is sensed by TLR3, which triggers the production of TNF-α. Similarly, MIF is rapidly released upon WNV infection and further promotes the production of TNF-α. Both inflammatory mediators favor WNV neuroinvasion by increasing the permeability of the blood–brain barrier (BBB).

4.2 IFN-γ

IFN-γ is a pleiotropic cytokine with multiple effects on the immune response. One of its major roles is the activation of antiviral mechanisms by the polarization and activation of T cells, enhancement of antigen presentation, and activation of macrophages and DCs (Schroder et al., 2004). IFN-γ is produced by CD8 + T cells and also by cells that mediate early (innate) antiviral responses such as NK cells and γδ T cells. Mice deficient in IFN-γ or in the IFN-γ receptor showed increased mortality rates and decreased survival time when compared to controls. The survival pattern of mice lacking IFN-γ production or signaling correlated with higher viremia and greater viral replication in peripheral tissues, which led to early and overall increase of WNV neuroinvasion (Shrestha et al., 2006b). In addition, it has been shown that treatment of primary DCs with IFN-γ greatly reduces WNV replication (Shrestha et al., 2006b), which further points towards a direct antiviral role of IFN-γ in WNV infection. Absence of IFN-γ does not significantly alter adaptive immune responses, and there is a cell-specific requirement for IFN-γ, as γδ T cells, but not CD8 T cells, require this cytokine to limit WNV dissemination (Wang et al., 2003; Shrestha et al., 2006b). Thus, it appears that the dominant antiviral role of IFN-γ falls within the innate response. Lastly, the action of IFN-γ in WNV infection does not seem to be restricted to immune defense mechanisms, as it has been also implicated in the development of the excitatory seizure pathways in the brain that become pathogenic in encephalitic WNV infection (Getts et al., 2007).

4.3 Other Inflammatory Cytokines

In addition to MIF and TNF-α, other innate inflammatory cytokines are induced during WNV infection. For example, IL-1β expression is upregulated in WNV-infected murine macrophages as soon as 12 h post infection (Shirato et al., 2006). In addition, upregulation of IL-1β was evident in neural tissue from WNV encephalitis patients as well as in WNV-infected cultured human glia (van Marle et al., 2007). Although it is not known whether the role of IL-1β in the CNS is protective or detrimental, peripheral IL-1β appears to be necessary for Langerhans cell migration, accumulation in the draining lymph nodes, and the initiation of lymph node shut-down in response to a cutaneous WNV infection (Byrne et al., 2001). Similarly to IL-1β, IL-6 is produced by macrophages and microglia promptly after WNV infection (Cheeran et al., 2005; Shirato et al., 2006). IL-6 serum levels were elevated in

WNV-infected mice at day 3 post-infection, and this response is partly dependent on TLR3 (Wang et al., 2004). However, when IL-6-deficient mice were challenged with WNV, they showed a mortality rate similar to that of control mice (Wang et al., 2004). Thus, not all the innate/inflammatory responses triggered by WNV may show a significant contribution to WNV pathogenesis.

5 Concluding Remarks

In the past decade much has been learned about the innate immune responses that shape WNV pathogenesis. This knowledge has emerged mainly from in vivo studies that employed mice deficient in innate immune molecules. Many of these innate responses, such as the one mediated by type I IFNs, as well as the $\gamma\delta$ T cells/IFN-γ response are clearly necessary not only for restraining initial WNV dissemination but also for orchestrating the adaptive responses that are responsible for WNV clearance and immunological memory. However, other WNV-induced innate responses like TLR3-mediated TNF-α production as well as MIF release have paradoxical consequences, as they are detrimental for the host owing to their effects on BBB permeability and subsequent WNV neuroinvasion (Fig. 1).

Because these deleterious responses appear during the early stages of infection, they arise as suggestive targets of therapeutic intervention strategies. For example, MIF pharmacologic inhibition has been proven to be feasible and efficacious in several experimental models of inflammation and shock (Lubetsky et al., 2002; Senter et al., 2002; Cvetkovic et al., 2005). Indeed, MIF inhibition has been found to protect from bacteria-induced shock in the same experimental infection models in which inhibition of downstream inflammatory factors showed no effect (Calandra et al., 2000). Studies on MIF gene polymorphisms have revealed the existence of high expression MIF alleles that are linked to increased susceptibility and severity of various inflammatory diseases (Baugh et al., 2002; Nohara et al., 2004; Mizue et al., 2005; Radstake et al., 2005). Investigating whether MIF polymorphisms influence the susceptibility to WNV encephalitis may be valuable for therapeutic and prognostic purposes as well as for assessing which individuals may benefit most from potential MIF-directed pharmacotherapies.

Obviously more research is needed to better understand how WNV infection is initially detected as well as what innate immunity mechanisms and pathways are activated upon WNV recognition. For example, although it is clear that TLR3 mediates WNV recognition, it is not known what specific cell type is the major contributor for peripheral TLR3-mediated cytokine responses (particularly TNF-α) to WNV infection. Along this line, the role of specific DC subsets during WNV infection in vivo has not been formally examined. A better understanding of how these cells react and orchestrate adaptive responses to WNV would be extremely useful for the design of future immunoprophylactic strategies. It remains to be elucidated whether other innate immune cell types that are often neglected when studying antiviral immunity contribute to WNV immunopathogenesis. In this sense, neutrophils have been shown to be able to degrade Japanese encephalitis virions (a closely related flavivirus) via triggering of the respiratory burst and generation of toxic radicals (Srivastava et al., 1999), and, interestingly, WNV meningoencephalitis is frequently associated with substantial cerebrospinal fluid (CSF) neutrophilia (Rawal et al., 2006; Tyler et al., 2006).

Another important issue is whether other canonical viral recognition receptors such as TLR7/8 play a significant role in sensing WNV infection as well. This question can be extended to nonconventional PRRs like carbohydrate-discriminating receptors (e.g., CLRs), which, because of their bona-fide interaction with WNV structural proteins, may modulate innate cell activation well before other WNV PAMPs such as ssRNA and dsRNA are exposed to recognition. In a broader sense, PRRs include any PAMP-recognizing molecule capable of triggering any type of antiviral response in leukocytes, including immediate, nongenomic effector functions such as phagocytosis or degranulation. The activation of nonconventional PRRs may underlie innate immune responses that may be pivotal for the control of WNV infection.

References

Abbott NJ. (2000) Inflammatory mediators and modulation of blood–brain barrier permeability. Cell Mol Neurobiol 20:131–147

Alexopoulou L, Holt AC, Medzhitov R, Flavell RA. (2001) Recognition of double-stranded RNA and activation of NF-kappaB by Toll-like receptor 3. Nature 413:732–738

Andoniou CE, Andrews DM, Degli-Esposti MA. (2006) Natural killer cells in viral infection: more than just killers. Immunol Rev 214:239–250

Arjona A et al. (2007) Abrogation of macrophage migration inhibitory factor decreases West Nile virus lethality by limiting viral neuroinvasion. J Clin Invest 117:3059–3066

Arndt U et al. (2002) Release of macrophage migration inhibitory factor and CXCL8/interleukin-8 from lung epithelial cells rendered necrotic by influenza A virus infection. J Virol 76:9298–9306

Bacher M, Eickmann M, Schrader J, Gemsa D, Heiske A. (2002) Human cytomegalovirus-mediated induction of MIF in fibroblasts. Virology 299:32–37

Baugh JA et al. (2002) A functional promoter polymorphism in the macrophage migration inhibitory factor (MIF) gene associated with disease severity in rheumatoid arthritis. Genes Immun 3:170–176

Bell JK, Askins J, Hall PR, Davies DR, Segal DM. (2006a) The dsRNA binding site of human Toll-like receptor 3. Proc Natl Acad Sci USA 103:8792–8797

Bell JK et al. (2006b) The molecular structure of the TLR3 extracellular domain. J Endotoxin Res 12:375–378

Ben-Nathan D, Huitinga I, Lustig S, van Rooijen N, Kobiler D. (1996) West Nile virus neuroinvasion and encephalitis induced by macrophage depletion in mice. Arch Virol 141:459–469

Boehme KW, Singh J, Perry ST, Compton T. (2004) Human cytomegalovirus elicits a coordinated cellular antiviral response via envelope glycoprotein B. J Virol 78:1202–1211

Byrne SN, Halliday GM, Johnston LJ, King NJ. (2001) Interleukin-1beta but not tumor necrosis factor is involved in West Nile virus-induced Langerhans cell migration from the skin in C57BL/6 mice. J Invest Dermatol 117:702–709

Calandra T, Roger T. (2003) Macrophage migration inhibitory factor: a regulator of innate immunity. Nat Rev Immunol 3:791–800

Calandra T et al. (2000) Protection from septic shock by neutralization of macrophage migration inhibitory factor. Nat Med 6:164–170

Caparros E et al. (2006) DC-SIGN ligation on dendritic cells results in ERK and PI3K activation and modulates cytokine production. Blood 107:3950–3958

Carding SR, Egan PJ. (2002) Gammadelta T cells: functional plasticity and heterogeneity. Nat Rev Immunol 2:336–345

Cheeran MC, Hu S, Sheng WS, Rashid A, Peterson PK, Lokensgard JR. (2005) Differential responses of human brain cells to West Nile virus infection. J Neurovirol 11:512–524

Chen LC et al. (2006) Correlation of serum levels of macrophage migration inhibitory factor with disease severity and clinical outcome in dengue patients. Am J Trop Med Hyg 74:142–147

Cheng Y, King NJ, Kesson AM. (2004) Major histocompatibility complex class I (MHC-I) induction by West Nile virus: involvement of 2 signaling pathways in MHC-I up-regulation. J Infect Dis 189:658–668

Choe J, Kelker MS, Wilson IA. (2005) Crystal structure of human toll-like receptor 3 (TLR3) ectodomain. Science 309:581–585

Cvetkovic I et al. (2005) Critical role of macrophage migration inhibitory factor activity in experimental autoimmune diabetes. Endocrinology 146:2942–2951

Daffis S, Samuel MA, Keller BC, Gale M Jr, Diamond MS. (2007) Cell-specific IRF-3 responses protect against West Nile virus infection by interferon-dependent and -independent mechanisms. PLoS Pathog 3:e106

David JR. (1966) Delayed hypersensitivity in vitro: its mediation by cell-free substances formed by lymphoid cell-antigen interaction. Proc Natl Acad Sci USA 56:72–77

Davis CW, Mattei LM, Nguyen HY, Ansarah-Sobrinho C, Doms RW, Pierson TC. (2006a) The location of asparagine-linked glycans on West Nile virions controls their interactions with CD209 (dendritic cell-specific ICAM-3 grabbing nonintegrin). J Biol Chem 281:37183–37194

Davis CW, Nguyen HY, Hanna SL, Sanchez MD, Doms RW, Pierson TC. (2006b) West Nile virus discriminates between DC-SIGN and DC-SIGNR for cellular attachment and infection. J Virol 80:1290–1301

Diebold SS, Kaisho T, Hemmi H, Akira S, Reis e Sousa C. (2004) Innate antiviral responses by means of TLR7-mediated recognition of single-stranded RNA. Science 303:1529–1531

Fredericksen BL, Gale M, Jr. (2006) West Nile virus evades activation of interferon regulatory factor 3 through RIG-I-dependent and -independent pathways without antagonizing host defense signaling. J Virol 80:2913–2923

Garcia-Tapia D, Loiacono CM, Kleiboeker SB. (2006) Replication of West Nile virus in equine peripheral blood mononuclear cells. Vet Immunol Immunopathol 110:229–244

Geijtenbeek TB et al. (2003) Mycobacteria target DC-SIGN to suppress dendritic cell function. J Exp Med 197:7–17

Getts DR et al. (2007) Role of IFN-gamma in an experimental murine model of West Nile virus-induced seizures. J Neurochem 103:1019–1030
Gilfoy FD, Mason PW. (2007) West Nile virus-induced interferon production is mediated by the double-stranded RNA-dependent protein kinase PKR. J Virol 81:11148–11158
Gitlin L et al.(2006) Essential role of mda-5 in type I IFN responses to polyriboinosinic:polyribocytidylic acid and encephalomyocarditis picornavirus. Proc Natl Acad Sci USA 103:8459–8464
Hayday AC. (2000) [gamma][delta] cells: a right time and a right place for a conserved third way of protection. Annu Rev Immunol 18:975–1026
Herzer K et al. (2003) Upregulation of major histocompatibility complex class I on liver cells by hepatitis C virus core protein via p53 and TAP1 impairs natural killer cell cytotoxicity. J Virol 77:8299–8309
Iwasaki A, Medzhitov R. (2004) Toll-like receptor control of the adaptive immune responses. Nat Immunol 5:987–995
Jin MS et al. (2007) Crystal structure of the TLR1-TLR2 heterodimer induced by binding of a tri-acylated lipopeptide. Cell 130:1071–1082
Johnston LJ, Halliday GM, King NJ. (1996) Phenotypic changes in Langerhans' cells after infection with arboviruses: a role in the immune response to epidermally acquired viral infection? J Virol 70:4761–4766
Johnston LJ, Halliday GM, King NJ. (2000) Langerhans cells migrate to local lymph nodes following cutaneous infection with an arbovirus. J Invest Dermatol 114:560–568
Kato H et al. (2006) Differential roles of MDA5 and RIG-I helicases in the recognition of RNA viruses. Nature 441:101–105
Kawai T, Akira S. (2006) Innate immune recognition of viral infection. Nat Immunol 7:131–137
Kelley TW, Prayson RA, Ruiz AI, Isada CM, Gordon SM. (2003) The neuropathology of West Nile virus meningoencephalitis. A report of two cases and review of the literature. Am J Clin Pathol 119:749–753
Kim HM et-al.(2007) Crystal structure of the TLR4-MD-2 complex with bound endotoxin antagonist Eritoran. Cell 130:906–917
King NJ, Kesson AM. (2003) Interaction of flaviviruses with cells of the vertebrate host and decoy of the immune response. Immunol Cell Biol 81:207–216
King NJ, Maxwell LE, Kesson AM. (1989) Induction of class I major histocompatibility complex antigen expression by West Nile virus on gamma interferon-refractory early murine trophoblast cells. Proc Natl Acad Sci USA 86:911–915
Kobayashi KS, Flavell RA. (2004) Shielding the double-edged sword: negative regulation of the innate immune system. J Leukoc Biol 75:428–433
Kreil TR, Eibl MM. (1996) Nitric oxide and viral infection: NO antiviral activity against a flavivirus in vitro, and evidence for contribution to pathogenesis in experimental infection in vivo. Virology 219:304–306
Kulkarni AB, Mullbacher A, Blanden RV. (1991) Functional analysis of macrophages, B cells and splenic dendritic cells as antigen-presenting cells in West Nile virus-specific murine T lymphocyte proliferation. Immunol Cell Biol 69 (Pt 2):71–80
Kurt-Jones EA et al. (2000) Pattern recognition receptors TLR4 and CD14 mediate response to respiratory syncytial virus. Nat Immunol 1:398–401
Lemaitre B, Nicolas E, Michaut L, Reichhart JM, Hoffmann JA. (1996) The dorsoventral regulatory gene cassette spatzle/Toll/cactus controls the potent antifungal response in Drosophila adults. Cell 86:973–983
Leng L, Bucala R. (2005) Macrophage migration inhibitory factor Crit Care Med 33:S475–477
Lin YL et al. (1997) Inhibition of Japanese encephalitis virus infection by nitric oxide: antiviral effect of nitric oxide on RNA virus replication. J Virol 71:5227–5235
Liu Y, King N, Kesson A, Blanden RV, Mullbacher A. (1988) West Nile virus infection modulates the expression of class I and class II MHC antigens on astrocytes in vitro. Ann N Y Acad Sci 540:483–485

Liu Y, King N, Kesson A, Blanden RV, Mullbacher A. (1989) Flavivirus infection up-regulates the expression of class I and class II major histocompatibility antigens on and enhances T cell recognition of astrocytes in vitro. J Neuroimmunol 21:157–168
Lubetsky JB et al. (2002) The tautomerase active site of macrophage migration inhibitory factor is a potential target for discovery of novel anti-inflammatory agents. J Biol Chem 277:24976–24982
Lund J, Sato A, Akira S, Medzhitov R, Iwasaki A. (2003) Toll-like receptor 9-mediated recognition of Herpes simplex virus-2 by plasmacytoid dendritic cells. J Exp Med 198:513–520
Lund JM et-al.(2004) Recognition of single-stranded RNA viruses by Toll-like receptor 7. Proc Natl Acad Sci USA 101:5598–5603
van Marle G et-al.(2007) West nile virus-induced neuroinflammation: glial infection and capsid protein-mediated neurovirulence. J Virol 81:10933–10949
Mizue Y et al. (2005) Role for macrophage migration inhibitory factor in asthma. Proc Natl Acad Sci USA 102:14410–14415
Morrissette N, Gold E, Aderem A. (1999) The macrophage – a cell for all seasons. Trends Cell Biol 9:199–201
Nohara H et al. (2004) Association of the -173 G/C polymorphism of the macrophage migration inhibitory factor gene with ulcerative colitis. J Gastroenterol 39:242–246
Pakozdi A et al. (2006) Macrophage migration inhibitory factor: a mediator of matrix metalloproteinase-2 production in rheumatoid arthritis. Arthritis Res Ther 8:R132
Parmar S, Platanias LC. (2005) Interferons. Cancer Treat Res 126:45–68
Petersen LR, Marfin AA. (2002) West Nile virus: a primer for the clinician. Ann Intern Med 137:173–179
Pierson TC et al. (2005) An infectious West Nile virus that expresses a GFP reporter gene. Virology 334:28–40
Pisarev VB, Shishkina EO, Larichev VF, Grigor'eva NV. (2003) Morphofunctional characteristics of antigen-presenting cells in lymph node in mice with experimental West Nile fever. Bull Exp Biol Med 135:293–295
Radstake TR et al. (2005) Correlation of rheumatoid arthritis severity with the genetic functional variants and circulating levels of macrophage migration inhibitory factor. Arthritis Rheum 52:3020–3029
Rawal A, Gavin PJ, Sturgis CD. (2006) Cerebrospinal fluid cytology in seasonal epidemic West Nile virus meningo-encephalitis. Diagn Cytopathol 34:127–129
Rios M et al. (2006) Monocytes-macrophages are a potential target in human infection with West Nile virus through blood transfusion. Transfusion 46:659–667
Saito T, Gale M Jr. (2007) Principles of intracellular viral recognition. Curr Opin Immunol 19:17–23
Samuel CE. (2002) Host genetic variability and West Nile virus susceptibility. Proc Natl Acad Sci USA 99:11555–11557
Samuel MA, Diamond MS. (2005) Alpha/beta interferon protects against lethal West Nile virus infection by restricting cellular tropism and enhancing neuronal survival. J Virol 79:13350–13361
Samuel MA, Diamond MS. (2006) Pathogenesis of West Nile Virus infection: a balance between virulence, innate and adaptive immunity, and viral evasion. J Virol 80:9349–9360
Samuel MA et al. (2006) PKR and RNase L contribute to protection against lethal West Nile Virus infection by controlling early viral spread in the periphery and replication in neurons. J Virol 80:7009–7019
Scherbik SV, Paranjape JM, Stockman BM, Silverman RH, Brinton MA. (2006) RNase L plays a role in the antiviral response to West Nile virus. J Virol 80:2987–2999
Schroder K, Hertzog PJ, Ravasi T, Hume DA. (2004) Interferon-gamma: an overview of signals, mechanisms and functions. J Leukoc Biol 75:163–189

Senter PD et al. (2002) Inhibition of macrophage migration inhibitory factor (MIF) tautomerase and biological activities by acetaminophen metabolites. Proc Natl Acad Sci USA 99:144–149

Shirato K, Miyoshi H, Kariwa H, Takashima I. (2006) The kinetics of proinflammatory cytokines in murine peritoneal macrophages infected with envelope protein-glycosylated or non-glycosylated West Nile virus. Virus Res 121:11–16

Shrestha B, Samuel MA, Diamond MS. (2006a) CD8 + T cells require perforin to clear West Nile virus from infected neurons. J Virol 80:119–129

Shrestha B et-al.(2006b) Gamma interferon plays a crucial early antiviral role in protection against West Nile virus infection. J Virol 80:5338–5348

Silva MC, Guerrero-Plat A, Gilfoy FD, Garofalo RP, Mason PW. (2007) Differential activation of human monocyte-derived and plasmacytoid dendritic cells by West Nile virus generated in different host cells. J Virol 81:13640–13648

Srivastava S, Khanna N, Saxena SK, Singh A, Mathur A, Dhole TN. (1999) Degradation of Japanese encephalitis virus by neutrophils. Int J Exp Pathol 80:17–24

Steinman RM, Hemmi H. (2006) Dendritic cells: translating innate to adaptive immunity. Curr Top Microbiol Immunol 311:17–58

Takahashi K, Ip WE, Michelow IC, Ezekowitz RA. (2006) The mannose-binding lectin: a prototypic pattern recognition molecule. Curr Opin Immunol 18:16–23

Theofilopoulos AN, Baccala R, Beutler B, Kono DH. (2005) Type I interferons (alpha/beta) in immunity and autoimmunity. Annu Rev Immunol 23:307–336

Town T, Jeng D, Alexopoulou L, Tan J, Flavell RA. (2006) Microglia recognize double-stranded RNA via TLR3. J Immunol 176:3804–3812

Tyler KL, Pape J, Goody RJ, Corkill M, Kleinschmidt-DeMasters BK. (2006) CSF findings in 250 patients with serologically confirmed West Nile virus meningitis and encephalitis. Neurology 66:361–365

Vargin VV, Semenov BF. (1986) Changes of natural killer cell activity in different mouse lines by acute and asymptomatic flavivirus infections. Acta Virol 30:303–308

Wang L, Das H, Kamath A, Bukowski JF. (2001) Human V gamma 2V delta 2 T cells produce IFN-gamma and TNF-alpha with an on/off/on cycling pattern in response to live bacterial products. J Immunol 167:6195–6201

Wang T et al. (2003) IFN-gamma-producing gamma delta T cells help control murine West Nile virus infection. J Immunol 171:2524–2531

Wang T, Town T, Alexopoulou L, Anderson JF, Fikrig E, Flavell RA. (2004) Toll-like receptor 3 mediates West Nile virus entry into the brain causing lethal encephalitis. Nat Med 10:1366–1373

Wang T et al. (2006) Gamma delta T cells facilitate adaptive immunity against West Nile virus infection in mice. J Immunol 177:1825–1832

Yamamoto M et-al.(2003) Role of adaptor TRIF in the MyD88-independent toll-like receptor signaling pathway. Science 301:640–643

Yin Z et al. (2000) Dominance of IL-12 over IL-4 in gamma delta T cell differentiation leads to default production of IFN-gamma: failure to down-regulate IL-12 receptor beta 2-chain expression. J Immunol 164:3056–3064

Yu X et al. (2007) Macrophage migration inhibitory factor induces MMP-9 expression in macrophages via the MEK-ERK MAP kinase pathway. J Interferon Cytokine Res 27:103–109

Zhang W, Yue B, Wang GQ, Lu SL. (2002) Serum and ascites levels of macrophage migration inhibitory factor, TNF-alpha and IL-6 in patients with chronic virus hepatitis B and hepatitis cirrhosis. Hepatobiliary Pancreat Dis Int 1:577–5801

9. Mechanisms of Complement Regulation of Infection by Flaviviruses

Erin Mehlhop and Michael S. Diamond

Abstract

The complement system is a family of more than 30 serum proteins and cell surface receptors that recognizes pathogen-associated molecular patterns, altered-self ligands, and immune complexes. Complement activation through the classical, lectin and alternative pathways triggers an array of functions including direct pathogen opsonization and/or lysis, and enhancement of B and T cell responses. Recent work has demonstrated that complement activation is necessary for the development of protective immune responses and mediates important antiviral effector functions of Flavivirus-specific antibody. In this review, we will examine the mechanisms and functions of complement activation and discuss our current understanding of the role of complement in protection against infection by Flaviviruses, with an emphasis on experiments with West Nile virus (WNV). In addition, we will discuss ongoing studies on the immune evasion mechanisms that WNV uses to limit complement activation.

Keywords

Complement, immunology flaviviruses, pathogenesis

1 Introduction

The genus *Flavivirus* is composed of 73 enveloped viruses containing ~11 kb single-stranded, positive-polarity RNA genomes (Lindenbach and Rice, 2006). Within this family, ~40 viruses are associated with severe human diseases including dengue (DENV), yellow fever (YFV), West Nile (WNV), Japanese encephalitis (JEV), and tick-borne encephalitis (TBE) viruses (Burke and Monath, 2001). Flaviviruses encode a single open reading frame that is translated in the cytoplasm as a polyprotein and cleaved into three structural (capsid (C), membrane (prM/M) and envelope (E)) and seven nonstructural (NS) proteins (Brinton, 2002; Nowak et al., 1989). Viral particles assemble at the endoplasmic reticulum, and mature 500 Å infectious virions are released by exocytosis (Elshuber et al., 2003; Guirakhoo et al., 1992; Stadler et al., 1997).

Antibodies against the envelope (E) protein protect from severe disease by Flaviviruses (Oliphant and Diamond, 2007; Roehrig et al., 2001). X-ray crystallographic analysis has defined three structural domains (domains I, II, and III) within the E protein (reviewed in Mukhopadhyay et al., 2005). Recent experiments suggest that potently neutralizing Flavivirus-specific antibodies localize within domain III (reviewed in Oliphant and Diamond, 2007). Although much insight has been gained into the structural and molecular basis of antibody neutralization of Flaviviruses, the role and significance of the in vivo effector functions associated with neutralizing antibody remain unclear. This review will discuss the mechanisms by which complement activation limits WNV infection and protects against disease. These advances have significant implications for the development and optimization of novel antibody-based therapies and targeted vaccines against Flaviviruses.

2 The Complement System

The complement system was originally described over 100 years ago as a heat-sensitive factor in plasma that "complemented" antibody-dependent lysis of bacteria and red blood cells (Bordet and Gengou, 1901). In the 1960s, biochemical advancements revealed the complexity of the interactions among individual components of this system (Muller-Eberhard, 1968; Porter and Reid, 1978). A central component of humoral innate immunity, the complement system is one of the oldest arms of the immune system, evolving over 1,300 million years ago by gene duplication (Dishaw et al., 2005; Nonaka and Kimura, 2006). Current evidence suggests that complement predates development of adaptive immune systems, as rudimentary complement systems exist even in invertebrates (Pinto et al., 2007). The mammalian complement system is a family of more than 30 serum proteins and cell surface receptors (Carroll, 2004) that interact in a serine protease cascade (Reid and Porter, 1981) leading to the release of proinflammatory peptides, attachment of opsonins, and the formation of a membrane attack complex (MAC). Complement activation plays key roles in (1) direct lysis of invading pathogens and pathogen-infected cells,

(2) neutralization by covalent opsonization with C1q and or C3 split products, (3) clearance by complement receptor-mediated phagocytosis, (4) leukocyte chemoattraction, and (5) priming of adaptive T and B cell responses (Fearon and Carroll, 2000; Volanakis, 2002). While the liver is the major source of many complement proteins, other cell types including monocytes, epithelial cells, endothelial cells, and cells of the CNS can synthesize complement components (Morgan and Gasque, 1997; Speth et al., 2002; van Beek et al., 2003).

Cloning and sequencing of the complement components has greatly augmented our understanding of individual proteins of the complement system. Identification of common structural motifs has allowed classification of active complement components into four functional groups (Table 1). (1) The collectins, as typified by C1q, contain a collagen-like coiled-coil domain with a globular C terminus (Kishore and Reid, 2000). The globular domain of collectins, which also include mannose-binding lectin (MBL), surfactant protein (SP)-A, SP-D, and conglutin, participates in glycoprotein recognition and binding. (2) The complement serine proteases are related to the chymotrypsin and trypsin enzyme subfamily (Gal et al., 2007; Perona and Craik, 1995). This group, which includes C1r, C1s, mannose-associated serine protease (MASP)-1, MASP-2, C2, factor (f)B, fD, and fI, contains an His-Asp-Ser catalytic triad. (3) The C3 family, including C3, C4, and C5, contains highly processed disulfide-linked multi-chain structures. C3 and C4 contain an internal thioester bond that upon activation allows C3b and C4b to bind covalently to target surfaces. (4) The terminal components C6–C9 all contain a perforin-like segment that is involved in Ca^{+2}-dependent membrane attack complex (MAC) lytic pore formation (Esser, 1994; Muller-Eberhard, 1986). Interaction of the components within these functional groups leads to complement activation, downstream signaling and effector functions.

2.1 Complement Activation

In general, complement activation occurs by proteolytic cleavage of complement components along three distinct activation pathways (Fig. 1). Different initiators trigger each of the three known activation pathways leading to the generation of C3 and C5 convertases. These enzyme complexes are responsible for anaphylatoxin release and MAC C5b-9 pore assembly. Pivotal to complement function, C3 convertases are required for the production of direct complement effector molecules

Table 1. Serum complement components

	Size[a]	Concentration[b]	Reference
Collectin family			
C1q	459	90–150	(Brasher, 1977; Nakamura et al., 1991; Sullivan et al., 1996)
MBL	576	0–5	(Carlsson et al., 2005; Schafranski et al., 2004)
SP-A	540	20–40	(Greene et al., 2002; Mason et al., 1998)
SP-D	510	40–90	(Greene et al., 2002; Mason et al., 1998)
Conglutin	428	40–70	(Akiyama et al., 1992)
Complement serine proteases			
C1r	87	~35	(Sullivan et al., 1996)
C1s	80	~30	(Sjoholm, 1975; Sullivan et al., 1996)
MASP-1	93	~6	(Knittel et al., 1997)
MASP-2	76	0.3–0.8	(Manuel et al., 2007)
C2	100	11–35	(Smith et al., 1984)
Factor B	90	75–250	(Oglesby et al., 1988)
Factor D	25	~0.7–1.2	(Hiemstra et al., 1989)
Factor I	88	~35	(Vyse et al., 1996)
C3 family			
C3	190	1,000–1,500	(Kohler and Muller-Eberhard, 1967)
C4	200	300–600	(Kohler and Muller-Eberhard, 1967)
C5	190	50–100	(Kohler and Muller-Eberhard, 1967)
Terminal components			
C6	100	20–80	(Wurzner et al., 1991)
C7	95	30–180	(Wurzner et al., 1991)
C8	150	55–80	(Benbassat et al., 1993; Nakamura et al., 1991; Schlesinger et al., 1992)
C9	71	30–90	(Benbassat et al., 1993; Nakamura et al., 1991)

[a]Size is expressed as kDa
[b]Serum concentration as determined in the references listed is expressed as μg/ml

and cleavage of fragments that initiate the later events in adaptive immune priming.

2.1.1 Classical Pathway

The classical pathway is initiated by C1 complex activation. This complex is composed of one molecule of the collectin C1q, and two molecules each of serine proteases C1r and C1s (Cooper, 1985; Perkins and Nealis, 1989). C1q consists of three polypeptide chains with unique

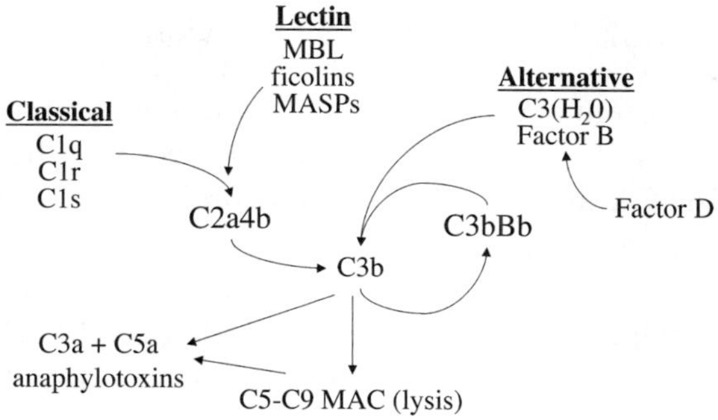

Figure 1. Complement activation pathways. The classical, lectin and alternative pathways can activate the serine protease complement activation cascade to induce C3b deposition, MAC formation and anaphylotoxin release. Details of activation cascades are described in the text.

globular head regions (A, B and C), and a collagen-like N-terminal tail (Kishore and Reid, 2000). The chains of C1q adopt a hexameric arrangement, as a trimer of AB heterodimers and CC homodimers. The hexameric composition of C1q may influence the recognition of activating ligands such as soluble immune complexes, DNA, C-reactive protein, serum amyloid P, bacterial lipopolysaccharide, viral glycoproteins, pentraxins, or cell surface SIGN-R1 (Kang et al., 2006; Kishore et al., 2004; Van Schravendijk and Dwek, 1982). C1q ligand binding induces a conformational change that displaces the C1 inhibitor and initiates auto-activation of C1r, and subsequent C1s activation (Sim and Reid, 1991).

To form the classical pathway C3 convertase, active C1s cleaves C4 into its C4a and C4b split products. This exposes the internal thioester bond within C4b, allowing covalent deposition on the activating membrane surface through hydroxyl or amino groups (Law and Dodds, 1997). Surface bound C4b can then bind C2, initiating C1s-mediated cleavage and formation of the classical pathway C3 convertase, C4b2a (Janeway, 2001). The C4b2a complex cleaves C3 within the α chain, exposing another reactive internal thioester bond on C3b necessary for covalent attachment to target surfaces. The binding of C3b back to the C3 convertase complex forms the classical pathway C5 convertase, C4b2a3b. C3 and C5 cleavage releases the chemotactic anaphylotoxins, C3a and C5a (Kohl, 2001). Classical pathway C3b activation and subsequent MAC formation have also been observed in the absence of C2 and C4 when high concentrations of immune complexes are present (May and Frank, 1973). Termed the C1-bypass pathway, complement activation through this route occurs in vivo and may contribute to immune-mediated arthritis (Wagner et al., 1999). Additionally, spontaneous low level classical pathway activation may occur continuously in the plasma (Manderson et al., 2001). This C1-tickover phenomenon requires the presence of antibody, C1q, C2, and C4 in serum and probably explains the labile nature of complement ex vivo. Whatever method of activation occurs, stimulation of the classical pathway cascade results in considerable downstream amplification, as some studies suggest that 240 C3b molecules are deposited on the activating surface for each C1q molecule (Ollert et al., 1994).

2.1.2 Lectin Pathway

Similar to the classical pathway, the lectin pathway is initiated primarily by mannose-binding lectin (MBL), a C-type serum lectin that recognizes terminal mannose, *N*-acetylglucosamine (GlcNAc), and glucose residues (Wallis, 2007). However, other serum lectins (human L-, H-, and M-ficolin, and mouse ficolin-A) also can trigger this pathway (Endo et al., 2005; Liu et al., 2005; Matsushita et al., 2000, 2002). Both MBL and ficolins contain a collagen-like domain, similar to C1q. All lectin pathway initiator molecules associate with MASP-1 and MASP-2, which have a similar domain structure to C1r and C1s (Sato et al., 1994). Upon carbohydrate recognition, MASP-2 auto-activates and cleaves C4 and C2 (Rossi et al., 2001; Thiel et al., 1997), resulting in the formation of the C4b2a C3 convertase. Similar to the classical pathway, the addition of C3b to the C3 convertase forms the C5 convertase, C4b2a3b. Interestingly, orthologs of human MBL (Takahashi et al., 2006) and ficolin (Sekine et al., 2001) have been identified in lamprey and sea squirts, suggesting that both MBL and ficolins contribute to the innate immune response of both vertebrates and invertebrates. In contrast, orthologs of C1q have been identified only in species that express immunoglobulin (Dodds and Matsushita, 2007), suggesting that classical pathway activation is probably derived from a lectin pathway precursor.

2.1.3 Alternative Pathway

In contrast to both the classical or lectin pathways, activation of the alternative pathway does not require a host encoded initiator molecule. Instead, activation of the alternative pathway often occurs by spontaneous hydrolysis of C3, known as "C3 tickover" (Nicol and

Lachmann, 1973). Hydrolysis of C3 exposes the reactive internal thioester bond in the absence of α chain cleavage (Isenman et al., 1981). Reactive C3 can then bind to fB, facilitating factor FD-mediated cleavage and formation of a fluid phase C3Bb C3 convertase (Pangburn and Muller-Eberhard, 1980; Pangburn et al., 1981). In solution, this fluid phase C3 convertase is highly labile. However, if an activating surface is present, newly cleaved C3b can initiate a positive-feedback loop by binding fB, facilitating further factor FD-mediated cleavage and alternative pathway C3 convertase assembly (Pangburn et al., 1983). The stability of the C3bBb convertase is greatly enhanced by the binding of properdin (Farries et al., 1988). Further cleavage of C3 amplifies alternative pathway activation and promotes assembly of the C3bBb3b C5 convertase. Importantly, alternative pathway amplification augments classical and lectin pathway initiated complement activation (Lutz and Jelezarova, 2006) and possibly causes pathogenic complement activation in a collagen antibody-mediated model of arthritis (Banda et al., 2006, 2007). Activation of complement by the alternative pathway has been demonstrated only in vertebrates where multiple control proteins have evolved to prevent damage to self (Dodds and Matsushita, 2007).

2.1.4 Terminal Pathway

All complement activation pathways converge at the terminal pathway, which is initiated by either classical or alternative pathway C5 convertase formation (Muller-Eberhard, 1986). This enzymatic complex cleaves C5 into the C5a anaphylatoxin (Kohl, 2001) and C5b. Although it lacks an internal thioester bond, cleaved C5b binds back to C3b on the activating surface, exposing a C6 binding site (Cooper and Muller-Eberhard, 1970; DiScipio et al., 1983). Binding of C6 stabilizes the complex and facilitates binding of C7 (Podack et al., 1978b). Upon C5b67 complex binding, C7 becomes highly amphiphilic with high affinity for phospholipid, anchoring the complex to the membrane lipid bilayer by hydrophobic interactions (Podack et al., 1978a; Preissner et al., 1985). Subsequent binding of C8 to the C5b67 complex allows for membrane insertion and association of a variable number of C9 molecules to form the mature lytic pore (Podack et al., 1982; Tschopp et al., 1985, 1986). Assembly of the MAC and subsequent target lysis represents one of the direct effector functions of complement activation.

2.2 Complement Receptors Link Activation to Adaptive Immune Responses

Recent studies clearly indicate that complement activation links innate immune recognition with adaptive immune stimulation. Soon after complement activation, proteolysis of C3b occurs. Cleavage abolishes further amplification of C3b deposition by the alternative pathway, or terminal pathway assembly. Proteolysis of C3b produces iC3b, C3c and C3dg fragments that are recognized by three superfamilies of complement receptors (CR) (Ross and Medof, 1985; van Lookeren Campagne et al., 2007): The CR1/CR2 family contains short consensus repeat (SCR) modules, the β_2 integrin family encodes CR3 and CR4, and the immunoglobulin Ig-superfamily contains the recently identified CRIg (Table 2). Additionally, complement anaphylotoxins C3a and C5a enhance inflammation and chemoattraction by binding the C3a-receptor (C3aR) and C5a-receptor (C5aR), respectively (Kohl, 2001).

CR2 (CD21) is the principal complement receptor responsible for enhancement of humoral immune responses (Carroll, 1998; Fearon and Carroll, 2000). A type I transmembrane glycoprotein, CR2 is expressed primarily on B lymphocytes and follicular dendritic cells (Molina et al., 1996; Reynes et al., 1985; Tedder et al., 1984). Binding of C3d, C3dg, or iC3b to CR2 (Weis et al., 1984) lowers the threshold for B cell activation by cross-linking the B cell receptor with the CD19/CD81/CR2 co-receptor complex (Matsumoto et al., 1991). Conjugation of C3d to hen egg lysozyme, influenza hemagglutinin, bovine viral diarrhea virus E2 glycoprotein or HIV-1 envelope protein increases the immunogenicity of the antigen upto 10,000-fold (Dempsey et al., 1996; Green et al., 2003;

Table 2. Complement receptors

Receptor	CD name	Ligand	Tissue distribution
CR1	CD35	C3b, C4b, C1q	Erythrocytes (human), B cells, follicular DC, PMN, monocytes, activated T cells
CR2	CD21	C3b, C3dg	B cells, follicular DC, activated T cells, thymocytes
CR3	CD11b/CD18	iC3b	Myeloid cells, NK, CD5 + B cells and resting T cells
CR4	CD11c/CD18	iC3b	Myeloid cells, tissue macrophages and dendritic cells, activated B cells
DAF	CD55	C3b	GPI anchored, found on all blood cells
MCP	CD49	C3b, C4b	Ubiquitously expressed
CD59		C8	GPI anchored, found in most tissues except CNS

Ross et al., 2000; Wang et al., 2004). Indeed, mice with a genetic deficiency in either C3 (Fischer et al., 1996) or CR2 (Ahearn et al., 1996; Croix et al., 1996; Molina et al., 1996) display significantly impaired humoral responses to T cell-dependent (TD) antigens. Additionally, expression of CR2 on germinal center follicular dendritic cells is required for B cell survival within the germinal center, affinity maturation, and the establishment of B cell memory (Barrington et al., 2002; Fischer et al., 1998; Wu et al., 2000).

CR1 (CD35) is a type I integral membrane protein that binds C3b, C4b, and C1q, and MBL (Fearon, 1980; Ghiran et al., 2000; Klickstein et al., 1997). This glycoprotein is expressed on all peripheral blood cells in humans with the exception of platelets, natural killer cells and most T cells (Fearon, 1980; Tedder et al., 1983). In primates, CR1 expression on erythrocytes contributes to immune complex clearance and transfer of C3b-opsonized antigens to splenic and hepatic macrophages (Bogers et al., 1992; Craig et al., 2000). This observation suggests that CR1 may have a role in augmenting antigen presentation and T cell activity. In mice, CR1 is expressed as an alternative splice product of the *Cr2* gene and is primarily expressed on B cells and follicular dendritic cells (Kinoshita et al., 1988; Kurtz et al., 1990; Molina et al., 1990). Profound defects in humoral immunity have been observed in *CR1/CR2*$^{-/-}$ mice (Ahearn et al., 1996; Chen et al., 2000; Croix et al., 1996; Molina et al., 1996), with little effect on T cell activity (Kopf et al., 2002; Suresh et al., 2003). CR1/CR2-mediated antigen trapping on follicular dendritic cells enhances antigen presentation to B cells, and is required for both primary and secondary humoral responses (Fang et al., 1998; Qin et al., 1998).

CR3 and CR4 are members of the β_2-integrin (CD18) family and mediate phagocytosis of iC3b-opsonized antigens. Expressed as heterodimers on most myeloid-lineage cells, CR3 and CR4 are formed by the non-covalent paring of α_M (CR3/CD11b) and α_X (CR4/CD11c) chains, respectively with a common β-chain (CD18) subunit (Springer, 1990). These integrins also mediate intercellular adhesion of phagocytes through the recognition of several distinct extracellular matrix and cell surface ligands. Expression of CR3 and CR4 on antigen presenting cells (APCs) may augment T cell activation by increasing antigen uptake and presentation. In the absence of complement T cell responsiveness to influenza virus, lymphocytic choriomeningitis virus (LCMV), Leishmania, and alloantigens is reduced (Kopf et al., 2002; Pratt et al., 2002; Stager et al., 2003; Suresh et al., 2003). Correspondingly, C3b opsonization augments antigen uptake (Rey-Millet et al., 1994; Villiers et al., 1996) and T cell stimulation (Arvieux et al., 1988; Jacquier-Sarlin et al., 1995; Rey-Millet et al., 1994). Covalent C3b modification targets antigen to specific MHC class II containing vesicles (Perrin-Cocon et al., 2004) and may increase lysosomal peptide-MHC stability (Serra et al., 1997), and the diversity of epitopes presented (Cretin et al., 2007). Alternatively, CR3 engagement by iC3b has also been suggested to suppress the T cell stimulatory capacity of some antigen presenting dendritic cells (Behrens et al., 2007; Luo et al., 2005). This effect is presumably due to up-regulation of IL-10 and down-regulation of IL-12 (Luo et al., 2005; Marth and Kelsall, 1997; Yoshida et al., 1998).

CRIg is a recently identified type I transmembrane receptor that contains a V-type immunoglobulin-like domain with homology to the B7-family of T cell costimulatory ligands (Helmy et al., 2006; Langnaese et al., 2000; Vogt et al., 2006). CRIg is expressed on a subset of tissue resident macrophages but is not found on circulating myeloid lineage cells (Helmy et al., 2006; Lee et al., 2006; Vogt et al., 2006). CRIg binds C3b and iC3b on the β-chain (Wiesmann et al., 2006) and localizes to recycling endosomes after internalization (Helmy et al., 2006); thus, CRIg may prevent local inflammation by constitutive removal of C3-opsonized targets. Because administration of exogenous soluble CRIg inhibits T-cell proliferation in vitro and in vivo (Vogt et al., 2006), CRIg is believed to negatively regulate T cell responses by as yet unidentified mechanisms.

2.3 Regulators of Complement Activation

Due to the potential for self-recognition and tissue damage, complement activation is tightly regulated. Uncontrolled complement activation has been implicated in rheumatoid arthritis, systemic lupus erythematosus, other autoimmune diseases (Colten and Rosen, 1992; Molina, 2002; Walport, 2001). To limit complement deposition, most membrane-bound and fluid phase complement regulatory proteins contain varying numbers of complement control protein (CCP) motifs (Hourcade et al., 1989). These globular repeating units of ~60 amino acids (also known as short consensus repeats) are composed of two disulfide bridges and a hydrophobic core (Barlow et al., 1992, 1993). CCP modules occur widely in many other proteins and mediate protein–protein and protein–carbohydrate interactions (Kirkitadze and Barlow, 2001).

Complement activation is regulated by four distinct mechanisms: (1) inhibition of collectin-associated proteases; (2) proteolytic inactivation of deposited C3b and C4b; (3) disassembly or decay of the C3 and

C5 convertases; and (4) prevention of MAC formation. C1-inhibitior is an acute phase serpin protein that inactivates C1r, C1s, MASP-1, and MASP-2 proteases by irreversibly binding to the active site of these proteases (Bos et al., 2002). Proteolytic inactivation of C3b and C4b is mediated by serum factor I and requires CCP-containing co-factors, such as membrane co-factor protein (MCP or CD46) and CR1 (CD35) (Krych-Goldberg and Atkinson, 2001; Seya and Atkinson, 1989). In contrast, glycosyl-phosphatidylinositol (GPI)-anchored decay accelerating factor (CD55) promotes the decay of classical and alternative C3 and C5 convertases (Fujita et al., 1987; Medof et al., 1984; Nicholson-Weller et al., 1982). In solution, factor H and C4 binding protein (C4BP) mediate both functions; accelerating the decay of the alternative (Kazatchkine et al., 1979; Weiler et al., 1976) and classical pathway (Scharfstein et al., 1978) C3 convertases respectively, and catalyzing factor I mediated enzymatic cleavage of C3b (Fujita et al., 1978; Whaley and Ruddy, 1976). Finally, C9 recruitment to the MAC and terminal complex lytic pore formation is regulated by the cell surface expression of GPI-anchored CD59 (Davies and Lachmann, 1993; Meri et al., 1990). CRIg also prevents MAC formation by inhibiting C5 association with the alternative pathway C5 convertase, C3bBb3b (Wiesmann et al., 2006). One should note that important functional differences in the distribution of membrane complement regulators between mice and humans exist: in mice, expression of CD46 is restricted to the testis (Tsujimura et al., 1998). Instead, Crry, a functional homolog of CD46 and CD55 that is absent in humans, is ubiquitously expressed (Kim et al., 1995; Molina, 2002).

3 Complement Regulation of Flavivirus Infection

Spontaneous or antibody-mediated complement activation can promote virus inactivation and elimination through several possible mechanisms (reviewed in Blue et al., 2004; Spear et al., 2001). Complement activation stimulates virus opsonization with the collectins C1q and MBL, as well as C3b, C4b, and their split products. Complement opsonization may prevent virus-receptor interaction and host cell entry, inhibit virus uncoating, and promote virus aggregation. Complement deposition on the surface of enveloped virions can also promote recognition by complement receptors, leading to enhanced B cell presentation, as well as phagocytosis. Finally, antibody-stimulated MAC formation can induce virolysis and/or lysis of virus-infected cells.

3.1 Flavivirus Infection Activates Complement

Flaviviruses directly trigger complement activation in vitro and in vivo. Increasing concentrations of baby rabbit complement or fresh mouse serum neutralized as much as 60% of a given dose of WNV in the absence of antibody (Mehlhop et al., 2005). Complement activation by Flaviviruses has also been described in vivo. C3 catabolism and production of C5a and sC5–9 during secondary DENV infection correlated with increased disease severity and development of dengue hemorrhagic fever (Avirutnan et al., 2006; Bokisch et al., 1973). Similarly, C3 and C4 consumption was observed in a mouse model of WNV infection (Mehlhop and Diamond, 2006) prior to the induction of a specific antibody response.

3.2 Complement Augments Antibody-Mediated Neutralization of Flaviviruses

Complement activation can augment antibody-mediated neutralization of several viruses, including influenza (Beebe et al., 1983; Mozdzanowska et al., 2006), HIV (Aasa-Chapman et al., 2005; Posner et al., 1992; Spear et al., 1993; Verity et al., 2006), respiratory syncytial (Baughman et al., 1968; Yoder et al., 2004), Varicella zoster (Beebe and Cooper, 1981; Grose et al., 1979; Schmidt and Lennette, 1975), Epstein–Barr (Nemerow et al., 1982; Sairenji et al., 1984), and herpes simplex (HSV) viruses (Lerner et al., 1974; Snyder et al., 1981; Wallis and Melnick, 1971). Only the first four components of the classical pathway are necessary to increase antibody-mediated neutralization of Newcastle disease virus (Linscott and Levinson, 1969), hepatitis C virus (Meyer et al., 2002), HIV (Sullivan et al., 1998), and HSV (Daniels et al., 1969, 1970). Recent studies indicate that complement-enhanced neutralization of influenza virus depends solely on C1q and correlates with IgG-subclasses that bind C1q efficiently (Feng et al., 2002; Mozdzanowska et al., 2006). By contrast, in vitro, complement-dependent virolysis appears to explain the enhanced antibody-dependent neutralization of HIV and parainfluenza virus (Spear et al., 1993; Vasantha et al., 1988). However, studies with human IgG1 antibodies against HIV suggest that complement effector functions may not be critical for augmenting antibody-mediated neutralization in vivo (Hessell et al., 2007).

Complement improves antibody-mediated neutralization of Flaviviruses. Complement augments immune serum-mediated neutralization of YFV, DENV, and Kunjin virus (Della-Porta and Westaway, 1977; Sabin, 1950; Spector and Tauraso, 1969) and monoclonal antibody-dependent neutralization of WNV (Mehlhop et al., 2005). Similarly, the protective efficacy of Flavivirus neutralizing antibodies in vivo correlates with IgG subclasses that efficiently fix complement (Schlesinger and Chapman, 1995). An IgG_{2a} subclass-switch variant of a YFV neutralizing IgG_1 mAb provided better protection against lethal challenge with the 17D YFV strain (Schlesinger and Chapman, 1995). However, while IgG_{2a} antibodies bind C1q more robustly and initiate classical pathway complement activation (Klaus et al., 1979), this isotype also has greater affinity for mouse Fc-γ receptor I (Fc-γRI, CD64) (Hulett and Hogarth, 1994; Ravetch and Kinet, 1991). Definitive evidence of the protective function of complement in vivo awaits protection studies with IgG subclass switch variants in complement and Fc-γR-deficient mice.

3.3 C1q Restricts Antibody-Dependent Enhancement of Flaviviruses

Fc-γR engagement by antibodies in vitro can paradoxically enhance replication of Flaviviruses (Gollins and Porterfield, 1984, 1985; Halstead and O'Rourke, 1977; Kliks, 1990; Kliks et al., 1989; Peiris et al., 1981; Peiris and Porterfield, 1979). At concentrations that do not reach the threshold necessary for neutralization, anti-Flavivirus antibodies enhance infection in cells expressing activating Fc-γR (Pierson et al., 2007). This phenomenon, also known as antibody-dependent enhancement of infection (ADE) is hypothesized to contribute to the pathogenesis of secondary DENV infection (Halstead, 2003), and possibly, the adverse effects following challenge of individuals immunized with some formalin-inactivated viral vaccines (Iankov et al., 2006; Ponnuraj et al., 2003; Porter et al., 1972; Prabhakar and Nathanson, 1981). Recent studies indicate that complement can restrict ADE by specific IgG against Flaviviruses. Complement limited ADE of WNV and DENV infection in Fc-γRIIA$^+$ (CD32A) cell lines and primary macrophages (Mehlhop et al., 2007; Yamanaka, 2007). Experiments with mouse sera deficient in individual complement components indicate that C1q is necessary and sufficient to restrict ADE of WNV infection in vitro. This effect was IgG subclass-dependent, as C1q restricted ADE by a human IgG_3 isotype-switch variant, but had little effect on $hIgG_2$ and $hIgG_4$ subclass variants (Mehlhop et al., 2007). These results correlate directly with the known affinity of human IgG subclasses for C1q (Bindon et al., 1988; Cooper, 1985). Interestingly, complement-dependent inhibition of DENV ADE may also require C3 activity

(Yamanaka, 2007). While these studies establish that complement restricts ADE by Flaviviruses in vitro, the precise inhibitory mechanisms at the cellular level remain unclear.

C1q may also limit Flavivirus ADE in vivo. Whereas enhancement of WNV infection was not observed after passive transfer of antiviral IgG_{2a} mAbs that bind C1q avidly in wild type mice, it was readily induced in $C1q^{-/-}$ mice (Mehlhop et al., 2007). This effect was epitope-dependent and IgG subclass-restricted. The ability of C1q to suppress ADE may explain some of the difficulties in consistently observing Flavivirus ADE in other animal models. Further investigation is necessary to define the links between complement restriction of ADE, Fc-γR specificity, and disease pathogenesis of Flaviviruses.

3.4 Complement C3 Enhances Flavivirus Infection via CR3

In cells that express CR3, antibody-dependent complement activation may enhance viral infection. In the presence of immune antibody, human complement enhanced HIV infection of $CR3^+$ dendritic cells approximately tenfold (Bajtay et al., 2004). Analogously, complement activation by antiviral IgM enhanced WNV infection of primary macrophages and monocytic cell lines (Cardosa et al., 1983, 1986). In these particular experiments, productive infection was not increased if a complement source was omitted. Consistent with this, blockade of Fc-γR had little inhibitory effect, whereas blockade of CR3 abrogated the complement and antibody-induced enhancement of WNV infection. Antibody-dependent complement-mediated enhancement of viral infection is restricted by antibody isotype, as complement did not enhance WNV infection of mouse macrophages in the presence of either antiviral IgG_1 or IgG_{2b} antibody (Mehlhop et al., 2007). Thus, under certain circumstances, antibody and complement-dependent opsonization of Flaviviruses by C3 split products increases infection of $CR3^+$ cells in an isotype-specific manner.

3.5 Complement Stimulates Adaptive Immunity to Flavivirus Infections

Complement limits viral dissemination indirectly by stimulating the adaptive immune responses to viruses. C3-deficient mice develop reduced antiviral antibody titers and increased viral burden following either influenza and vesicular stomatitis virus infection (Kopf et al., 2002; Ochsenbein et al., 1999). Similarly, mice lacking myeloid-derived

C3 or C4 showed impaired primary and secondary IgG responses to HSV-1 (Da Costa et al., 1999; Gadjeva et al., 2002; Verschoor et al., 2001, 2003). A deficiency in CR2 also restricts secondary IgG responses against HSV-1 (Da Costa et al., 1999), suggesting that C3d ligation of CR2 is essential for the establishment and/or maintenance of antiviral memory B cells. Complement activation is also required for optimal antiviral T cell function. Expansion of influenza and lymphocytic choriomeningitis virus (LCMV)-specific cytolytic $CD8^+$ T cells requires C3, but not CR2 (Kopf et al., 2002; Suresh et al., 2003), suggesting that complement activation triggers antiviral T cell function independent of its role in humoral responses.

The priming functions of complement are also critical for protection against WNV infection. C3-deficient mice are significantly more susceptible to lethal WNV infection with greater viral burden and reduced antiviral antibody titers (Mehlhop et al., 2005). An analogous phenotype was observed in CR2-deficicent mice (Mehlhop et al., 2005) suggesting that C3 protects against lethal WNV disease by stimulating antiviral humoral immunity. Subsequent infection studies of mice lacking C1q, C4, factor D, or factor B indicate that all pathways of complement activation orchestrate protection against WNV infection (Mehlhop and Diamond, 2006). However, each complement activation pathway appears to exert independent protective effects in response to WNV infection. Humoral IgM responses to WNV probably depend upon activation of C4 by the lectin recognition pathway. In the absence of C1q and classical pathway activation, normal IgM responses to WNV were observed (Mehlhop and Diamond, 2006). However, WNV-specific IgM titers were significantly reduced in the absence of C4, suggesting lectin pathway involvement in IgM induction. Normal WNV-specific IgM and IgG responses were observed in factor $B^{-/-}$ mice, consistent with studies observing no defect in B cell priming in the absence of alternative pathway activation (Matsumoto et al., 1997). In contrast, the lectin and alternative pathways appear necessary for efficient T cell priming as $C4^{-/-}$, factor $B^{-/-}$, and factor $D^{-/-}$ mice exhibited reduced WNV-specific $CD8^+$ T cell responses (Mehlhop and Diamond, 2006). The T cell defects in $C4^{-/-}$ mice may be indirect as depressed IgM responses could affect viral opsonization and antigen presentation. Interestingly, no defect in adaptive immune priming was observed in the absence of classical pathway activity. As $C1q^{-/-}$ mice still exhibited enhanced susceptibility to WNV infection, C1q-mediated antibody effector functions probably control WNV infection in vivo.

4 Immune Evasion of Complement by Flaviviruses

As discussed above, several complement regulatory proteins limit pathological deposition of complement on host cells. These complement regulators promote cleavage of C3b and C4b, dismantle C3 and C5 convertases, and inhibit membrane attack complex formation. Several viruses have co-opted these complement regulatory functions to limit complement-dependent targeting and inflammation (reviewed in Finlay and McFadden, 2006; Tortorella et al., 2000). For example, during budding, HIV, human T cell leukemia virus, and human cytomegalovirus incorporate CD46, CD55 and CD59 onto their surface, thus, limiting complement recognition or attack (Saifuddin et al., 1995; Spear et al., 1995). In contrast, some large DNA viruses have "stolen" and modified host genes with complement regulatory activity. Vaccinia, cowpox, and variola viruses encode a secreted C4BP homolog that has co-factor and C3 convertase decay accelerating activity (Kotwal et al., 1990; McKenzie et al., 1992; Miller et al., 1997; Sahu et al., 1998; Shchelkunov, 1995). Herpesvirus family members encode cell-surface associated complement regulators that perform analogous functions (Fodor et al., 1995; Friedman et al., 1996; Kapadia et al., 2002; Kostavasili et al., 1997; Neipel et al., 1997). Finally, HIV encodes glycoproteins that recruit factor H to protect from complement-dependent lysis (Pinter et al., 1995; Stoiber et al., 1995, 1996). Thus, viruses from several families have developed independent means of attenuating complement activation to enhance viral infectivity.

Recent evidence suggests that WNV NS1 has immune evasion function and protects against complement activation by binding the negative regulator factor H (Chung et al., 2007). Factor H is an abundant soluble protein that sustains factor I-mediated cleavage of C3b and inactivates the alternative pathway C3 convertase (reviewed in Zipfel et al., 2002). Co-immunoprecipitation experiments demonstrate that soluble NS1 binds to factor H in human serum or in its purified form and degrades C3b in solution (Chung et al., 2007). Additionally, cell surface NS1 limits C3b deposition and C5b-9 MAC formation. Thus, extracellular or cell surface NS1 may minimize immune system targeting of WNV by decreasing complement activation in solution and on the surface of infected cells. These data appear to contradict earlier studies which suggested that DENV NS1 may activate complement (Brandt et al., 1970) and studies indicating that NS1 levels in serum correlate with complement consumption and severe dengue disease (Alcon et al., 2002; Avirutnan et al., 2006; Brandt et al., 1970; Libraty et al., 2002; Young et al., 2000). However, subsequent studies indicate that DENV NS1 by itself does not activate complement (P. Avirutnan, J.P. Atkinson, and M.S. Diamond, unpublished results). Future studies are necessary to determine if factor H or other complement components interact similarly with NS1 from other Flaviviruses.

5 Concluding Remarks

Recent studies have established that activation of the complement system has a critical role in protection against WNV. Following infection, complement activation primes adaptive immune responses and modulates the effector functions of Flavivirus-specific antibodies in an IgG subclass-specific manner. As a response, Flaviviruses have evolved novel strategies to limit complement activation. It is likely that the balance between complement activation and attenuation plays a crucial role in determining the outcome of a productive infection. A more detailed molecular and structural analysis of complement interactions with Flaviviruses may foster the development of novel therapeutic strategies to control infection.

References

Aasa-Chapman, M. M., Holuigue, S., Aubin, K., Wong, M., Jones, N. A., Cornforth, D., Pellegrino, P., Newton, P., Williams, I., Borrow, P., and McKnight, A. (2005). Detection of antibody-dependent complement-mediated inactivation of both autologous and heterologous virus in primary human immunodeficiency virus type 1 infection. J Virol *79*, 2823–2830.

Ahearn, J. M., Fischer, M. B., Croix, D., Goerg, S., Ma, M., Xia, J., Zhou, X., Howard, R. G., Rothstein, T. L., and Carroll, M. C. (1996). Disruption of the Cr2 locus results in a reduction in B-1a cells and in an impaired B cell response to T-dependent antigen. Immunity *4*, 251–262.

Alcon, S., Talarmin, A., Debruyne, M., Falconar, A., Deubel, V., and Flamand, M. (2002). Enzyme-linked immunosorbent assay specific to Dengue virus type 1 nonstructural protein NS1 reveals circulation of the antigen in the blood during the acute phase of disease in patients experiencing primary or secondary infections. J Clin Microbiol *40*, 376–381.

Arvieux, J., Yssel, H., and Colomb, M. G. (1988). Antigen-bound C3b and C4b enhance antigen-presenting cell function in activation of human T-cell clones. Immunology *65*, 229–235.

Avirutnan, P., Punydee, N., Noisakran, S., Komoltri, C., Thiemmeca, S., Auethavornanan, K., Jairungsri, A., Kanlaya, R., Tangthawornchaikul, N., Puttikhunt, C., et-al.. (2006). Vascular leakage in severe dengue virus infections: a potential role for the nonstructural viral protein NS1 and complement. J Infect Dis *193*, 1078–1088.

Bajtay, Z., Speth, C., Erdei, A., and Dierich, M. P. (2004). Cutting edge: productive HIV-1 infection of dendritic cells via complement receptor type 3 (CR3, CD11b/CD18). J Immunol *173*, 4775–4778.

Banda, N. K., Thurman, J. M., Kraus, D., Wood, A., Carroll, M. C., Arend, W. P., and Holers, V. M. (2006). Alternative complement pathway activation is essential for inflammation and joint destruction in the passive transfer model of collagen-induced arthritis. J Immunol *177*, 1904–1912.

Banda, N. K., Takahashi, K., Wood, A. K., Holers, V. M., and Arend, W. P. (2007). Pathogenic complement activation in collagen antibody-induced arthritis in mice requires amplification by the alternative pathway. J Immunol *179*, 4101–4109.

Barlow, P. N., Norman, D. G., Steinkasserer, A., Horne, T. J., Pearce, J., Driscoll, P. C., Sim, R. B., and Campbell, I. D. (1992). Solution structure of the fifth repeat of factor H: a second example of the complement control protein module. Biochemistry *31*, 3626–3634.

Barlow, P. N., Steinkasserer, A., Norman, D. G., Kieffer, B., Wiles, A. P., Sim, R. B., and Campbell, I. D. (1993). Solution structure of a pair of complement modules by nuclear magnetic resonance. J Mol Biol *232*, 268–284.

Barrington, R. A., Pozdnyakova, O., Zafari, M. R., Benjamin, C. D., and Carroll, M. C. (2002). B lymphocyte memory: role of stromal cell complement and FcgammaRIIB receptors. J Exp Med *196*, 1189–1199.

Baughman, R. H., Fenters, J. D., Marquis, G. S., Jr., and Holper, J. C. (1968). Effect of complement and viral filtration on the neutralization of respiratory syncytial virus. Appl Microbiol *16*, 1076–1080.

Beebe, D. P., and Cooper, N. R. (1981). Neutralization of vesicular stomatitis virus (VSV) by human complement requires a natural IgM antibody present in human serum. J Immunol *126*, 1562–1568.

Beebe, D. P., Schreiber, R. D., and Cooper, N. R. (1983). Neutralization of influenza virus by normal human sera: mechanisms involving antibody and complement. J Immunol *130*, 1317–1322.

Behrens, E. M., Sriram, U., Shivers, D. K., Gallucci, M., Ma, Z., Finkel, T. H., and Gallucci, S. (2007). Complement receptor 3 ligation of dendritic cells suppresses their stimulatory capacity. J Immunol *178*, 6268–6279.

Bindon, C. I., Hale, G., Bruggemann, M., and Waldmann, H. (1988). Human monoclonal IgG isotypes differ in complement activating function at the level of C4 as well as C1q. J Exp Med *168*, 127–142.

Blue, C. E., Spiller, O. B., and Blackbourn, D. J. (2004). The relevance of complement to virus biology. Virology *319*, 176–184.

Bogers, W. M., Stad, R. K., Van Es, L. A., and Daha, M. R. (1992). Both Kupffer cells and liver endothelial cells play an important role in the clearance of IgA and IgG immune complexes. Res Immunol *143*, 219–224.

Bokisch, V. A., Top, F. H., Jr., Russell, P. K., Dixon, F. J., and Muller-Eberhard, H. J. (1973). The potential pathogenic role of complement in dengue hemorrhagic shock syndrome. N Engl J Med *289*, 996–1000.

Bordet, J., and Gengou, O. (1901). Sur l'existence de substances sensibilisatrices dans la plupart des serum antimicrobiens. Ann Inst Pasteur *15*, 289–302.

Bos, I. G., Hack, C. E., and Abrahams, J. P. (2002). Structural and functional aspects of C1-inhibitor. Immunobiology *205*, 518–533.

Brandt, W. E., Chiewslip, D., Harris, D. L., and Russell, P. K. (1970). Partial purification and characterization of a dengue virus soluble complement-fixing antigen. J Immunol *105*, 1565–1568.

Brinton, M. A. (2002). The molecular biology of West Nile virus: a new invader of the western hemisphere. Annu Rev Microbiol *56*, 371–402.

Burke, D. S., and Monath, T. P. (2001). Flaviviruses. In Fields Virology, P. M. Howley, ed. (Philadelphia, Lippincott Williams &Wilkins), pp. 1043–1125.

Cardosa, M. J., Porterfield, J. S., and Gordon, S. (1983). Complement receptor mediates enhanced flavivirus replication in macrophages. J Exp Med *158*, 258–263.

Cardosa, M. J., Gordon, S., Hirsch, S., Springer, T. A., and Porterfield, J. S. (1986). Interaction of West Nile virus with primary murine macrophages: role of cell activation and receptors for antibody and complement. J Virol *57*, 952–959.

Carroll, M. C. (1998). The role of complement and complement receptors in induction and regulation of immunity. Annu Rev Immunol *16*, 545–568.

Carroll, M. C. (2004). The complement system in regulation of adaptive immunity. Nat Immunol *5*, 981–986.

Chen, Z., Koralov, S. B., Gendelman, M., Carroll, M. C., and Kelsoe, G. (2000). Humoral immune responses in *Cr2*$^{-/-}$ mice: enhanced affinity maturation but impaired antibody persistence. J Immunol *164*, 4522–4532.

Chung, K. M., Thompson, B. S., Fremont, D. H., and Diamond, M. S. (2007). Antibody recognition of cell surface-associated NS1 triggers Fc-gamma receptor-mediated phagocytosis and clearance of West Nile virus-infected cells. J Virol *81*, 9551–9555.

Colten, H. R., and Rosen, F. S. (1992). Complement deficiencies. Annu Rev Immunol *10*, 809–834.

Cooper, N. R. (1985). The classical complement pathway: activation and regulation of the first complement component. Adv Immunol *37*, 151–216.

Cooper, N. R., and Muller-Eberhard, H. J. (1970). The reaction mechanism of human C5 in immune hemolysis. J Exp Med *132*, 775–793.

Craig, M. L., Bankovich, A. J., McElhenny, J. L., and Taylor, R. P. (2000). Clearance of anti-double-stranded DNA antibodies: the natural immune complex clearance mechanism. Arthritis Rheum *43*, 2265–2275.

Cretin, F. C., Serra, V. A., Villiers, M. B., Laharie, A. M., Marche, P. N., and Gabert, F. M. (2007). C3b complexation diversifies naturally processed T cell epitopes. Mol Immunol *44*, 2893–2899.

Croix, D. A., Ahearn, J. M., Rosengard, A. M., Han, S., Kelsoe, G., Ma, M., and Carroll, M. C. (1996). Antibody response to a T-dependent antigen requires B cell expression of complement receptors. J Exp Med *183*, 1857–1864.

Da Costa, X. J., Brockman, M. A., Alicot, E., Ma, M., Fischer, M. B., Zhou, X., Knipe, D. M., and Carroll, M. C. (1999). Humoral response to herpes simplex virus is complement-dependent. Proc Natl Acad Sci U S A *96*, 12708–12712.

Daniels, C. A., Borsos, T., Rapp, H. J., Snyderman, R., and Notkins, A. L. (1969). Neutralization of sensitized virus by the fourth component of complement. Science *165*, 508–509.

Daniels, C. A., Borsos, T., Rapp, H. J., Snyderman, R., and Notkins, A. L. (1970). Neutralization of sensitized virus by purified components of complement. Proc Natl Acad Sci U S A *65*, 528–535.

Davies, A., and Lachmann, P. J. (1993). Membrane defence against complement lysis: the structure and biological properties of CD59. Immunol Res *12*, 258–275.

Della-Porta, A. J., and Westaway, E. G. (1977). Immune response in rabbits to virion and non-virion antigens of the Flavivirus kunjin. Infect Immun *15*, 874–882.

Dempsey, P. W., Allison, M. E., Akkaraju, S., Goodnow, C. C., and Fearon, D. T. (1996). C3d of complement as a molecular adjuvant: bridging innate and acquired immunity. Science *271*, 348–350.

DiScipio, R. G., Smith, C. A., Muller-Eberhard, H. J., and Hugli, T. E. (1983). The activation of human complement component C5 by a fluid phase C5 convertase. J Biol Chem *258*, 10629–10636.

Dishaw, L. J., Smith, S. L., and Bigger, C. H. (2005). Characterization of a C3-like cDNA in a coral: phylogenetic implications. Immunogenetics *57*, 535–548.

Dodds, A. W., and Matsushita, M. (2007). The phylogeny of the complement system and the origins of the classical pathway. Immunobiology *212*, 233–243.

Elshuber, S., Allison, S. L., Heinz, F. X., and Mandl, C. W. (2003). Cleavage of protein prM is necessary for infection of BHK-21 cells by tick-borne encephalitis virus. J Gen Virol *84*, 183–191.

Endo, Y., Nakazawa, N., Liu, Y., Iwaki, D., Takahashi, M., Fujita, T., Nakata, M., and Matsushita, M. (2005). Carbohydrate-binding specificities of mouse ficolin A, a splicing variant of ficolin A and ficolin B and their complex formation with MASP-2 and sMAP. Immunogenetics *57*, 837–844.

Esser, A. F. (1994). The membrane attack complex of complement. Assembly, structure and cytotoxic activity. Toxicology *87*, 229–247.

Fang, Y., Xu, C., Fu, Y. X., Holers, V. M., and Molina, H. (1998). Expression of complement receptors 1 and 2 on follicular dendritic cells is necessary for the generation of a strong antigen-specific IgG response. J Immunol *160*, 5273–5279.

Farries, T. C., Lachmann, P. J., and Harrison, R. A. (1988). Analysis of the interactions between properdin, the third component of complement (C3), and its physiological activation products. Biochem J *252*, 47–54.

Fearon, D. T. (1980). Identification of the membrane glycoprotein that is the C3b receptor of the human erythrocyte, polymorphonuclear leukocyte, B lymphocyte, and monocyte. J Exp Med *152*, 20–30.

Fearon, D. T., and Carroll, M. C. (2000). Regulation of B lymphocyte responses to foreign and self-antigens by the CD19/CD21 complex. Annu Rev Immunol *18*, 393–422.

Feng, J. Q., Mozdzanowska, K., and Gerhard, W. (2002). Complement component C1q enhances the biological activity of influenza virus hemagglutinin-specific antibodies depending on their fine antigen specificity and heavy-chain isotype. J Virol *76*, 1369–1378.

Finlay, B. B., and McFadden, G. (2006). Anti-immunology: evasion of the host immune system by bacterial and viral pathogens. Cell *124*, 767–782.

Fischer, M. B., Ma, M., Goerg, S., Zhou, X., Xia, J., Finco, O., Han, S., Kelsoe, G., Howard, R. G., Rothstein, T. L., et-al. (1996). Regulation of the B cell response to T-dependent antigens by classical pathway complement. J Immunol *157*, 549–556.

Fischer, M. B., Goerg, S., Shen, L., Prodeus, A. P., Goodnow, C. C., Kelsoe, G., and Carroll, M. C. (1998). Dependence of germinal center B cells on expression of CD21/CD35 for survival. Science *280*, 582–585.

Fodor, W. L., Rollins, S. A., Bianco-Caron, S., Rother, R. P., Guilmette, E. R., Burton, W. V., Albrecht, J. C., Fleckenstein, B., and Squinto, S. P. (1995). The complement control protein homolog of herpesvirus saimiri regulates serum complement by inhibiting C3 convertase activity. J Virol *69*, 3889–3892.

Friedman, H. M., Wang, L., Fishman, N. O., Lambris, J. D., Eisenberg, R. J., Cohen, G. H., and Lubinski, J. (1996). Immune evasion properties of herpes simplex virus type 1 glycoprotein gC. J Virol *70*, 4253–4260.

Fujita, T., Gigli, I., and Nussenzweig, V. (1978). Human C4-binding protein. II. Role in proteolysis of C4b by C3b-inactivator. J Exp Med *148*, 1044–1051.

Fujita, T., Inoue, T., Ogawa, K., Iida, K., and Tamura, N. (1987). The mechanism of action of decay-accelerating factor (DAF). DAF inhibits the assembly of C3 convertases by dissociating C2a and Bb. J Exp Med *166*, 1221–1228.

Gadjeva, M., Verschoor, A., Brockman, M. A., Jezak, H., Shen, L. M., Knipe, D. M., and Carroll, M. C. (2002). Macrophage-derived complement component C4 can restore humoral immunity in C4-deficient mice. J Immunol *169*, 5489–5495.

Gal, P., Barna, L., Kocsis, A., and Zavodszky, P. (2007). Serine proteases of the classical and lectin pathways: similarities and differences. Immunobiology *212*, 267–277.

Ghiran, I., Barbashov, S. F., Klickstein, L. B., Tas, S. W., Jensenius, J. C., and Nicholson-Weller, A. (2000). Complement receptor 1/CD35 is a receptor for mannan-binding lectin. J Exp Med *192*, 1797–1808.

Gollins, S. W., and Porterfield, J. S. (1984). Flavivirus infection enhancement in macrophages: radioactive and biological studies on the effect of antibody on viral fate. J Gen Virol *65* (Pt 8), 1261–1272.

Gollins, S. W., and Porterfield, J. S. (1985). Flavivirus infection enhancement in macrophages: an electron microscopic study of viral cellular entry. J Gen Virol *66* (Pt 9), 1969–1982.

Green, T. D., Montefiori, D. C., and Ross, T. M. (2003). Enhancement of antibodies to the human immunodeficiency virus type 1 envelope by using the molecular adjuvant C3d. J Virol *77*, 2046–2055.

Grose, C., Edmond, B. J., and Brunell, P. A. (1979). Complement-enhanced neutralizing antibody response to varicella-zoster virus. J Infect Dis *139*, 432–437.

Guirakhoo, F., Bolin, R. A., and Roehrig, J. T. (1992). The Murray Valley encephalitis virus prM protein confers acid resistance to virus particles and alters the expression of epitopes within the R2 domain of E glycoprotein. Virology *191*, 921–931.

Halstead, S. B. (2003). Neutralization and antibody-dependent enhancement of dengue viruses. Adv Virus Res *60*, 421–467.

Halstead, S. B., and O'Rourke, E. J. (1977). Antibody-enhanced dengue virus infection in primate leukocytes. Nature *265*, 739–741.

Helmy, K. Y., Katschke, K. J., Jr., Gorgani, N. N., Kljavin, N. M., Elliott, J. M., Diehl, L., Scales, S. J., Ghilardi, N., and van Lookeren Campagne, M. (2006). CRIg: a macrophage complement receptor required for phagocytosis of circulating pathogens. Cell *124*, 915–927.

Hessell, A. J., Hangartner, L., Hunter, M., Havenith, C. E., Beurskens, F. J., Bakker, J. M., Lanigan, C. M., Landucci, G., Forthal, D. N., Parren, P. W., et-al. (2007). Fc receptor but not complement binding is important in antibody protection against HIV. Nature *449*, 101–104.

Hourcade, D., Holers, V. M., and Atkinson, J. P. (1989). The regulators of complement activation (RCA) gene cluster. Adv Immunol *45*, 381–416.

Hulett, M. D., and Hogarth, P. M. (1994). Molecular basis of Fc receptor function. Adv Immunol *57*, 1–127.

Iankov, I. D., Pandey, M., Harvey, M., Griesmann, G. E., Federspiel, M. J., and Russell, S. J. (2006). Immunoglobulin g antibody-mediated enhancement of measles virus infection can bypass the protective antiviral immune response. J Virol *80*, 8530–8540.

Isenman, D. E., Kells, D. I., Cooper, N. R., Muller-Eberhard, H. J., and Pangburn, M. K. (1981). Nucleophilic modification of human complement protein C3: correlation of conformational changes with acquisition of C3b-like functional properties. Biochemistry *20*, 4458–4467.

Jacquier-Sarlin, M. R., Gabert, F. M., Villiers, M. B., and Colomb, M. G. (1995). Modulation of antigen processing and presentation by covalently linked complement C3b fragment. Immunology *84*, 164–170.

Janeway, C. A. (2001). Innate immunity: the complement system and innate immunity. In Immunobiology: The Immune System in Health and Disease, C. A. Janeway, P. Travers, M. J. Walport, and M. Schlomchik, eds. (New York, Garland Science), pp. xxiii, 823.

Kang, Y. S., Do, Y., Lee, H. K., Park, S. H., Cheong, C., Lynch, R. M., Loeffler, J. M., Steinman, R. M., and Park, C. G. (2006). A dominant complement fixation pathway for pneumococcal polysaccharides initiated by SIGN-R1 interacting with C1q. Cell *125*, 47–58.

Kapadia, S. B., Levine, B., Speck, S. H., and Virgin, H. W. T. (2002). Critical role of complement and viral evasion of complement in acute, persistent, and latent gamma-herpesvirus infection. Immunity *17*, 143–155.

Kazatchkine, M. D., Fearon, D. T., and Austen, K. F. (1979). Human alternative complement pathway: membrane-associated sialic acid regulates the competition between B and beta1 H for cell-bound C3b. J Immunol *122*, 75–81.

Kim, Y. U., Kinoshita, T., Molina, H., Hourcade, D., Seya, T., Wagner, L. M., and Holers, V. M. (1995). Mouse complement regulatory protein Crry/p65 uses the specific mechanisms of both human decay-accelerating factor and membrane cofactor protein. J Exp Med *181*, 151–159.

Kinoshita, T., Takeda, J., Hong, K., Kozono, H., Sakai, H., and Inoue, K. (1988). Monoclonal antibodies to mouse complement receptor type 1 (CR1). Their use in a distribution study showing that mouse erythrocytes and platelets are CR1-negative. J Immunol *140*, 3066–3072.

Kirkitadze, M. D., and Barlow, P. N. (2001). Structure and flexibility of the multiple domain proteins that regulate complement activation. Immunol Rev *180*, 146–161.

Kishore, U., and Reid, K. B. (2000). C1q: structure, function, and receptors. Immunopharmacology *49*, 159–170.

Kishore, U., Gaboriaud, C., Waters, P., Shrive, A. K., Greenhough, T. J., Reid, K. B., Sim, R. B., and Arlaud, G. J. (2004). C1q and tumor necrosis factor superfamily: modularity and versatility. Trends Immunol *25*, 551–561.

Klaus, G. G., Pepys, M. B., Kitajima, K., and Askonas, B. A. (1979). Activation of mouse complement by different classes of mouse antibody. Immunology *38*, 687–695.

Klickstein, L. B., Barbashov, S. F., Liu, T., Jack, R. M., and Nicholson-Weller, A. (1997). Complement receptor type 1 (CR1, CD35) is a receptor for C1q. Immunity *7*, 345–355.

Kliks, S. (1990). Antibody-enhanced infection of monocytes as the pathogenetic mechanism for severe dengue illness. AIDS Res Hum Retroviruses *6*, 993–998.

Kliks, S. C., Nisalak, A., Brandt, W. E., Wahl, L., and Burke, D. S. (1989). Antibody-dependent enhancement of dengue virus growth in human monocytes as a risk factor for dengue hemorrhagic fever. Am J Trop Med Hyg *40*, 444–451.

Kohl, J. (2001). Anaphylatoxins and infectious and non-infectious inflammatory diseases. Mol Immunol *38*, 175–187.

Kopf, M., Abel, B., Gallimore, A., Carroll, M., and Bachmann, M. F. (2002). Complement component C3 promotes T-cell priming and lung migration to control acute influenza virus infection. Nat Med *8*, 373–378.

Kostavasili, I., Sahu, A., Friedman, H. M., Eisenberg, R. J., Cohen, G. H., and Lambris, J. D. (1997). Mechanism of complement inactivation by glycoprotein C of herpes simplex virus. J Immunol *158*, 1763–1771.

Kotwal, G. J., Isaacs, S. N., McKenzie, R., Frank, M. M., and Moss, B. (1990). Inhibition of the complement cascade by the major secretory protein of vaccinia virus. Science *250*, 827–830.

Krych-Goldberg, M., and Atkinson, J. P. (2001). Structure-function relationships of complement receptor type 1. Immunol Rev *180*, 112–122.

Kurtz, C. B., O'Toole, E., Christensen, S. M., and Weis, J. H. (1990). The murine complement receptor gene family. IV. Alternative splicing of Cr2 gene transcripts predicts two distinct gene products that share homologous domains with both human CR2 and CR1. J Immunol *144*, 3581–3591.

Langnaese, K., Colleaux, L., Kloos, D. U., Fontes, M., and Wieacker, P. (2000). Cloning of Z39Ig, a novel gene with immunoglobulin-like domains located on human chromosome X. Biochim Biophys Acta *1492*, 522–525.

Law, S. K., and Dodds, A. W. (1997). The internal thioester and the covalent binding properties of the complement proteins C3 and C4. Protein Sci *6*, 263–274.

Lee, M. Y., Kim, W. J., Kang, Y. J., Jung, Y. M., Kang, Y. M., Suk, K., Park, J. E., Choi, E. M., Choi, B. K., Kwon, B. S., and Lee, W. H. (2006). Z39Ig is expressed on macrophages and may mediate inflammatory reactions in arthritis and atherosclerosis. J Leukoc Biol *80*, 922–928.

Lerner, A. M., Shippey, M. J., and Crane, L. R. (1974). Serologic responses to herpes simplex virus in rabbits: complement-requiring neutralizing, conventional neutralizing, and passive hemagglutinating antibodies. J Infect Dis *129*, 623–636.

Libraty, D. H., Young, P. R., Pickering, D., Endy, T. P., Kalayanarooj, S., Green, S., Vaughn, D. W., Nisalak, A., Ennis, F. A., and Rothman, A. L. (2002). High circulating levels of the dengue virus nonstructural protein NS1 early in dengue illness correlate with the development of dengue hemorrhagic fever. J Infect Dis *186*, 1165–1168.

Lindenbach, B. D., and Rice, C. M. (2006). Flaviviridae: the viruses and their replication. In Field's Virology, D. M. Knipe, and P. M. Howley, eds. (Philadelphia, Lippincott Williams & Wilkins), pp. 991–1041.

Linscott, W. D., and Levinson, W. E. (1969). Complement components required for virus neutralization by early immunoglobulin antibody. Proc Natl Acad Sci U S A *64*, 520–527.

Liu, Y., Endo, Y., Iwaki, D., Nakata, M., Matsushita, M., Wada, I., Inoue, K., Munakata, M., and Fujita, T. (2005). Human M-ficolin is a secretory protein that activates the lectin complement pathway. J Immunol *175*, 3150–3156.

Luo, X., Liu, L., Tang, N., Lu, K. Q., McCormick, T. S., Kang, K., and Cooper, K. D. (2005). Inhibition of monocyte-derived dendritic cell differentiation and interleukin-12 production by complement iC3b via a mitogen-activated protein kinase signalling pathway. Exp Dermatol *14*, 303–310.

Lutz, H. U., and Jelezarova, E. (2006). Complement amplification revisited. Mol Immunol *43*, 2–12.

Manderson, A. P., Pickering, M. C., Botto, M., Walport, M. J., and Parish, C. R. (2001). Continual low-level activation of the classical complement pathway. J Exp Med *194*, 747–756.

Marth, T., and Kelsall, B. L. (1997). Regulation of interleukin-12 by complement receptor 3 signaling. J Exp Med *185*, 1987–1995.

Matsumoto, A. K., Kopicky-Burd, J., Carter, R. H., Tuveson, D. A., Tedder, T. F., and Fearon, D. T. (1991). Intersection of the complement and immune systems: a signal transduction complex of the B lymphocyte-containing complement receptor type 2 and CD19. J Exp Med *173*, 55–64.

Matsumoto, M., Fukuda, W., Circolo, A., Goellner, J., Strauss-Schoenberger, J., Wang, X., Fujita, S., Hidvegi, T., Chaplin, D. D., and Colten, H. R. (1997). Abrogation of the alternative complement pathway by targeted deletion of murine factor B. Proc Natl Acad Sci U S A *94*, 8720–8725.

Matsushita, M., Endo, Y., and Fujita, T. (2000). Cutting edge: complement-activating complex of ficolin and mannose-binding lectin-associated serine protease. J Immunol *164*, 2281–2284.

Matsushita, M., Kuraya, M., Hamasaki, N., Tsujimura, M., Shiraki, H., and Fujita, T. (2002). Activation of the lectin complement pathway by H-ficolin (Hakata antigen). J Immunol *168*, 3502–3506.

May, J. E., and Frank, M. M. (1973). A new complement-mediated cytolytic mechanism – the C1-bypass activation pathway. Proc Natl Acad Sci U S A *70*, 649–652.

McKenzie, R., Kotwal, G. J., Moss, B., Hammer, C. H., and Frank, M. M. (1992). Regulation of complement activity by vaccinia virus complement-control protein. J Infect Dis *166*, 1245–1250.

Medof, M. E., Kinoshita, T., and Nussenzweig, V. (1984). Inhibition of complement activation on the surface of cells after incorporation of decay-accelerating factor (DAF) into their membranes. J Exp Med *160*, 1558–1578.

Mehlhop, E., and Diamond, M. S. (2006). Protective immune responses against West Nile virus are primed by distinct complement activation pathways. J Exp Med *203*, 1371–1381.

Mehlhop, E., Whitby, K., Oliphant, T., Marri, A., Engle, M., and Diamond, M. S. (2005). Complement activation is required for induction of a protective antibody response against West Nile virus infection. J Virol *79*, 7466–7477.

Mehlhop, E., Ansarah-Sobrinho, C., Johnson, S., Engle, M., Fremont, D. H., Pierson, T. C., and Diamond, M. S. (2007). C1q inhibits antibody-dependent enhancement of flavivirus infection in vitro and in vivo in an IgG subclass specific manner. Cell Host Microbe *2*, 417–426.

Meri, S., Morgan, B. P., Davies, A., Daniels, R. H., Olavesen, M. G., Waldmann, H., and Lachmann, P. J. (1990). Human protectin (CD59), an 18,000–20,000 MW complement lysis restricting factor, inhibits C5b-8 catalysed insertion of C9 into lipid bilayers. Immunology *71*, 1–9.

Meyer, K., Basu, A., Przysiecki, C. T., Lagging, L. M., Di Bisceglie, A. M., Conley, A. J., and Ray, R. (2002). Complement-mediated enhancement of antibody function for neutralization of pseudotype virus containing hepatitis C virus E2 chimeric glycoprotein. J Virol *76*, 2150–2158.

Miller, C. G., Shchelkunov, S. N., and Kotwal, G. J. (1997). The cowpox virus-encoded homolog of the vaccinia virus complement control protein is an inflammation modulatory protein. Virology *229*, 126–133.

Molina, H. (2002). The murine complement regulator Crry: new insights into the immunobiology of complement regulation. Cell Mol Life Sci *59*, 220–229.

Molina, H., Kinoshita, T., Inoue, K., Carel, J. C., and Holers, V. M. (1990). A molecular and immunochemical characterization of mouse CR2. Evidence for a single gene model of mouse complement receptors 1 and 2. J Immunol *145*, 2974–2983.

Molina, H., Holers, V. M., Li, B., Fung, Y., Mariathasan, S., Goellner, J., Strauss-Schoenberger, J., Karr, R. W., and Chaplin, D. D. (1996). Markedly impaired humoral immune response in mice deficient in complement receptors 1 and 2. Proc Natl Acad Sci U S A *93*, 3357–3361.

Morgan, B. P., and Gasque, P. (1997). Extrahepatic complement biosynthesis: where, when and why? Clin Exp Immunol *107*, 1–7.

Mozdzanowska, K., Feng, J., Eid, M., Zharikova, D., and Gerhard, W. (2006). Enhancement of neutralizing activity of influenza virus-specific antibodies by serum components. Virology *352*, 418–426.

Mukhopadhyay, S., Kuhn, R. J., and Rossmann, M. G. (2005). A structural perspective of the flavivirus life cycle. Nat Rev Microbiol *3*, 13–22.

Muller-Eberhard, H. J. (1968). Chemistry and reaction mechanisms of complement. Adv Immunol *8*, 1–80.

Muller-Eberhard, H. J. (1986). The membrane attack complex of complement. Annu Rev Immunol *4*, 503–528.

Neipel, F., Albrecht, J. C., and Fleckenstein, B. (1997). Cell-homologous genes in the Kaposi's sarcoma-associated rhadinovirus human herpesvirus 8: determinants of its pathogenicity? J Virol *71*, 4187–4192.

Nemerow, G. R., Jensen, F. C., and Cooper, N. R. (1982). Neutralization of Epstein–Barr virus by nonimmune human serum. Role of cross-reacting antibody to herpes simplex virus and complement. J Clin Invest *70*, 1081–1091.

Nicholson-Weller, A., Burge, J., Fearon, D. T., Weller, P. F., and Austen, K. F. (1982). Isolation of a human erythrocyte membrane glycoprotein with decay-accelerating activity for C3 convertases of the complement system. J Immunol *129*, 184–189.

Nicol, P. A., and Lachmann, P. J. (1973). The alternate pathway of complement activation. The role of C3 and its inactivator (KAF). Immunology *24*, 259–275.

Nonaka, M., and Kimura, A. (2006). Genomic view of the evolution of the complement system. Immunogenetics *58*, 701–713.

Nowak, T., Farber, P. M., and Wengler, G. (1989). Analyses of the terminal sequences of West Nile virus structural proteins and of the in vitro translation of these proteins allow the proposal of a complete scheme of the proteolytic cleavages involved in their synthesis. Virology *169*, 365–376.

Ochsenbein, A. F., Pinschewer, D. D., Odermatt, B., Carroll, M. C., Hengartner, H., and Zinkernagel, R. M. (1999). Protective T cell-independent antiviral antibody responses are dependent on complement. J Exp Med *190*, 1165–1174.

Oliphant, T., and Diamond, M. S. (2007). The molecular basis of antibody-mediated neutralization of West Nile virus. Expert Opin Biol Ther *7*, 885–892.

Ollert, M. W., Kadlec, J. V., David, K., Petrella, E. C., Bredehorst, R., and Vogel, C. W. (1994). Antibody-mediated complement activation on nucleated cells. A quantitative analysis of the individual reaction steps. J Immunol *153*, 2213–2221.

Pangburn, M. K., and Muller-Eberhard, H. J. (1980). Relation of putative thioester bond in C3 to activation of the alternative pathway and the binding of C3b to biological targets of complement. J Exp Med *152*, 1102–1114.

Pangburn, M. K., Schreiber, R. D., and Muller-Eberhard, H. J. (1981). Formation of the initial C3 convertase of the alternative complement pathway. Acquisition of C3b-like activities by spontaneous hydrolysis of the putative thioester in native C3. J Exp Med *154*, 856–867.

Pangburn, M. K., Schreiber, R. D., and Muller-Eberhard, H. J. (1983). C3b deposition during activation of the alternative complement pathway and the effect of deposition on the activating surface. J Immunol *131*, 1930–1935.

Peiris, J. S., and Porterfield, J. S. (1979). Antibody-mediated enhancement of Flavivirus replication in macrophage-like cell lines. Nature *282*, 509–511.

Peiris, J. S., Gordon, S., Unkeless, J. C., and Porterfield, J. S. (1981). Monoclonal anti-Fc receptor IgG blocks antibody enhancement of viral replication in macrophages. Nature *289*, 189–191.

Perkins, S. J., and Nealis, A. S. (1989). The quaternary structure in solution of human complement subcomponent C1r2C1s2. Biochem J *263*, 463–469.

Perona, J. J., and Craik, C. S. (1995). Structural basis of substrate specificity in the serine proteases. Protein Sci *4*, 337–360.

Perrin-Cocon, L. A., Villiers, C. L., Salamero, J., Gabert, F., and Marche, P. N. (2004). B cell receptors and complement receptors target the antigen to distinct intracellular compartments. J Immunol *172*, 3564–3572.

Pierson, T., Xu, Q., Nelson, S., Oliphant, T., Nybakken, G., Fremont, D., and Diamond, M. (2007). The stoichiometry of antibody-mediated neutralization and enhancement of West Nile virus infection. Cell Host Microbe *1*, 135–145.

Pinter, C., Siccardi, A. G., Lopalco, L., Longhi, R., and Clivio, A. (1995). HIV glycoprotein 41 and complement factor H interact with each other and share functional as well as antigenic homology. AIDS Res Hum Retroviruses *11*, 971–980.

Pinto, M. R., Melillo, D., Giacomelli, S., Sfyroera, G., and Lambris, J. D. (2007). Ancient origin of the complement system: emerging invertebrate models. Adv Exp Med Biol *598*, 372–388.

Podack, E. R., Biesecker, G., Kolb, W. P., and Muller-Eberhard, H. J. (1978a). The C5b-6 complex: reaction with C7, C8, C9. J Immunol *121*, 484–490.

Podack, E. R., Kolb, W. P., and Muller-Eberhard, H. J. (1978b). The C5b-6 complex: formation, isolation, and inhibition of its activity by lipoprotein and the S-protein of human serum. J Immunol *120*, 1841–1848.

Podack, E. R., Tschoop, J., and Muller-Eberhard, H. J. (1982). Molecular organization of C9 within the membrane attack complex of complement. Induction of circular C9 polymerization by the C5b-8 assembly. J Exp Med *156*, 268–282.

Ponnuraj, E. M., Springer, J., Hayward, A. R., Wilson, H., and Simoes, E. A. (2003). Antibody-dependent enhancement, a possible mechanism in augmented pulmonary disease of respiratory syncytial virus in the Bonnet monkey model. J Infect Dis *187*, 1257–1263.

Porter, R. R., and Reid, K. B. (1978). The biochemistry of complement. Nature *275*, 699–704.

Porter, D. D., Larsen, A. E., and Porter, H. G. (1972). The pathogenesis of Aleutian disease of mink. II. Enhancement of tissue lesions following the administration of a killed virus vaccine or passive antibody. J Immunol *109*, 1–7.

Posner, M. R., Elboim, H. S., Cannon, T., Cavacini, L., and Hideshima, T. (1992). Functional activity of an HIV-1 neutralizing IgG human monoclonal antibody: ADCC and complement-mediated lysis. AIDS Res Hum Retroviruses *8*, 553–558.

Prabhakar, B. S., and Nathanson, N. (1981). Acute rabies death mediated by antibody. Nature *290*, 590–591.

Pratt, J. R., Basheer, S. A., and Sacks, S. H. (2002). Local synthesis of complement component C3 regulates acute renal transplant rejection. Nat Med *8*, 582–587.

Preissner, K. T., Podack, E. R., and Muller-Eberhard, H. J. (1985). The membrane attack complex of complement: relation of C7 to the metastable membrane binding site of the intermediate complex C5b-7. J Immunol *135*, 445–451.

Qin, D., Wu, J., Carroll, M. C., Burton, G. F., Szakal, A. K., and Tew, J. G. (1998). Evidence for an important interaction between a complement-derived CD21 ligand on follicular dendritic cells and CD21 on B cells in the initiation of IgG responses. J Immunol *161*, 4549–4554.

Ravetch, J. V., and Kinet, J. P. (1991). Fc receptors. Annu Rev Immunol *9*, 457–492.

Reid, K. B., and Porter, R. R. (1981). The proteolytic activation systems of complement. Annu Rev Biochem *50*, 433–464.

Rey-Millet, C. A., Villiers, C. L., Gabert, F. M., Chesne, S., and Colomb, M. G. (1994). C3b covalently associated to tetanus toxin modulates TT processing and presentation by U937 cells. Mol Immunol *31*, 1321–1327.

Reynes, M., Aubert, J. P., Cohen, J. H., Audouin, J., Tricottet, V., Diebold, J., and Kazatchkine, M. D. (1985). Human follicular dendritic cells express CR1, CR2, and CR3 complement receptor antigens. J Immunol *135*, 2687–2694.

Roehrig, J. T., Staudinger, L. A., Hunt, A. R., Mathews, J. H., and Blair, C. D. (2001). Antibody prophylaxis and therapy for flavivirus encephalitis infections. Ann N Y Acad Sci *951*, 286–297.

Ross, G. D., and Medof, M. E. (1985). Membrane complement receptors specific for bound fragments of C3. Adv Immunol *37*, 217–267.

Ross, T. M., Xu, Y., Bright, R. A., and Robinson, H. L. (2000). C3d enhancement of antibodies to hemagglutinin accelerates protection against influenza virus challenge. Nat Immunol *1*, 127–131.

Rossi, V., Cseh, S., Bally, I., Thielens, N. M., Jensenius, J. C., and Arlaud, G. J. (2001). Substrate specificities of recombinant mannan-binding lectin-associated serine proteases-1 and -2. J Biol Chem *276*, 40880–40887.

Sabin, A. B. (1950). The dengue group of viruses and its family relationships. Bacteriol Rev *14*, 225–232.

Sahu, A., Isaacs, S. N., Soulika, A. M., and Lambris, J. D. (1998). Interaction of vaccinia virus complement control protein with human complement proteins: factor I-mediated degradation of C3b to iC3b1 inactivates the alternative complement pathway. J Immunol *160*, 5596–5604.

Saifuddin, M., Parker, C. J., Peeples, M. E., Gorny, M. K., Zolla-Pazner, S., Ghassemi, M., Rooney, I. A., Atkinson, J. P., and Spear, G. T. (1995). Role of virion-associated glycosylphosphatidylinositol-linked proteins CD55 and CD59 in complement resistance of cell line-derived and primary isolates of HIV-1. J Exp Med *182*, 501–509.

Sairenji, T., Sullivan, J. L., and Humphreys, R. E. (1984). Complement-dependent, Epstein–Barr virus-neutralizing antibody appearing early in the sera of patients with infectious mononucleosis. J Infect Dis *149*, 763–768.

Sato, T., Endo, Y., Matsushita, M., and Fujita, T. (1994). Molecular characterization of a novel serine protease involved in activation of the complement system by mannose-binding protein. Int Immunol *6*, 665–669.

Scharfstein, J., Ferreira, A., Gigli, I., and Nussenzweig, V. (1978). Human C4-binding protein. I. Isolation and characterization. J Exp Med *148*, 207–222.

Schlesinger, J. J., and Chapman, S. (1995). Neutralizing F(ab')2 fragments of protective monoclonal antibodies to yellow fever virus (YF) envelope protein fail to protect mice against lethal YF encephalitis. J Gen Virol *76 (Pt 1)*, 217–220.

Schmidt, N. J., and Lennette, E. H. (1975). Neutralizing antibody responses to varicella-zoster virus. Infect Immun *12*, 606–613.

Sekine, H., Kenjo, A., Azumi, K., Ohi, G., Takahashi, M., Kasukawa, R., Ichikawa, N., Nakata, M., Mizuochi, T., Matsushita, M., et-al. (2001). An ancient lectin-dependent complement system in an ascidian: novel lectin isolated from the plasma of the solitary ascidian, Halocynthia roretzi. J Immunol *167*, 4504–4510.

Serra, V. A., Cretin, F., Pepin, E., Gabert, F. M., and Marche, P. N. (1997). Complement C3b fragment covalently linked to tetanus toxin increases lysosomal sodium dodecyl sulfate-stable HLA-DR dimer production. Eur J Immunol *27*, 2673–2679.

Seya, T., and Atkinson, J. P. (1989). Functional properties of membrane cofactor protein of complement. Biochem J *264*, 581–588.

Shchelkunov, S. N. (1995). Functional organization of variola major and vaccinia virus genomes. Virus Genes *10*, 53–71.

Sim, R. B., and Reid, K. B. (1991). C1: molecular interactions with activating systems. Immunol Today *12*, 307–311.

Snyder, D. B., Myrup, A. C., and Dutta, S. K. (1981). Complement requirement for virus neutralization by antibody and reduced serum complement levels associated with experimental equine herpesvirus 1 infection. Infect Immun *31*, 636–640.

Spear, G. T., Takefman, D. M., Sullivan, B. L., Landay, A. L., and Zolla-Pazner, S. (1993). Complement activation by human monoclonal antibodies to human immunodeficiency virus. J Virol *67*, 53–59.

Spear, G. T., Lurain, N. S., Parker, C. J., Ghassemi, M., Payne, G. H., and Saifuddin, M. (1995). Host cell-derived complement control proteins CD55 and CD59 are incorporated into the virions of two unrelated enveloped viruses. Human T cell leukemia/lymphoma virus type I (HTLV-I) and human cytomegalovirus (HCMV). J Immunol *155*, 4376–4381.

Spear, G. T., Hart, M., Olinger, G. G., Hashemi, F. B., and Saifuddin, M. (2001). The role of the complement system in virus infections. Curr Top Microbiol Immunol *260*, 229–245.

Spector, S. L., and Tauraso, N. M. (1969). Yellow fever virus. II. Factors affecting the plaque neutralization test. Appl Microbiol *18*, 736–743.

Speth, C., Dierich, M. P., and Gasque, P. (2002). Neuroinvasion by pathogens: a key role of the complement system. Mol Immunol *38*, 669–679.

Springer, T. A. (1990). Adhesion receptors of the immune system. Nature *346*, 425–434.

Stadler, K., Allison, S. L., Schalich, J., and Heinz, F. X. (1997). Proteolytic activation of tick-borne encephalitis virus by furin. J Virol *71*, 8475–8481.

Stager, S., Alexander, J., Kirby, A. C., Botto, M., Rooijen, N. V., Smith, D. F., Brombacher, F., and Kaye, P. M. (2003). Natural antibodies and complement are endogenous adjuvants for vaccine-induced CD8 + T-cell responses. Nat Med *9*, 1287–1292.

Stoiber, H., Ebenbichler, C., Schneider, R., Janatova, J., and Dierich, M. P. (1995). Interaction of several complement proteins with gp120 and gp41, the two envelope glycoproteins of HIV-1. AIDS *9*, 19–26.

Stoiber, H., Pinter, C., Siccardi, A. G., Clivio, A., and Dierich, M. P. (1996). Efficient destruction of human immunodeficiency virus in human serum by inhibiting the protective action of complement factor H and decay accelerating factor (DAF, CD55). J Exp Med *183*, 307–310.

Sullivan, B. L., Takefman, D. M., and Spear, G. T. (1998). Complement can neutralize HIV-1 plasma virus by a C5-independent mechanism. Virology *248*, 173–181.

Suresh, M., Molina, H., Salvato, M. S., Mastellos, D., Lambris, J. D., and Sandor, M. (2003). Complement component 3 is required for optimal expansion of CD8 T cells during a systemic viral infection. J Immunol *170*, 788–794.

Takahashi, M., Iwaki, D., Matsushita, A., Nakata, M., Matsushita, M., Endo, Y., and Fujita, T. (2006). Cloning and characterization of mannose-binding lectin from lamprey (Agnathans). J Immunol *176*, 4861–4868.

Tedder, T. F., Fearon, D. T., Gartland, G. L., and Cooper, M. D. (1983). Expression of C3b receptors on human be cells and myelomonocytic cells but not natural killer cells. J Immunol *130*, 1668–1673.

Tedder, T. F., Clement, L. T., and Cooper, M. D. (1984). Expression of C3d receptors during human B cell differentiation: immunofluorescence analysis with the HB-5 monoclonal antibody. J Immunol *133*, 678–683.

Thiel, S., Vorup-Jensen, T., Stover, C. M., Schwaeble, W., Laursen, S. B., Poulsen, K., Willis, A. C., Eggleton, P., Hansen, S., Holmskov, U., et-al. (1997). A second serine protease associated with mannan-binding lectin that activates complement. Nature *386*, 506–510.

Tortorella, D., Gewurz, B. E., Furman, M. H., Schust, D. J., and Ploegh, H. L. (2000). Viral subversion of the immune system. Annu Rev Immunol *18*, 861–926.

Tschopp, J., Podack, E. R., and Muller-Eberhard, H. J. (1985). The membrane attack complex of complement: C5b-8 complex as accelerator of C9 polymerization. J Immunol *134*, 495–499.

Tschopp, J., Masson, D., and Stanley, K. K. (1986). Structural/functional similarity between proteins involved in complement- and cytotoxic T-lymphocyte-mediated cytolysis. Nature *322*, 831–834.

Tsujimura, A., Shida, K., Kitamura, M., Nomura, M., Takeda, J., Tanaka, H., Matsumoto, M., Matsumiya, K., Okuyama, A., Nishimune, Y., et-al.. (1998). Molecular cloning of a murine homologue of membrane cofactor protein (CD46): preferential expression in testicular germ cells. Biochem J *330* (Pt 1), 163–168.

van Beek, J., Elward, K., and Gasque, P. (2003). Activation of complement in the central nervous system: roles in neurodegeneration and neuroprotection. Ann N Y Acad Sci *992*, 56–71.

van Lookeren Campagne, M., Wiesmann, C., and Brown, E. J. (2007). Macrophage complement receptors and pathogen clearance. Cell Microbiol *9*, 2095–2102.

Van Schravendijk, M. R., and Dwek, R. A. (1982). Interaction of C1q with DNA. Mol Immunol *19*, 1179–1187.

Vasantha, S., Coelingh, K. L., Murphy, B. R., Dourmashkin, R. R., Hammer, C. H., Frank, M. M., and Fries, L. F. (1988). Interactions of a nonneutralizing IgM antibody and complement in parainfluenza virus neutralization. Virology *167*, 433–441.

Verity, E. E., Williams, L. A., Haddad, D. N., Choy, V., O'Loughlin, C., Chatfield, C., Saksena, N. K., Cunningham, A., Gelder, F., and McPhee, D. A. (2006). Broad neutralization and complement-mediated lysis of HIV-1 by PEHRG214, a novel caprine anti-HIV-1 polyclonal antibody. AIDS *20*, 505–515.

Verschoor, A., Brockman, M. A., Knipe, D. M., and Carroll, M. C. (2001). Cutting edge: myeloid complement C3 enhances the humoral response to peripheral viral infection. J Immunol *167*, 2446–2451.

Verschoor, A., Brockman, M. A., Gadjeva, M., Knipe, D. M., and Carroll, M. C. (2003). Myeloid C3 determines induction of humoral responses to peripheral herpes simplex virus infection. J Immunol *171*, 5363–5371.

Villiers, M. B., Villiers, C. L., Jacquier-Sarlin, M. R., Gabert, F. M., Journet, A. M., and Colomb, M. G. (1996). Covalent binding of C3b to tetanus toxin: influence on uptake/internalization of antigen by antigen-specific and non-specific B cells. Immunology *89*, 348–355.

Vogt, L., Schmitz, N., Kurrer, M. O., Bauer, M., Hinton, H. I., Behnke, S., Gatto, D., Sebbel, P., Beerli, R. R., Sonderegger, I., et-al. (2006). VSIG4, a B7 family-related protein, is a negative regulator of T cell activation. J Clin Invest *116*, 2817–2826.

Volanakis, J. E. (2002). The role of complement in innate and adaptive immunity. Curr Top Microbiol Immunol *266*, 41–56.

Wagner, E., Platt, J. L., Howell, D. N., Marsh, H. C., Jr., and Frank, M. M. (1999). IgG and complement-mediated tissue damage in the absence of C2: evidence of a functionally active C2-bypass pathway in a guinea pig model. J Immunol *163*, 3549–3558.

Wallis, R. (2007). Interactions between mannose-binding lectin and MASPs during complement activation by the lectin pathway. Immunobiology *212*, 289–299.

Wallis, C., and Melnick, J. L. (1971). Herpesvirus neutralization: the role of complement. J Immunol *107*, 1235–1242.

Walport, M. J. (2001). Complement. First of two parts. N Engl J Med *344*, 1058–1066.

Wang, L., Sunyer, J. O., and Bello, L. J. (2004). Fusion to C3d enhances the immunogenicity of the E2 glycoprotein of type 2 bovine viral diarrhea virus. J Virol *78*, 1616–1622.

Weiler, J. M., Daha, M. R., Austen, K. F., and Fearon, D. T. (1976). Control of the amplification convertase of complement by the plasma protein beta1H. Proc Natl Acad Sci U S A *73*, 3268–3272.

Weis, J. J., Tedder, T. F., and Fearon, D. T. (1984). Identification of a 145,000 Mr membrane protein as the C3d receptor (CR2) of human B lymphocytes. Proc Natl Acad Sci U S A *81*, 881–885.

Whaley, K., and Ruddy, S. (1976). Modulation of the alternative complement pathways by beta 1 H globulin. J Exp Med *144*, 1147–1163.

Wiesmann, C., Katschke, K. J., Yin, J., Helmy, K. Y., Steffek, M., Fairbrother, W. J., McCallum, S. A., Embuscado, L., DeForge, L., Hass, P. E., and van Lookeren Campagne, M. (2006). Structure of C3b in complex with CRIg gives insights into regulation of complement activation. Nature *444*, 217–220.

Wu, X., Jiang, N., Fang, Y. F., Xu, C., Mao, D., Singh, J., Fu, Y. X., and Molina, H. (2000). Impaired affinity maturation in Cr2−/− mice is rescued by adjuvants without improvement in germinal center development. J Immunol *165*, 3119–3127.

Yamanaka, A., Kosugi, S., and Konishi, E. (2008) Infection-enhancing and-neutralizing activities of mouse monoclonal antibodies against dengue type 2 and 4 viruses are controlled by complement levels. J Virol 82, 927–937.

Yoder, S. M., Zhu, Y., Ikizler, M. R., and Wright, P. F. (2004). Role of complement in neutralization of respiratory syncytial virus. J Med Virol *72*, 688–694.

Yoshida, Y., Kang, K., Berger, M., Chen, G., Gilliam, A. C., Moser, A., Wu, L., Hammerberg, C., and Cooper, K. D. (1998). Monocyte induction of IL-10 and down-regulation of IL-12 by iC3b deposited in ultraviolet-exposed human skin. J Immunol *161*, 5873–5879.

Young, P. R., Hilditch, P. A., Bletchly, C., and Halloran, W. (2000). An antigen capture enzyme-linked immunosorbent assay reveals high levels of the dengue virus protein NS1 in the sera of infected patients. J Clin Microbiol *38*, 1053–1057.

Zipfel, P. F., Skerka, C., Hellwage, J., Jokiranta, S. T., Meri, S., Brade, V., Kraiczy, P., Noris, M., and Remuzzi, G. (2002). Factor H family proteins: on complement, microbes and human diseases. Biochem Soc Trans *30*, 971–978.

10. Antibody-Mediated Neutralization of West Nile Virus: Factors that Govern Neutralization Potency

CHRISTIANE A. JOST AND THEODORE C. PIERSON

Abstract

Flaviviruses are complex immunogens that elicit antibodies of varying specificity and with a spectrum of functional properties. Flavivirus virions are covered by a dense array of envelope (E) proteins that mediate steps of the virus entry pathway and are a primary target of neutralizing antibodies. The development of virus-specific antibodies is a critical aspect of protection against flavivirus infection and a major goal of ongoing efforts to produce vaccines against flaviviruses of clinical importance, such as the West Nile virus. In this chapter, we will review current models that describe how antibodies engage flaviviruses and block infection. Recent insight into the relationships that govern where antibodies bind virions and how this impacts the potency and mechanisms of neutralization of antibodies have been driven in part by insights from the structural biology of flaviviruses. The factors that define antibody potency will be discussed with a focus on the stoichiometric requirements for neutralization.

1 Introduction

Flaviviruses are a group of RNA viruses of global significance due to their widespread distribution and ability to cause a variety of severe diseases in humans (Rice, 1996). West Nile virus (WNV) is a mosquito-borne member of this genus and the etiologic agent of West Nile encephalitis. WNV is endemic in parts of Africa, Australia, Europe, Asia, and the Middle East, and has been responsible for periodic outbreaks of encephalitis in humans and horses. The introduction of WNV into North America in 1999 and its rapid spread across the United States into Canada, Mexico, and the Caribbean identifies this virus as an emerging pathogen of clinical and economic significance for the Western Hemisphere (reviewed in Mackenzie et al., 2004). Seroprevalence studies indicate that most WNV infections of humans are subclinical. Clinically apparent infections range from a febrile illness (West Nile fever) to more severe and potentially fatal neurologic disease (Hayes et al., 2005). Roughly 1,000 severe WNV related illnesses are reported in the United States each year (Sejvar, 2007). Currently, no WNV vaccine has been approved for use in humans, and treatment of infection is supportive.

The development of a small animal model of WNV infection and disease has allowed for significant advances in the understanding of WNV pathogenesis and immunity (reviewed in Samuel and Diamond, 2006). Altogether, these studies highlight the importance of an intact innate and adaptive immune system for protection against the flavivirus infection, including the development of a virus-specific antibody response. B cell- and antibody-deficient strains of mice have higher viral titers in the central nervous system (CNS) and experience greater mortality following WNV infection as compared to wild type mice (Diamond et al., 2003a,b; Halevy et al., 1994). In addition to the capacity of antibodies to directly neutralize WNV infection, studies with Fc- and complement receptor deficient strains highlight the importance of Fc-dependent effector functions in protection and immunity (Chung et al., 2007; Oliphant et al., 2005, 2006). Passive transfer of WNV-specific antibodies or immune sera protects animals from lethal challenge with WNV (Ben-Nathan et al., 2003; Camenga et al., 1974; Diamond et al., 2003a; Engle and Diamond, 2003; Henchal et al., 1988; Roehrig et al., 2001; Tesh et al., 2002; Wang et al., 2001). WNV-specific antibodies can also protect from disease when administered after infection. The therapeutic potential of WNV-specific antibodies is supported by the apparent successful clinical use of WNV-immune immunoglobulin preparations in humans with neurological symptoms (Agrawal and Petersen, 2003; Haley et al., 2003; Hamdan et al., 2002; Shimoni et al., 2001). The development of a humanized neutralizing monoclonal antibody (mAb) as a WNV therapeutic is underway (Morrey et al., 2006, 2007).

2 Flavivirus Virion Structure

Flaviviruses are small (~50 nm diameter) spherical virions covered by a dense array of 180 envelope (E) proteins that orchestrate the processes of virus entry, assembly, and budding (Heinz and Allison, 2000). The E protein is also a primary target for neutralizing antibodies (discussed below, and in detail by Throsby et al., this volume). Flaviviruses bind cells via interactions between virions and largely uncharacterized cellular factors (reviewed in Anderson, 2003), are internalized in a clathrin-dependent process (Chu and Ng, 2004; Gollins and Porterfield, 1985) and penetrate the host cell membrane of the early endosome in a pH-dependent fashion (Krishnan et al., 2007; Stiasny and Heinz, 2006; van der Schaar et al., 2007). Viral membrane fusion is driven by conformational changes in the E protein, and results in the introduction of a roughly 11 kb viral genomic RNA of positive sense polarity into the cytoplasm. The single open reading frame of the viral genome is translated as a polyprotein that is subsequently cleaved into three structural proteins (capsid (C), pre-membrane (prM), and E protein) and seven nonstructural proteins by viral and host cell proteases. Viral RNA replication occurs in the cytoplasm of infected cells and is reviewed by Brinton, this volume (reviewed by Brinton, 2002).

Flaviviruses assemble at and bud into the endoplasmic reticulum (ER) as immature virions from where they traffic through and mature in the secretory pathway (Mackenzie and Westaway, 2001). Cryoelectron microscopy studies of the structure of immature dengue virus (DENV), yellow fever virus (YFV), and WNV reveal sixty spikes of E protein trimers arranged on the surface of the virion with icosahedral symmetry (Fig. 1a) (Zhang et al., 2003, 2007). On immature virions the prM protein is positioned at the tip of each E protein of the trimer, masking the fusion loop at the distal end of each E protein. Presumably, the role of prM in this context is to regulate when E proteins become competent for fusion by preventing low-pH induced changes in the oligomerization state of E proteins that may otherwise occur during viral egress through acidic compartments of the secretory pathway (Guirakhoo et al., 1991; Heinz et al., 1994).

During transit through the *trans*-Golgi network the immature virion undergoes a maturation process characterized by cleavage of the prM protein by the cellular serine protease furin (Guirakhoo et al., 1991; Heinz et al., 1994; Stadler et al., 1997). This cleavage catalyzes a significant rearrangement between E proteins on the virion, and the release of an approximately 20 kDa "pr" portion of the protein; the small membrane anchored M peptide (~8 kDa) remains associated with the virion. The E proteins of mature flavivirus virions exist as 90 antiparallel dimers that lay flat against the virion surface and are arranged with $T = 3$ pseudo-icosahedral symmetry, resulting in a relatively smooth spherical virion (Fig. 1b) (Kuhn et al., 2002; Mukhopadhyay et al., 2003). This orientation is in striking contrast to the immature flavivirus virion and

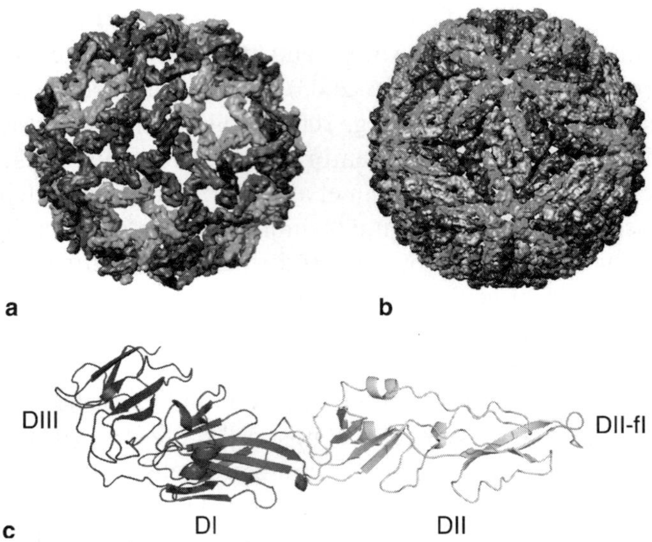

Figure 1. Structure of the flavivirus E protein and its organization on the mature virion. (**a**) Cryoelectron reconstruction of the WNV immature virion. Image was created using Chimera (http://www.cgl.ucsf.edu/chimera/). (**b**) Cryoelectron reconstruction of the WNV mature virion illustrating the arrangement of E proteins into a $T = 3$ pseudo-icosahedron. Individual E proteins are colored according to their proximity to the 2-, 3-, or 5-fold symmetry axes (*blue*, *green* and *red* respectively). Image obtained using the viperdb Virus Particle Explorer (http://viperdb.scripps.edu/) (Shepherd et al., 2006). (**c**) Ribbon diagram of a WNV E protein monomer with domains II, I and III shown as *yellow*, *red* and *blue* ribbons respectively. The fusion loop at the tip of DII (DII-fl) is shown in *green*. (*See Color Plates*)

the perpendicular envelope projections of the highly characterized type I group of fusion glycoproteins (e.g. Influenza HA, HIV-1 gp120) (reviewed in Colman and Lawrence, 2003). From the perspective of the antibody, the pseudo-icosahedral structure increases the complexity of the antigenic surface of the flavivirus virion: each E protein is present in one of three chemically distinct environments defined by their proximity to the 5-, 3- or 2-fold symmetry axis of the virion (Fig. 1b; shown in green, red, and blue, respectively).

3 The Structure of the Envelope Protein and Epitopes Recognized by Neutralizing Antibodies

The E protein of flaviviruses is the primary target for neutralizing antibodies, although antibodies that recognize prM have also been identified (AbuBakar et al., 1997; Colombage et al., 1998; Falconar, 1999;

Pincus et al., 1992; Throsby et al., 2006). Antibodies specific for the nonstructural protein NS1 are commonly observed (Shu et al., 2000). While capable of protecting animals from lethal challenge, these antibodies do not directly neutralize virus infection and are discussed elsewhere (Alcon-LePoder et al., 2006; Chung et al., 2006). The structure of the E protein has been solved for several flaviviruses using X-ray crystallography. Altogether, these studies reveal that the ectodomain of the E protein is composed of three distinct domains connected by short flexible linkers (Fig. 1c) (Kanai et al., 2006; Modis et al., 2003, 2004; Nybakken et al., 2006; Rey et al., 1995; Zhang et al., 2004). Domain III (DIII) is an immunoglobulin-like fold located at the carboxy-terminus of the E protein ectodomain and is thought to play a role in the attachment of virions to cells (Crill and Roehrig, 2001; Rey et al., 1995; Volk et al., 2004). Domain II is composed of two elongated finger-like structures that mediate oligomerization of E proteins on the virion during the structural transitions that drive membrane fusion. Importantly, this domain also contains a highly conserved fusion loop at the distal end that is thought to insert itself into the host cell membrane during the low pH-catalyzed fusion process (Bressanelli et al., 2004; Modis et al., 2004). Domain I (DI) is a central β-barrel domain that connects DII and DIII. The E protein ectodomain is connected to the viral membrane via a stem anchor composed of two amphipathic helices, and anchored there by two anti-parallel transmembrane domains (Allison et al., 1999).

Epitope mapping studies have identified more than twelve structurally distinct epitopes distributed throughout the three domains of the E protein (Beasley and Barrett, 2002; Crill and Chang, 2004; Heinz et al., 1983; Mandl et al., 1989; Nybakken et al., 2005; Oliphant et al., 2005, 2006; Roehrig et al., 1983, 1998; Sanchez et al., 2005). Antibodies that bind these epitopes display varying degrees of neutralization potency in vitro and efficiency in vivo. The most potent WNV-specific mAbs characterized to date bind the upper lateral surface of DIII (DIII lateral ridge: DIII-lr) (Beasley and Barrett, 2002; Oliphant et al., 2005; Sanchez et al., 2005). The structural basis for recognition of the DIII-lr epitope by the potently neutralizing murine mAb E16 has been defined with high resolution by both structural and biochemical approaches, with excellent agreement between the two methods (Nybakken et al., 2005; Oliphant et al., 2005). The footprint of E16 on DIII includes four discontinuous loops, with important contacts between the antibody and residues S306, K307, T330, and T332, in agreement with the finding that mutation of these residues allows escape from neutralization. (Beasley and Barrett, 2002). While the DIII-lr is structurally conserved among flaviviruses, sequence variation between distantly related flaviviruses restricts recognition of this epitope to type-specific antibodies. Antibodies that bind the structurally analogous epitope on DIII of DENV (Gromowski and Barrett, 2007; Hiramatsu et al., 1996; Roehrig et al., 1998; Serafin and Aaskov, 2001; Sukupolvi-Petty et al., 2007; Thullier et al., 2001), Japanese Encephalitis virus (JEV) (Cecilia et al., 1988; Cecilia and Gould, 1991), YFV (Ryman et al., 1998), and Murray Valley Encephalitis virus (MVEV) (Hall et al., 1990) have been characterized.

While antibodies directed against the DIII-lr epitope are characterized by potent neutralizing activity in vitro, their role in natural infection in vivo is less clear. The functional contribution of DIII-lr antibodies in the polyclonal responses of mice, horses, and humans has been investigated using virions that contain a single point mutation (T332) in the DIII-lr epitope (Oliphant et al., 2007; Sanchez et al., 2005; Nelson et al., in press). These functional studies suggest that the contribution of antibodies specific for the DIII-lr epitope to the overall neutralizing activity of a polyclonal response is modest. Indeed, analysis of the antibody repertoire of convalescent WNV immune subjects indicate that less than 10% of B-cells produce antibodies specific for DIII (Throsby et al., 2006). Instead, these studies identify the fusion loop at the tip of DII as a major epitope in the human polyclonal response to WNV (Throsby et al., 2006). The fusion loop (DII-fl) is a highly conserved stretch of amino acids (residues 98–110) (Fig. 1c; in green) shared between distantly related flaviviruses (Allison et al., 2001). Thus, DII-fl-specific antibodies are highly cross-reactive (Crill and Chang, 2004; Crill et al., 2007; Goncalvez et al., 2004; Oliphant et al., 2006; Roehrig et al., 1998; Stiasny et al., 2006; reviewed in Roehrig, 2003). In comparison to mAbs that recognize the DIII-lr, DII-fl antibodies are characterized by modest potency in vitro, and protect only at relatively high concentrations in vivo (Oliphant et al., 2006; Stiasny et al., 2006).

Antibodies specific for epitopes in DI and regions of DIII and DII outside the DIII-lr and DII-fl epitope, respectively, have been characterized (Beasley and Barrett, 2002; Crill and Chang, 2004; Heinz et al., 1983; Nybakken et al., 2005; Oliphant et al., 2005, 2006; Roehrig et al., 1983, 1998; Sanchez et al., 2005; Stiasny et al., 2006; Nelson et al., in press). Typically, these antibodies are also markedly less potent in vitro and in vivo as compared to DIII-lr antibodies (Oliphant et al., 2006; Stiasny et al., 2006). One interesting exception is a recently characterized DENV-specific mAb that binds a determinant on the lateral ridge of DI (DI-lr) and potently neutralizes DENV infection (Lai et al., 2007).

4 The Stoichiometry of Flavivirus Neutralization

The neutralization of animal viruses by antibody has been the subject of study and debate for decades (Burnet et al., 1937, 2001; Della-Porta and Westaway, 1978). At least two models of the requirements for antibody-mediated neutralization have been proposed. "Single hit" models of neutralization describe the inactivation of virus infectivity following the docking of a single antibody molecule onto the virion (Dulbecco et al., 1956). The experimental basis for this model was the observation that the kinetics of neutralization for some viruses appear to be of first order (at 37°C), and do not display an initial lag phase indicative of a requirement for binding of more than a single antibody molecule. As virions typically display many copies of the viral antigens recognized by antibody, single hit models suggest that not all epitopes are functionally equivalent, and that neutralization occurs following engagement of critical sites on the virion (Dulbecco et al., 1956). In contrast, "multiple hit" models of neutralization suggest that inactivation of viral infectivity occurs following engagement of the virion by more than a single antibody molecule (Burnet et al., 1937; Della-Porta and Westaway, 1978). The central concepts of this model are that the fate of the virion is determined by the number of antibodies docked onto it at critical steps of the virus entry pathway, and that the process of antibody binding to the virus particle does not in itself result in inactivating changes on the virion (neutralization is reversible) (Burnet et al., 1937). Several lines of evidence suggest that the neutralization of WNV is a multiple hit phenomenon in which virion inactivation occurs when the number of antibodies bound to individual virion exceeds a required threshold (discussed below).

4.1 Epitope Accessibility Governs the Neutralization Potency of Antibodies

The number of antibodies bound to WNV determines the infectivity of individual virions. At least two factors define the requirements for a particular antibody to bind the virion with a stoichiometry sufficient for neutralization: antibody affinity and the accessibility of the epitope on the virion. The strength of binding between antibody and viral antigen (antibody affinity) determines the fraction of epitopes on the virion that are bound by antibody at any given concentration of antibody (epitope occupancy) (reviewed in Klasse and Sattentau, 2002). Thus, in some instances differences in the affinity of antibody-virion interactions may explain differences in antibody potency.

Epitope accessibility is a critical factor that governs the neutralization potency of antibodies and shapes the occupancy requirements for neutralization. As neutralization requires engagement of the virion with a stoichiometry that exceeds a required threshold (defined below), the number of antibodies docked on the virion at any concentration of antibody depends on the affinity of the interaction and the total number of accessible epitopes on the virion (Fig. 2). While infectious flavivirus virions are thought to uniformly incorporate 180 E proteins, not all of the epitopes on the surface of the virion are equally accessible for binding. The pseudo-icosahedral organization of the virion displays E proteins in

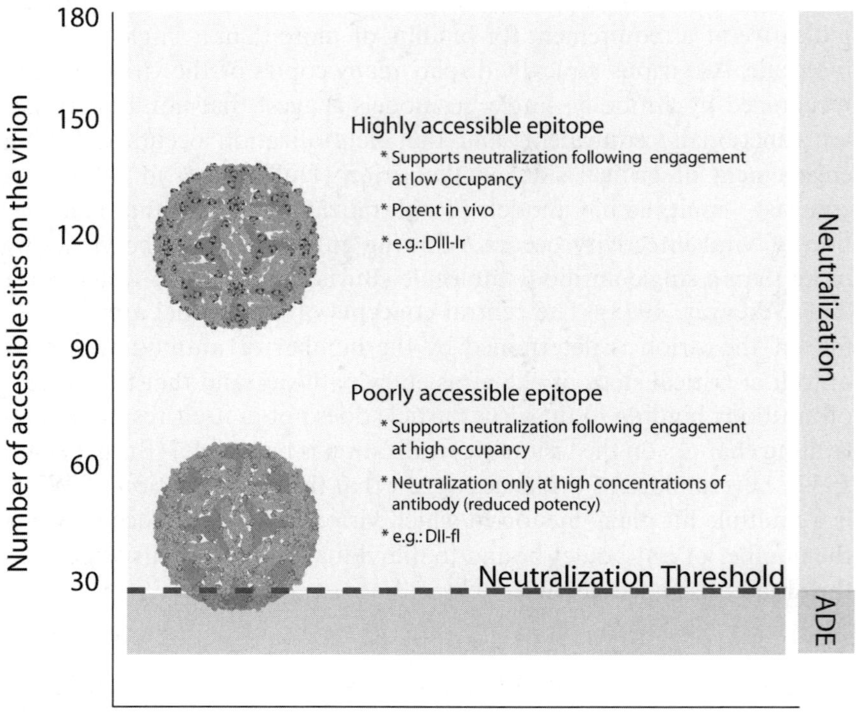

Figure 2. Relationship between epitope accessibility and the occupancy requirement for neutralization. The accessibility of epitopes recognized by two antibodies on the mature WNV virion is illustrated using molecular modeling: residues that comprise each determinant are illustrated as solid spheres (*pink*). E proteins are colored according to their proximity to the 2-, 3- or 5-fold symmetry axes (*blue*, *green* and *yellow* respectively). The number of accessible binding sites is indicated schematically on the *left*, whereas the threshold for neutralization is indicated as a *red line* (modeled as 30 mAbs based on studies with WNV DIII-lr-specific mAbs) (Pierson et al., 2007). To exceed the threshold requirements for neutralization, only a fraction of highly accessible determinants must be simultaneously occupied by antibody (a low occupancy requirement). In contrast, for poorly exposed epitopes (fewer total accessible sites) a significantly greater percentage of accessible epitopes must be bound to achieve the same number of antibodies docked on the average virion (a high occupancy requirement). Not all epitopes appear to exist on the average virion at levels that exceed this threshold. (*See Color Plates*)

one of three distinct chemical environments defined by their proximity to the 2-, 3-, or 5-fold axes of symmetry (Kuhn et al., 2002; Mukhopadhyay et al., 2003; Oliphant et al., 2006; Pierson et al., 2007; Stiasny et al., 2006). Epitopes in each of these environments may be differentially accessible for antibody binding due to steric constraints imposed by adjacent E proteins on the virion (Nybakken et al., 2005; Oliphant et al., 2006; Pierson et al., 2007; Stiasny et al., 2006). As a result, the number of sites on the virion available for binding by antibodies may be significantly different among structurally distinct epitopes, or those associated with different symmetry axes.

If the neutralization threshold is relatively constant for antibodies, irrespective of their specificity, differences in epitope accessibility will govern the occupancy requirements for neutralization. Antibodies specific for highly accessible structures (such as the DIII-lr) can exceed the stoichiometric threshold required for neutralization by binding the virion at relatively low occupancy (Gromowski and Barrett, 2007; Pierson et al., 2007). In contrast, antibodies that recognize poorly exposed or "cryptic" determinants need to bind a large fraction of them in order to achieve neutralization. Finally, not all epitopes may exist on the average virion with a frequency that exceeds the threshold required for neutralization. Engagement at these epitopes will not support neutralization, even at saturating concentrations of antibody (Oliphant et al., 2006; Pierson et al., 2007; Nelson et al., in press).

4.2 The Stoichiometry of WNV Neutralization

The stoichiometric requirements for WNV have been estimated using mAbs that bind the DIII-lr. Structural and molecular modeling studies of E16 complexed with its epitope on the mature virion indicate that it is unable to engage all 180 E proteins, as E proteins proximal to the fivefold axis of symmetry cannot be bound due to steric constraints (Kaufmann et al., 2006; Nybakken et al., 2005). Thus, a total of 120 epitopes for E16 exist on mature WNV. The fraction of these epitopes bound by E16 when WNV is neutralized was determined using two independent quantitative approaches. Dose response studies performed with E16 revealed relatively steep neutralization profiles and a capacity to block infection at picomolar concentrations of antibody (Oliphant et al., 2005; Pierson et al., 2007). That the concentration of E16 required for neutralizing WNV was lower than the affinity of this mAb for virions suggests that not all the epitopes on the virion must be occupied by antibody when the virion is neutralized. Analysis of the occupancy requirements for neutralization by mAbs that bind the DIII-lr epitope indicated that half the virions in the population are neutralized when roughly 20–30% of the epitopes on the virion were occupied by mAb. Furthermore, significant neutralization (>99%) was observed when only a modest fraction (60%) of accessible epitopes was engaged. This low occupancy requirement for neutralization was confirmed using a genetic approach in which the number of epitopes on individual virions was manipulated (Pierson et al., 2007). Together, these studies suggest that neutralization of WNV occurs when the virion is bound by approximately 30 mAbs.

The stoichiometric requirements for neutralization of several viruses (picornaviruses (Che et al., 1998; Colonno et al., 1989; Icenogle et al., 1983), papilloma virus (Booy et al., 1998; Roden et al., 1994), influenza (Taylor et al., 1987), HIV-1 (Klasse and Moore, 1996; Schonning et al., 1999; Yang et al., 2005), rabies virus (Flamand et al., 1993) have been investigated. The number of antibodies required to neutralize different types of virions varies dramatically. For example, as few as four mAbs are reported to be required for neutralization of poliovirus (Icenogle et al., 1983), whereas inhibition of rabies virus involves more than 200 IgG molecules (Flamand et al., 1993). However, Burton and colleagues have proposed the unifying concept that the number of antibodies required for neutralization can be explained as a function of the surface area of the virion (Burton et al., 2001). Extrapolating from studies of the stoichiometric requirements for neutralization for unrelated virus types, they suggest that the number of antibodies required for neutralization is a function of the size of the virus particle. Interestingly, estimates for the number of antibodies required to neutralize WNV are in agreement with the predictions of this "coating theory" (Klasse and Burton, 2007; Pierson et al., 2007). Why 30 mAbs are required for neutralization of WNV is not presently clear. The binding footprint resulting from the monovalent engagement of 30 antibodies on the surface of the virion is quite small (less than 10%) (Nybakken et al., 2005; Pierson et al., 2007). Thus, the stoichiometric requirements for neutralization may reflect the steric impact of the relatively large size and space occupied by the entire antibody molecule.

5 Factors that Modulate Antibody Potency

Cryo-electron microscopic reconstructions of flaviviruses provide important models that enable the analysis of epitope accessibility in the context of the intact virion (Kaufmann et al., 2006; Kuhn et al., 2002;

Nybakken et al., 2005; Oliphant and Diamond, 2007; Pierson et al., 2007; Stiasny et al., 2006). Many antibodies recognize epitopes predicted to be poorly accessible on the mature virion (e.g. the DII-fl), yet show neutralizing activity in vitro and in vivo (Oliphant et al., 2006; Stiasny et al., 2006; Nelson et al., in press). How antibodies engage poorly accessible epitopes on virions with a stoichiometry that permits neutralization is difficult to reconcile. However, one caveat to the molecular modeling studies based on these reconstructions is that the models reflect the average state of the virion at the conditions used to prepare the virions for analysis. At physiologic temperatures, flaviviruses may be quite dynamic with epitopes transiently and reversibly exposed for antibody recognition. The structural dynamics of other enveloped and non-enveloped viruses have been investigated and provide precedents for antibody or drug interactions with structures that are not predicted to be accessible on the virion based on static models of virion structure (Burrage et al., 2000; Lewis et al., 1998; Li et al., 1994; Reisdorph et al., 2003). Several characteristics of the E protein structure and arrangement(s) suggest that they may exist in multiple dynamic states on the average virion including: (1) the rotational flexibility conferred by the hinge regions between each domain of the E protein (Modis et al., 2003, 2004; Nybakken et al., 2006; Rey et al., 1995; Zhang et al., 2004), (2) the large reorganization of E proteins on the virion that occur during maturation and low pH-mediated fusion events (Stiasny et al., 2007b, and references within), and (3) the capacity of the virion to expand during changes in E protein conformation (Kuhn et al., 2002). The structure of a neutralizing DENV mAb bound to an "inaccessible" determinant was recently solved (Lok et al., 2008). That this epitope can be bound by antibody at 37°C, but not at 4°C suggests that differences in epitope accessibility occur between these two temperatures. The unique arrangement of E proteins on the virion when complexed with the antibody suggests that it stabilizes the E proteins in an otherwise transient configuration not predicted by static models (Lok et al., 2008). Altogether, this important study highlights dynamic aspects that contribute to epitope accessibility and the antigenic complexity of the surface of the virion.

5.1 Impact of Virion Maturation of Antibody-Mediated Neutralization

Analysis of the functional properties of many antibodies suggests that flavivirus virions may be heterogenous with respect to their sensitivity to neutralization (Nelson et al., in press). In some instances, incubation of

WNV with increasing concentrations of mAb reveals a fraction of virions insensitive to neutralization even at concentrations of antibody that permit saturation (Oliphant et al., 2007; Pierson et al., 2007; Nelson et al., in press). A similar "resistant fraction" was observed when the number of epitopes on the average WNV virion was reduced using genetic methods, suggesting that the resistance of individual virions to neutralization can be explained by a failure to display an epitope with a frequency that exceeds the neutralization threshold (Pierson et al., 2007). The impact of virion maturation on neutralization potency and the existence of individual virions at different stages of the maturation process provide a biochemical rationale for the resistant fraction (Nelson et al., in press).

Relatively little is known about the relationship between the extent of prM cleavage during the maturation process and the acquisition of infectivity of individual virions. Virions may acquire the potential to infect cells following prM-cleavage-mediated activation of a relatively small fraction of the E proteins on the virion (Guirakhoo et al., 1991, 1992; Heinz et al., 1994; Randolph and Stollar, 1990; Randolph et al., 1990; Stadler et al., 1997), as observed for the analogous E2 protein of alphaviruses (Berglund et al., 1993; Salminen et al., 1992). Partially mature virions containing at least some uncleaved prM protein are present in bulk virus populations and are infectious (Fig. 3a) (Guirakhoo et al., 1991). Antibodies specific for structures predicted to be poorly accessible on mature virions (e.g. DII-fl) have significantly less potency when assayed on relatively homogenous populations of mature virions, as compared to the heterogeneous population of virions released from cells (Nelson et al., in press). These antibodies do not fail to bind mature virions, nor do they do so with significantly reduced affinity. Instead, the maturation process is thought to reduce the number of antibodies that may simultaneously bind the virion to levels that do not exceed the threshold required for neutralization even at full occupancy (Fig. 3b). Changes in E protein epitope accessibility associated with maturation may reflect dynamic aspects of virion structure that are not evident from static models of virion structure. Partially mature virions that contain uncleaved prM may be more dynamic than the mature virus with respect to the number or stability of alternate structural conformations that E proteins can attain as compared to fully mature virions. For antibodies that bind epitopes, which require virtually complete occupancy to achieve neutralization, even modest changes in epitope accessibility can significantly affect the outcome of antibody binding.

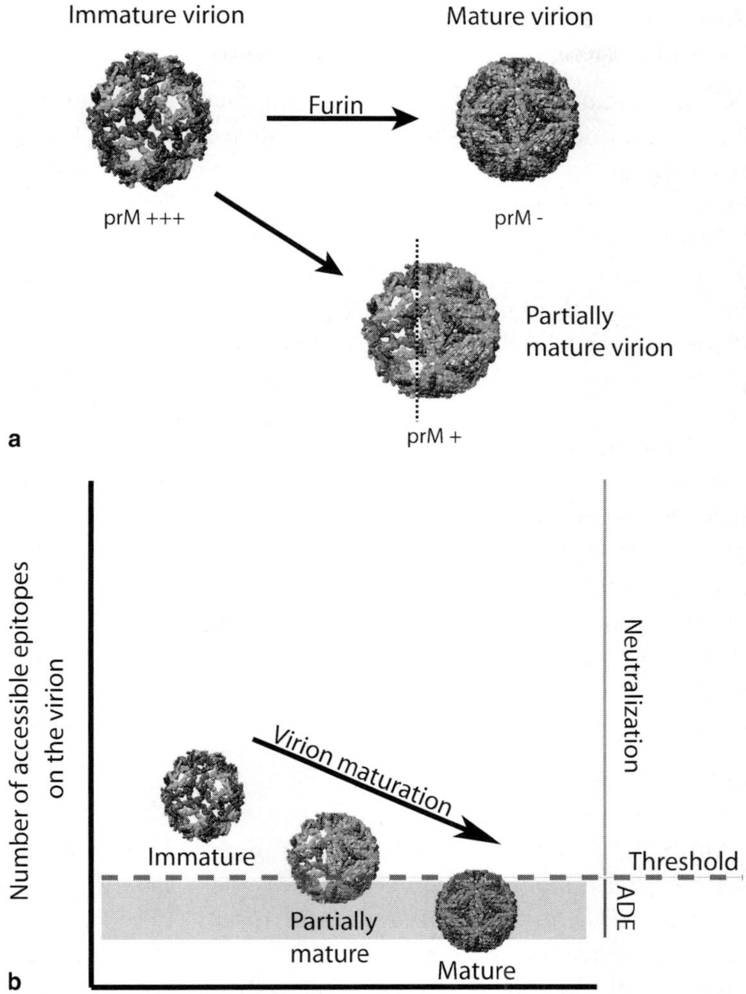

Figure 3. Relationship between virus maturation and neutralization. (**a**) During virion egress from the cell, the prM protein of immature virions is cleaved by the cellular serine protease furin (Stadler et al., 1997). Cleavage results in a significant reorganization of E proteins on the surface of the virion into the mature form of the virus characterized by a herringbone pseudo-icosahedral arrangement of E proteins (Heinz et al., 1994). However, not all prM proteins may be cleaved during this process. Partially mature virions that contain prM are present in bulk virus populations and are infectious (Davis et al. 2006; Guirakhoo et al., 1992). (**b**) The virion maturation process significantly impacts sensitivity to neutralization by antibodies predicted to bind poorly accessible structures. Mature virions are poorly neutralized by mAbs that bind several structurally distinct "cryptic" epitopes, as well as the polyclonal response of some human WNV vaccines. The reduction in neutralization sensitivity reflects a decrease in the number of accessible epitopes, which in turn reduces the number of antibodies that may simultaneously bind the virion. (*See Color Plates*)

5.2 Complement Augments the Neutralization Potency of Antibodies

Complement has been shown to augment the neutralization potency of antibodies using several virus systems, including flaviviruses (Aasa-Chapman et al., 2005; Beebe et al., 1983; Della-Porta and Westaway, 1978; Feng et al., 2002; Mehlhop et al., 2005; Meyer et al., 2002; Mozdzanowska et al., 2006; Spector and Tauraso, 1969).Studies using animal models highlight the importance of an intact complement system for humoral immunity against WNV (Mehlhop and Diamond, 2006; Mehlhop et al., 2005). Virus opsonization by classical pathway complement components C1q, C4b, and C3b may promote the formation of C5b-C9 membrane attack components (MAC) that result in direct lysis of the virion (reviewed in Volanakis, 2002). The efficiency of this process on viruses may be limited by the small surface area of viral membrane exposed in the context of the mature virion. More directly, complement may augment the neutralization potency of antibodies directly by modulating the occupancy requirements for neutralization: increasing antibody affinity or increasing the steric effects of bound antibody may result in a more efficient blockade of virus attachment or fusion (Feng et al., 2002; Meyer et al., 2002).

6 Mechanisms of Neutralization

Antibodies have the potential to neutralize the infectivity of flaviviruses by interfering with several steps of the virus entry pathway including attachment, internalization, and fusion (Fig. 4). Antibody-mediated neutralization of several viruses has been reported to occur by blocking the attachment of viruses to the cell (reviewed in Klasse and Sattentau, 2002). Perhaps the most well characterized example is antibodies that block the binding of the HIV-1 envelope protein to the CD4 receptor or CCR5 co-receptor on T cells (reviewed in Pantophlet and Burton, 2006). While the cellular factors involved remain unclear, antibodies may block flavivirus infection by inhibiting the interactions between virions and the cell surface during the attachment step. mAbs specific for DIII have been shown in some studies to block infection at this stage (He et al., 1995). Several lines of indirect evidence suggest that DIII plays an important role in virus attachment including: (1) DIII protrudes the farthest from the surface of the virion (Mukhopadhyay et al., 2005); (2) many of the mutations that impact tropism or virulence map to DIII (Holzmann et al., 1990; Jennings et al., 1994; Jiang et al., 1993; Mandl et al., 2000); and (3) soluble forms of DIII can block infection

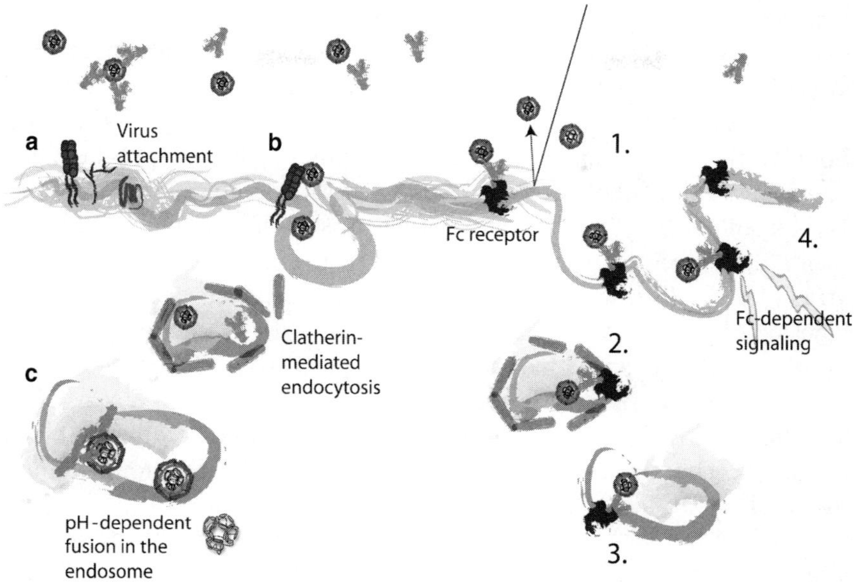

Figure 4. Overview of the virus life cycle and its interaction with antibodies. Flaviviruses bind cells through largely uncharacterized cellular factors (Anderson, 2003; Davis et al., 2006) (**a**), are internalized in a clathrin-dependent process (Chu and Ng, 2004; Gollins and Porterfield, 1985) (**b**), and fuse with the host cell membranes of the early endosome in a pH-dependent fashion (Krishnan et al., 2007; van der Schaar et al., 2007) (**c**). Antibodies have the potential to reduce the efficiency of any of these steps including blocking attachment (He et al., 1995) or by inhibiting fusion with the endosomal membrane (Gollins and Porterfield, 1986; Nybakken et al., 2005; Stiasny et al., 2007a). Antibody-dependent enhancement describes the increase in virus entry and replication in the presence of non-neutralizing quantities of antibody (Halstead, 2003; Halstead and O'Rourke, 1977). ADE is dependent upon the interactions of virion immune complexes with Fc- or complement receptors (Cardosa et al., 1983, 1986; Peiris et al., 1981; Peiris and Porterfield, 1979). As the stable attachment of virus particles to cells is a major factor that limits infection, ADE may function by increasing the efficiency and durability of virus attachment to cells (Gollins and Porterfield, 1984) (1). This in turn may result in more efficient internalization into the cell, and/or the trafficking of virus to sites in the endocytic pathway from which post-fusion events may be more efficient (2 and 3). Furthermore, Fc-receptor-mediated signaling events may play a significant role in ADE by blunting the host-cell response to infection (Rodrigo et al., 2006) (4). (*See Color Plates*)

(Chu et al., 2005). Thus, blockade of the binding step is an attractive model for the neutralizing mechanism of DIII-specific mAbs.

mAbs also have the potential to neutralize infectivity at steps downstream of binding, perhaps by inhibiting the conformational changes in the E protein associated with membrane fusion. Pioneering electron microscopy studies by Gollins and Porterfield suggest that West Nile virions complexed with neutralizing quantities of antibody can be internalized by target cells, suggesting a post-attachment mechanism of neutralization (Gollins and Porterfield, 1986). Neutralization by the WNV DIII-specific mAb E16 also occurs at a post-attachment step of the viral entry pathway (Nybakken et al., 2005). Structural and cryoelectron reconstruction analysis suggest that docking of this antibody to the virion imposes steric constraints on the low-pH-mediated rearrangements of the E proteins that drive fusion (Kaufmann et al., 2006; Nybakken et al., 2005). The ability of mAbs to directly block flavivirus fusion has recently been demonstrated using tick-borne encephalitis virus (TBEV) (Stiasny et al., 2007a). Using a cell-free system in which labeled flavivirus virions are induced to fuse with synthetic lipid membranes, mAbs to some but not all epitopes on the TBEV E protein directly blocked fusion. Using similar approaches, the WNV DIII-specific mAb E16 has also been shown to directly and completely block fusion (B. Thompson, J. Smit, M. Diamond, and D. Fremont, unpublished observations).

While specific mechanisms of neutralization can readily be demonstrated using individual in vitro approaches, neutralization by a single antibody may occur via multiple mechanisms that operate simultaneously, depending on how many antibodies bind to the virion. For example, the WNV-specific mAb E16 neutralizes infection at a relatively low occupancy of epitopes on the virion (Pierson et al., 2007). This inhibition is thought to occur primarily by blocking the conformational changes in E protein required for fusion as discussed above. However, when individual virions are coated with saturating quantities of E16, attachment blockade may also occur (Nybakken et al., 2005). Furthermore, antibodies that neutralize infection only at full occupancy may block fusion in some experimental contexts, but may not neutralize infection in the endosome via this mechanism as the concentration of antibody is low and falls below that required for saturation and inhibitory function.

7 Antibody Dependent Enhancement of Infection

Antibodies are an important component of protective immune responses to flavivirus infection (reviewed in Diamond et al., 2003a). However, under some circumstances antibodies may exacerbate disease (Halstead, 2003). Interest in the pathogenic role of antibody in flavivirus infection arises in part from the complexities of DENV pathogenesis. Four antigenically related serotypes of DENV (serotypes 1–4) circulate in nature and are transmitted to humans through the bite of the *Aedes aegypti* mosquito (Calisher et al., 1989; Goldwasser and Davies, 1953). Infection from any of these viruses typically results in a self-limited flu-like illness called dengue fever, from which the infected individual recovers and is thereafter immune to re-infection with the same serotype of DENV. However, more severe and potentially fatal clinical manifestations of dengue infection have been reported (dengue hemorrhagic fever (DHF) or dengue shock syndrome (DSS)) (Chambers et al., 1990). Several prospective clinical studies strongly suggest that prior exposure to DENV is a significant risk factor for severe dengue disease (Alvarez et al., 2006; Endy et al., 2004; Stephens et al., 2002; Vaughn et al., 2000).

While the factors responsible for the severe complications of secondary DENV infection are not completely understood and are probably complex, one hypothesis is that antibodies elicited by primary infection not only fail to protect the host from a secondary infection by a heterologous serotype, but also promote enhanced viral infectivity in vivo (Halstead, 2003). Antibody-dependent enhancement of infection (ADE) describes the dramatic increase in infection of Fc-receptor or complement receptor-bearing cells in the presence of non-neutralizing concentrations of antibody or immune sera (Halstead and O'Rourke, 1977). The most direct link between ADE and the clinical outcome of DENV infection comes from investigations of the unusually large number of severe cases of disease following primary infection observed in infants during the first year of life. At birth, DENV-specific passively acquired antibodies are present at a relatively high concentration and exhibit neutralizing activity in vitro. However, as the infant ages, the maternally acquired antibody wanes to levels that are no longer protective, do not neutralize virus and enhance virus infection in vitro (Kliks et al., 1988; Nguyen et al., 2004; Simmons et al., 2007). The waning antibody titers of infants to levels that support ADE in vitro parallels the risk of DSS following primary DENV infection during the first year of life. Indeed, passive transfer of antibody can promote increased viral burden in animal models of DENV infection (Goncalvez et al., 2007; Halstead, 1979; Lai et al., 2007).

ADE has been demonstrated in vitro using a large number of different viruses, including WNV (reviewed in Takada and Kawaoka, 2003). However, the role of ADE during WNV infection in vivo, if any, is not yet clear. While passive transfer studies using a murine model of WNV infection demonstrate enhanced viral replication in the spleen, this did not translate into a change in disease phenotype or survival (Mehlhop and Diamond, 2006). However, ADE may prove to be important in the context of vaccination and cross-reactive responses with other viruses of the JEV antigenic complex. Mice immunized with UV-inactivated MVEV had significantly reduced average survival time and higher mortality when challenged with sublethal or lethal doses of JEV (Lobigs et al., 2003).

These experiments suggest that the immunity induced by some MVEV vaccine formulations enhance JEV. Whether cross-reactive responses could modulate WNV pathogenesis under some circumstances is unknown.

7.1 The Stoichiometry of ADE

Neutralization of flavivirus infection requires engagement of the virion with a stoichiometry that exceeds the threshold for neutralization. Studies with more than 100 WNV mAbs indicate that all antibodies that bind the virion have the potential to enhance infectivity of Fcγ-bearing cells, albeit with varying efficiency (Oliphant et al., 2006; Pierson et al., 2007). The phenomena of neutralization and ADE are related simply by the number of antibodies bound to the virion, which in turn is controlled by antibody affinity and epitope accessibility as described in detail above. Antibodies that bind the virion with low affinity, or recognize poorly accessible epitopes, promote ADE at higher concentrations relative to those that efficiently bind the virion via interactions with highly accessible determinants. Importantly, some epitopes elicit antibodies that may enhance infection in vitro even at saturating concentrations. Other factors that impact the potential for antibodies to promote ADE include isotype, glycosylation state (Nimmerjahn and Ravetch, 2008), and the presence of additional serum factors such as complement (Iankov et al., 2006; Mehlhop et al., 2007; Yamanaka et al., 2008; see review by Mehlhop and Diamond, this volume). While the upper limit for the number of antibodies bound to the virion capable of promoting infection is defined by the stoichiometric requirements of neutralization (Pierson et al., 2007), the minimal number of antibodies capable of promoting ADE is less clear. Using mixtures of two variants of the mAb E16 that differ only in their capacity to bind Fc-receptors, it has been proposed that between 15 and 30 mAbs are required for ADE (Pierson et al., 2007). The requirement for more than a single antibody may reflect a role for Fcγ-receptor cross-linking during ADE, or a requirement for more than a single antibody-Fcγ-receptor interaction during stable attachment of the virion to the target cell.

7.2 Mechanisms of ADE

The cell biology of ADE is not completely understood. ADE has been described as an opsonic phenomenon in which antibody functions as a bridge between the virion and Fcγ- or complement receptors on the target cells (Cardosa et al., 1983, 1986; Gollins and Porterfield, 1984).

Non-neutralizing concentrations of antibody increase the efficiency of virus binding to cells, and promote more efficient internalization of opsonized viruses (Gollins and Porterfield, 1984, 1985). Antibodies specific to either Fcγ- or complement receptors can block ADE in vitro (Peiris et al., 1981; Peiris and Porterfield, 1979). ADE is most dramatically demonstrated using cell types that do not bind virions efficiently in the absence of antibody (Boonnak et al., 2008; Pierson et al., 2007). Conversely, ADE may not significantly increase infection of cells expressing Fcγ-receptors that have other mechanisms for efficiently capturing virions, as suggested by studies of DENV with dendritic cells that express high levels of the attachment factor CD209 (Boonnak et al., 2008; Tassaneetrithep et al., 2003). In addition, the mechanism of ADE may also include processes that occur downstream of virus attachment. Fcγ-receptor signaling has been suggested to play a role in ADE (Rodrigo et al., 2006). Furthermore, DENV was shown to directly interfere with the host's innate immune response following Fcγ-receptor mediated entry by modulating expression of inflammatory cytokines (Chareonsirisuthigul et al., 2007; Yang et al., 2001). Suppression of antiviral transcription factors STAT-1 and NF-κB in macrophages following enhanced Ross River virus infection has also been demonstrated (Mahalingam and Lidbury, 2002).

8 Implications for Vaccine Development

Advances in the structural biology of flaviviruses, coupled with studies of the neutralization potency of antibodies using quantitative in vitro tests and animal models of WNV infection, have defined in biochemical terms the characteristics associated with potent neutralizing antibodies. The most potent mAbs characterized to date, block infection at relatively low occupancy of epitopes on the virion (Fig. 2) (Gromowski and Barrett, 2007; Pierson et al., 2007). This allows neutralization at relatively low concentrations of antibody, or via low affinity interactions. By comparison, neutralization by antibodies that recognize epitopes predicted to be less accessible for binding requires engagement of a large fraction of epitopes on the virion. The occupancy requirements for different antibodies reflect a requirement to engage a larger fraction of a less abundant epitope in order to reach the threshold required for neutralization.

The types of antibodies elicited by vaccines may prove to be an important predictor of durable protection or vaccine safety. Eliciting high affinity antibodies may not be sufficient. Antibodies that recognize cryptic determinants are typically less potent due to the large occupancy requirements for neutralization irrespective of antibody affinity. Furthermore, they are also more likely to promote enhancement on Fcγ-receptor bearing cells. In instances where the abundance of accessible epitopes on individual virions does not exceed the threshold for neutralization, enhancement of infection is possible even at saturating concentrations of antibody. Importantly, preliminary data suggest that the antibody repertoire of naturally infected humans is composed in the majority of antibodies specific for the largely cryptic fusion loop (Throsby et al., 2006). Eliciting antibodies against the DIII-lr, which are extremely potent in vitro and in vivo, may require the inclusion of genetically tailored immunogens in vaccines against WNV and other flaviviruses.

References

Aasa-Chapman, M.M., Holuigue, S., Aubin, K., Wong, M., Jones, N.A., Cornforth, D., Pellegrino, P., Newton, P., Williams, I., Borrow, P., et al. (2005). Detection of antibody-dependent complement-mediated inactivation of both autologous and heterologous virus in primary human immunodeficiency virus type 1 infection. J Virol *79*, 2823–2830.

AbuBakar, S., Azmi, A., Mohamed-Saad, N., Shafee, N., and Chee, H.Y. (1997). Antibody responses of dengue fever patients to dengue 2 (New Guinea C strain) viral proteins. Malaysian J Pathol *19*, 41–51.

Agrawal, A.G., and Petersen, L.R. (2003). Human immunoglobulin as a treatment for West Nile virus infection. J Infect Dis *188*, 1–4.

Alcon-LePoder, S., Sivard, P., Drouet, M.T., Talarmin, A., Rice, C., and Flamand, M. (2006). Secretion of flaviviral non-structural protein NS1: from diagnosis to pathogenesis. Novartis Foundation Symposium *277*, 233–247; discussion 247–253.

Allison, S.L., Stiasny, K., Stadler, K., Mandl, C.W., and Heinz, F.X. (1999). Mapping of functional elements in the stem-anchor region of tick-borne encephalitis virus envelope protein E. J Virol *73*, 5605–5612.

Allison, S.L., Schalich, J., Stiasny, K., Mandl, C.W., and Heinz, F.X. (2001). Mutational evidence for an internal fusion peptide in flavivirus envelope protein E. J Virol *75*, 4268–4275.

Alvarez, M., Rodriguez-Roche, R., Bernardo, L., Vazquez, S., Morier, L., Gonzalez, D., Castro, O., Kouri, G., Halstead, S.B., and Guzman, M.G. (2006). Dengue hemorrhagic fever caused by sequential dengue 1–3 virus infections over a long time interval: Havana epidemic, 2001–2002. Am J Trop Med Hyg *75*, 1113–1117.

Anderson, R. (2003). Manipulation of cell surface macromolecules by flaviviruses. Adv Virus Res *59*, 229–274.

Beasley, D.W., and Barrett, A.D. (2002). Identification of neutralizing epitopes within structural domain III of the West Nile virus envelope protein. J Virol *76*, 13097–13100.

Beebe, D.P., Schreiber, R.D., and Cooper, N.R. (1983). Neutralization of influenza virus by normal human sera: mechanisms involving antibody and complement. J Immunol *130*, 1317–1322.

Ben-Nathan, D., Lustig, S., Tam, G., Robinzon, S., Segal, S., and Rager-Zisman, B. (2003). Prophylactic and therapeutic efficacy of human intravenous immunoglobulin in treating West Nile virus infection in mice. J Infect Dis *188*, 5–12.

Berglund, P., Sjoberg, M., Garoff, H., Atkins, G.J., Sheahan, B.J., and Liljestrom, P. (1993). Semliki Forest virus expression system: production of conditionally infectious recombinant particles. Bio/Technology *11*, 916–920.

Boonnak, K., Slike, B.M., Burgess, T.H., Mason, R.M., Wu, S.J., Sun, P., Porter, K., Rudiman, I.F., Yuwono, D., Puthavathana, P., et al. (2008). Role of dendritic cells in antibody dependent enhancement of dengue infection. J Virol *82*, 3939–3951.

Booy, F.P., Roden, R.B., Greenstone, H.L., Schiller, J.T., and Trus, B.L. (1998). Two antibodies that neutralize papillomavirus by different mechanisms show distinct binding patterns at 13 A resolution. J Mol Biol *281*, 95–106.

Bressanelli, S., Stiasny, K., Allison, S.L., Stura, E.A., Duquerroy, S., Lescar, J., Heinz, F.X., and Rey, F.A. (2004). Structure of a flavivirus envelope glycoprotein in its low-pH-induced membrane fusion conformation. EMBO J *23*, 728–738.

Brinton, M.A. (2002). The molecular biology of West Nile virus: a new invader of the western hemisphere. Annu Rev Microbiol *56*, 371–402.

Burnet, F.M., Keogh, E.V., and Lush, D. (1937). The immunological reactions of the filterable viruses. Aust J Exp Biol Med Sci *15*, 227–368.

Burrage, T., Kramer, E., and Brown, F. (2000). Structural differences between foot-and-mouth disease and poliomyelitis viruses influence their inactivation by aziridines. Vaccine *18*, 2454–2461.

Burton, D.R., Saphire, E.O., and Parren, P.W. (2001). A model for neutralization of viruses based on antibody coating of the virion surface. Curr Top Microbiol Immunol *260*, 109–143.

Calisher, C.H., Karabatsos, N., Dalrymple, J.M., Shope, R.E., Porterfield, J.S., Westaway, E.G., and Brandt, W.E. (1989). Antigenic relationships between flaviviruses as determined by cross-neutralization tests with polyclonal antisera. J Gen Virol *70* (Pt 1), 37–43.

Camenga, D.L., Nathanson, N., and Cole, G.A. (1974). Cyclophosphamide-potentiated West Nile viral encephalitis: relative influence of cellular and humoral factors. J Infect Dis *130*, 634–641.

Cardosa, M.J., Porterfield, J.S., and Gordon, S. (1983). Complement receptor mediates enhanced flavivirus replication in macrophages. J Exp Med *158*, 258–263.

Cardosa, M.J., Gordon, S., Hirsch, S., Springer, T.A., and Porterfield, J.S. (1986). Interaction of West Nile virus with primary murine macrophages: role of cell activation and receptors for antibody and complement. J Virol *57*, 952–959.

Cecilia, D., and Gould, E.A. (1991). Nucleotide changes responsible for loss of neuroinvasiveness in Japanese encephalitis virus neutralization-resistant mutants. Virology *181*, 70–77.

Cecilia, D., Gadkari, D.A., Kedarnath, N., and Ghosh, S.N. (1988). Epitope mapping of Japanese encephalitis virus envelope protein using monoclonal antibodies against an Indian strain. J Gen Virol *69* (Pt 11), 2741–2747.

Chambers, T.J., Hahn, C.S., Galler, R., and Rice, C.M. (1990). Flavivirus genome organization, expression, and replication. Annu Rev Microbiol *44*, 649–688.

Chareonsirisuthigul, T., Kalayanarooj, S., and Ubol, S. (2007). Dengue virus (DENV) antibody-dependent enhancement of infection upregulates the production of anti-inflammatory cytokines, but suppresses anti-DENV free radical and pro-inflammatory cytokine production, in THP-1 cells. J Gen Virol *88*, 365–375.

Che, Z., Olson, N.H., Leippe, D., Lee, W.M., Mosser, A.G., Rueckert, R.R., Baker, T.S., and Smith, T.J. (1998). Antibody-mediated neutralization of human rhinovirus 14 explored by means of cryoelectron microscopy and X-ray crystallography of virus-Fab complexes. J Virol *72*, 4610–4622.

Chu, J.J., and Ng, M.L. (2004). Infectious entry of West Nile virus occurs through a clathrin-mediated endocytic pathway. J Virol *78*, 10543–10555.

Chu, J.J., Rajamanonmani, R., Li, J., Bhuvanakantham, R., Lescar, J., and Ng, M.L. (2005). Inhibition of West Nile virus entry by using a recombinant domain III from the envelope glycoprotein. J Gen Virol *86*, 405–412.

Chung, K.M., Nybakken, G.E., Thompson, B.S., Engle, M.J., Marri, A., Fremont, D.H., and Diamond, M.S. (2006). Antibodies against West Nile virus nonstructural protein NS1 prevent lethal infection through Fc gamma receptor-dependent and -independent mechanisms. J Virol *80*, 1340–1351.

Chung, K.M., Thompson, B.S., Fremont, D.H., and Diamond, M.S. (2007). Antibody recognition of cell surface-associated NS1 triggers Fc-gamma receptor-mediated phagocytosis and clearance of West Nile virus-infected cells. J Virol *81*, 9551–9555.

Colman, P.M., and Lawrence, M.C. (2003). The structural biology of type I viral membrane fusion. Nat Rev Mol Cell Biol *4*, 309–319.

Colombage, G., Hall, R., Pavy, M., and Lobigs, M. (1998). DNA-based and alphavirus-vectored immunisation with prM and E proteins elicits long-lived and protective immunity against the flavivirus, Murray Valley encephalitis virus. Virology *250*, 151–163.

Colonno, R.J., Callahan, P.L., Leippe, D.M., Rueckert, R.R., and Tomassini, J.E. (1989). Inhibition of rhinovirus attachment by neutralizing monoclonal antibodies and their Fab fragments. J Virol *63*, 36–42.

Crill, W.D., and Chang, G.J. (2004). Localization and characterization of flavivirus envelope glycoprotein cross-reactive epitopes. J Virol *78*, 13975–13986.

Crill, W.D., and Roehrig, J.T. (2001). Monoclonal antibodies that bind to domain III of dengue virus E glycoprotein are the most efficient blockers of virus adsorption to Vero cells. J Virol *75*, 7769–7773.

Crill, W.D., Trainor, N.B., and Chang, G.J. (2007). A detailed mutagenesis study of flavivirus cross-reactive epitopes using West Nile virus-like particles. J Gen Virol *88*, 1169–1174.

Davis, C.W., Nguyen, H.Y., Hanna, S.L., Sanchez, M.D., Doms, R.W., and Pierson, T.C. (2006). West Nile virus discriminates between DC-SIGN and DC-SIGNR for cellular attachment and infection. J Virol *80*, 1290–1301.

Della-Porta, A.J., and Westaway, E.G. (1978). A multi-hit model for the neutralization of animal viruses. J Gen Virol *38*, 1–19.

Diamond, M.S., Shrestha, B., Marri, A., Mahan, D., and Engle, M. (2003a). B cells and antibody play critical roles in the immediate defense of disseminated infection by West Nile encephalitis virus. J Virol *77*, 2578–2586.

Diamond, M.S., Sitati, E.M., Friend, L.D., Higgs, S., Shrestha, B., and Engle, M. (2003b). A critical role for induced IgM in the protection against West Nile virus infection. J Exp Med *198*, 1853–1862.

Dulbecco, R., Vogt, M., and Strickland, A.G. (1956). A study of the basic aspects of neutralization of two animal viruses, western equine encephalitis virus and poliomyelitis virus. Virology *2*, 162–205.

Endy, T.P., Nisalak, A., Chunsuttitwat, S., Vaughn, D.W., Green, S., Ennis, F.A., Rothman, A.L., and Libraty, D.H. (2004). Relationship of preexisting dengue virus (DV) neutralizing antibody levels to viremia and severity of disease in a prospective cohort study of DV infection in Thailand. J Infect Dis *189*, 990–1000.

Engle, M.J., and Diamond, M.S. (2003). Antibody prophylaxis and therapy against West Nile virus infection in wild-type and immunodeficient mice. J Virol *77*, 12941–12949.

Falconar, A.K. (1999). Identification of an epitope on the dengue virus membrane (M) protein defined by cross-protective monoclonal antibodies: design of an improved epitope sequence based on common determinants present in both envelope (E and M) proteins. Arch Virol *144*, 2313–2330.

Feng, J.Q., Mozdzanowska, K., and Gerhard, W. (2002). Complement component C1q enhances the biological activity of influenza virus hemagglutinin-specific antibodies depending on their fine antigen specificity and heavy-chain isotype. J Virol *76*, 1369–1378.

Flamand, A., Raux, H., Gaudin, Y., and Ruigrok, R.W. (1993). Mechanisms of rabies virus neutralization. Virology *194*, 302–313.

Goldwasser, R.A., and Davies, A.M. (1953). Transmission of a West Nile-like virus by *Aedes aegypti*. Trans R Soc Trop Med Hyg *47*, 336–337.

Gollins, S.W., and Porterfield, J.S. (1984). Flavivirus infection enhancement in macrophages: radioactive and biological studies on the effect of antibody on viral fate. J Gen Virol *65* (Pt 8), 1261–1272.

Gollins, S.W., and Porterfield, J.S. (1985). Flavivirus infection enhancement in macrophages: an electron microscopic study of viral cellular entry. J Gen Virol *66* (Pt 9), 1969–1982.

Gollins, S.W., and Porterfield, J.S. (1986). A new mechanism for the neutralization of enveloped viruses by antiviral antibody. Nature *321*, 244–246.

Goncalvez, A.P., Men, R., Wernly, C., Purcell, R.H., and Lai, C.J. (2004). Chimpanzee Fab fragments and a derived humanized immunoglobulin G1 antibody that efficiently cross-neutralize dengue type 1 and type 2 viruses. J Virol *78*, 12910–12918.

Goncalvez, A.P., Engle, R.E., St Claire, M., Purcell, R.H., and Lai, C.J. (2007). Monoclonal antibody-mediated enhancement of dengue virus infection in vitro and in vivo and strategies for prevention. Proc Natl Acad Sci U S A *104*, 9422–9427.

Gromowski, G.D., and Barrett, A.D. (2007). Characterization of an antigenic site that contains a dominant, type-specific neutralization determinant on the envelope protein domain III (ED3) of dengue 2 virus. Virology *366*, 349–360.

Guirakhoo, F., Heinz, F.X., Mandl, C.W., Holzmann, H., and Kunz, C. (1991). Fusion activity of flaviviruses: comparison of mature and immature (prM-containing) tick-borne encephalitis virions. J Gen Virol *72* (Pt 6), 1323–1329.

Guirakhoo, F., Bolin, R.A., and Roehrig, J.T. (1992). The Murray Valley encephalitis virus prM protein confers acid resistance to virus particles and alters the expression of epitopes within the R2 domain of E glycoprotein. Virology *191*, 921–931.

Halevy, M., Akov, Y., Ben-Nathan, D., Kobiler, D., Lachmi, B., and Lustig, S. (1994). Loss of active neuroinvasiveness in attenuated strains of West Nile virus: pathogenicity in immunocompetent and SCID mice. Arch Virol *137*, 355–370.

Haley, M., Retter, A.S., Fowler, D., Gea-Banacloche, J., and O'Grady, N.P. (2003). The role for intravenous immunoglobulin in the treatment of West Nile virus encephalitis. Clin Infect Dis *37*, e88–e90.

Hall, R.A., Kay, B.H., Burgess, G.W., Clancy, P., and Fanning, I.D. (1990). Epitope analysis of the envelope and non-structural glycoproteins of Murray Valley encephalitis virus. J Gen Virol *71* (Pt 12), 2923–2930.

Halstead, S.B. (1979). In vivo enhancement of dengue virus infection in rhesus monkeys by passively transferred antibody. J Infect Dis *140*, 527–533.

Halstead, S.B. (2003). Neutralization and antibody-dependent enhancement of dengue viruses. Adv Virus Res *60*, 421–467.

Halstead, S.B., and O'Rourke, E.J. (1977). Dengue viruses and mononuclear phagocytes. I. Infection enhancement by non-neutralizing antibody. J Exp Med *146*, 201–217.

Hamdan, A., Green, P., Mendelson, E., Kramer, M.R., Pitlik, S., and Weinberger, M. (2002). Possible benefit of intravenous immunoglobulin therapy in a lung transplant recipient with West Nile virus encephalitis. Transpl Infect Dis *4*, 160–162.

Hayes, E.B., Sejvar, J.J., Zaki, S.R., Lanciotti, R.S., Bode, A.V., and Campbell, G.L. (2005). Virology, pathology, and clinical manifestations of West Nile virus disease. Emerg Infect Dis *11*, 1174–1179.

Heinz, F.X., and Allison, S.L. (2000). Structures and mechanisms in flavivirus fusion. Adv Virus Res *55*, 231–269.

Heinz, F.X., Berger, R., Tuma, W., and Kunz, C. (1983). A topological and functional model of epitopes on the structural glycoprotein of tick-borne encephalitis virus defined by monoclonal antibodies. Virology *126*, 525–537.

Heinz, F.X., Stiasny, K., Puschner-Auer, G., Holzmann, H., Allison, S.L., Mandl, C.W., and Kunz, C. (1994). Structural changes and functional control of the tick-borne encephalitis virus glycoprotein E by the heterodimeric association with protein prM. Virology *198*, 109–117.

Henchal, E.A., Henchal, L.S., and Schlesinger, J.J. (1988). Synergistic interactions of anti-NS1 monoclonal antibodies protect passively immunized mice from lethal challenge with dengue 2 virus. J Gen Virol *69* (Pt 8), 2101–2107.

He, R.T., Innis, B.L., Nisalak, A., Usawattanakul, W., Wang, S., Kalayanarooj, S., and Anderson, R. (1995). Antibodies that block virus attachment to Vero cells are a major component of the human neutralizing antibody response against dengue virus type 2. J Med Virol *45*, 451–461.

Hiramatsu, K., Tadano, M., Men, R., and Lai, C.J. (1996). Mutational analysis of a neutralization epitope on the dengue type 2 virus (DEN2) envelope protein: monoclonal antibody resistant DEN2/DEN4 chimeras exhibit reduced mouse neurovirulence. Virology *224*, 437–445.

Holzmann, H., Heinz, F.X., Mandl, C.W., Guirakhoo, F., and Kunz, C. (1990). A single amino acid substitution in envelope protein E of tick-borne encephalitis virus leads to attenuation in the mouse model. J Virol *64*, 5156–5159.

Iankov, I.D., Pandey, M., Harvey, M., Griesmann, G.E., Federspiel, M.J., and Russell, S.J. (2006). Immunoglobulin g antibody-mediated enhancement of measles virus infection can bypass the protective antiviral immune response. J Virol *80*, 8530–8540.

Icenogle, J., Shiwen, H., Duke, G., Gilbert, S., Rueckert, R., and Anderegg, J. (1983). Neutralization of poliovirus by a monoclonal antibody: kinetics and stoichiometry. Virology *127*, 412–425.

Jennings, A.D., Gibson, C.A., Miller, B.R., Mathews, J.H., Mitchell, C.J., Roehrig, J.T., Wood, D.J., Taffs, F., Sil, B.K., Whitby, S.N., et-al. (1994). Analysis of a yellow fever virus isolated from a fatal case of vaccine-associated human encephalitis. J Infect Dis *169*, 512–518.

Jiang, W.R., Lowe, A., Higgs, S., Reid, H., and Gould, E.A. (1993). Single amino acid codon changes detected in louping ill virus antibody-resistant mutants with reduced neurovirulence. J Gen Virol *74* (Pt 5), 931–935.

Kanai, R., Kar, K., Anthony, K., Gould, L.H., Ledizet, M., Fikrig, E., Marasco, W.A., Koski, R.A., and Modis, Y. (2006). Crystal structure of west nile virus envelope glycoprotein reveals viral surface epitopes. J Virol *80*, 11000–11008.

Kaufmann, B., Nybakken, G.E., Chipman, P.R., Zhang, W., Diamond, M.S., Fremont, D.H., Kuhn, R.J., and Rossmann, M.G. (2006). West Nile virus in complex with the Fab fragment of a neutralizing monoclonal antibody. Proc Natl Acad Sci U S A *103*, 12400–12404.

Klasse, P.J., and Burton, D.R. (2007). Antibodies to West Nile virus: a double-edged sword. Cell Host Microbe *1*, 87–89.

Klasse, P.J., and Moore, J.P. (1996). Quantitative model of antibody- and soluble CD4-mediated neutralization of primary isolates and T-cell line-adapted strains of human immunodeficiency virus type 1. J Virol *70*, 3668–3677.

Klasse, P.J., and Sattentau, Q.J. (2002). Occupancy and mechanism in antibody-mediated neutralization of animal viruses. J Gen Virol *83*, 2091–2108.

Kliks, S.C., Nimmanitya, S., Nisalak, A., and Burke, D.S. (1988). Evidence that maternal dengue antibodies are important in the development of dengue hemorrhagic fever in infants. Am J Trop Med Hyg *38*, 411–419.

Krishnan, M.N., Sukumaran, B., Pal, U., Agaisse, H., Murray, J.L., Hodge, T.W., and Fikrig, E. (2007). Rab 5 is required for the cellular entry of dengue and West Nile viruses. J Virol *81*, 4881–4885.

Kuhn, R.J., Zhang, W., Rossmann, M.G., Pletnev, S.V., Corver, J., Lenches, E., Jones, C.T., Mukhopadhyay, S., Chipman, P.R., Strauss, E.G., et-al. (2002). Structure of dengue virus: implications for flavivirus organization, maturation, and fusion. Cell *108*, 717–725.

Lai, C.J., Goncalvez, A.P., Men, R., Wernly, C., Donau, O., Engle, R.E., and Purcell, R.H. (2007). Epitope determinants of a chimpanzee dengue virus type 4 (DENV-4)-neutralizing antibody and protection against DENV-4 challenge in mice and rhesus monkeys by passively transferred humanized antibody. J Virol *81*, 12766–12774.

Lewis, J.K., Bothner, B., Smith, T.J., and Siuzdak, G. (1998). Antiviral agent blocks breathing of the common cold virus. Proc Natl Acad Sci U S A *95*, 6774–6778.

Li, Q., Yafal, A.G., Lee, Y.M., Hogle, J., and Chow, M. (1994). Poliovirus neutralization by antibodies to internal epitopes of VP4 and VP1 results from reversible exposure of these sequences at physiological temperature. J Virol *68*, 3965–3970.

Lobigs, M., Pavy, M., and Hall, R. (2003). Cross-protective and infection-enhancing immunity in mice vaccinated against flaviviruses belonging to the Japanese encephalitis virus sero-complex. Vaccine *21*, 1572–1579.

Lok, S.M., Kostyuchenko, V., Nybakken, G.E., Holdaway, H.A., Battisti, A.J., Sukupolvi-Petty, S., Sedlak, D., Fremont, D.H., Chipman, P.R., Roehrig, J.T., et-al. (2008). Binding of a neutralizing antibody to dengue virus alters the arrangement of surface glycoproteins. Nat Struct Mol Biol *15*, 312–317.

Mackenzie, J.M., and Westaway, E.G. (2001). Assembly and maturation of the flavivirus Kunjin virus appear to occur in the rough endoplasmic reticulum and along the secretory pathway, respectively. J Virol *75*, 10787–10799.

Mackenzie, J.S., Gubler, D.J., and Petersen, L.R. (2004). Emerging flaviviruses: the spread and resurgence of Japanese encephalitis, West Nile and dengue viruses. Nat Med *10*, S98–S109.

Mahalingam, S., and Lidbury, B.A. (2002). Suppression of lipopolysaccharide-induced antiviral transcription factor (STAT-1 and NF-kappa B) complexes by antibody-dependent enhancement of macrophage infection by Ross River virus. Proc Natl Acad Sci U S A *99*, 13819–13824.

Mandl, C.W., Guirakhoo, F., Holzmann, H., Heinz, F.X., and Kunz, C. (1989). Antigenic structure of the flavivirus envelope protein E at the molecular level, using tick-borne encephalitis virus as a model. J Virol *63*, 564–571.

Mandl, C.W., Allison, S.L., Holzmann, H., Meixner, T., and Heinz, F.X. (2000). Attenuation of tick-borne encephalitis virus by structure-based site-specific mutagenesis of a putative flavivirus receptor binding site. J Virol *74*, 9601–9609.

Mehlhop, E., and Diamond, M.S. (2006). Protective immune responses against West Nile virus are primed by distinct complement activation pathways. J Exp Med *203*, 1371–1381.

Mehlhop, E., Whitby, K., Oliphant, T., Marri, A., Engle, M., and Diamond, M.S. (2005). Complement activation is required for induction of a protective antibody response against West Nile virus infection. J Virol *79*, 7466–7477.

Mehlhop, E., Ansarah-Sobrinho, C., Johnson, S., Engle, M., Fremont, D.H., Pierson, T.C., and Diamond, M.S. (2007). Complement protein C1q inhibits antibody-dependent enhancement of flavivirus infection in an IgG subclass-specific manner. Cell Host Microbe *2*, 417–426.

Meyer, K., Basu, A., Przysiecki, C.T., Lagging, L.M., Di Bisceglie, A.M., Conley, A.J., and Ray, R. (2002). Complement-mediated enhancement of antibody function for neutralization of pseudotype virus containing hepatitis C virus E2 chimeric glycoprotein. J Virol *76*, 2150–2158.

Modis, Y., Ogata, S., Clements, D., and Harrison, S.C. (2003). A ligand-binding pocket in the dengue virus envelope glycoprotein. Proc Natl Acad Sci U S A *100*, 6986–6991.

Modis, Y., Ogata, S., Clements, D., and Harrison, S.C. (2004). Structure of the dengue virus envelope protein after membrane fusion. Nature *427*, 313–319.

Morrey, J.D., Siddharthan, V., Olsen, A.L., Roper, G.Y., Wang, H., Baldwin, T.J., Koenig, S., Johnson, S., Nordstrom, J.L., and Diamond, M.S. (2006). Humanized monoclonal antibody against West Nile virus envelope protein administered after neuronal infection protects against lethal encephalitis in hamsters. J Infect Dis *194*, 1300–1308.

Morrey, J.D., Siddharthan, V., Olsen, A.L., Wang, H., Julander, J.G., Hall, J.O., Li, H., Nordstrom, J.L., Koenig, S., Johnson, S., et-al. (2007). Defining limits of treatment with humanized neutralizing monoclonal antibody for West Nile virus neurological infection in a hamster model. Antimicrob Agents Chemother *51*, 2396–2402.

Mozdzanowska, K., Feng, J., Eid, M., Zharikova, D., and Gerhard, W. (2006). Enhancement of neutralizing activity of influenza virus-specific antibodies by serum components. Virology *352*, 418–426.

Mukhopadhyay, S., Kim, B.S., Chipman, P.R., Rossmann, M.G., and Kuhn, R.J. (2003). Structure of West Nile virus. Science *302*, 248.

Mukhopadhyay, S., Kuhn, R.J., and Rossmann, M.G. (2005). A structural perspective of the flavivirus life cycle. Nat Rev Microbiol *3*, 13–22.

Nguyen, T.H., Lei, H.Y., Nguyen, T.L., Lin, Y.S., Huang, K.J., Le, B.L., Lin, C.F., Yeh, T.M., Do, Q.H., Vu, T.Q., et-al. (2004). Dengue hemorrhagic fever in infants: a study of clinical and cytokine profiles. J Infect Dis *189*, 221–232.

Nimmerjahn, F., and Ravetch, J.V. (2008). Fcgamma receptors as regulators of immune responses. Nat Rev *8*, 34–47.

Nybakken, G.E., Oliphant, T., Johnson, S., Burke, S., Diamond, M.S., and Fremont, D.H. (2005). Structural basis of West Nile virus neutralization by a therapeutic antibody. Nature *437*, 764–769.

Nybakken, G.E., Nelson, C.A., Chen, B.R., Diamond, M.S., and Fremont, D.H. (2006). Crystal structure of the West Nile virus envelope glycoprotein. J Virol *80*, 11467–11474.

Oliphant, T., and Diamond, M.S. (2007). The molecular basis of antibody-mediated neutralization of West Nile virus. Expert Opin Biol Ther *7*, 885–892.

Oliphant, T., Engle, M., Nybakken, G.E., Doane, C., Johnson, S., Huang, L., Gorlatov, S., Mehlhop, E., Marri, A., Chung, K.M., et-al. (2005). Development of a humanized monoclonal antibody with therapeutic potential against West Nile virus. Nat Med *11*, 522–530.

Oliphant, T., Nybakken, G.E., Engle, M., Xu, Q., Nelson, C.A., Sukupolvi-Petty, S., Marri, A., Lachmi, B.E., Olshevsky, U., Fremont, D.H., et-al. (2006). Antibody recognition and neutralization determinants on domains I and II of West Nile virus envelope protein. J Virol *80*, 12149–12159.

Oliphant, T., Nybakken, G.E., Austin, S.K., Xu, Q., Bramson, J., Loeb, M., Throsby, M., Fremont, D.H., Pierson, T.C., and Diamond, M.S. (2007). Induction of epitope-specific neutralizing antibodies against West Nile virus. J Virol *81*, 11828–11839.

Pantophlet, R., and Burton, D.R. (2006). GP120: target for neutralizing HIV-1 antibodies. Annu Rev Immunol *24*, 739–769.

Peiris, J.S., and Porterfield, J.S. (1979). Antibody-mediated enhancement of Flavivirus replication in macrophage-like cell lines. Nature *282*, 509–511.

Peiris, J.S., Gordon, S., Unkeless, J.C., and Porterfield, J.S. (1981). Monoclonal anti-Fc receptor IgG blocks antibody enhancement of viral replication in macrophages. Nature *289*, 189–191.

Pierson, T.C., Xu, Q., Nelson, S., Oliphant, T., Nybakken, G.E., Fremont, D.H., and Diamond, M.S. (2007). The stoichiometry of antibody-mediated neutralization and enhancement of West Nile virus infection. Cell Host Microbe *1*, 135–145.

Pincus, S., Mason, P.W., Konishi, E., Fonseca, B.A., Shope, R.E., Rice, C.M., and Paoletti, E. (1992). Recombinant vaccinia virus producing the prM and E proteins of yellow fever virus protects mice from lethal yellow fever encephalitis. Virology *187*, 290–297.

Randolph, V.B., and Stollar, V. (1990). Low pH-induced cell fusion in flavivirus-infected *Aedes albopictus* cell cultures. J Gen Virol *71* (Pt 8), 1845–1850.

Randolph, V.B., Winkler, G., and Stollar, V. (1990). Acidotropic amines inhibit proteolytic processing of flavivirus prM protein. Virology *174*, 450–458.

Reisdorph, N., Thomas, J.J., Katpally, U., Chase, E., Harris, K., Siuzdak, G., and Smith, T.J. (2003). Human rhinovirus capsid dynamics is controlled by canyon flexibility. Virology *314*, 34–44.

Rey, F.A., Heinz, F.X., Mandl, C., Kunz, C., and Harrison, S.C. (1995). The envelope glycoprotein from tick-borne encephalitis virus at 2 A resolution. Nature *375*, 291–298.

Rice, C.M. (1996). Flaviviridae: the viruses and their replication. In Fields Virology, B.N. Fields, P.M. Knipe, and P.M. Howley, eds. (Philadelphia, Lippincott-Raven), pp. 931–959.

Roden, R.B., Weissinger, E.M., Henderson, D.W., Booy, F., Kirnbauer, R., Mushinski, J.F., Lowy, D.R., and Schiller, J.T. (1994). Neutralization of bovine papillomavirus by antibodies to L1 and L2 capsid proteins. J Virol *68*, 7570–7574.

Rodrigo, W.W., Jin, X., Blackley, S.D., Rose, R.C., and Schlesinger, J.J. (2006). Differential enhancement of dengue virus immune complex infectivity mediated by signaling-competent and signaling-incompetent human Fcgamma RIA (CD64) or FcgammaRIIA (CD32). J Virol *80*, 10128–10138.

Roehrig, J.T. (2003). Antigenic structure of flavivirus proteins. Adv Virus Res *59*, 141–175.

Roehrig, J.T., Mathews, J.H., and Trent, D.W. (1983). Identification of epitopes on the E glycoprotein of Saint Louis encephalitis virus using monoclonal antibodies. Virology *128*, 118–126.

Roehrig, J.T., Bolin, R.A., and Kelly, R.G. (1998). Monoclonal antibody mapping of the envelope glycoprotein of the dengue 2 virus, Jamaica. Virology *246*, 317–328.

Roehrig, J.T., Staudinger, L.A., Hunt, A.R., Mathews, J.H., and Blair, C.D. (2001). Antibody prophylaxis and therapy for flavivirus encephalitis infections. Ann N Y Acad Sci *951*, 286–297.

Ryman, K.D., Ledger, T.N., Campbell, G.A., Watowich, S.J., and Barrett, A.D. (1998). Mutation in a 17D-204 vaccine substrain-specific envelope protein epitope alters the pathogenesis of yellow fever virus in mice. Virology *244*, 59–65.

Salminen, A., Wahlberg, J.M., Lobigs, M., Liljestrom, P., and Garoff, H. (1992). Membrane fusion process of Semliki Forest virus. II: Cleavage-dependent reorganization of the spike protein complex controls virus entry. J Cell Biol *116*, 349–357.

Samuel, M.A., and Diamond, M.S. (2006). Pathogenesis of West Nile virus infection: a balance between virulence, innate and adaptive immunity, and viral evasion. J Virol *80*, 9349–9360.

Sanchez, M.D., Pierson, T.C., McAllister, D., Hanna, S.L., Puffer, B.A., Valentine, L.E., Murtadha, M.M., Hoxie, J.A., and Doms, R.W. (2005). Characterization of neutralizing antibodies to West Nile virus. Virology *336*, 70–82.

Schonning, K., Lund, O., Lund, O.S., and Hansen, J.E. (1999). Stoichiometry of monoclonal antibody neutralization of T-cell line-adapted human immunodeficiency virus type 1. J Virol *73*, 8364–8370.

Sejvar, J.J. (2007). The long-term outcomes of human West Nile virus infection. Clin Infect Dis *44*, 1617–1624.

Serafin, I.L., and Aaskov, J.G. (2001). Identification of epitopes on the envelope (E) protein of dengue 2 and dengue 3 viruses using monoclonal antibodies. Arch Virol *146*, 2469–2479.

Shepherd, C.M., Borelli, I.A., Lander, G., Natarajan, P., Siddavanahalli, V., Bajaj, C., Johnson, J.E., BrooksIII, C.L., , and Reddy, V.S. (2006). VIPERdb: a relational database for structural virology. Nucleic Acids Res *34*, D386–D389.

Shimoni, Z., Niven, M.J., Pitlick, S., and Bulvik, S. (2001). Treatment of West Nile virus encephalitis with intravenous immunoglobulin. Emerg Infect Dis *7*, 759.

Shu, P.Y., Chen, L.K., Chang, S.F., Yueh, Y.Y., Chow, L., Chien, L.J., Chin, C., Lin, T.H., and Huang, J.H. (2000). Dengue NS1-specific antibody responses: isotype distribution and serotyping in patients with dengue fever and dengue hemorrhagic fever. J Med Virol *62*, 224–232.

Simmons, C.P., Chau, T.N., Thuy, T.T., Tuan, N.M., Hoang, D.M., Thien, N.T., Lien le, B., Quy, N.T., Hieu, N.T., Hien, T.T., et-al. (2007). Maternal antibody and viral factors in the pathogenesis of dengue virus in infants. J Infect Dis *196*, 416–424.

Spector, S.L., and Tauraso, N.M. (1969). Yellow fever virus. II. Factors affecting the plaque neutralization test. Appl Microbiol *18*, 736–743.

Stadler, K., Allison, S.L., Schalich, J., and Heinz, F.X. (1997). Proteolytic activation of tick-borne encephalitis virus by furin. J Virol *71*, 8475–8481.

Stephens, H.A., Klaythong, R., Sirikong, M., Vaughn, D.W., Green, S., Kalayanarooj, S., Endy, T.P., Libraty, D.H., Nisalak, A., Innis, B.L., et-al. (2002). HLA-A and -B allele associations with secondary dengue virus infections correlate with disease severity and the infecting viral serotype in ethnic Thais. Tissue Antigens *60*, 309–318.

Stiasny, K., and Heinz, F.X. (2006). Flavivirus membrane fusion. J Gen Virol *87*, 2755–2766.

Stiasny, K., Kiermayr, S., Holzmann, H., and Heinz, F.X. (2006). Cryptic properties of a cluster of dominant flavivirus cross-reactive antigenic sites. J Virol *80*, 9557–9568.

Stiasny, K., Brandler, S., Kossl, C., and Heinz, F.X. (2007a). Probing the flavivirus membrane fusion mechanism by using monoclonal antibodies. J Virol *81*, 11526–11531.

Stiasny, K., Kossl, C., Lepault, J., Rey, F.A., and Heinz, F.X. (2007b). Characterization of a structural intermediate of flavivirus membrane fusion. PLoS Pathog*3*, e20.

Sukupolvi-Petty, S., Austin, S.K., Purtha, W.E., Oliphant, T., Nybakken, G.E., Schlesinger, J.J., Roehrig, J.T., Gromowski, G.D., Barrett, A.D., Fremont, D.H., (2007). Type- and sub-complex-specific neutralizing antibodies against domain III of dengue virus type-2 envelope protein recognize adjacent epitopes. J Virol.

Takada, A., and Kawaoka, Y. (2003). Antibody-dependent enhancement of viral infection: molecular mechanisms and in vivo implications. Rev Med Virol *13*, 387–398.

Tassaneetrithep, B., Burgess, T.H., Granelli-Piperno, A., Trumpfheller, C., Finke, J., Sun, W., Eller, M.A., Pattanapanyasat, K., Sarasombath, S., Birx, D.L., et-al. (2003). DC-SIGN (CD209) mediates dengue virus infection of human dendritic cells. J Exp Med *197*, 823–829.

Taylor, H.P., Armstrong, S.J., and Dimmock, N.J. (1987). Quantitative relationships between an influenza virus and neutralizing antibody. Virology *159*, 288–298.

Tesh, R.B., Arroyo, J., Travassos Da Rosa, A.P., Guzman, H., Xiao, S.Y., and Monath, T.P. (2002). Efficacy of killed virus vaccine, live attenuated chimeric virus vaccine, and passive immunization for prevention of West Nile virus encephalitis in hamster model. Emerg Infect Dis *8*, 1392–1397.

Throsby, M., Geuijen, C., Goudsmit, J., Bakker, A.Q., Korimbocus, J., Kramer, R.A., Clijsters-van der Horst, M., de Jong, M., Jongeneelen, M., Thijsse, S., et-al. (2006). Isolation and characterization of human monoclonal antibodies from individuals infected with West Nile Virus. J Virol *80*, 6982–6992.

Thullier, P., Demangel, C., Bedouelle, H., Megret, F., Jouan, A., Deubel, V., Mazie, J.C., and Lafaye, P. (2001). Mapping of a dengue virus neutralizing epitope critical for the infectivity of all serotypes: insight into the neutralization mechanism. J Gen Virol *82*, 1885–1892.

van der Schaar, H.M., Rust, M.J., Waarts, B.L., van der Ende-Metselaar, H., Kuhn, R.J., Wilschut, J., Zhuang, X., and Smit, J.M. (2007). Characterization of the early events in dengue virus cell entry by biochemical assays and single-virus tracking. J Virol *81*, 12019–12028.

Vaughn, D.W., Green, S., Kalayanarooj, S., Innis, B.L., Nimmannitya, S., Suntayakorn, S., Endy, T.P., Raengsakulrach, B., Rothman, A.L., Ennis, F.A., et-al. (2000). Dengue viremia titer, antibody response pattern, and virus serotype correlate with disease severity. J Infect Dis *181*, 2–9.

Volanakis, J.E. (2002). The role of complement in innate and adaptive immunity. Curr Top Microbiol Immunol *266*, 41–56.

Volk, D.E., Beasley, D.W., Kallick, D.A., Holbrook, M.R., Barrett, A.D., and Gorenstein, D.G. (2004). Solution structure and antibody binding studies of the envelope protein domain III from the New York strain of West Nile virus. J Biol Chem *279*, 38755–38761.

Wang, T., Anderson, J.F., Magnarelli, L.A., Wong, S.J., Koski, R.A., and Fikrig, E. (2001). Immunization of mice against West Nile virus with recombinant envelope protein. J Immunol *167*, 5273–5277.

Yamanaka, A., Kosugi, S., and Konishi, E. (2008). Infection-enhancing and -neutralizing activities of mouse monoclonal antibodies against dengue type 2 and 4 viruses are controlled by complement levels. J Virol *82*, 927–937.

Yang, K.D., Yeh, W.T., Yang, M.Y., Chen, R.F., and Shaio, M.F. (2001). Antibody-dependent enhancement of heterotypic dengue infections involved in suppression of IFNgamma production. J Med Virol *63*, 150–157.

Yang, X., Kurteva, S., Lee, S., and Sodroski, J. (2005). Stoichiometry of antibody neutralization of human immunodeficiency virus type 1. J Virol *79*, 3500–3508.

Zhang, Y., Corver, J., Chipman, P.R., Zhang, W., Pletnev, S.V., Sedlak, D., Baker, T.S., Strauss, J.H., Kuhn, R.J., and Rossmann, M.G. (2003). Structures of immature flavivirus particles. EMBO J *22*, 2604–2613.

Zhang, Y., Zhang, W., Ogata, S., Clements, D., Strauss, J.H., Baker, T.S., Kuhn, R.J., and Rossmann, M.G. (2004). Conformational changes of the flavivirus E glycoprotein. Structure *12*, 1607–1618.

Zhang, Y., Kaufmann, B., Chipman, P.R., Kuhn, R.J., and Rossmann, M.G. (2007). Structure of immature West Nile virus. J Virol *81*, 6141–6145.

11. Structural Basis of Antibody Protection Against West Nile Virus

RICHARD J. KUHN

Abstract

The structural basis for antibody-mediated neutralization of West Nile virus infection has been investigated by a variety of biophysical and biochemical approaches that have begun to reveal the underlying mechanisms involved in this process. Crystallographic studies of E protein complexed with neutralizing monoclonal antibody Fab fragments have provided detailed information about the residues that are key to the process of antibody–epitope interaction. These studies have been mated to lower resolution cryo-electron microscopy studies that provide structures of bound Fab fragment to native flavivirus virions. The results suggest that antibody-mediated virus neutralization occurs in several different ways that impact the ability of the virus to go forward in its life cycle. Furthermore, these studies have highlighted the dynamic processes that flaviviruses undergo as part of their life cycle. Collectively, these studies suggest new approaches to the design of next generation flavivirus vaccines.

Keywords

Cryo-electron microscopy, X-ray crystallography, neutralization, icosahedral, virion

1 Introduction

West Nile virus (WNV) is an enveloped virus in the Flaviviridae family of positive strand RNA viruses (Lindenbach et al., 2007). Although the three genera have different sets of structural proteins, recent observations suggest that their virions share predominant structural features (Mukhopadhyay et al., 2005). A major characteristic found in all of these viruses is the presence of a lipid bilayer derived from the host. In the case of the flavivirus genus, in which WNV is a member, the viral membrane comes from either an insect or a vertebrate host. Embedded in this bilayer are the two envelope proteins found in flaviviruses, the envelope (E) and membrane (M) proteins. These proteins function to protect the virus during extracellular transit and to promote attachment and entry of the virus into host cells.

A large number of monoclonal antibodies specifically directed against WNV have been described (Beasley and Barrett, 2002; Oliphant et al., 2005, 2007; Throsby et al., 2006). The majority of these antibodies react with the major surface protein E, and most strongly neutralizing WNV-specific antibodies recognize determinants of domain III on this protein. However, there are protective antibodies against domains I and II, and several neutralizing antibodies that recognize epitopes that are hidden on the surface of the virion. To understand the mechanisms of antibody-mediated protection, especially given the cryptic nature of some of the neutralizing epitopes, a structural perspective of the antibody–virus complex is necessary. Although high-resolution structures are now available for several flavivirus E protein–Fab complexes, binding should also be examined based on the arrangement of the E protein within the virion. Such structures can be predicted in some cases based on the footprint of the antibody on the E protein. However, it is now clear that flavivirus particles undergo considerable dynamic motion. This so-called breathing suggests that antibodies might bind to transient conformational and translational states, and these can be captured for structural analyses. Cryo-electron microscopy (cryo-EM) and image reconstruction have been used to examine the structure of virus particles in complex antibody Fabs and will be the focus of the following discussion.

2 Structure of West Nile Virus

2.1 The Viral Proteins

The composition of the flavivirus particle consists of three structural proteins, the RNA genome, and the host-derived lipid bilayer membrane. The three structural proteins are synthesized from the amino terminal end of the viral polyprotein and are the capsid protein (C), the membrane protein (M), and the envelope protein (E) (Rice et al., 1985). The M and E proteins contain transmembrane domains and are anchored in the lipid membrane. M is initially translated as a precursor protein known as prM, and forms a heterodimer with the E protein (Heinz et al., 1994). Maturation occurs in the late Golgi by a furin cleavage that releases the pr peptide from the particle and results in an infectious flavivirus virion (Stadler et al., 1997). Although the structure of M has not been described, recent studies have determined the structure of the precursor component, pr, and its association with the E protein (Li et al., 2008). The structures of the WNV capsid protein, as well as the dengue protein, have shown it to consist of four alpha helices and the protein is arranged as a dimer (Dokland et al., 2004; Ma et al., 2004). Despite several genetic and structural studies, the relationship between the capsid protein structure and its role in assembly and in the structure of the virus particle has not been established (Kofler et al., 2002, 2003; Patkar et al., 2007).

The structure of the E protein from several flaviviruses has been reported. In a groundbreaking study, Rey et al. (1995) determined the structure of the E protein ectodomain (minus the stem and anchor regions) from tick-borne encephalitis virus. The structure revealed a three-domain protein that was associated as a dimer in the crystal structure. The dimer structure could also be identified by trypsin treatment of virions to produce soluble ectodomains (Allison et al., 1995b). Domain I, found at the N terminus, was referred to as the central domain based on its location relative to domains II and III. Intimately associated with domain I was domain II, also referred to as the dimerization domain, which consisted primarily of beta strands that formed an extended structure away from domains I and III. At the distal tip of domain II, a short but conserved hydrophobic sequence that has been referred to as the "fusion loop" was located (Allison et al., 2001). This sequence has been suggested to penetrate the target membrane under conditions of low pH-induced fusion. The polypeptide chain travels back through domain I on its way to form domain III, which follows an immunoglobulin C-like fold. Domain III has been implicated in receptor attachment in several studies and is a target for many neutralizing antibodies described for flaviviruses. Since the report of the tick-borne encephalitis virus E protein structure in 1995, numerous additional E protein structures including WNV E have been reported (Kanai et al., 2006; Modis et al., 2003, 2005; Nybakken et al., 2006; Zhang et al., 2004).

A remarkable finding based on the structure, and on the location of antigenic sites, was that the E protein dimer was arranged with its long axis parallel to the viral membrane (Rey et al., 1995). Thus, in contrast with HIV and influenza, the flavivirus particle was not predicted to have protruding spikes, and roughly, half of the E protein would be facing toward the membrane and protected from antibodies. Based on the size of the E protein dimer and the diameter of the virion, it was predicted that 180 copies of the E protein would be arranged together in the form of a $T = 3$ icosahedron (Kuhn and Rossmann, 1995). However, it was apparent that radical transformational rearrangements would be required to convert the E homodimer into the homotrimer that was identified as the functional state of the E protein during fusion (Allison et al., 1995a; Bressanelli et al., 2004; Modis et al., 2004).

2.2 The Flavivirus Virion

The first high-resolution image of an intact flavivirus particle was obtained from subviral particles of tick-borne encephalitis virus (Ferlenghi et al., 2001). During normal flavivirus infections, noninfectious particles are released that lack the capsid protein and the genome RNA (Konishi et al., 1992; Mason et al., 1991). The majority of these particles are smaller than native virions (~315 Å in diameter), and can also be produced by heterologous expression of the prM and E glycoproteins. Ferlenghi et al. (2001) produced a cryo-EM reconstruction of a recombinant subviral particle from tick-borne encephalitis virus that confirmed the orientation of the E protein dimer relative to the lipid bilayer. The E proteins were organized in a classic $T = 1$ arrangement, and this allowed the authors to propose a $T = 3$ model for the native virion. This proposed arrangement immediately suggested the rearrangement of the E proteins that would be required for formation of the fusion-competent homotrimer.

The structure of the native dengue 2 virion was subsequently solved using cryo-EM and reconstruction techniques (Kuhn et al., 2002). Like the subviral particle, the E protein dimers were arranged parallel to the viral membrane giving the surface of the virion a smooth appearance. Although there were 180 copies of the E protein in the virion, the arrangement of the E dimers did not comply with predicted icosahedral quasisymmetry. Instead, the E proteins were arranged in a parallel series of three dimers oriented in a herringbone pattern (see Fig. 1). In this orientation, each of the three E proteins within the icosahedral asymmetric unit was in a unique chemical environment. This organization suggested that binding of antibodies and receptor molecules to the E protein would differ radically depending upon the site of attachment. Subsequent studies were able to improve the resolution of the dengue virus so that the stem and anchor regions of the E and M proteins were also seen (Zhang et al., 2003a). These studies demonstrated that only a minor portion of the M protein is solvent accessible beyond the E protein shell and is probably disordered. It is likely that only 15–20 amino acids

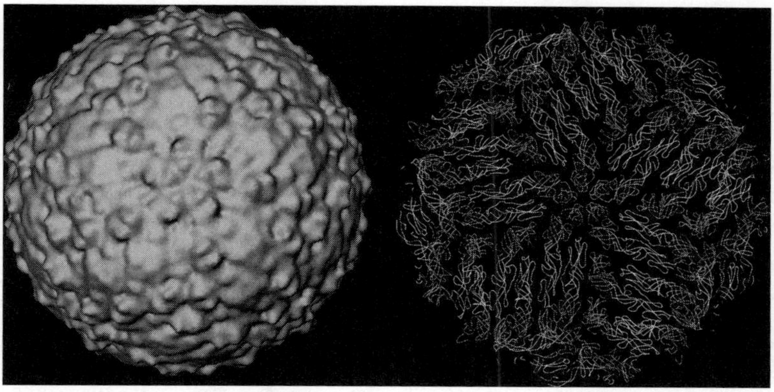

Figure 1. The structure of West Nile virus. The structure of WNV was determined by cryo-EM and image reconstruction. The *left panel* shows a surface-rendered view of the virion looking down the icosahedral fivefold axis. The *right panel* shows the arrangement of the E protein after X-ray-derived coordinates were computationally fitted into the cryo-EM density. The view is identical to that of the *left panel*. The E protein is color coded according to domain with domain I in *red*, domain II in *yellow*, and domain III in *blue*. (*See Color Plates*)

of M protrude beyond the holes formed between the two opposing E proteins in the dimer. The role of this M sequence is obscure, but the protein does contain two transmembrane domains that are rigid and clearly visible in the cryo-EM reconstruction. Likewise, the E protein contains two transmembrane domains that do not penetrate the inner bilayer and do not appear to have substantial contact with the nucleocapsid core. The E protein also contains the stem region that consists of two alpha helices linked by a conserved sequence region. These helices appear to be intimately associated with the bilayer, but their role in the flavivirus life cycle has not been established.

The structure of the WNV has also been solved using cryo-EM and image reconstruction techniques (Mukhopadhyay et al., 2003). The structure does not differ significantly from the dengue virus structure, but there are apparent differences when the E glycoprotein is fitted into each virus structure and difference maps are generated. These differences confirm the distinct sites of glycosylation between the two viral E proteins and the insertion of five amino acids in E in WNV relative to dengue 2 virus.

2.3 Immature WNV Particles

To understand the assembly pathway of flaviviruses, cryo-EM reconstructions have been carried out on dengue 2 and WNV immature particles (Zhang et al., 2003b, 2007). Initially, these particles were isolated by treatment of cells with ammonium chloride, which raises the pH of the Golgi and prevents cleavage of the precursor prM by furin. However, it was shown with WNV that collecting infected cell supernatants and analyzing the resultant particles, one could identify larger spiky particles that were immature and were identical to the particles released after ammonium chloride treatment. The immature particles were larger than native particles (600 vs. 500 Å) and displayed trimers of prM–E heterodimers that formed the spike-like projections. The orientation of the E protein in these particles had the distal end of domain II, containing the fusion peptide, oriented away from the membrane. The majority of the prM density was positioned on top of the fusion peptide, with three of the prM–E molecules coming together at the tip. Recent studies have now shown that immature particles within the low pH environment (pH 6.0) of the trans-Golgi network, have an appearance more similar to the native virions, but with the pr moiety of prM sitting on top of the fusion peptide (Yu et al., 2008). Furthermore, these studies have shown that pr is not released immediately after cleavage but requires the particle to be exposed to a neutral pH to affect the release. The structure of the pr moiety is consistent with these observations and the interactions with the fusion peptide and other residues of the E protein (Li et al., 2008). The role of the immature particle in eliciting and modulating an immune response is unclear, but it is likely that immature virus particles are released from infected cells during a normal infection.

3 Structural Basis of West Nile Virus Neutralization

3.1 WNV Antibody Neutralization

The E protein has long been recognized as the target for the majority of neutralizing antibodies. Prior to the X-ray structure of the E protein, mapping of neutralization sites using genetic and biochemical approaches produced a fairly accurate map of the epitopes on the dengue and the tick-borne encephalitis E proteins (Mandl et al., 1989; Roehrig et al., 1998). More recently, Diamond and colleagues generated a large panel of WNV monoclonal antibodies against the E protein and mapped the epitopes by a yeast surface display method (Oliphant et al., 2005). One antibody was studied in great detail, in part, because of its ability to protect mice against a lethal challenge with WNV. This monoclonal antibody termed E16 recognizes residues on the lateral surface of domain III as determined by the yeast surface display (Nybakken et al., 2005). These residues comprise an epitope that is also recognized by human convalescent sera from individuals who had recovered from WNV infection.

The mechanism by which E16 functions in WNV neutralization was investigated using several biochemical approaches (Nybakken et al., 2005). The ability of the antibody to block virus attachment to Vero cells was assessed using a binding assay, which compared the activity of specific and nonspecific monoclonals. This assay was done on ice to prevent entry of virus into the cells, and virus binding was determined by quantifying the amount of viral RNA associated with the cells. E16 along with another domain III-specific monoclonal partially blocked attachment of virus to cells but not nearly as well as other antibodies such as E53 and E60. To further identify the key step in entry that is blocked, Nybakken and colleagues carried out pre- and postadsorption antibody inhibition assays using E16 and E53. In this assay, E53 was effective in blocking virus only when added in the preadsorption step, whereas E16 worked whether added pre- or postadsorption, suggesting that it blocks at a step after virus attachment to the cells.

3.2 Structure of the E16 Fab–Domain III Complex

The structural basis for the binding of E16 on the surface of the WNV E protein was elaborated by producing crystals of the E16 Fab fragment complexed with domain III (Nybakken et al., 2005). The X-ray structure that was solved to 2.5 Å resolution showed the interaction of E16 with four discontinuous segments of the immunoglobulin-fold of domain III (Fig. 2). The results were consistent with the yeast surface display mapping, and the four segments of domain III were the N-terminal residues (302–309), and three of the beta-strand-connecting loops, BC (residues 330–333), DE (residues 365–368), and FG (residues 389–391). As expected, the binding surface exhibits a high degree of complementarity and has a surface area of 1,550Å$^{(2)}$. Residues that were implicated by yeast display as highly critical for interaction, Ser306, Lys 307, Thr330, and Thr332, are in the interface region and highlight the nature of the interaction. Perhaps not surprisingly, the E16 epitope is well conserved among WNV isolates, and E16 is able to neutralize isolates of WNV from lineage I and II.

The X-ray crystallographic structure of E16 bound to domain III of the WNV E protein is consistent with results obtained with other neutralizing antibodies bound to viral epitopes in terms of the binding surface, the nature of the contact residues, and the residues that are critical for binding and therefore neutralization escape. However, given the

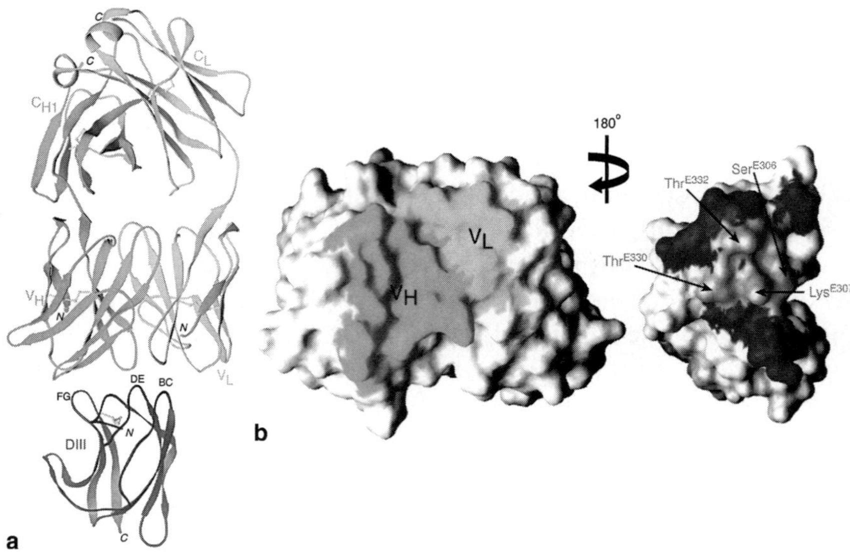

Figure 2. The X-ray structure of Fab E16 in complex with the West Nile virus domain III. (**a**) A ribbon diagram of the complex with the Fab heavy chain shown in *green* and the light chain shown in *cyan*; domain III is shown in *blue*. The loops on domain III that interact with E16 are identified. The free N and C termini are also shown. (**b**) The molecular surfaces of the Fab (*left*) and domain III are shown using the same color scheme. Residues identified by the yeast display are identified in domain III and the region is colored *magenta*. This figure was provided courtesy of Daved Fremont and reprinted with permission of Nature Publishing Group. (*See Color Plates*)

nonequivalent nature of the E protein distribution in the flavivirus virion, it was important to interrogate the predicted binding of the E16 antibody to the surface of the WNV virion. It should be noted that the neutralization affects of E16 were observed with both intact antibody as well as with Fab fragments. This makes intuitive sense, as it is unlikely that most antibodies will bind to virions in a bivalent manner, and thus their neutralization should be dependent on monovalent binding to the target epitope. The predicted binding of E16 onto the surface of the E protein in the context of virus was accomplished by docking the crystal structure of the Fab–domain III complex into the cryo-EM density map of WNV. Since there are three molecules of the E protein and domain III within the asymmetric unit, full occupancy would dictate that all three sites were occupied and saturation of the particle with E16 Fab would require 180 molecules. However, docking clearly showed that binding of E16 at sites that surrounded the fivefold axes would be sterically occluded. That is, the close proximity of domain IIIs at the fivefold axis would prevent the close approach and docking of Fab to the epitope, even though the epitope itself was solvent exposed. Binding of E16 to the threefold and to the twofold E proteins would be permitted, and thus it was predicted that 120 E16 Fab's would bind at saturation to the WNV particle.

3.3 Structure of E16 Complexed with Virus

To determine whether the prediction based on the X-ray structure of the complex and modeling into the virion was accurate, complexes of intact WNV and E16 Fab fragment were generated using an excess of antibody, and cryo-EM was carried out (Kaufmann et al., 2006). The reconstruction of the complex was compared to the native WNV particle and a difference map was calculated. This map provided the position of each Fab fragment bound to the virus under the saturating conditions used for complex formation. As predicted, two Fab molecules bound in the asymmetric unit, with the E protein closest to the fivefold axis devoid of bound antibody. This is shown in Fig. 3 looking down the fivefold axis. Two things are apparent in this image reconstruction: (1) a significant amount of the E protein is solvent accessible, particularly in domains I and II, and could be bound by other antibodies or other

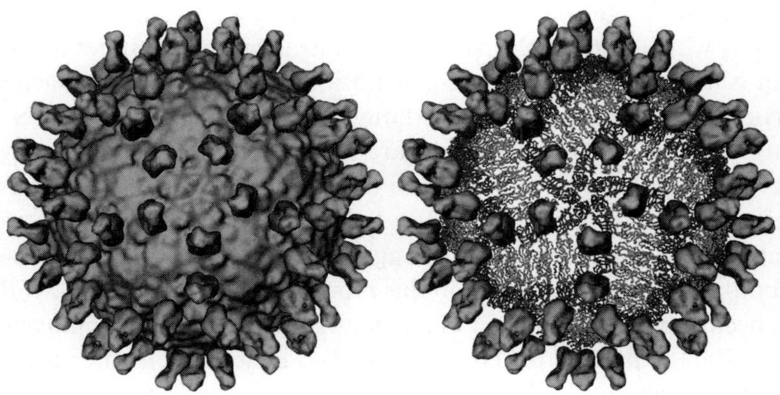

Figure 3. The structure of West Nile virus in complex with Fab E16. The structure of WNV in complex with Fab E16 was determined by cryo-EM and image reconstruction. The *left panel* shows a surface-rendered view of the virion looking down the icosahedral fivefold axis. The virus particle is colored as in Fig. 1 (*green*). The Fab fragment is denoted by the *gray* color. The *right panel* shows the arrangement of the E protein after X-ray-derived coordinates were computationally fitted into the cryo-EM density. The view is identical to that of the *left panel*, with the Fab fragments shown in *gray*. Note the availability of domain IIIs around the fivefold axis. (*See Color Plates*)

molecules such as cell surface receptors; and (2) domain III around the fivefold axis is also accessible and can bind to receptor molecules, perhaps in an oligomeric dependent manner. Minor differences in the orientation of elbow angle between the constant and variable domains of the Fab were apparent between the X-ray and cryo-EM structures. This is probably the result of more severe constraints due to crystal packing that are not present in the cryo-EM complex. Importantly, there was no evidence of antibody-induced conformational change in the E protein or other components of the virus particle. In other words, the difference map that was computed only showed the Fab fragments and no alterations in the virion structure.

The observed densities of the virus and the Fab fragment are similar in intensity arguing for approximately a one-to-one ratio or a 100% occupancy of the Fabs into the 120 available binding sites. However, results from Pierson's laboratory (see the chapter by Jost and Pierson) suggest that far less than Fab–virus saturation is required for neutralization of virus infectivity (Pierson et al., 2007). Experiments that correlate occupancy of Fab with neutralization suggest that approximately 30 E16 Fabs are required to neutralize WNV infectivity in cell culture.

3.4 Implications for Function

Several pieces of data can now be assembled to suggest a mechanism by which E16 functions in virus neutralization. First, let us assess what is needed for attachment, penetration, and fusion of the flavivirus particle into the infected cell. Although the entry receptor for WNV is not known, it is likely that there are attachment receptors that facilitate binding of the virion onto the host cell surface, followed by two-dimensional diffusion that greatly facilitates the interaction between the virus and the receptor. For dengue virus, DC-SIGN and mannose receptor have been implicated in this process and for WNV, DC-SIGNR has been suggested (Davis et al., 2006; Miller et al., 2008; Pokidysheva et al., 2006; Tassaneetrithep et al., 2003). Particles then enter through clathrin-coated pits and proceed into early endosomes (van der Schaar et al., 2007). Upon entering the late endosomal compartment, the virions are exposed to low pH, and the E proteins rearrange from their homodimer state to a homotrimer state that is required for membrane fusion (Allison et al., 1995a; Bressanelli et al., 2004; Modis et al., 2004). This initiates the process by which the viral and cellular membranes fuse and the viral capsid and genome are released into the cytoplasm. The detailed steps required for transition of the native virion into a fusion-competent particle are not understood yet at a structural level. However, this transition is likely to involve substantial conformational and translational movements of the E protein relative to the virus surface.

Prior data demonstrated that E16 does not abolish attachment and penetration of virus to cells. The cryo-EM structure of the complex supports this and shows that there are substantial amounts of all three domains that can be accessed by cellular molecules for the purpose of attachment. A variety of experiments including, most recently, single particle tracking (van der Schaar, Wilschut, Diamond, Smit, personal communication), suggest that WNV bound with neutralizing amounts of E16 can enter cells, but is blocked at a step prior to the membrane fusion. Although the prefusion state of the virion immediately prior to membrane insertion of E is not known, biochemical and structural results identify an E homotrimer as the oligomeric form necessary for fusion. To achieve this structure requires a substantial change in the organization of the E protein from its native configuration in the virion. A model has been proposed for this presumed state of the prefusion virus that resembles a $T = 3$ icosahedral virus and is based on the $T = 1$ subviral particles of tick-borne encephalitis virus that have been described using cryo-EM (Kuhn et al., 2002). In this hypothetical arrangement, the E proteins have rearranged into radially extended homotrimers with solvent exposed membranes. The particle expands as the E proteins change their oligomeric state and reorient with their fusion peptides pointing away from the viral membrane. To achieve this proposed state, significant rotational and translational movements occur with the E protein, particularly those molecules closest to the twofold and threefold symmetry axes. With E16 Fab bound to domain IIIs at each of these locations, steric opposition would prevent the E proteins from completing this required rearrangement. This argument is further enhanced by the suggestion of Nybakken et al. (2005) that E16 binds to the linker region between the domains I and III of the E protein, the same region that undergoes a reorientation of 70° as domain III packs up against domain II in the dimer to trimer transition. Thus, the mechanism of E16 neutralization is that bound antibody prevents the conformational changes that are required for activation of the fusion-promoting state of the virion. The observation that only 30 antibodies are required for neutralization suggests that this block may be cooperative, and that binding of antibodies in one region of the particle may influence other regions undergoing conformational changes.

4 Mechanisms of Action of Other Antibodies

4.1 Neutralization of Dengue Virus by 1A1D-2

Neutralizing antibodies are predominantly directed against domain III; however, there are also epitopes on domains II and I of the E protein that elicit neutralizing antibodies (Beasley and Barrett, 2002; Crill and Chang, 2004; Crill et al., 2007; Oliphant et al., 2005; Roehrig et al., 1998; Stiasny et al., 2006). Interestingly, many of these antibodies are not as strongly neutralizing and some are capable of cross-protection with other flaviviruses. Some of these antibodies are predicted to bind to sites that are cryptic and are not available for binding based on the structure of the native flavivirus particles that have been demonstrated by cryo-EM and reconstruction techniques. A recent structure of dengue-specific neutralization antibody in the complex with the virus reveals an alternative mechanism for neutralization and highlights the important contributions that structural analyses can make in understanding of antibody-mediated virus neutralization (Lok et al., 2008).

The monoclonal 1A1D-2 strongly neutralizes dengue serotypes 1, 2, and 3 and unlike the mechanism described above for E16, it does this primarily by interfering with virus attachment to host cells (Roehrig et al., 1998). Initial attempts at forming Fab–virus complexes using the conditions that worked well for the E16–WNV failed to produce cryo-EM reconstruction results that were acceptable, as there was very limited binding of Fab to the particles. Solving the X-ray structure of the 1A1D-2 Fab in the complex with the dengue 2 domain III suggested a possible explanation for the failure to form complexes under the conditions used for cryo-EM (Lok et al., 2008). The binding surface of 1A1D-2 is distinct from the E16 footprint, and consists of residues from the L1, L2, H1, H2, and H3 complementary determining regions. However, the epitope is partially occluded in the native virus structure in each of the three unique positions of the E protein (18% of the binding surface is buried). This would prevent the antibody from binding to the epitope, a result that appeared to be in conflict with the known biological data. Since there appeared to be partial binding of the Fab to the dengue particle when complexed at room temperature, an elevated temperature was used to encourage additional binding. Raising the temperature to 37°C was sufficient for a large number of particles to bind Fab and cryo-EM and reconstructions were carried out. Remarkably, the organization of the E protein in the particle when bound by Fab was dramatically different from the native herringbone arrangement. Again, two of the three unique E molecules were bound by the 1A1D-2 Fab, but each of the E proteins had undergone some translational movements. The two E molecules within the asymmetric unit that had bound Fab underwent substantial movements, whereas the unbound E monomer closest to the twofold axis moved very little.

What created this altered arrangement of the E proteins on the surface of the virion? The observation that binding of Fab appeared to be temperature-dependent suggested that particle dynamics might be involved. Previous studies with Flock House virus and with rhinovirus demonstrated that the particles underwent significant changes in the organization of their capsid proteins such that sites or regions of these proteins that normally were inaccessible to proteases became transiently available (Bothner et al., 1998; Lewis et al., 1998). Factors that stabilized (antiviral compounds), or destabilized (elevated temperature) the virions influenced the sensitivity of the capsid proteins to the action of proteases. Such "breathing" could explain the ability of antibodies to recognize cryptic sites on flavivirus E proteins. Elevated temperature increased the probability of the E protein having the 1A1D-2 epitope accessible, and binding of one or more Fabs to the site locked the E proteins into this unusual configuration. Presumably, this arrangement would not support the attachment of virions to the target cells. Analysis of WNV mutants resistant to antibody-mediated neutralization has also suggested dynamic mobility in the E protein and this leads to long-range communication between distinct structural elements and residues (Maillard et al., 2008).

5 Structural Insights into Antibody-Mediated Flavivirus Neutralization

The two examples described provide two different mechanisms for neutralization. In the E16 case, binding of the antibody traps the particle so that future conformational states necessary for entry are prohibited. For 1A1D-2, the antibody is trapping an existing but transient conformational state of the virus particle, and locks it into a configuration that prevents attachment to the cell. In both cases, viral dynamics, of which nothing is really known, appears to be an important mechanistic aspect of antibody-mediated neutralization. The ability of antibodies to recognize cryptic, or hidden sites on the E protein, has important implications for explaining a large number of previously described antibodies, as well as in developing next generation vaccines. Sites that are hidden tend to be in regions on the interior of the protein that have structural or functional roles. These epitopes might be less likely or able to mutate in the presence of antibody selection, as opposed to surface accessible epitopes. Furthermore, these sites might be more conserved across the flaviviruses, suggesting that such antibodies might be cross-reactive, although less potently neutralizing. Future structural and biophysical studies of flavivirus–antibody complexes will be important to investigate these observations and define the correlates of potent and cross-protective antibody responses against WNV and other flaviviruses.

References

Allison SL, Schalich J, Stiasny K et al. (1995a) Oligomeric rearrangement of tick-borne encephalitis virus envelope proteins induced by an acidic pH. J Virol 69:695–700

Allison SL, Stadler K, Mandl CW et al. (1995b) Synthesis and secretion of recombinant tick-borne encephalitis virus protein E in soluble and particulate form. J Virol 69:5816–5820

Allison SL, Schalich J, Stiasny K et al. (2001) Mutational evidence for an internal fusion peptide in flavivirus envelope protein E. J Virol 75:4268–4275

Beasley DW, Barrett AD (2002) Identification of neutralizing epitopes within structural domain III of the West Nile virus envelope protein. J Virol 76:13097–13100

Bothner B, Dong XF, Bibbs L et al. (1998) Evidence of viral capsid dynamics using limited proteolysis and mass spectrometry. J Biol Chem 273:673–676

Bressanelli S, Stiasny K, Allison SL et al. (2004) Structure of a flavivirus envelope glycoprotein in its low-pH-induced membrane fusion conformation. EMBO J 23:1–11

Crill WD, Chang GJ (2004) Localization and characterization of flavivirus envelope glycoprotein cross-reactive epitopes. J Virol 78:13975–13986

Crill WD, Trainor NB, Chang GJ (2007) A detailed mutagenesis study of flavivirus cross-reactive epitopes using West Nile virus-like particles. J Gen Virol 88:1169–1174

Davis CW, Nguyen HY, Hanna SL et al. (2006) West Nile virus discriminates between DC-SIGN and DC-SIGNR for cellular attachment and infection. J Virol 80:1290–1301

Dokland T, Walsh M, Mackenzie JM et al. (2004) West Nile virus core protein; tetramer structure and ribbon formation. Structure 12:1157–1163

Ferlenghi I, Clarke M, Ruttan T et al. (2001) Molecular organization of a recombinant subviral particle from tick-borne encephalitis. Mol Cell 7:593–602

Heinz FX, Stiasny K, Puschner-Auer G et al. (1994) Structural changes and functional control of the tick-borne encephalitis virus glycoprotein E by the heterodimeric association with protein prM. Virology 198:109–117

Kanai R, Kar K, Anthony K et al. (2006) Crystal structure of West Nile virus envelope glycoprotein reveals viral surface epitopes. J Virol 80:11000–11008

Kaufmann B, Nybakken GE, Chipman PR et al. (2006) West Nile virus in complex with the Fab fragment of a neutralizing monoclonal antibody. Proc Natl Acad Sci USA 103:12400–12404

Kofler RM, Heinz FX, Mandl CW (2002) Capsid protein C of tick-borne encephalitis virus tolerates large internal deletions and is a favorable target for attenuation of virulence. J Virol 76:3534–3543

Kofler RM, Leitner A, O'Riordain G et al. (2003) Spontaneous mutations restore the viability of tick-borne encephalitis virus mutants with large deletions in protein C. J Virol 77:443–451

Konishi E, Pincus S, Paoletti E et al. (1992) Mice immunized with a subviral particle containing the Japanese encephalitis virus prM/M and E proteins are protected from lethal JEV infection. Virology 188:714–720

Kuhn RJ, Rossmann MG (1995) Virology. When it's better to lie low [news; comment]. Nature 375:275–276

Kuhn RJ, Zhang W, Rossmann MG et al. (2002) Structure of dengue virus: implications for flavivirus organization, maturation, and fusion. Cell 108:717–725

Lewis JK, Bothner B, Smith TJ et al. (1998) Antiviral agent blocks breathing of the common cold virus. Proc Natl Acad Sci USA 95:6774–6778

Li L, Lok S-M, Yu I-M et al. (2008) Structure of the flavivirus precursor membrane-envelope protein complex and its implication for maturation. Science, 319:1830–1834

Lindenbach BD, Thiel H-J, Rice CM (2007) Flaviviridae: the viruses and their replication. In: Knipe DM, Howley PM (eds) Fields virology. Lippincott Williams & Wilkins, Philadelphia, pp 1001–1022

Lok S-M, Kostyuchenko VA, Nybakken GE et al. (2008) Binding of a neutralizing antibody to dengue virus resulted in an altered surface glycoproteins arrangement. Nat Struct Mol Biol 15:312–317

Ma L, Jones CT, Groesch TD et al. (2004) Solution structure of dengue virus capsid protein reveals another fold. Proc Natl Acad Sci USA 101:3414–3419

Maillard RA, Jordan M, Beasley DW et al. (2008) Long range communication in the envelope protein domain III and its effect on the resistance of West Nile virus to antibody-mediated neutralization. J Biol Chem 283:613–622

Mandl CW, Guirakhoo F, Holzmann H et-al. (1989) Antigenic structure of the flavivirus envelope protein E at the molecular level, using tick-borne encephalitis virus as a model. J Virol 63:564–571

Mason PW, Pincus S, Fournier MJ et al. (1991) Japanese encephalitis virus-vaccinia recombinants produce particulate forms of the structural membrane proteins and induce high levels of protection against lethal JEV infection. Virology 180:294–305

Miller JL, deWet BJM, Martinez-Pomares L et al. (2008) The mannose receptor mediates dengue virus infection of macrophages. PLoS Pathogens 4:1–11

Modis Y, Ogata S, Clements D et al. (2003) A ligand-binding pocket in the dengue virus envelope glycoprotein. Proc Natl Acad Sci USA 100:6986–6991

Modis Y, Ogata S, Clements D et al. (2004) Structure of the Dengue virus envelope protein after fusion. Nature 427:313–319

Modis Y, Ogata S, Clements D et al. (2005) Variable surface epitopes in the crystal structure of dengue virus type 3 envelope glycoprotein. J Virol 79:1223–1231

Mukhopadhyay S, Kim B-S, Chipman PR et al. (2003) Structure of West Nile virus. Science 303:248

Mukhopadhyay S, Kuhn RJ, Rossmann MG (2005) A structural perspective of the Flavivirus life cycle. Nat Rev Microbiol 3:13–22

Nybakken GE, Oliphant T, Johnson S et al. (2005) Structural basis of West Nile virus neutralization by a therapeutic antibody. Nature 437:764–769

Nybakken GE, Nelson CA, Chen BR et al. (2006) Crystal structure of the West Nile virus envelope glycoprotein. J Virol 80:11467–11474

Oliphant T, Engle M, Nybakken GE et al. (2005) Development of a humanized monoclonal antibody with therapeutic potential against West Nile virus. Nat Med 11:522–530

Oliphant T, Nybakken GE, Austin SK et al. (2007) Induction of epitope-specific neutralizing antibodies against West Nile Virus. J Virol 81:11828–11839

Patkar CG, Jones CT, Chang Y-H et al. (2007) Functional requirements of the yellow fever virus capsid protein. J Virol 81:6471–6481

Pierson TC, Xu Q, Nelson S et al. (2007) The stoichiometry of antibody-mediated neutralization and enhancement of West Nile virus infection. Cell Host Microbe 1:135–145

Pokidysheva E, Zhang Y, Battisti AJ et al. (2006) Cryo-EM reconstruction of dengue virus in complex with the carbohydrate recognition domain of DC-SIGN. Cell 124:485–493

Rey FA, Heinz FX, Mandl C et al. (1995) The envelope glycoprotein from tick-borne encephalitis virus at 2 Å resolution. Nature (London) 375:291–298

Rice CM, Lenches EM, Eddy SR et al. (1985) Nucleotide sequence of yellow fever virus: implications for flavivirus gene expression and evolution. Science 229:726–733

Roehrig JR, Bolin RA, Kelly RG (1998) Monoclonal antibody mapping of the envelope glycoprotein of the dengue 2 virus, Jamaica. Virology 246:317–328

Stadler K, Allison SL, Schalich J et al. (1997) Proteolytic activation of tick-borne encephalitis virus by furin. J Virol 71:8475–8481

Stiasny K, Kiermayr S, Holzmann H et al. (2006) Cryptic properties of a cluster of dominant flavivirus cross-reactive antigenic sites. J Virol 80:9557–9568

Tassaneetrithep B, Burgess TH, Granelli-Piperno A et al. (2003) DC-SIGN (CD209) mediates dengue virus infection of human dendritic cells. J Exp Med 197:823–829

Throsby M, Geuijen C, Goudsmit J et al. (2006) Isolation and Characterization of Human Monoclonal Antibodies from Individuals Infected with West Nile Virus. J Virol 80:6982–6992

van der Schaar HM, Rust MJ, Waarts BL et al. (2007) Characterization of the early events in dengue virus cell entry by biochemical assays and single-virus tracking. J Virol 81:12019–12028

Yu I-M, Zhang W, Holdaway HA et al. (2008) Structure of immature dengue virus at low pH primes proteolytic maturation. Science, 319:1834–1837

Zhang W, Chipman PR, Corver J et al. (2003a) Visualization of membrane protein domains by cryo-electron microscopy of dengue virus. Nat Struct Biol 10:907–912

Zhang Y, Corver J, Chipman PR et al. (2003b) Structures of immature flavivirus particles. EMBO J 22:2604–2613

Zhang Y, Zhang W, Ogata S et al. (2004) Conformational changes of the flavivirus E glycoprotein. Structure 12:1607–1618

Zhang Y, Kaufmann B, Chipman PR et al. (2007) Structure of immature West Nile Virus. J Virol 81:6141–6145

12. Molecular Mechanisms of Flaviviral Membrane Fusion

Yorgo Modis and Vinod Nayak

Abstract

Enveloped viruses rely on transmembrane fusion proteins to fuse the viral membrane to the host-cell membrane and deliver the viral genome into the cytoplasm for replication. Although the structures and evolutionary origins of viral fusion proteins vary widely, all fusion proteins use the same physical principles and topology to drive membrane fusion. First, exposure of a hydrophobic fusion anchor allows them to insert into the host-cell membrane. Conserved hydrophobic residues in the fusion anchor penetrate part way into the outer bilayer leaflet of the host-cell membrane. The fusion protein then folds back on itself, directing the C-terminal viral transmembrane anchor toward the fusion loop. This fold-back forces the host-cell membrane (held by the fusion loop) and the viral membrane (held by the C-terminal transmembrane anchor) against one another until they fuse. In West Nile virus and other flaviviruses this fold-back in the fusion protein, E, is triggered by the reduced pH of an endosome, is accompanied by the assembly of E monomers into trimers, and occurs by domain rearrangement rather than by an extensive refolding of secondary structure. The rearrangement releases a large amount of energy, which is used to exert a bending force on the apposed viral and cellular membranes, propelling them toward each other and, eventually, causing them to fuse. The conserved regions of E that are responsible for driving membrane fusion are attractive targets for antiviral therapies.

Keywords

Virus cell entry, envelope protein, membrane fusion, fusion loop, conformational change, fusion inhibitor

1 Introduction

Flaviviruses and other enveloped viruses acquire a lipid bilayer membrane when they bud across the plasma membrane or the membrane of the endoplasmic reticulum (ER) during virion assembly (Lindenbach and Rice, 2001; Schlesinger and Schlesinger, 2001). During infection, the viral membrane must fuse to the host-cell membrane to deliver the viral genome into the cytoplasm for replication (Fig. 1). The fusion of the viral and host-cell membranes is therefore one of the key molecular events during viral entry. Adjacent membranes do not fuse spontaneously. Membrane fusion requires considerable energy – on the order of 100 kJ mol^{-1} (or 40 kT). Most of this energy is used to generate a force that is strong enough to bend the two membranes toward each other until they are separated by only a few Ångstroms (Kozlov and Chernomordik, 1998; Kuzmin et al., 2001). In flaviviruses, the envelope (E) protein anchored in the viral membrane exerts this force during a pH-induced conformational rearrangement (Modis et al., 2004). In addition to their role as fusion proteins, flavivirus E proteins are also responsible for cellular attachment of the virus by binding to a receptor on the cell surface and are targets for antibody neutralization.

Fusion proteins of enveloped viruses fall into two structural classes. The influenza virus hemagglutinin (HA) is the prototype of class I fusion proteins (Skehel and Wiley, 2000), which encompass those of other orthomyxoviruses and paramyxoviruses, retroviruses, filoviruses, coronaviruses, and herpesviruses. Class II fusion proteins are a structurally and evolutionarily a distinct class of proteins found in the flaviviruses,

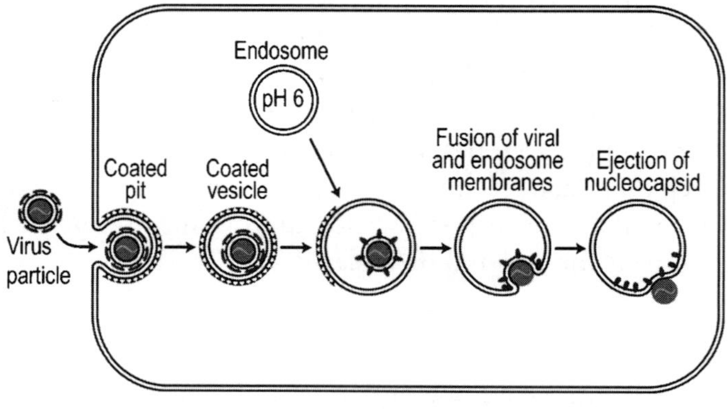

Figure 1. Cell entry of flaviviruses. Virus particles bind target cells through a surface receptor, which is linked to the clathrin-dependent endocytic pathway. Internalized vesicles fuse with endosomal compartments. The reduced pH of these compartments promotes conformational rearrangements in the viral envelope proteins that catalyze the fusion of the host-cell and viral membranes. Upon membrane fusion, the viral genome is delivered into cytoplasm. (*See Color Plates*)

including West Nile, dengue and yellow fever viruses, and alphaviruses, such as Semliki Forest, Sindbis, and the equine encephalitis viruses. Hepatitis C has a relatively similar genomic organization to the flaviviruses, and therefore most likely relies on a class II fusion protein as well. Crystal structures of several class I and class II fusion proteins (including West Nile E) before and after their fusogenic conformational rearrangements have provided a detailed molecular understanding of the fusion mechanism (Baker et al., 1999; Bullough et al., 1994; Caffrey et al., 1998; Chan et al., 1997; Chen et al., 1999; Fass et al., 1996; Kanai et al., 2006; Kobe et al., 1999; Lescar et al., 2001; Malashkevich et al., 1998; Modis et al., 2003, 2004, 2005; Nybakken et al., 2006; Rey et al., 1995; Rosenthal et al., 1998; Tan et al., 1997; Weissenhorn et al., 1997, 1998; Wilson et al., 1981; Xu et al., 2004a, b; Yin et al., 2005, 2006; Zhao et al., 2000). The structures show that the two classes of fusion proteins have completely different structural folds, and that fusion proteins from both classes nevertheless remarkably use the same physical principles and general topology to drive membrane fusion. First, the fusion protein inserts a hydrophobic fusion anchor partway into the outer bilayer leaflet of the host-cell membrane. The fusion anchor is either an N-terminal peptide (Gething et al., 1978), as in influenza and HIV (Gallaher, 1987), or an internal loop, as in SARS coronavirus (Supekar et al., 2004), avian sarcoma leucosis virus (Cheng et al., 2004) and all class II enveloped viruses including flaviviruses (Allison et al., 2001). Second, the fusion protein folds back on itself, directing the (C-terminal) viral transmembrane anchor toward the fusion anchor. This fold-back forces the host-cell membrane (held by the fusion anchor) and the viral membrane (held by the C-terminal transmembrane anchor) against each other, resulting in fusion of the two membranes. In this chapter, we describe our current paradigm of how West Nile E and other flaviviral fusion proteins drive viral membrane fusion, based on the existing structural and biochemical data. We also review how this knowledge may be translated into antiflaviviral therapies.

2 Overall Architecture of Flaviviral Membrane Fusion Proteins

In flaviviruses, membrane fusion is catalyzed by the envelope protein, E. Three-dimensional structures of four flaviviral E proteins in their native, or prefusion state have been determined at near atomic resolution (Table 1) (Kanai et al., 2006; Modis et al., 2003, 2005; Nybakken et al., 2006; Rey et al., 1995; Zhang et al., 2004). Figure 2 compares the three-domain structures of the E proteins of West Nile

Table 1. Flavivirus E fusion protein crystal structures and corresponding electron cryomicroscopy structures (listed in chronological order of publication)

Virus	Fusion state	Oligomeric state in solution	References
Tick-borne encephalitis	Prefusion	Dimer	Rey et al. (1995)
Dengue type 2	Prefusion	Dimer	Modis et al. (2003) and Zhang et al. (2003a, b, 2004)
Dengue type 2	Postfusion	Trimer	Modis et al. (2004)
Dengue type 3	Prefusion	Dimer	Modis et al. (2005)
Tick-borne encephalitis	Postfusion	Trimer	Bressanelli et al. (2004)
West Nile	Prefusion	Monomer	Kanai et al. (2006) and Nybakken et al. (2006)

virus (Kanai et al., 2006) and dengue virus (Modis et al., 2003). As expected from sequence identities of over 37% across the flavivirus genus, and of up to 49% between dengue and West Nile viruses, flaviviral E proteins share a common molecular architecture and the structures of the individual domains of West Nile and dengue E are very similar. Domain I, an eight-stranded β-barrel, organizes the structure. Two long insertions between pairs of consecutive β-strands in domain I form the elongated domain II, which bears the fusion anchor, or fusion loop, at its tip (Figs. 2 and 4). Domain II contains 12 β-strands and two α-helices. Domain III is an IgC-like module, with ten β-strands. However, unlike dengue and tick-borne encephalitis E proteins, West Nile E does not form head-to-tail dimers and is a monomer in solution (and in the crystalline form). The monomeric state of West Nile E in solution is mainly due to the different relative orientation of domain II with respect to domains I and III. Indeed, domain II participates in all of the dimer contacts in the dimeric dengue and TBE E structures, and the orientation of domain II in West Nile E is not compatible with dimer formation (Fig. 2).

Many of the antigenic sites on E map to domain III, as do most of the structural determinants of virulence and tropism (Rey et al., 1995; Sanchez et al., 2005). This observation, and the widespread occurrence of immunoglobulin modules in cell-adhesion proteins, suggests that domain III participates in attachment to a cellular receptor (Rey et al., 1995). Indeed, positively charged patches on the surface of domain III in dengue virus have been suggested to promote attachment by binding heparan sulfate on the cell surface (Chen et al., 1997). Also, while E proteins have very similar overall structures, they differ in the length and structure of surface-exposed loops, which may affect the specificity of receptor binding (Crill and Roehrig 2001; Hung et al., 2004; Rey et al., 1995). However, despite these hints on the basis of cellular attachment, a cellular

Color Plates

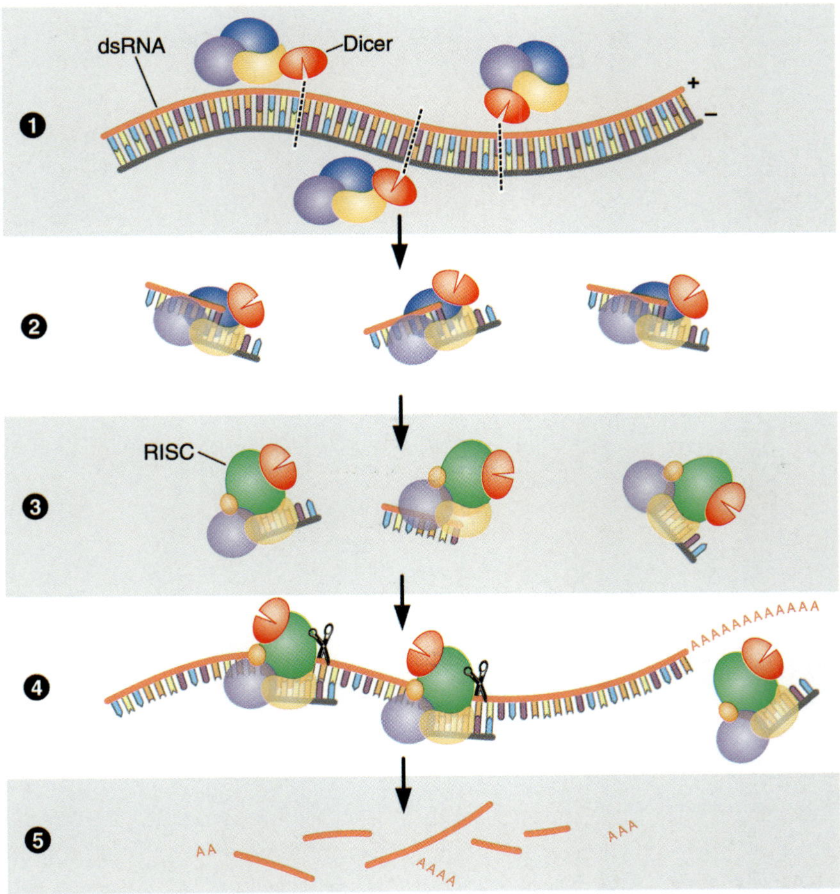

Chapter 3 Figure 2. RNA interference in mosquitors. (1) Exogenously-introduced double-stranded (ds) RNA is recognized by the non-sequence specific endonuclease Dicer-2 and cleaved into ~21 bp small interfering (si) RNA. (2) Dicer-2 and R2D2 in combination load the siRNA into the RNA-induced silencing complex (RISC; 3), which retains the "guide" RNA strand. (4) Fully– assembled RISC, including Argonaute-2 as well as Dicer-2 and R2D2, uses the RNA guide sequence to identify and base-pair to complementary sequences on mRNA and Argonaute-2 cleaves the mRNA at the center of the complementary sequence. (5) The degraded mRNA is no longer functional for translation (Blair et al., 2006).

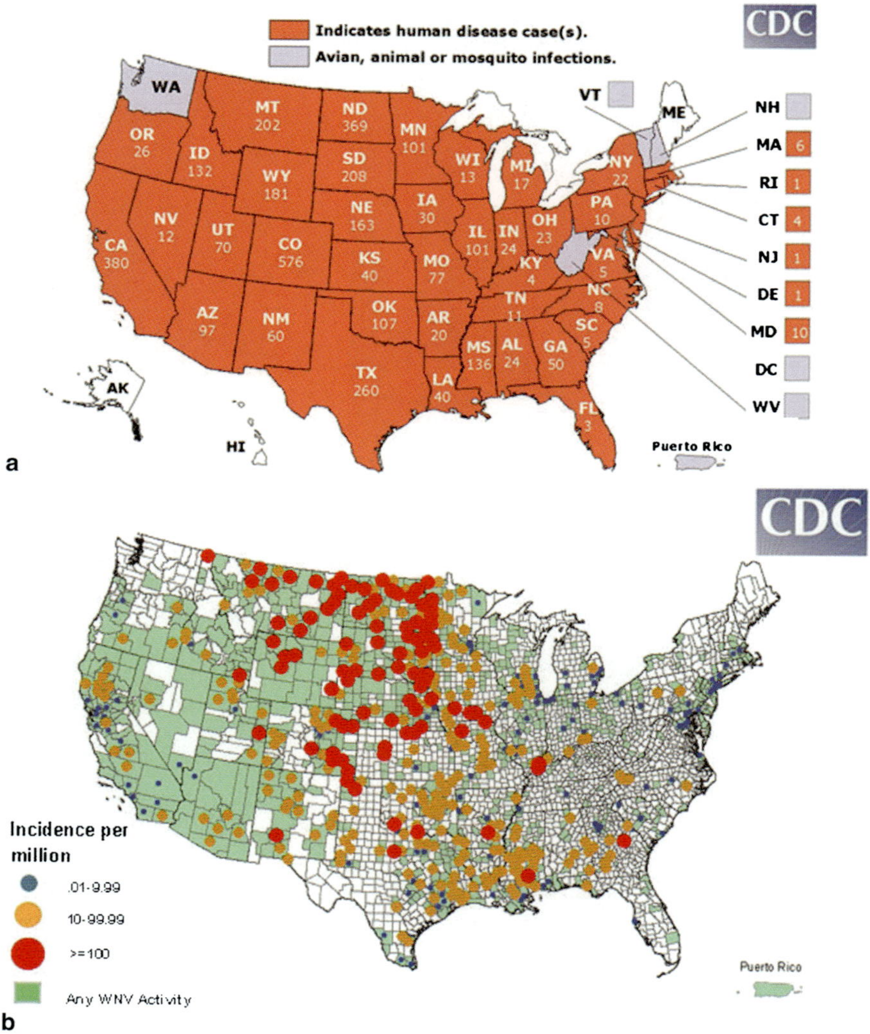

Chapter 4 Figure 1. West Nile virus cases in 2007. (**a**) Number of WNV cases by state reported to the CDC for the 2007 season. (**b**) The incidence of neuroinvasive WNV infection during the year 2007. Values are updated as of November 7, 2007 (modified from CDC at http://www.cdc.gov/ncidod/dvbid/westnile/).

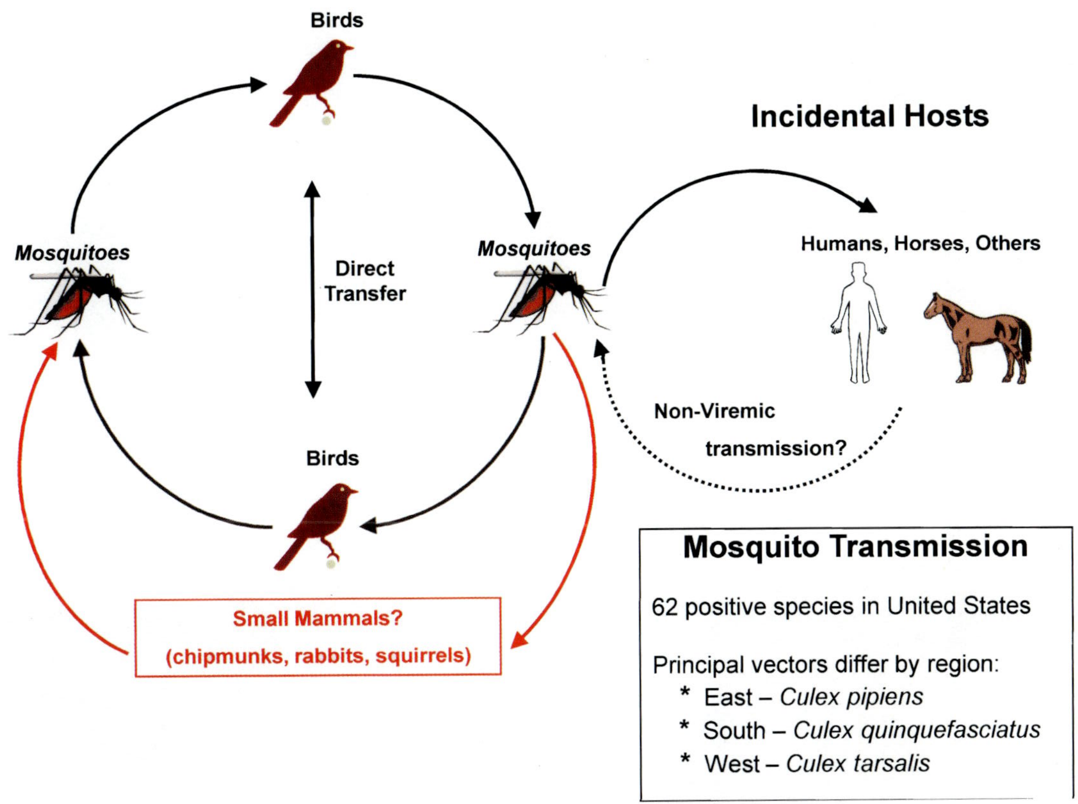

Chapter 6 Figure 2. The West Nile virus transmission cycle.

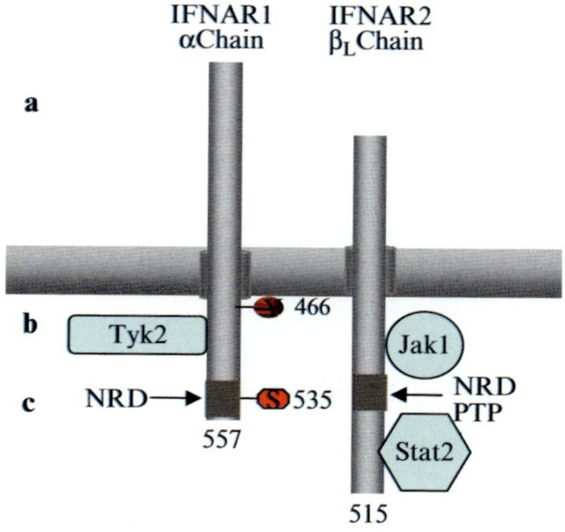

Chapter 7 Figure 2. Inhibition of IFN signal transduction by WNV. The figure depicts the binding sites for the Janus kinases Tyk2 and Jak1 and the transcription factor Stat2 on cytoplasmic domains of the two chains of the IFNα/βR. Potential targets of WNV to inhibit IFN-induced phosphorylation of Tyk2 and Jak1 could include (**a**) Blocking ligand binding, inhibiting receptor dimerization, or destruction of a receptor subunit; (**b**) Inhibition of JAK kinase activation or activity, or activation or relocalization of a phosphatase to the receptor complex; (**c**) Activation of the negative regulatory domain (NRD), perhaps by phoshphorylation of Ser_{535}. PTP, protein tyrosine phosphatase.

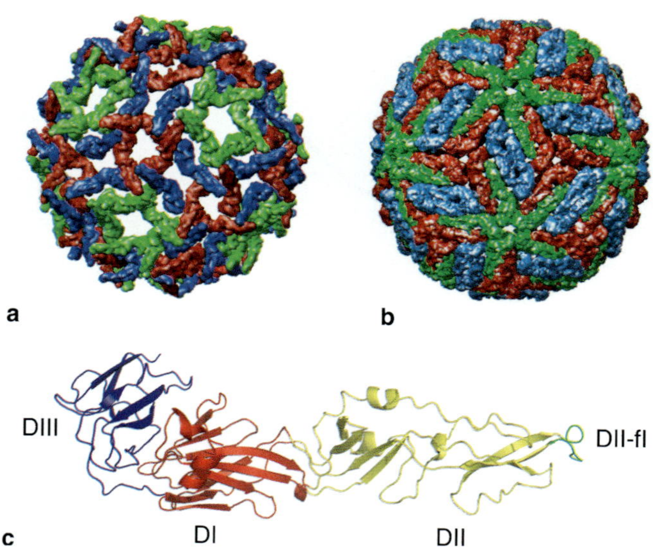

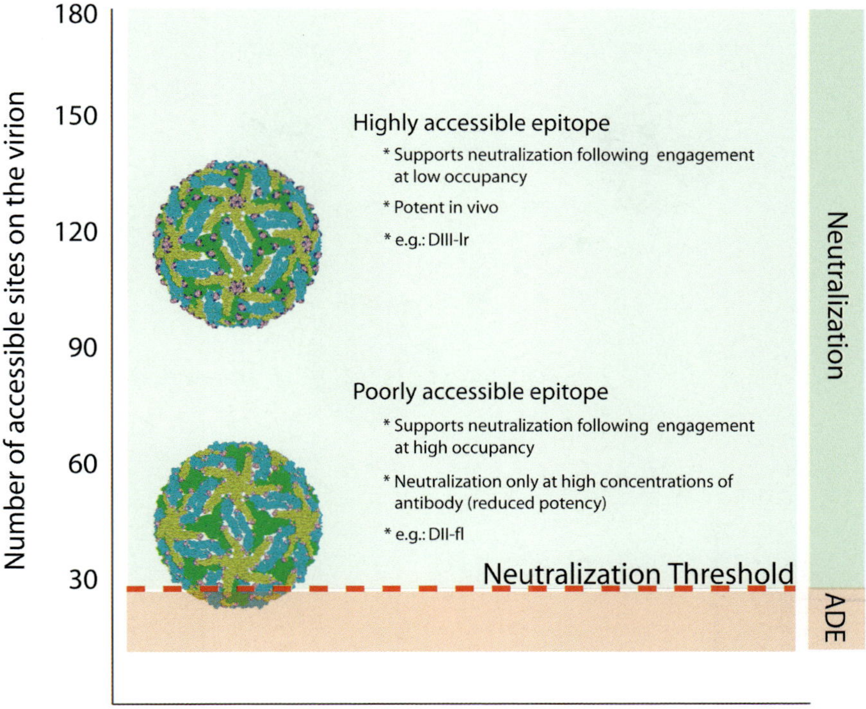

Chapter 10 Figure 2. Relationship between epitope accessibility and the occupancy requirement for neutralization. The accessibility of epitopes recognized by two antibodies on the mature WNV virion is illustrated using molecular modeling: residues that comprise each determinant are illustrated as solid spheres (*pink*). E proteins are colored according to their proximity to the 2-, 3- or 5-fold symmetry axes (*blue*, *green* and *yellow* respectively). The number of accessible binding sites is indicated schematically on the *left*, whereas the threshold for neutralization is indicated as a *red line* (modeled as 30 mAbs based on studies with WNV DIII-lr-specific mAbs) (Pierson et al., 2007). To exceed the threshold requirements for neutralization, only a fraction of highly accessible determinants must be simultaneously occupied by antibody (a low occupancy requirement). In contrast, for poorly exposed epitopes (fewer total accessible sites) a significantly greater percentage of accessible epitopes must be bound to achieve the same number of antibodies docked on the average virion (a high occupancy requirement). Not all epitopes appear to exist on the average virion at levels that exceed this threshold.

◄

Chapter 10 Figure 1. Structure of the flavivirus E protein and its organization on the mature virion. (**a**) Cryoelectron reconstruction of the WNV immature virion. Image was created using Chimera (http://www.cgl.ucsf.edu/chimera/). (**b**) Cryoelectron reconstruction of the WNV mature virion illustrating the arrangement of E proteins into a $T = 3$ pseudo-icosahedron. Individual E proteins are colored according to their proximity to the 2-, 3-, or 5-fold symmetry axes (*blue*, *green* and *red* respectively). Image obtained using the viperdb Virus Particle Explorer (http://viperdb.scripps.edu/) (Shepherd et al., 2006). (**c**) Ribbon diagram of a WNV E protein monomer with domains II, I and III shown as *yellow*, *red* and *blue* ribbons respectively. The fusion loop at the tip of DII (DII-fl) is shown in *green*.

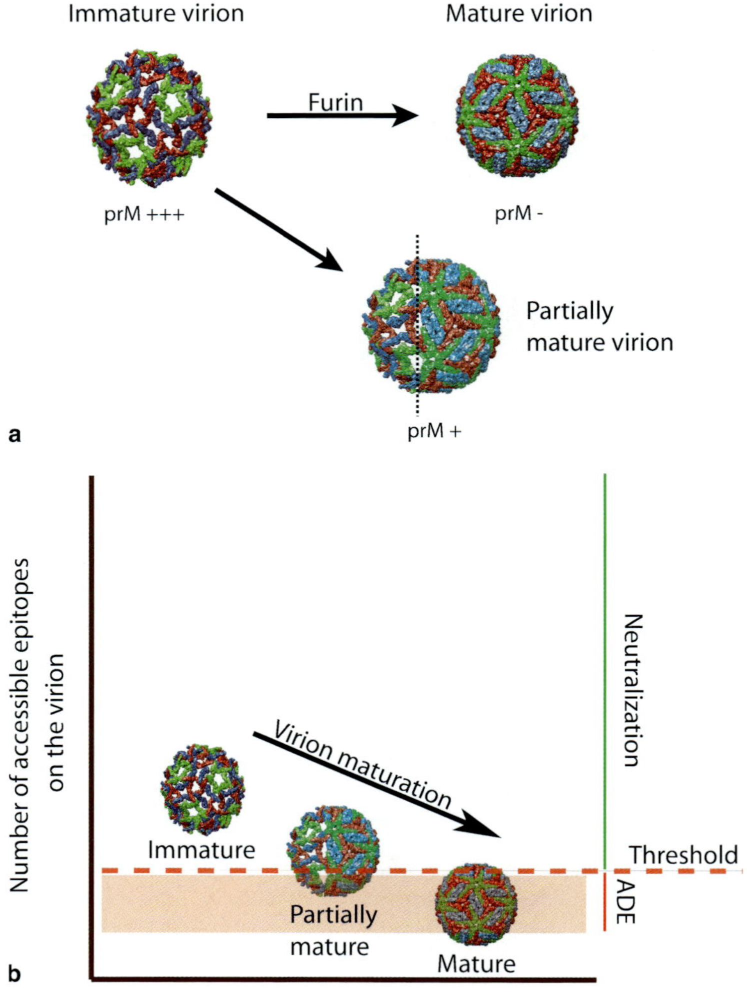

Chapter 10 Figure 3. Relationship between virus maturation and neutralization. (**a**) During virion egress from the cell, the prM protein of immature virions is cleaved by the cellular serine protease furin (Stadler et al., 1997). Cleavage results in a significant reorganization of E proteins on the surface of the virion into the mature form of the virus characterized by a herring-bone pseudo-icosahedral arrangement of E proteins (Heinz et al., 1994). However, not all prM proteins may be cleaved during this process. Partially mature virions that contain prM are present in bulk virus populations and are infectious (Davis et al. 2006; Guirakhoo et al., 1992). (**b**) The virion maturation process significantly impacts sensitivity to neutralization by antibodies predicted to bind poorly accessible structures. Mature virions are poorly neutralized by mAbs that bind several structurally distinct "cryptic" epitopes, as well as the polyclonal response of some human WNV vaccines. The reduction in neutralization sensitivity reflects a decrease in the number of accessible epitopes, which in turn reduces the number of antibodies that may simultaneously bind the virion.

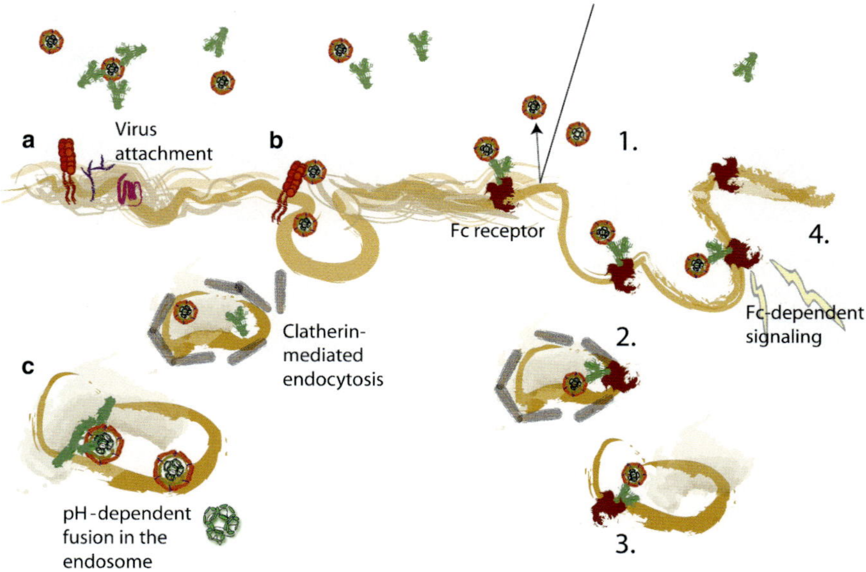

Chapter 10 Figure 4. Overview of the virus life cycle and its interaction with antibodies. Flaviviruses bind cells through largely uncharacterized cellular factors (Anderson, 2003; Davis et al., 2006) (**a**), are internalized in a clathrin-dependent process (Chu and Ng, 2004; Gollins and Porterfield, 1985) (**b**), and fuse with the host cell membranes of the early endosome in a pH-dependent fashion (Krishnan et al., 2007; van der Schaar et al., 2007) (**c**). Antibodies have the potential to reduce the efficiency of any of these steps including blocking attachment (He et al., 1995) or by inhibiting fusion with the endosomal membrane (Gollins and Porterfield, 1986; Nybakken et al., 2005; Stiasny et al., 2007a). Antibody-dependent enhancement describes the increase in virus entry and replication in the presence of non-neutralizing quantities of antibody (Halstead, 2003; Halstead and O'Rourke, 1977). ADE is dependent upon the interactions of virion immune complexes with Fc- or complement receptors (Cardosa et al., 1983, 1986; Peiris et al., 1981; Peiris and Porterfield, 1979). As the stable attachment of virus particles to cells is a major factor that limits infection, ADE may function by increasing the efficiency and durability of virus attachment to cells (Gollins and Porterfield, 1984) (1). This in turn may result in more efficient internalization into the cell, and/or the trafficking of virus to sites in the endocytic pathway from which post-fusion events may be more efficient (2 and 3). Furthermore, Fc-receptor-mediated signaling events may play a significant role in ADE by blunting the host-cell response to infection (Rodrigo et al., 2006) (4).

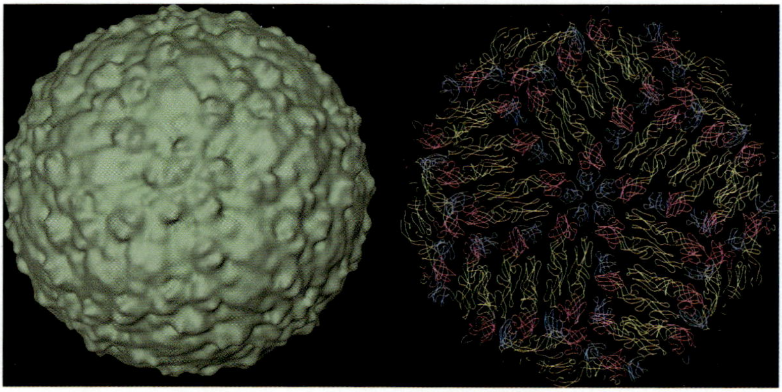

Chapter 11 Figure 1. The structure of West Nile virus. The structure of WNV was determined by cryo-EM and image reconstruction. The *left panel* shows a surface-rendered view of the virion looking down the icosahedral fivefold axis. The *right panel* shows the arrangement of the E protein after X-ray-derived coordinates were computationally fitted into the cryo-EM density. The view is identical to that of the *left panel*. The E protein is color coded according to domain with domain I in *red*, domain II in *yellow*, and domain III in *blue*.

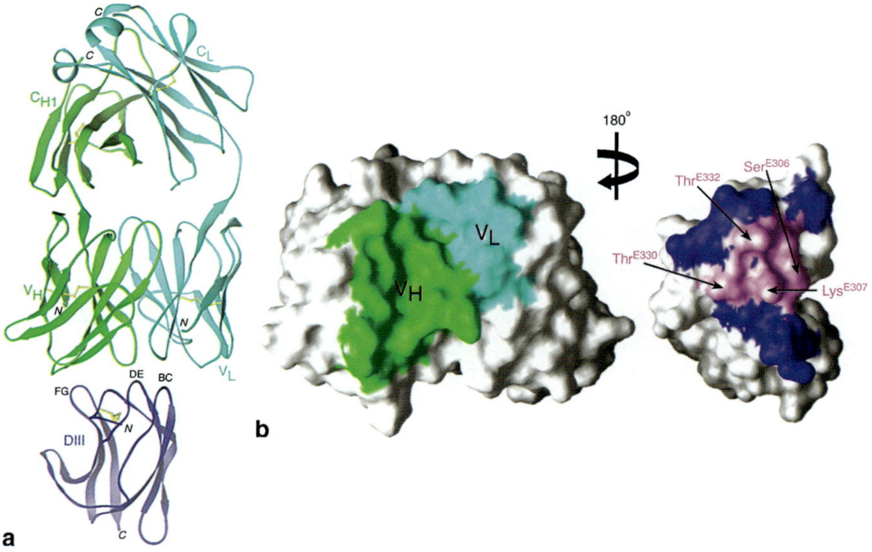

Chapter 11 Figure 2. The X-ray structure of Fab E16 in complex with the West Nile virus domain III. (**a**) A ribbon diagram of the complex with the Fab heavy chain shown in *green* and the light chain shown in *cyan*; domain III is shown in *blue*. The loops on domain III that interact with E16 are identified. The free N and C termini are also shown. (**b**) The molecular surfaces of the Fab (*left*) and domain III are shown using the same color scheme. Residues identified by the yeast display are identified in domain III and the region is colored *magenta*. This figure was provided courtesy of Daved Fremont and reprinted with permission of Nature Publishing Group.

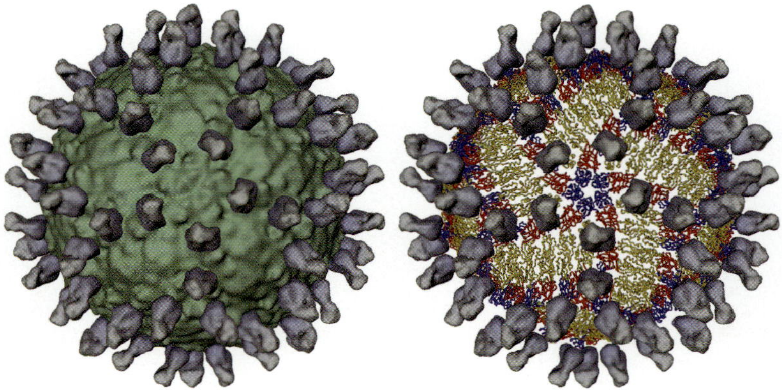

Chapter 11 Figure 3. The structure of West Nile virus in complex with Fab E16. The structure of WNV in complex with Fab E16 was determined by cryo-EM and image reconstruction. The *left panel* shows a surface-rendered view of the virion looking down the icosahedral fivefold axis. The virus particle is colored as in Fig. 1 (*green*). The Fab fragment is denoted by the *gray* color. The *right panel* shows the arrangement of the E protein after X-ray-derived coordinates were computationally fitted into the cryo-EM density. The view is identical to that of the *left panel*, with the Fab fragments shown in *gray*. Note the availability of domain IIIs around the fivefold axis.

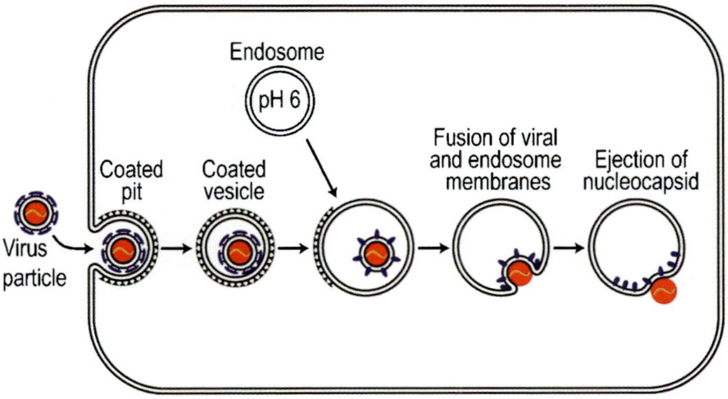

Chapter 12 Figure 1. Cell entry of flaviviruses. Virus particles bind target cells through a surface receptor, which is linked to the clathrin-dependent endocytic pathway. Internalized vesicles fuse with endosomal compartments. The reduced pH of these compartments promotes conformational rearrangements in the viral envelope proteins that catalyze the fusion of the host-cell and viral membranes. Upon membrane fusion, the viral genome is delivered into cytoplasm.

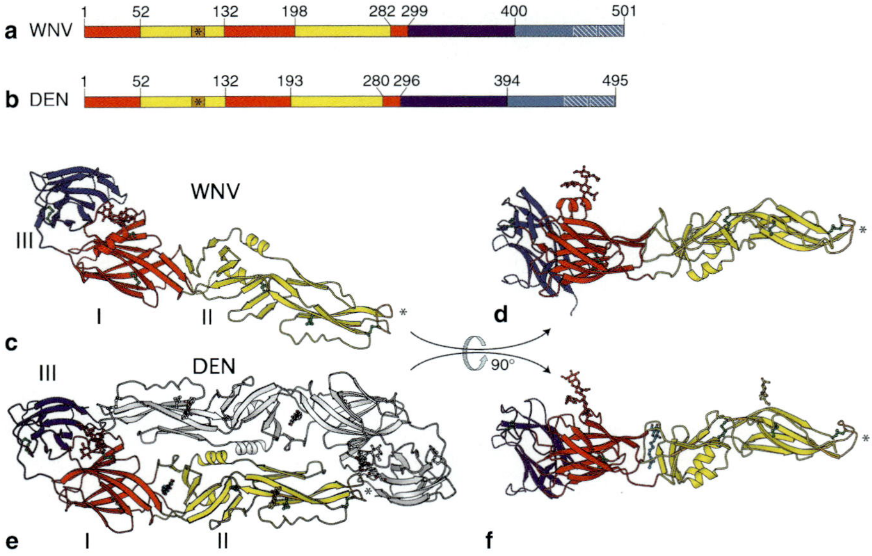

Chapter 12 Figure 2. Representative flavivirus fusion protein structures. (**a**) The three domains of the fusion protein of West Nile virus, E: domain I (residues 1–51, 132–197, 282–298), domain II (residues 52–131, 198–281), domain III (residues 299–501). A 53-residue "stem" (*cyan*) links the ectodomains to a two-helix C-terminal transmembrane anchor (*hatched areas*). (**b**) The three domains of dengue type 2 E: domain I (residues 1–51, 132–192, 280–295), domain II (residues 52–131, 193–279), domain III (residues 296–495). A 53-residue stem links the ectodomains to the transmembrane anchor. (**c**) Crystal structure of West Nile E in the prefusion conformation (Kanai et al., 2006). The relative orientations of domains I and II are slightly different from those observed in the mature virus particle (Kanai et al., 2006; Mukhopadhyay et al., 2003). The fusion loops in (**a**)–(**f**) are marked with an *asterisk*. (**d**) View rotated 90° relative to (**c**). (**e**) Crystal structure of dengue type 2 E in the prefusion conformation (Modis et al., 2003), as observed in the mature virus particle (Zhang et al., 2003a). A second subunit of E, forming the dimer found on the viral surface and in solution, is shown in *light gray*. (**f**) View rotated 90° relative to (**e**), with the second subunit omitted for clarity.

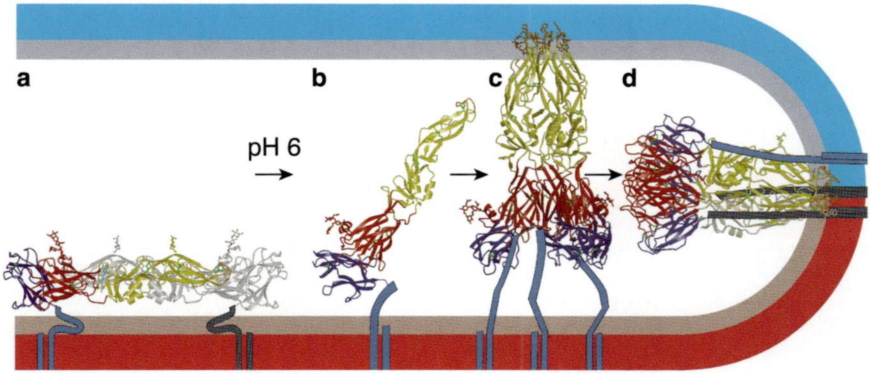

Chapter 12 Figure 3. Pre- and postfusion structures of dengue E, and proposed intermediates. (**a**) A dimer of dengue E (Modis et al., 2003) molecules in the prefusion conformation as found on the viral surface, viewed perpendicular to the viral membrane. The fusion loop is buried in the dimer interface. The outer (proximal) bilayer leaflets of the cellular and viral membranes are shown to scale as blue and red bands, respectively. The thin outer layer within each leaflet represents the polar headgroup layer, and the thicker inner layer represents the hydrocarbon layer. The stem-anchor segments are absent from the crystal structure, but are represented here schematically as rods in the viral membrane. (**b**) Upon acidification of the solute in the endosome, domain II rotates 15–30° about a hinge in the domain II–domain I interface. This exposes the fusion loop, which then inserts into the host-cell membrane. (**c**) Insertion of the fusion loop into the target membrane leads to domain II self association to form trimers. The panel shows a hypothetical West Nile E trimer constructed by superimposing domain II of the West Nile E monomeric prefusion structure onto the dengue E trimeric postfusion structure (Modis et al., 2004). This intermediate bridges the viral and cellular membranes. (**d**) The postfusion, trimeric structure of dengue E. After insertion of their fusion loops into the target membrane, the fusion proteins form trimers and fold back on themselves, bringing the fusion loops close to the C-terminal transmembrane anchors.

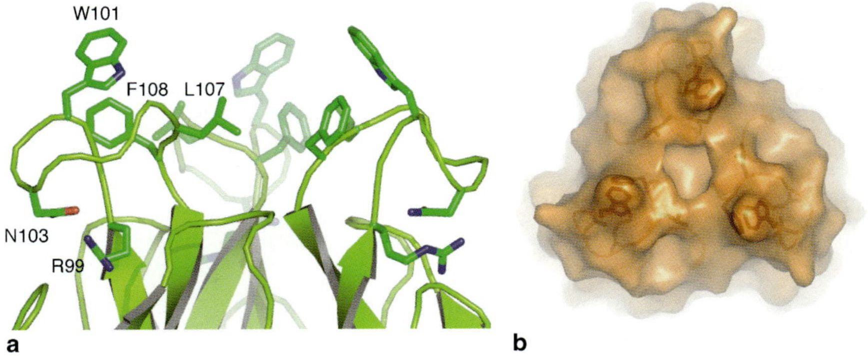

Chapter 12 Figure 4. Close-up of the aromatic anchor formed by the fusion loop in dengue virus E. (**a**) In flaviviruses, the fusion loop has a rigid structure, with three conserved hydrophobic residues (Trp, Phe, and Leu) protruding at the tip of domain II. These three residues insert into the hydrocarbon layer of the target cell membrane. The three fusion loops in the dengue E postfusion trimer (Modis et al., 2004) are shown, with the residues of the one of the protomers labeled. (**b**) The clustered fusion loops form a nonpolar, bowl-shaped apex, with the three conserved hydrophobic residues at the rim of the bowl, shown here in surface representation.

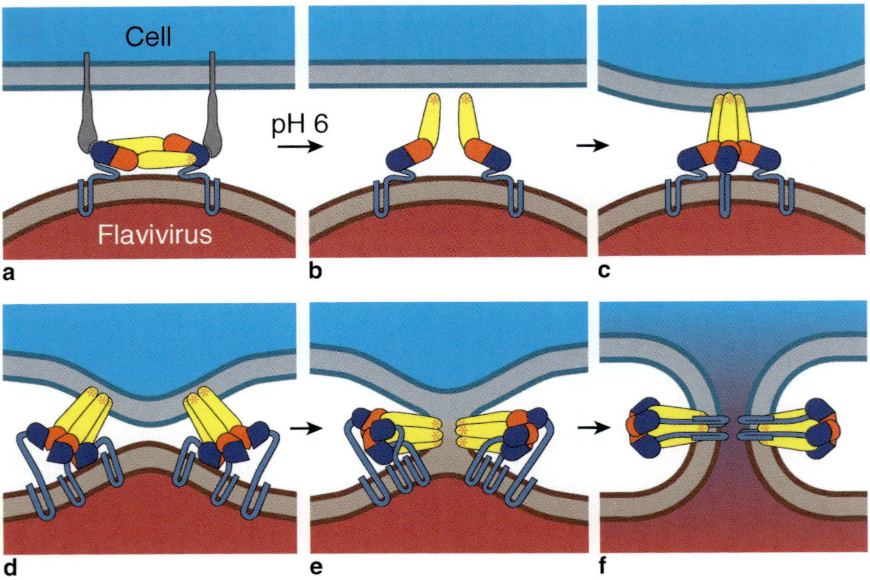

Chapter 12 Figure 5. Proposed fusion mechanism for fusion mediated by class II fusion proteins. (**a**) The virus binds to a receptor on the cell surface. In flaviviruses, the fusion protein E binds the receptor. Following attachment, the virus is internalized to an endosome. (**b**) Acidic pH in the endosome causes domain II to hinge outward from the virion surface, exposing the fusion loop, and allowing E monomers to rearrange laterally in the plane of the membrane. (**c**) The fusion loop inserts into the hydrocarbon layer of the host-cell membrane, promoting trimer formation. (**d**) Formation of trimer contacts spreads from the fusion loop at the tip of the trimer, to the base of the trimer. The protein folds back on itself, directing the fusion loop toward the C-terminal transmembrane anchor. Energy release by this refolding bends the apposed membranes. (**e**) Creation of additional trimer contacts between the stem anchor and domain II leads first to hemifusion and then (**f**) to formation of a lipidic fusion pore.

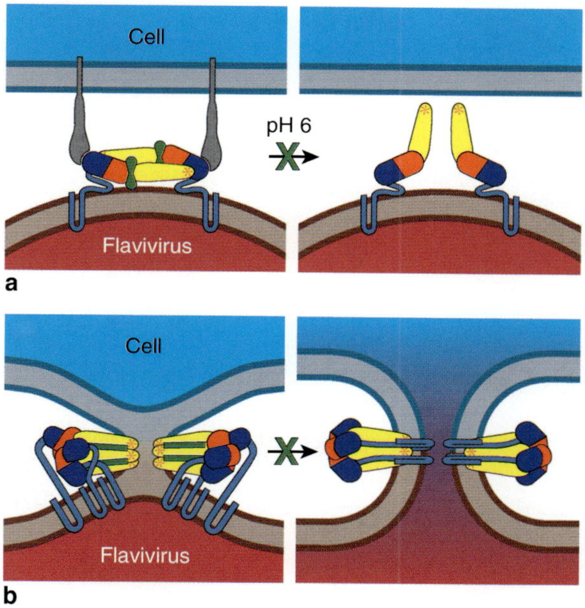

Chapter 12 Figure 6. Fusion inhibition strategies. (**a**) The discovery of a ligand-binding pocket at the interface between domains I and II in dengue virus E (Modis et al., 2003), just at the locus of a hinge motion required for fusion, suggests that compounds inserted in the pocket might hinder the hinge motion and hence inhibit the fusion transition. This approach would block the first step in the fusion mechanism (Fig. 5a, b). (**b**) Peptides corresponding to the stem region of the fusion protein may inhibit viral entry by binding to the trimeric core of the protein in its postfusion conformation, and interfering with the folding back against it of the fusion protein's own stem (Modis et al., 2004). An analogous strategy has been successful with HIV gp41 (Baldwin et al., 2003; Kilby et al., 1998). This approach would block the last step in the fusion mechanism (Fig. 5e, f). Antibody and vaccine-based strategies also offer promise in the treatment of flaviviral infection by inhibiting membrane fusion (see text).

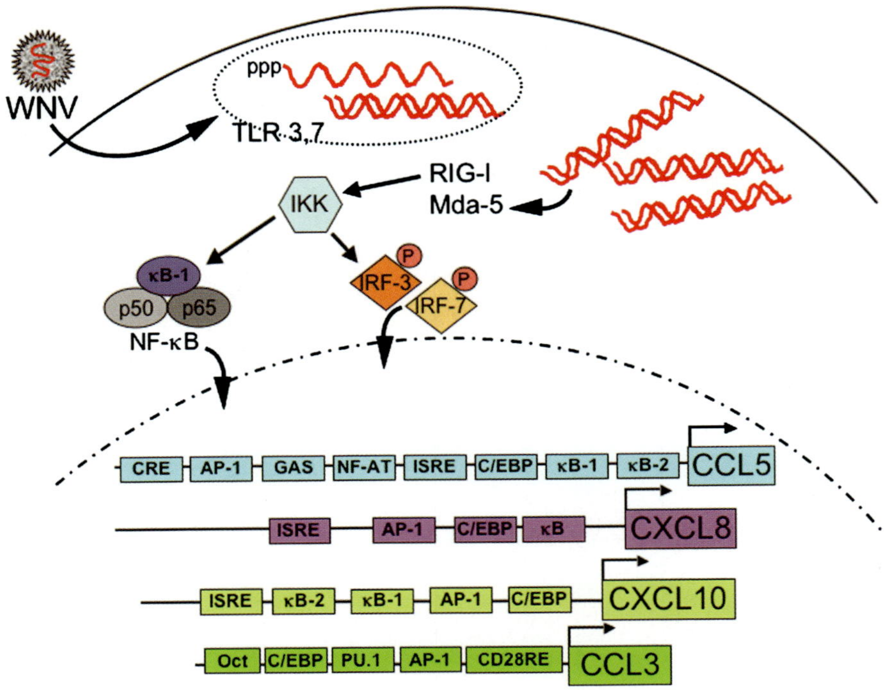

Chapter 15 Figure 1. Viral RNA sensing pathways lead to chemokine expression during WNV infection. Replication of WNV within infected cells leads to generation of 5′-triphosphate viral RNA moieties that trigger the activation of several PRRs including endosomally based toll-like receptors (TLR)-3 and -7 and cytoplasmic dsRNA sensors, retinoic acid-inducible gene I (RIG-I) and melanoma-differentiation-associated gene 5 (MDA-5). RIG-1/MDA5-5 activates the inhibitor of IκB kinase (IKK), which leads to both the activation of the transcription factors interferon response factor (IRF)-3 and -7 and the dissociation of κB-1 from NF-κB. These transcription factors, in conjunction with those induced by cytokine-mediated pathways, migrate into the nucleus (*dotted line*) and lead to expression of proinflammatory chemokines CCL5, CXCL8, CXCL10, and CCL3. *Abbreviations: Cre* cyclic adenosine monophosphate response element, *AP-1* activator protein-1, *GAS* IFN-γ activation site, *NF-AT* nuclear factor of activated T cells, *ISRE* IFN-stimulated response element, *C/EBP* CCAAT enhancer-binding proteins, *κB-1*, *-2* NF-κB binding sites, *CD28RE* CD28 response element.

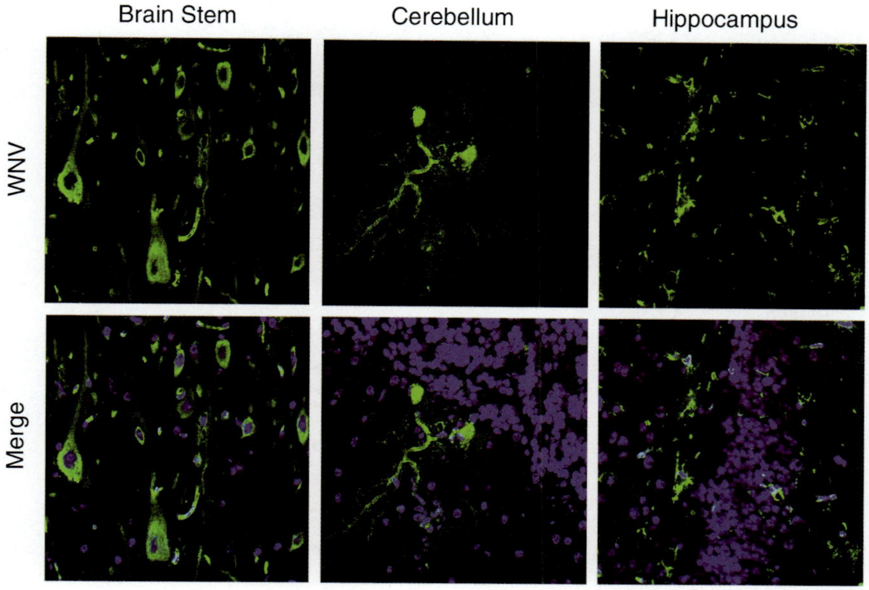

Chapter 17 Figure 1. WNV antigen staining in infected CNS tissues. Mice were inoculated intradermally with WNV and immunohistochemistry was performed on brains at day 10 after infection for WNV antigen (*green*) and nuclear staining (*blue*). WNV was detected at high levels in neurons throughout the brain stem and hippocampus and in the Purkinje layer of the cerebellum.

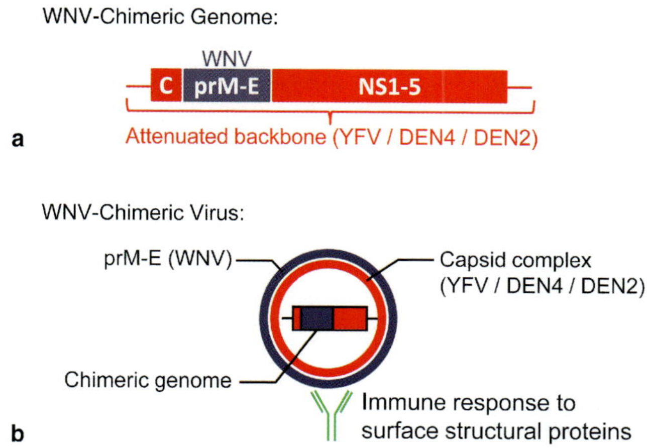

Chapter 20 Figure 1. Chimeric live attenuated vaccines. (**a**) The genes coding for the structural proteins prM and E of West Nile virus (WNV) are used to replace the corresponding genes in the attenuated host virus backbone. Host viruses used to develop WNV chimeras include attenuated strains of YFV, DEN4 and DEN2. (**b**) The chimeric virus consists of the WNV prM and E proteins and the nucleocapsid of the host virus with the encapsidated chimeric genome. The surface structural proteins of the virus elicit a specific immune response to WNV.

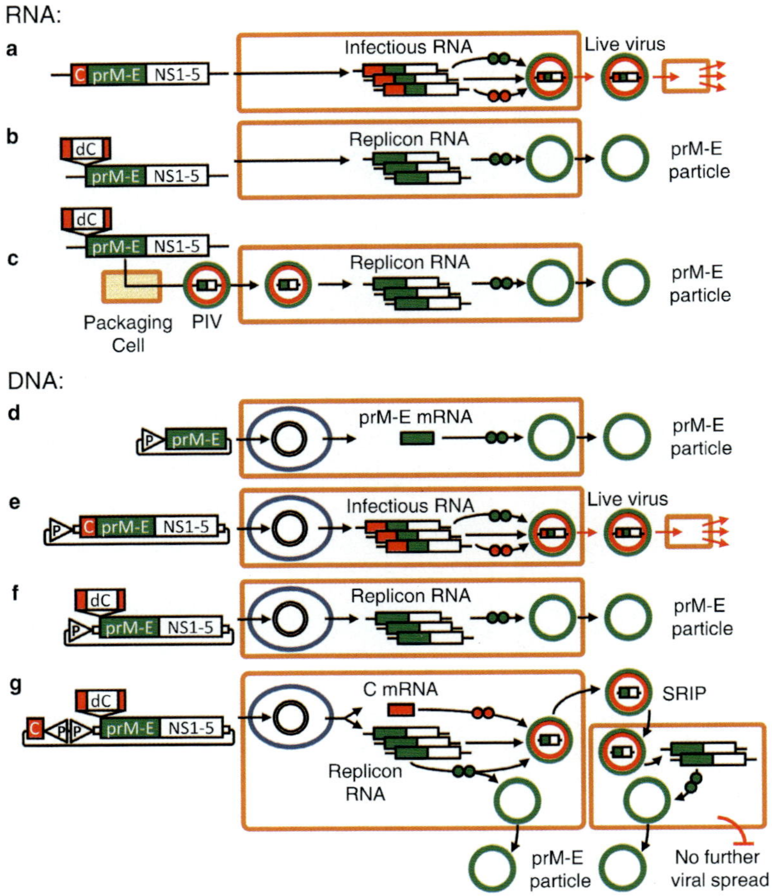

Chapter 20 Figure 2. Nucleic acid-based vaccines in development. *RNA:* (**a**) Infectious RNA; the full-length infectious RNA when delivered into cells results in generation of live virus which spreads to other cells. (**b**) Non-infectious RNA: a large internal deletion in the capsid gene in the full-length RNA prevents the packaging of replicon RNA into viral particles. The two structural proteins prM and E form secreted immunogenic prM–E particles, which elicit potent antibody response. (**c**) Pseudo-infectious viruses (PIV): a non-infectious RNA with large deletion in capsid gene is first introduced into a packaging cell expressing structural protein to generate PIVs. PIVs are then used to immunize recipients where they infect cells and deliver the capsid-deleted RNA to produce highly immunogenic secreted prM–E particles. *DNA:* (**d**) prM–E expressing vector: plasmid DNA encoding only the prM and E genes driven by a mammalian expression promoter (P) transcribes prM–E mRNA which leads to the formation and secretion of prM–E particles. (**e**) Infectious DNA: the full-length infectious RNA is transcribed by cellular RNA polymerase II from the mammalian expression promoter and produce infectious virus which spreads to other cells. (**f**) Non-infectious DNA: capsid-deleted self-replicating RNA is transcribed by RNA polymerase II and produces secreted prM–E particles. (**g**) Single-round infectious particles (SRIPs): a single plasmid DNA with two mammalian expression promoters placed into back-to-back orientation directs transcription of two RNAs: (1) capsid-deleted self-replicating RNA and (2) capsid mRNA. The former is packaged into SRIPs by the three structural proteins. PrM–E particles are also secreted from these cells. Secreted SRIPs infect and deliver the replicon RNA to other cells to further stimulate the host immune response by expressing viral non-structural proteins and secreted prM–E. The absence of capsid gene expression in SRIP-infected cells effectively terminates the infectious cycle.

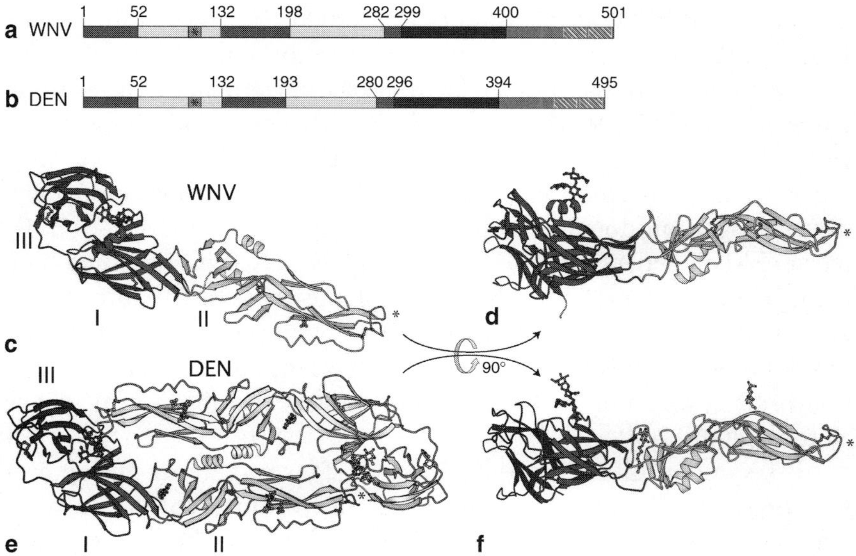

Figure 2. Representative flavivirus fusion protein structures. (**a**) The three domains of the fusion protein of West Nile virus, E: domain I (residues 1–51, 132–197, 282–298), domain II (residues 52–131, 198–281), domain III (residues 299–501). A 53-residue "stem" (*cyan*) links the ectodomains to a two-helix C-terminal transmembrane anchor (*hatched areas*). (**b**) The three domains of dengue type 2 E: domain I (residues 1–51, 132–192, 280–295), domain II (residues 52–131, 193–279), domain III (residues 296–495). A 53-residue stem links the ectodomains to the transmembrane anchor. (**c**) Crystal structure of West Nile E in the prefusion conformation (Kanai et al., 2006). The relative orientations of domains I and II are slightly different from those observed in the mature virus particle (Kanai et al., 2006; Mukhopadhyay et al., 2003). The fusion loops in (**a**)–(**f**) are marked with an *asterisk*. (**d**) View rotated 90° relative to (**c**). (**e**) Crystal structure of dengue type 2 E in the prefusion conformation (Modis et al., 2003), as observed in the mature virus particle (Zhang et al., 2003a). A second subunit of E, forming the dimer found on the viral surface and in solution, is shown in *light gray*. (**f**) View rotated 90° relative to (**e**), with the second subunit omitted for clarity. (*See Color Plates*)

receptor that specifically recognizes a protein epitope on an envelope protein of a class II enveloped virus has yet to be conclusively identified.

Flavivirus E proteins have either one or two N-linked glycosylation sites. West Nile virus and dengue virus use these glycans to attach to the surface of dendritic cells. In West Nile virus, the single glycan on E (at residue 154) is recognized by the C-type lectin DC-SIGNR (Davis et al., 2006). In the case of dengue virus, a different glycan (at residue 67) is recognized by a related lectin, DC-SIGN (Navarro-Sanchez et al., 2003; Tassaneetrithep et al., 2003). The basis for the differential recognition of the glycans on West Nile and dengue virions by different dendritic cell receptors may be due to a match in the spacing of the respective glycans on the viral surface with the spacing of the carbohydrate recognition domains of the receptors (Kanai et al., 2006; Mitchell et al., 2001). Indeed, DC-SIGN appears to require two adjacent glycans approximately 18 Å apart for optimal binding, as illustrated by the binding pattern of DC-SIGN on dengue virus particles (Pokidysheva et al., 2006). In contrast, the spacing between glycans on the surface of West Nile virus is 50 Å. This is close to the 54-Å separation between carbohydrate recognition domains in the crystal structure of the DC-SIGNR tetramer (Feinberg et al., 2005), suggesting that DC-SIGNR could bind multiple (up to four) viral glycans simultaneously. In support of this notion, the specificity of carbohydrate recognition of DC-SIGN and DC-SIGNR depends largely on whether glycans can bind all four carbohydrate recognition domains simultaneously (Feinberg et al., 2005; Mitchell et al., 2001).

It is important to note that all the crystal structures of fusion proteins determined so far, from both classes and regardless of their conformational state, lack the C-terminal viral membrane anchor. This anchor consists of one or two transmembrane helices. The crystallized species should therefore be referred to strictly as soluble fragments of the ectodomains of the full-length fusion protein. Furthermore, all available crystal structures of class II fusion proteins also lack the "stem" region (Allison et al., 1999), a 30–55 amino acid linker between domain III and the C-terminal transmembrane anchor (Figs. 2a, b and 3). As will be discussed below, the stem region plays a key role in the final stages of membrane fusion. Its function is analogous to that of the "outer helix" in class I fusion proteins (Skehel and Wiley, 2000).

3 Maturation and Priming of Fusion-Competent Virions

All viral membrane fusion proteins rely on a proteolytic cleavage event to become primed to respond to the environmental conditions appropriate for fusion. These conditions are usually induced by the acidic pH of an endosome (Fig. 1), but for some class I enveloped viruses, such as HIV, co-receptor binding is required instead. In contrast to class I fusion proteins, however, class II fusion proteins rely on a priming proteolytic cleavage that does not cleave the fusion protein itself. Instead, class II proteins associate with a second, "protector" protein, called prM in flaviviruses (or E2 in alphaviruses). PrM is cleaved by furin-like proteases when immature virus particles assembled in the ER reach the trans-Golgi network (Stadler et al., 1997). The cleavage of prM to its mature product, M, releases a conformational constraint on E (the fusion

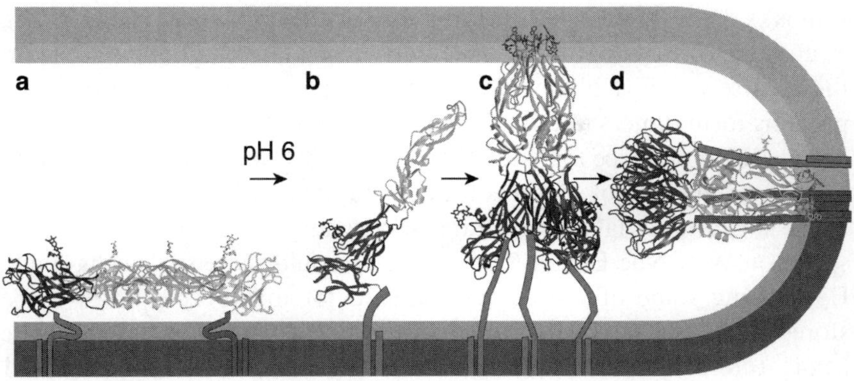

Figure 3. Pre- and postfusion structures of dengue E, and proposed intermediates. (**a**) A dimer of dengue E (Modis et al., 2003) molecules in the prefusion conformation as found on the viral surface, viewed perpendicular to the viral membrane. The fusion loop is buried in the dimer interface. The outer (proximal) bilayer leaflets of the cellular and viral membranes are shown to scale as blue and red bands, respectively. The thin outer layer within each leaflet represents the polar headgroup layer, and the thicker inner layer represents the hydrocarbon layer. The stem-anchor segments are absent from the crystal structure, but are represented here schematically as rods in the viral membrane. (**b**) Upon acidification of the solute in the endosome, domain II rotates 15–30° about a hinge in the domain II–domain I interface. This exposes the fusion loop, which then inserts into the host-cell membrane. (**c**) Insertion of the fusion loop into the target membrane leads to domain II self association to form trimers. The panel shows a hypothetical West Nile E trimer constructed by superimposing domain II of the West Nile E monomeric prefusion structure onto the dengue E trimeric postfusion structure (Modis et al., 2004). This intermediate bridges the viral and cellular membranes. (**d**) The postfusion, trimeric structure of dengue E. After insertion of their fusion loops into the target membrane, the fusion proteins form trimers and fold back on themselves, bringing the fusion loops close to the C-terminal transmembrane anchors. (*See Color Plates*)

protein), which allows E to reach its mature, prefusion conformation in a large rearrangement on the viral surface. The mature virus particles are then released from the host cell by exocytosis. In the mature conformation, the fusion protein is primed to respond to acidic pH and induce membrane fusion with a further conformational rearrangement.

Structures from electron cryomicroscopy of both immature and mature flavivirus particles provide a detailed picture of the rearrangement that accompanies maturation in these viruses (Kuhn et al., 2002; Mukhopadhyay et al., 2003; Zhang et al., 2003a, b). Unlike alphaviruses, which retain the $T = 4$ icosahedral packing of their envelope proteins during maturation (Ferlenghi et al., 1998; Mancini et al., 2000), flaviviruses undergo a dramatic rearrangement in the organization of their E proteins on the viral surface during maturation. Indeed, when the protector protein prM is cleaved in flaviviruses, the fusion protein

E breaks the $T = 3$ icosahedral symmetry of the immature virion (Zhang et al., 2003b) to adopt an unusual icosahedral herringbone pattern in the mature virion (Kuhn et al., 2002; Zhang et al., 2002). The E fusion proteins form dimers in the mature virion, including in West Nile virus, despite the preference for West Nile E to adopt a monomeric conformation in the absence of icosahedral constraints (see Fig. 2). This has led to the suggestion that icosahedral assembly might impose some physical strain on West Nile E, and that this strain could serve to "spring-load" E, allowing some of the energy required for membrane fusion to be stored in the metastable mature virus particle (Kanai et al., 2006). The key feature of the maturation process, however, is that cleavage of prM allows the fusion loop of E to reposition itself so that it is poised to insert into the host-cell membrane in response to acidification in the endosome. Mature virus particles are therefore infectious (Elshuber et al., 2003; Stadler et al., 1997), unlike immature virions (Guirakhoo et al., 1991, 1992), which are insensitive to pH. The flavivirus fusion loop is shielded from the viral surface in mature virions by E-E dimer contacts (Figs. 3a and 5a).

4 The Fusogenic Conformational Rearrangement

The three-dimensional structures of three class II fusion proteins (dengue, tick-borne encephalitis, and Semliki Forest viruses) in their postfusion states reveal striking differences from the prefusion forms (Fig. 3), and suggest a molecular mechanism for membrane fusion (see Fig. 5) (Bressanelli et al., 2004; Gibbons et al., 2004; Modis et al., 2004). Like class I fusion proteins, flaviviral E proteins and other class II fusion proteins are homotrimers in their postfusion conformations. E proteins form trimers from monomers on the viral surface, however, whereas class I proteins are trimeric in their prefusion state (Skehel and Wiley, 2000). A comparison of the pre- and postfusion states of influenza HA – the only example in its class where both structures are known for the same protein – shows that, as in class II fusion proteins, nearly all of the trimer contacts in the postfusion state are formed during the fusogenic conformational rearrangement.

Unlike influenza HA, which undergoes extensive refolding during membrane fusion, the three domains of class II fusion proteins retain most of their folded structures (Fig. 3). Instead, the domains undergo major rearrangements in their relative orientations, through flexion of the interdomain linkers. Domain III undergoes the most significant displacement in the fusion transition. It rotates by approximately 70°, and its center of mass shifts by 30–40 Å toward domain II. This shift brings the C terminus of domain III about 40 Å closer to the fusion loop, located at the tip of domain II. Domain II rotates 15–30° with respect to domain I about a hinge region near the interface with domain I (Modis et al., 2003). Mutations in this region affect the pH threshold of fusion in various flaviviruses (Beasley and Aaskov, 2001; Cecilia and Gould, 1991; Hasegawa et al., 1992; Hurrelbrink and McMinn, 2001; Lee et al., 1997; Monath et al., 2002). These conformational rearrangements place the end of domain III – and the beginning of the stem region that links domain III to the C-terminal viral transmembrane anchor – pointing toward the fusion loop (Fig. 3d) (Gibbons et al., 2004; Modis et al., 2004). A deep channel between domains II of adjacent subunits in the trimer extends from the C terminus of the crystallized fragment to the three clustered fusion loops at the tip of the trimer, in the flaviviral postfusion structures. In the full-length fusion proteins, it is thought that the stem binds in this channel in an extended, but mainly α-helical conformation (Modis et al., 2004; Zhang et al., 2003a). This proposed conformation for the stem places the viral transmembrane anchor in the immediate vicinity of the fusion loop, just as in the postfusion conformation of class I viral fusion proteins.

The fusion transition in flaviviruses is irreversible. The structural rearrangements just described may impart irreversibility by contributing a high barrier to the initiation of trimerization and an even higher barrier to the dissociation of postfusion trimers once formed. Moreover, many new polar and nonpolar contacts are formed during the fusion transition, in several different areas, mostly near the threefold axis of the trimer. The total surface buried in the dengue E trimer is 15,000 $Å^2$ (Modis et al., 2004), nearly four times more than is buried in the dengue prefusion E dimer. The stem, which is missing from currently available crystal structures, most likely forms additional contacts with the core trimer structure (Figs. 3 and 5). The stem does indeed promote trimer assembly in vitro, even in the absence of liposomes (Allison et al., 1999).

5 The Flaviviral Fusion Loop

In both class I and class II enveloped viruses, the process of viral membrane fusion begins with the exposure of a fusion anchor, and its subsequent insertion into the host-cell membrane. Fusion anchors from both viral classes vary in length but are in general rich in glycines and hydrophobic residues, particularly aromatic residues such as tryptophan or phenylalanine. Viruses from different genera rarely have significant levels of sequence identity in their fusion proteins. The fusion anchor in class I fusion proteins – the "fusion peptide" – is a region of approximately 20 residues at or near the N terminus of the fusion protein. The crystal structure of the parainfluenza virus 5 fusion (F) protein in its prefusion form reveals the fusion peptide wedged between two subunits of the protein, in a partly extended, partly β-sheet and partly α-helical conformation (Yin et al., 2006). Structural studies on influenza HA in its postfusion conformation using NMR and other spectroscopic techniques show that the fusion peptide is mostly α-helical in character and that its structure changes only subtly as it inserts partway into the outer leaflet of the host-cell lipid bilayer (Dubovskii et al., 2000; Han et al., 2001). The fusion peptide is disordered or absent in all of the currently available postfusion class I protein crystal structures.

In contrast, the crystal structures of class II fusion proteins in postfusion conformations offer direct views of fusion anchors – in this case, fusion loops – as they insert into a target membrane (Fig. 4) (Bressanelli et al., 2004; Gibbons et al., 2004; Modis et al., 2004). Like the class I fusion peptide, the class II fusion loop penetrates only partway into the hydrocarbon layer of the target membrane. Exposed carbonyls and charged residues prevent the fusion loop from penetrating further than 6 Å (Bressanelli et al., 2004; Modis et al., 2004). In flaviviruses, the fusion loop adopts a conserved and tightly folded conformation, which is stabilized by a disulfide bond (Fig. 4a). The structure of the fusion

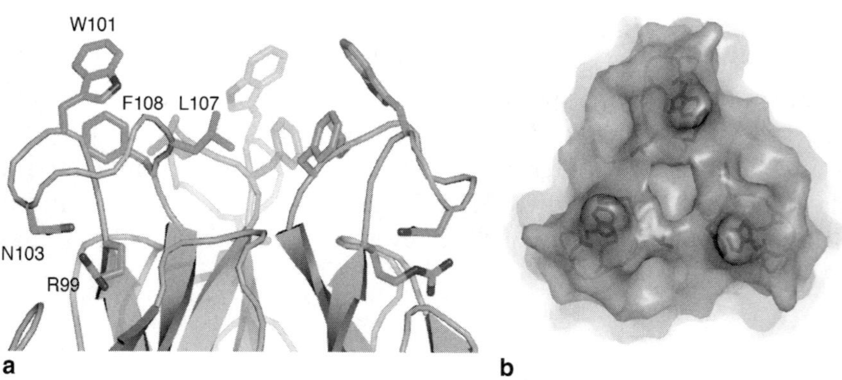

Figure 4. Close-up of the aromatic anchor formed by the fusion loop in dengue virus E. (**a**) In flaviviruses, the fusion loop has a rigid structure, with three conserved hydrophobic residues (Trp, Phe, and Leu) protruding at the tip of domain II. These three residues insert into the hydrocarbon layer of the target cell membrane. The three fusion loops in the dengue E postfusion trimer (Modis et al., 2004) are shown, with the residues of the one of the protomers labeled. (**b**) The clustered fusion loops form a nonpolar, bowl-shaped apex, with the three conserved hydrophobic residues at the rim of the bowl, shown here in surface representation. (*See Color Plates*)

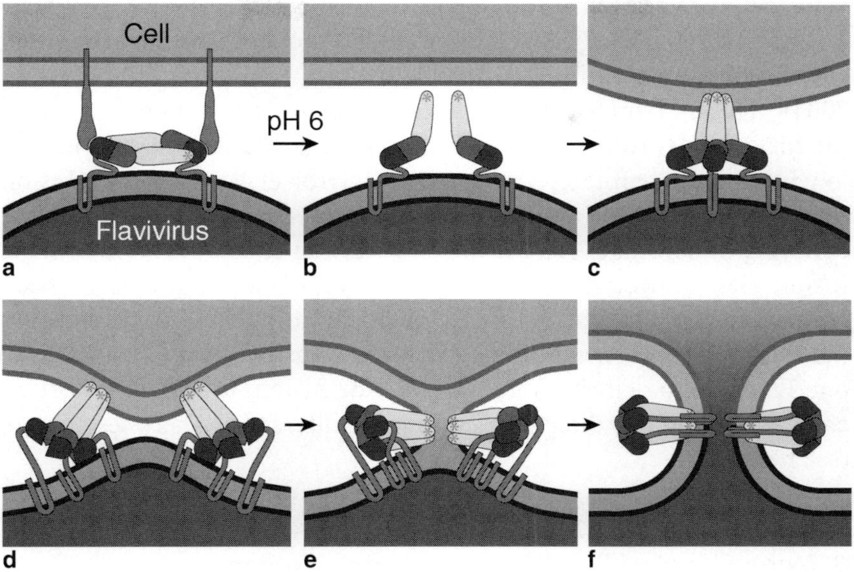

Figure 5. Proposed fusion mechanism for fusion mediated by class II fusion proteins. (**a**) The virus binds to a receptor on the cell surface. In flaviviruses, the fusion protein E binds the receptor. Following attachment, the virus is internalized to an endosome. (**b**) Acidic pH in the endosome causes domain II to hinge outward from the virion surface, exposing the fusion loop, and allowing E monomers to rearrange laterally in the plane of the membrane. (**c**) The fusion loop inserts into the hydrocarbon layer of the host-cell membrane, promoting trimer formation. (**d**) Formation of trimer contacts spreads from the fusion loop at the tip of the trimer, to the base of the trimer. The protein folds back on itself, directing the fusion loop toward the C-terminal transmembrane anchor. Energy release by this refolding bends the apposed membranes. (**e**) Creation of additional trimer contacts between the stem anchor and domain II leads first to hemifusion and then (**f**) to formation of a lipidic fusion pore. (*See Color Plates*)

loop is essentially identical in the pre- and postfusion conformations of the protein, suggesting that membrane insertion has no effect on the structure of the fusion loop. During the fusion transition, three conserved hydrophobic residues in the fusion loop (a tryptophan, leucine, and pheylalanine) become exposed on the molecular surface. Three fusion loops end up tightly clustered at the tip of the trimer in the postfusion conformation, where they form a crater-like surface with a hydrophobic rim (Fig. 4). Electron cryomicroscopy (Modis et al., 2004) and mutagenesis studies (Allison et al., 2001) confirm that these hydrophobic, mostly aromatic residues on the crater rim insert into the host-cell membrane, acting as an "aromatic anchor" for the fusion protein. The concave shape of the crater is thought to be important in generating distortions or perturbations in the host-cell membrane (Modis et al., 2004), which are required for fusion with the viral membrane (Tamm et al., 2002).

Semliki Forest virus E1 (a class II fusion protein) forms irregular clusters, or "rosettes," consisting of 40–60 postfusion trimers through contacts between fusion loops in adjacent trimers (Gibbons et al., 2004). This is reminiscent of influenza virus HA, which aggregates into rosettes through interactions between the fusion peptide, at low pH and after proteolytic activation (Ruigrok et al., 1988). This fusion loop/peptide clustering may provide a mechanism for the direct coupling of several E1/HA trimers to work in concert around a single fusion site. However, flavivirus E protein trimers have never been observed to form clusters or rosettes, and current evidence suggests that although the energetics of membrane fusion requires more than one E trimer to deliver enough energy for fusion, the trimers work in concert without any direct protein-protein interactions between trimers (see below).

6 Mechanism of Flaviviral Membrane Fusion

Combined with previous knowledge, the structures of flavivirus fusion proteins in their postfusion states (Bressanelli et al., 2004; Modis et al., 2004) have elucidated how conformational changes in these proteins drive membrane fusion. The structures confirm two major principles of membrane fusion machineries: (1) the fusion protein must insert an anchor into each of the two membranes to be fused and (2) the protein folds back on itself in a thermodynamically favorable conformational rearrangement that drives membrane fusion by forcing the two anchors into close proximity of one another.

In the current model, viral membrane fusion proceeds as follows (Fig. 5). First, receptor binding by the envelope protein (which in flaviviruses is also the fusion protein) leads to clathrin-mediated endocytosis of the virus (Figs. 1 and 5a). When the virus reaches an endosomal compartment, the reduced pH of the lumen (~pH 6) causes an initial conformational rearrangement that exposes the previously buried fusion loop (Zhang et al., 2003a), at the tip of domain II. In flaviviruses, domains I and II flex relative to each other by 30° (Modis et al., 2004). This hinge motion causes domain II, and therefore the fusion loop, to swing away from the viral surface and toward the host-cell membrane (Fig. 5b). The notion that the domain I-domain II interface acts as a hinge early in the fusion transition is supported by the observation that mutations at this interface alter the pH threshold of fusion in various flaviviruses (Beasley and Aaskov, 2001; Cecilia and Gould, 1991; Hasegawa et al., 1992;

Hurrelbrink and McMinn 2001; Lee et al., 1997; Modis et al., 2003; Monath et al., 2002). As domain II swings away from the viral surface, constraints imposed by the tight packing of E on the viral surface are released, allowing E to diffuse freely in the plane of the viral membrane. The stem may provide enough flexibility at this stage to allow the E ectodomains to extend away from the membrane (Fig. 4b, c).

The second key step in the fusion process is insertion of the exposed fusion loop into the host-cell membrane (Fig. 5c). Flavivirus E proteins probably insert their fusion loops as monomers, but membrane insertion quickly promotes trimerization of the fusion loops (Stiasny et al., 2002), by lateral rearrangement of E monomers in the plane of the viral membrane. The resulting trimeric prefusion intermediate (Figs. 4c and 5c) bridges the host-cell and viral membranes, anchored by its fusion loops in the former and by the viral transmembrane anchors in the latter. This proposed intermediate is analogous to the "prehairpin" intermediate postulated for class I viral fusion mechanisms (Chan and Kim, 1998).

After the fusion loops have inserted into the host-cell membrane, formation of trimer contacts spreads from the fusion loops at the tip of the trimer to domain I at the base of the trimer. Domain II rotates and shifts, folding the stem and C-terminal anchor back toward the fusion loop (Fig. 5d), and burying additional protein surfaces. Energy released by this refolding drives the two membranes to bend toward one another (Baker et al., 1999; Melikyan et al., 2000b; Russell et al., 2001), as the C-terminal anchor is forced closer to the fusion loop, forming apposing nipples in the membranes (Fig. 5d) (Kuzmin et al., 2001). Fusion-loop insertion is expected to induce bilayer curvature as lipid molecules are laterally displaced in the leaflet. Such a curvature would stabilize the lateral surfaces of the nipples. The concave shape of the crater-like surface formed by the fusion loops at the trimer tip may also destabilize membranes, as has been postulated for fusion peptides in class I fusion proteins (Tamm et al., 2002). Based on the amount of energy required to deform lipid bilayers, the concerted action of at least three trimers is needed around the fusion site of enveloped viruses to provide sufficient energy to form nipples in the membranes. In the case of influenza, fusion requires at least three HA trimers (Danieli et al., 1996), and is more likely to be driven by rings, or "rosettes," of 6–8 trimers (Blumenthal et al., 1996). The concerted refolding of each trimer in the ring allows the energy released by each refolding to be combined to generate a total force that is great enough to create the necessary distortions in the host-cell and viral membranes (Kozlov and Chernomordik, 1998;

Kuzmin et al., 2001). It is unclear exactly how many trimers are needed to drive membrane fusion in flaviviruses, or how their conformational changes are coupled. The clustering of fusion loops may provide a mechanism for the direct coupling of several E1 trimers to work in concert around a single fusion site in alphaviruses, but such clustering has not yet been observed in flaviviruses. It is possible that coupling occurs through the membrane: only when several trimers fold back in concert can they overcome the resistance of the membrane to deformation and reach their final, stable postfusion conformation (Fig. 5d–f).

As the fusion transition proceeds, the stem anneals onto the core of the trimer, along a channel that spans the length of domain II, at the interface between adjacent subunits (Figs. 3d and 5d–f). The annealing of the stem onto the domain II forces the fusion loop and the viral transmembrane anchor closer and closer, until the proximal leaflets of the two membranes fuse to form a "hemifusion stalk" (Fig. 5e). Hemifusion is an obligatory intermediate of membrane fusion in general (Kozlov and Chernomordik, 1998; Kuzmin et al., 2001; Razinkov et al., 1999). Figure 5e illustrates the need for shallow penetration of the viral fusion anchor into the host-cell membrane: assuming several trimers act in concert around a single fusion site, fusion anchors from different trimers cannot insert beyond the outer (proximal) leaflet of the lipid bilayer without colliding with each other. This constraint on the length of the fusion anchor holds true for both class I fusion peptides and class II fusion loops.

Hemifusion stalks can briefly "flicker" open into narrow fusion pores and then return to the hemifused state (Razinkov et al., 1999). To prevent transient fusion pores from closing, the stem must complete its annealing onto the core of the trimer, and the C-terminal transmembrane anchor must migrate into the pore (Fig. 5f). Indeed, the transition from hemifusion stalk to full fusion pore appears to require that the viral transmembrane anchor span the membrane completely, in all biological membrane fusion systems. This requirement was clearly demonstrated in an experiment in which the C-terminal transmembrane anchor of influenza HA was replaced with a glycosylphophatidylinositol (GPI) lipid anchor (Kemble et al., 1994; Melikyan et al., 1995; Nussler et al., 1997), or with a half-length α-helical anchor (Armstrong et al., 2000). In both of these truncated HA mutants, the fusion reaction stalls at the stage of hemifusion. Other viral fusion proteins and cellular SNARE fusion proteins also require at least one transmembrane anchor (Bagai and Lamb, 1996; Dutch and Lamb, 2001; Januszeski et al., 1997; McNew et al., 1997, 2000, 2000a; Saifee et al., 1998; West et al., 2001). In flaviviruses, the C-terminal anchor is also essential to resolve fusion intermediates into a fully fused pore. Upon completion of fusion, the E trimer reaches the conformation seen in the postfusion crystal structures of dengue and tick-borne encephalitis E (Bressanelli et al., 2004; Modis et al., 2004). The stems (not present in the structures) are thought to dock along the surface of domains II, with the fusion loops and transmembrane anchors lying next to each other in the fused membrane (Fig. 5f).

Some viral fusion proteins require a specific lipid composition to catalyze membrane fusion. Alphavirus E1, for example, can only fuse membranes that contain cholesterol and sphingolipids (Nieva et al., 1994). Mutations in different regions of the Semliki Forest virus E1 lower its dependence on cholesterol and/or sphingolipids for membrane fusion (Chanel-Vos and Kielian 2004; Vashishtha et al., 1998); however, the molecular basis for this requirement is still not well understood (Chatterjee et al., 2002). In flaviviruses, cholesterol facilitates fusion, but neither cholesterol nor sphingolipids are essential for fusion (Stiasny et al., 2003).

7 Strategies for Fusion Inhibition

Many flaviviruses are important human pathogens including dengue, hepatitis C, yellow fever, Japanese encephalitis, and tick-borne encephalitis viruses in addition to West Nile virus (Burke and Monath, 2001). For most of these viruses, there are no specific treatments for infection, their control by vaccination has proved elusive, and the number of infections is on the rise. The three-dimensional structures of flavivirus fusion proteins described above suggest novel strategies for inhibiting viral entry by blocking membrane fusion. One such strategy stems from the discovery in dengue E of a long, tapering channel lined with hydrophobic side chains (Modis et al., 2003). In the crystal structure, the channel is occupied by a molecule of *n*-octyl-β-d-glucoside, a nonionic detergent. In the absence of detergent, a β-hairpin covering the channel swings shut, closing up the channel (Modis et al., 2003). The location of this ligand-binding pocket at the domain I–domain II interface coincides with that of mutations affecting the pH threshold of fusion in various flaviviruses (Beasley and Aaskov, 2001; Cecilia and Gould, 1991; Hasegawa et al., 1992; Hurrelbrink and McMinn, 2001; Lee et al., 1997; Monath et al., 2002). Most of these mutations affect side chains lining the ligand-binding pocket. The structure of dengue virus E in the postfusion conformation shows that this region acts as a hinge between domains I and II during the fusogenic conformational rearrangement (see above) (Modis et al., 2004). The existence of a ligand-binding pocket just at the locus of a hinge suggests that compounds, which bind tightly to this position, may hinder the conformational changes required for membrane fusion (Fig. 6a). Such small molecules may have a similar mechanism of action as some of the well-studied anti-picornavirus compounds (e.g., disoxaril, pleconaril), which block a concerted structural transition in the icosahedral assembly by binding in a hydrophobic canyon on the viral surface (Smith et al., 1986). Alternatively, small molecules that pry open the β-hairpin in the pocket may inhibit infection by facilitating the fusogenic conformational change, causing premature triggering. Knowing the structure of the binding pocket with a bound ligand should guide efforts to design derivative ligands with higher affinities for use as inhibitors of flavivirus membrane fusion.

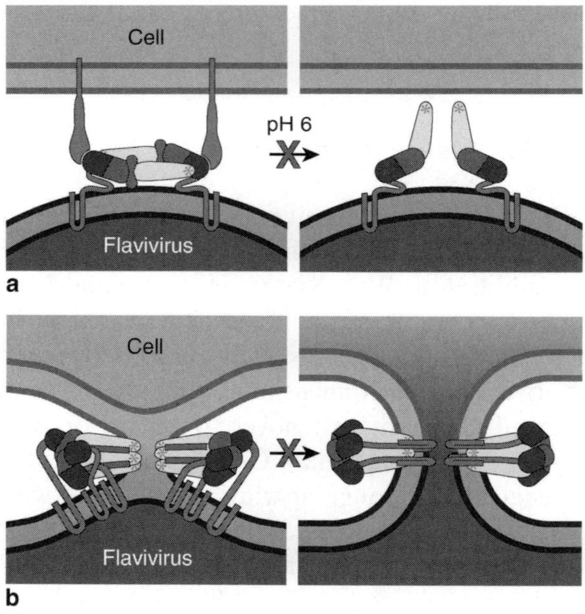

Figure 6. Fusion inhibition strategies. (**a**) The discovery of a ligand-binding pocket at the interface between domains I and II in dengue virus E (Modis et al., 2003), just at the locus of a hinge motion required for fusion, suggests that compounds inserted in the pocket might hinder the hinge motion and hence inhibit the fusion transition. This approach would block the first step in the fusion mechanism (Fig. 5a, b). (**b**) Peptides corresponding to the stem region of the fusion protein may inhibit viral entry by binding to the trimeric core of the protein in its postfusion conformation, and interfering with the folding back against it of the fusion protein's own stem (Modis et al., 2004). An analogous strategy has been successful with HIV gp41 (Baldwin et al., 2003; Kilby et al., 1998). This approach would block the last step in the fusion mechanism (Fig. 5e, f). Antibody and vaccine-based strategies also offer promise in the treatment of flaviviral infection by inhibiting membrane fusion (see text). (*See Color Plates*)

The postfusion structures of dengue (Modis et al., 2004) and Semliki Forest (Gibbons et al., 2004) viruses suggest a second possible strategy for fusion inhibition, related to an approach successful in developing the HIV fusion inhibitor T-20, or enfurvirtide (Baldwin et al., 2003; Kilby et al., 1998). Peptides corresponding to the stem region of the gp41 fusion protein inhibit HIV entry by binding to the trimeric, N-terminal "inner core" of the protein and interfering with the folding back against it of the stem and C-terminal viral transmembrane anchor. The expected annealing of the stem into a deep channel in flavivirus E proteins during the fusion transition (Figs. 3c, d and 5d–f) suggests that an analogous strategy may be successful with flaviviruses. Peptides derived from stem sequences could block completion of the fusogenic conformational change, by competing with the intramolecular stem for interaction with surfaces on domain II, at the trimer interface (Fig. 6b). Stem-like peptides or peptidomimetic compounds could thus inhibit viral membrane fusion in flaviviruses, and other class II enveloped viruses, by preventing the final folding back of the fusion protein, which is required to drive the viral and host-cell membranes together to the point of fusion. Indeed, a 33-residue peptide from the dengue E stem inhibits infection of cells in culture by both dengue and West Nile viruses at $<25\ \mu M$ concentrations (Hrobowski et al., 2005).

A third strategy for fusion inhibition in flaviviruses is to use an antibody to prevent the fusogenic conformational change from proceeding to completion. Indeed the conventional view that neutralizing antibodies against flavivirus E proteins inhibit virus infection by blocking cellular attachment (Rey et al., 1995; Sanchez et al., 2005) was recently challenged when it was found that the domain III-specific antibody E16 protected mice from a lethal dose of West Nile virus even when the antibody was administered several days postinfection (Oliphant et al., 2005). E16 is a poor inhibitor of attachment and was therefore proposed to sterically interfere with a postattachment step – most likely one of the fusogenic domain rearrangements described above (Nybakken et al., 2005). This mechanism of antibody neutralization presents the advantage that it is independent of the receptor used by the virus to attach and enter the cell, which is particularly important given the apparent ability of flaviviruses to use multiple attachment receptors (Davis et al., 2006; Navarro-Sanchez et al., 2003; Tassaneetrithep et al., 2003). Hence, a promising avenue in vaccine design is to raise antibodies against E protein antigens that either directly participate in fusion, such as the highly conserved fusion loop (Goncalvez et al., 2004; Oliphant et al., 2006; Stiasny et al., 2006), or that preclude the fusogenic domain rearrangement from proceeding to completion, such as the E16 epitope on domain III.

References

Allison SL, Stiasny K, Stadler K et al. (1999) Mapping of functional elements in the stem-anchor region of tick-borne encephalitis virus envelope protein E. J Virol 73:5605–5612
Allison SL, Schalich J, Stiasny K et al. (2001) Mutational evidence for an internal fusion peptide in flavivirus envelope protein E. J Virol 75:4268–4275
Armstrong RT, Kushnir AS, White JM (2000) The transmembrane domain of influenza hemagglutinin exhibits a stringent length requirement to support the hemifusion to fusion transition. J Cell Biol 151:425–437
Bagai S, Lamb RA (1996) Truncation of the COOH-terminal region of the paramyxovirus SV5 fusion protein leads to hemifusion but not complete fusion. J Cell Biol 135:73–84
Baker KA, Dutch RE, Lamb RA et al. (1999) Structural basis for paramyxovirus-mediated membrane fusion. Mol Cell 3:309–319
Baldwin CE, Sanders RW, Berkhout B (2003) Inhibiting HIV-1 entry with fusion inhibitors. Curr Med Chem 10:1633–1642
Beasley DW, Aaskov JG (2001) Epitopes on the dengue 1 virus envelope protein recognized by neutralizing IgM monoclonal antibodies. Virology 279:447–458
Blumenthal R, Sarkar DP, Durell S et al. (1996) Dilation of the influenza hemagglutinin fusion pore revealed by the kinetics of individual cell-cell fusion events. J Cell Biol 135:63–71
Bressanelli S, Stiasny K, Allison SL et al. (2004) Structure of a flavivirus envelope glycoprotein in its low-pH-induced membrane fusion conformation. EMBO J 23:728–738
Bullough PA, Hughson FM, Skehel JJ et-al. (1994) Structure of influenza haemagglutinin at the pH of membrane fusion. Nature 371:37–43
Burke DS, Monath TP (2001) Flaviviruses. In: Knipe DM, Howley PM (eds) Fields virology. Lippincott Williams and Wilkins, Philadelphia, pp 1043–1125
Caffrey M, Cai M, Kaufman J et al. (1998) Three-dimensional solution structure of the 44 kDa ectodomain of SIV gp41. EMBO J 17:4572–4584
Cecilia D, Gould EA (1991) Nucleotide changes responsible for loss of neuroinvasiveness in Japanese encephalitis virus neutralization-resistant mutants. Virology 181:70–77
Chan DC, Kim PS (1998) HIV entry and its inhibition. Cell 93:681–684
Chan DC, Fass D, Berger JM et al. (1997) Core structure of gp41 from the HIV envelope glycoprotein. Cell 89:263–273
Chanel-Vos C, Kielian M (2004) A conserved histidine in the ij loop of the Semliki Forest virus E1 protein plays an important role in membrane fusion. J Virol 78:13543–13552
Chatterjee PK, Eng CH, Kielian M (2002) Novel mutations that control the sphingolipid and cholesterol dependence of the Semliki Forest virus fusion protein. J Virol 76:12712–12722
Chen Y, Maguire T, Hileman RE et al. (1997) Dengue virus infectivity depends on envelope protein binding to target cell heparan sulfate. Nat Med 3:866–871
Chen J, Skehel JJ, Wiley DC (1999) N- and C-terminal residues combine in the fusion-pH influenza hemagglutinin HA(2) subunit to form an N cap that terminates the triple-stranded coiled coil. Proc Natl Acad Sci USA 96:8967–8972
Cheng SF, Wu CW, Kantchev EA et al. (2004) Structure and membrane interaction of the internal fusion peptide of avian sarcoma leukosis virus. Eur J Biochem 271:4725–4736
Crill WD, Roehrig JT (2001) Monoclonal antibodies that bind to domain III of dengue virus E glycoprotein are the most efficient blockers of virus adsorption to Vero cells. J Virol 75:7769–7773
Danieli T, Pelletier SL, Henis YI et al. (1996) Membrane fusion mediated by the influenza virus hemagglutinin requires the concerted action of at least three hemagglutinin trimers. J Cell Biol 133:559–569
Davis CW, Nguyen HY, Hanna SL et al. (2006) West Nile virus discriminates between DC-SIGN and DC-SIGNR for cellular attachment and infection. J Virol 80:1290–1301
Dubovskii PV, Li H, Takahashi S et al. (2000) Structure of an analog of fusion peptide from hemagglutinin. Protein Sci 9:786–798

Dutch RE, Lamb RA (2001) Deletion of the cytoplasmic tail of the fusion protein of the paramyxovirus simian virus 5 affects fusion pore enlargement. J Virol 75:5363–5369

Elshuber S, Allison SL, Heinz FX et al. (2003) Cleavage of protein prM is necessary for infection of BHK-21 cells by tick-borne encephalitis virus. J Gen Virol 84:183–191

Fass D, Harrison SC, Kim PS (1996) Retrovirus envelope domain at 1.7 angstrom resolution. Nat Struct Biol 3:465–469

Feinberg H, Guo Y, Mitchell DA et al. (2005) Extended neck regions stabilize tetramers of the receptors DC-SIGN and DC-SIGNR. J Biol Chem 280:1327–1335

Ferlenghi I, Gowen B, de Haas F et al. (1998) The first step: activation of the Semliki Forest virus spike protein precursor causes a localized conformational change in the trimeric spike. J Mol Biol 283:71–81

Gallaher WR (1987) Detection of a fusion peptide sequence in the transmembrane protein of human immunodeficiency virus. Cell 50:327–328

Gething MJ, White JM, Waterfield MD (1978) Purification of the fusion protein of Sendai virus: analysis of the NH2-terminal sequence generated during precursor activation. Proc Natl Acad Sci USA 75:2737–2740

Gibbons DL, Vaney MC, Roussel A et al. (2004) Conformational change and protein-protein interactions of the fusion protein of Semliki Forest virus. Nature 427:320–325

Goncalvez AP, Purcell RH, Lai CJ (2004) Epitope determinants of a chimpanzee Fab antibody that efficiently cross-neutralizes dengue type 1 and type 2 viruses map to inside and in close proximity to fusion loop of the dengue type 2 virus envelope glycoprotein. J Virol 78:12919–12928

Guirakhoo F, Heinz FX, Mandl CW et al. (1991) Fusion activity of flaviviruses: comparison of mature and immature (prM-containing) tick-borne encephalitis virions. J Gen Virol 72 (Pt 6):1323–1329

Guirakhoo F, Bolin RA, Roehrig JT (1992) The Murray Valley encephalitis virus prM protein confers acid resistance to virus particles and alters the expression of epitopes within the R2 domain of E glycoprotein. Virology 191:921–931

Han X, Bushweller JH, Cafiso DS et al. (2001) Membrane structure and fusion-triggering conformational change of the fusion domain from influenza hemagglutinin. Nat Struct Biol 8:715–720

Hasegawa H, Yoshida M, Shiosaka T et al. (1992) Mutations in the envelope protein of Japanese encephalitis virus affect entry into cultured cells and virulence in mice. Virology 191:158–165

Hrobowski YM, Garry RF, Michael SF (2005) Peptide inhibitors of dengue virus and West Nile virus infectivity. Virol J 2:49

Hung JJ, Hsieh MT, Young MJ et al. (2004) An external loop region of domain III of dengue virus type 2 envelope protein is involved in serotype-specific binding to mosquito but not mammalian cells. J Virol 78:378–388

Hurrelbrink RJ, McMinn PC (2001) Attenuation of Murray Valley encephalitis virus by site-directed mutagenesis of the hinge and putative receptor-binding regions of the envelope protein. J Virol 75:7692–7702

Januszeski MM, Cannon PM, Chen D et al. (1997) Functional analysis of the cytoplasmic tail of Moloney murine leukemia virus envelope protein. J Virol 71:3613–3619

Kanai R, Kar K, Anthony K et al. (2006) Crystal structure of West Nile virus envelope glycoprotein reveals viral surface epitopes. J Virol 80:11000–11008

Kemble GW, Danieli T, White JM (1994) Lipid-anchored influenza hemagglutinin promotes hemifusion, not complete fusion. Cell 76:383–391

Kilby JM, Hopkins S, Venetta TM et al. (1998) Potent suppression of HIV-1 replication in humans by T-20, a peptide inhibitor of gp41-mediated virus entry. Nat Med 4:1302–1307

Kobe B, Center RJ, Kemp BE et al. (1999) Crystal structure of human T cell leukemia virus type 1 gp21 ectodomain crystallized as a maltose-binding protein chimera reveals structural evolution of retroviral transmembrane proteins. Proc Natl Acad Sci USA 96:4319–4324

Kozlov MM, Chernomordik LV (1998) A mechanism of protein-mediated fusion: coupling between refolding of the influenza hemagglutinin and lipid rearrangements. Biophys J 75:1384–1396

Kuhn RJ, Zhang W, Rossmann MG et al. (2002) Structure of dengue virus: implications for flavivirus organization, maturation, and fusion. Cell 108:717–725

Kuzmin PI, Zimmerberg J, Chizmadzhev YA et al. (2001) A quantitative model for membrane fusion based on low-energy intermediates. Proc Natl Acad Sci USA 98:7235–7240

Lee E, Weir RC, Dalgarno L (1997) Changes in the dengue virus major envelope protein on passaging and their localization on the three-dimensional structure of the protein. Virology 232:281–290

Lescar J, Roussel A, Wien MW et al. (2001) The Fusion glycoprotein shell of Semliki Forest virus: an icosahedral assembly primed for fusogenic activation at endosomal pH. Cell 105:137–148

Lindenbach BD, Rice CM (2001) *Flaviviridae*: the viruses and their replication. In: Knipe DM, Howley PM (eds) Fields virology. Lippincott Williams and Wilkins, Philadelphia, pp 991–1041

Malashkevich VN, Chan DC, Chutkowski CT et al. (1998) Crystal structure of the simian immunodeficiency virus (SIV) gp41 core: conserved helical interactions underlie the broad inhibitory activity of gp41 peptides. Proc Natl Acad Sci USA 95:9134–9139

Mancini EJ, Clarke M, Gowen BE et al. (2000) Cryo-electron microscopy reveals the functional organization of an enveloped virus, Semliki Forest virus. Mol Cell 5:255–266

McNew JA, Weber T, Parlati F et al. (2000) Close is not enough: SNARE-dependent membrane fusion requires an active mechanism that transduces force to membrane anchors. J Cell Biol 150:105–117

Melikyan GB, White JM, Cohen FS (1995) GPI-anchored influenza hemagglutinin induces hemifusion to both red blood cell and planar bilayer membranes. J Cell Biol 131:679–691

Melikyan GB, Jin H, Lamb RA et-al. (1997) The role of the cytoplasmic tail region of influenza virus hemagglutinin in formation and growth of fusion pores. Virology 235:118–128

Melikyan GB, Markosyan RM, Brener SA et al. (2000a) Role of the cytoplasmic tail of ecotropic moloney murine leukemia virus Env protein in fusion pore formation. J Virol 74:447–455

Melikyan GB, Markosyan RM, Hemmati H et al. (2000b) Evidence that the transition of HIV-1 gp41 into a six-helix bundle, not the bundle configuration, induces membrane fusion. J Cell Biol 151:413–423

Mitchell DA, Fadden AJ, Drickamer K (2001) A novel mechanism of carbohydrate recognition by the C-type lectins DC-SIGN and DC-SIGNR. Subunit organization and binding to multivalent ligands. J Biol Chem 276:28939–28945

Modis Y, Ogata S, Clements D et al. (2003) A ligand-binding pocket in the dengue virus envelope glycoprotein. Proc Natl Acad Sci USA 100:6986–6991

Modis Y, Ogata S, Clements D et al. (2004) Structure of the dengue virus envelope protein after membrane fusion. Nature 427:313–319

Modis Y, Ogata S, Clements D et al. (2005) Variable surface epitopes in the crystal structure of dengue virus type 3 envelope glycoprotein. J Virol 79:1223–1231

Monath TP, Arroyo J, Levenbook I et al. (2002) Single mutation in the flavivirus envelope protein hinge region increases neurovirulence for mice and monkeys but decreases viscerotropism for monkeys: relevance to development and safety testing of live, attenuated vaccines. J Virol 76:1932–1943

Mukhopadhyay S, Kim BS, Chipman PR et al. (2003) Structure of West Nile virus. Science 302:248

Navarro-Sanchez E, Altmeyer R, Amara A et al. (2003) Dendritic-cell-specific ICAM3-grabbing non-integrin is essential for the productive infection of human dendritic cells by mosquito-cell-derived dengue viruses. EMBO Rep 4:1–6

Nieva JL, Bron R, Corver J et al. (1994) Membrane fusion of Semliki Forest virus requires sphingolipids in the target membrane. EMBO J 13:2797–2804
Nussler F, Clague MJ, Herrmann A (1997) Meta-stability of the hemifusion intermediate induced by glycosylphosphatidylinositol-anchored influenza hemagglutinin. Biophys J 73:2280–2291
Nybakken GE, Oliphant T, Johnson S et al. (2005) Structural basis of West Nile virus neutralization by a therapeutic antibody. Nature 437:764–769
Nybakken GE, Nelson CA, Chen BR et al. (2006) Crystal structure of the West Nile virus envelope glycoprotein. J Virol 80:11467–11474
Oliphant T, Engle M, Nybakken GE et al. (2005) Development of a humanized monoclonal antibody with therapeutic potential against West Nile virus. Nat Med 11:522–530
Oliphant T, Nybakken GE, Engle M et al. (2006) Antibody recognition and neutralization determinants on domains I and II of West Nile virus envelope protein. J Virol 80:12149–12159
Pokidysheva E, Zhang Y, Battisti AJ et al. (2006) Cryo-EM reconstruction of dengue virus in complex with the carbohydrate recognition domain of DC-SIGN. Cell 124:485–493
Razinkov VI, Melikyan GB, Cohen FS (1999) Hemifusion between cells expressing hemagglutinin of influenza virus and planar membranes can precede the formation of fusion pores that subsequently fully enlarge. Biophys J 77:3144–3151
Rey FA, Heinz FX, Mandl C et al. (1995) The envelope glycoprotein from tick-borne encephalitis virus at 2 A resolution. Nature 375:291–298
Rosenthal PB, Zhang X, Formanowski F et al. (1998) Structure of the haemagglutinin-esterase-fusion glycoprotein of influenza C virus. Nature 396:92–96
Ruigrok RW, Aitken A, Calder LJ et al. (1988) Studies on the structure of the influenza virus haemagglutinin at the pH of membrane fusion. J Gen Virol 69 (Pt 11):2785–2795
Russell CJ, Jardetzky TS, Lamb RA (2001) Membrane fusion machines of paramyxoviruses: capture of intermediates of fusion. EMBO J 20:4024–4034
Saifee O, Wei L, Nonet ML (1998) The Caenorhabditis elegans unc-64 locus encodes a syntaxin that interacts genetically with synaptobrevin. Mol Biol Cell 9:1235–1252
Sanchez MD, Pierson TC, McAllister D et al. (2005) Characterization of neutralizing antibodies to West Nile virus. Virology 336:70–82
Schlesinger S, Schlesinger MJ (2001) *Togaviridae*: the viruses and their replication. In: Knipe DM, Howley PM (eds) Fields virology. Lippincott Williams and Wilkins, Philadelphia, pp 895–916
Skehel JJ, Wiley DC (2000) Receptor binding and membrane fusion in virus entry: the influenza hemagglutinin. Annu Rev Biochem 69:531–569
Smith TJ, Kremer MJ, Luo M et al. (1986) The site of attachment in human rhinovirus 14 for antiviral agents that inhibit uncoating. Science 233:1286–1293
Stadler K, Allison SL, Schalich J et al. (1997) Proteolytic activation of tick-borne encephalitis virus by furin. J Virol 71:8475–8481
Stiasny K, Allison SL, Schalich J et al. (2002) Membrane interactions of the tick-borne encephalitis virus fusion protein E at low pH. J Virol 76:3784–3790
Stiasny K, Koessl C, Heinz FX (2003) Involvement of lipids in different steps of the flavivirus fusion mechanism. J Virol 77:7856–7862
Stiasny K, Kiermayr S, Holzmann H et al. (2006) Cryptic properties of a cluster of dominant flavivirus cross-reactive antigenic sites. J Virol 80:9557–9568
Supekar VM, Bruckmann C, Ingallinella P et al. (2004) Structure of a proteolytically resistant core from the severe acute respiratory syndrome coronavirus S2 fusion protein. Proc Natl Acad Sci USA 101:17958–17963
Tamm LK, Han X, Li Y et al. (2002) Structure and function of membrane fusion peptides. Biopolymers 66:249–260
Tan K, Liu J, Wang J et al. (1997) Atomic structure of a thermostable subdomain of HIV-1 gp41. Proc Natl Acad Sci USA 94:12303–12308
Tassaneetrithep B, Burgess TH, Granelli-Piperno A et al. (2003) DC-SIGN (CD209) mediates dengue virus infection of human dendritic cells. J Exp Med 197:823–829

Vashishtha M, Phalen T, Marquardt MT et al. (1998) A single point mutation controls the cholesterol dependence of Semliki Forest virus entry and exit. J Cell Biol 140:91–99

Weissenhorn W, Dessen A, Harrison SC et al. (1997) Atomic structure of the ectodomain from HIV-1 gp41. Nature 387:426–430

Weissenhorn W, Carfi A, Lee KH et al. (1998) Crystal structure of the Ebola virus membrane fusion subunit, GP2, from the envelope glycoprotein ectodomain. Mol Cell 2:605–616

West JT, Johnston PB, Dubay SR et al. (2001) Mutations within the putative membrane-spanning domain of the simian immunodeficiency virus transmembrane glycoprotein define the minimal requirements for fusion, incorporation, and infectivity. J Virol 75:9601–9612

Wilson IA, Skehel JJ, Wiley DC (1981) Structure of the haemagglutinin membrane glycoprotein of influenza virus at 3 A resolution. Nature 289:366–373

Xu Y, Liu Y, Lou Z et al. (2004a) Structural basis for coronavirus-mediated membrane fusion. Crystal structure of mouse hepatitis virus spike protein fusion core. J Biol Chem 279:30514–30522

Xu Y, Lou Z, Liu Y et al. (2004b) Crystal structure of severe acute respiratory syndrome coronavirus spike protein fusion core. J Biol Chem 279:49414–49419

Yin HS, Paterson RG, Wen X et al. (2005) Structure of the uncleaved ectodomain of the paramyxovirus (hPIV3) fusion protein. Proc Natl Acad Sci USA 102:9288–9293

Yin HS, Wen X, Paterson RG et al. (2006) Structure of the parainfluenza virus 5 F protein in its metastable, prefusion conformation. Nature 439:38–44

Zhang W, Mukhopadhyay S, Pletnev SV et al. (2002) Placement of the structural proteins in Sindbis virus. J Virol 76:11645–11658

Zhang W, Chipman PR, Corver J et al. (2003a) Visualization of membrane protein domains by cryo-electron microscopy of dengue virus. Nat Struct Biol 10:907–912

Zhang Y, Corver J, Chipman PR et al. (2003b) Structures of immature flavivirus particles. EMBO J 22:2604–2613

Zhang Y, Zhang W, Ogata S et al. (2004) Conformational changes of the flavivirus E glycoprotein. Structure (Camb) 12:1607–1618

Zhao X, Singh M, Malashkevich VN et al. (2000) Structural characterization of the human respiratory syncytial virus fusion protein core. Proc Natl Acad Sci USA 97:14172–14177

13. CD4$^+$ and CD8$^+$ T-Cell Immune Responses in West Nile Virus Infection

MARIO LOBIGS, ARNO MÜLLBACHER, AND MATTHIAS REGNER

Abstract

CD4+ and CD8+ T cells, in combination with the innate and humoral immune responses, are critical for recovery of mice from infection with West Nile virus. Mouse models of West Nile virus encephalitis are therefore excellently suited to investigate the role of T cells in the balance between viral clearance and CNS disease following infection with a cytopathic and neurotropic virus. Here we review the in vitro properties of West Nile virus-immune T cell responses, their in vivo disease ameliorating and potentiating effects, and the contribution of the different T cell effector functions to disease outcome.

Keywords

flavivirus; viral encephalitis; cellular immunity

1 Role of T Cells in Viral Infection: An Overview

A role for T cells in recovery from viral infection in humans has been suspected from the empirical observation that individuals with genetically impaired T-cell development are highly susceptible to viral infections, while B-cell deficiency generally leads to uncontrollable bacterial infections (Fulginiti et al., 1968). Thus, an understanding of the antiviral T-cell response in humans is vital to advance our ability to prophylactically and therapeutically intervene in clinical infections.

The mouse, *Mus musculus*, has become over the past 100 years the animal model of choice for biomedical research (Fenner, 1982). As for viral disease models in mice, two have been the most extensively studied and have provided fundamental insights into host/viral pathogen interactions: Mouse pox, infection with the natural mouse pathogen ectromelia virus, proved an excellent model for an acute viral infection and its pathogenesis was elegantly described (Fenner, 1949). The role of cytolytic T cells in this infection model as essential mediators in recovery was established by Blanden in the early 1970s (Blanden, 1974). Infection of mice with lymphocytic choriomeningitis virus is the second model, which has provided enormous basic understanding on the role of T cells in persistent viral infections and has become the principal model for elucidating virus-induced immunopathology (Doherty and Zinkernagel, 1974; Lehmann-Grube, 1982). Recent studies into the role of immune T cells in flavivirus infection showed that both beneficial (recovery) as well as detrimental aspects (immunopathology) are observed as a result of T-cell function during encephalitic episodes (Camenga et al., 1974; Licon Luna et al., 2002; Müllbacher et al., 2003; Wang et al., 2003b, 2004b; Samuel et al., 2007) Accordingly, both classical mouse models are of relevance to our appreciation of T-cell-mediated immunity in flavivirus infection.

Phenotypically, T cells have been subdivided on the basis of their antigen receptor usage (α/β vs. γ/δ), their co-receptor expression (CD4 vs. CD8) or their functionality (cytotoxic vs. helper). Functionally, T cells can belong to either of the antigen receptor usage or co-receptor expressing subsets, although cytolytic T (Tc) cells are predominantly of $CD8^+$ and helper T (Th) of $CD4^+$ phenotype. Both α/β and γ/δ T cells can express cytolytic activity as well as release cytokines (Carding and Egan, 2002). While γ/δ T cells recognize non-classical MHC antigens in the mouse as part of the innate immune response, α/β T cells recognize MHC class I or II plus peptide antigen as mediators of adaptive immunity (Rudolph et al., 2006). Members of all the above subsets of T cells have been reported to play a role in the immune response to WNV, albeit not always in a protective capacity (see below).

Early research on T-cell immunity in flavivirus infection focused on the Tc cell responses. Gajdosova et al. (1980, 1981) first obtained cytotoxicity of ex vivo derived splenocytes from mice immunized with one of several flaviviruses, including a European isolate of WNV. This early report already demonstrated a broad cross-reactivity of effector cells on targets infected with distantly related members of the genus, flavivirus. A more detailed analysis of the Tc cell response to flavivirus infection was undertaken by Kesson et al. (1987, 1988) using WNV, strain *Sarafend*. One of the reasons for the choice of WNV *Sarafend* as the preferred experimental model for investigating Tc cell responses in flavivirus infection over other related viruses, such as Kunjin virus (now classified as a strain of WNV; Scherret et al., 2001), Japanese encephalitis virus (JEV) or Murray Valley encephalitis virus (MVEV), was that it induced a substantially stronger Tc cell response in most mouse strains as measured by in vitro target cell lysis using the ^{51}Cr release assay, than the other viruses (Hill et al., 1992, 1993) for reasons as yet poorly understood.

More recent advances in our understanding of T-cell immunity of both helper and cytotoxic phenotype against WNV became possible with the use of vaccinia virus expression vectors encoding regions of the WNV polyprotein for the mapping and characterization of T-cell determinants (Parrish et al., 1991) and the use of knock-out mouse strains for the evaluation of the contribution of effector functions of the cellular immune responses in recovery from infection or disease exacerbation (see below).

2 WNV-Immune CD4+ and CD8+ T-Cell Responses, In Vitro

2.1 The CD4+ T-Cell Response

Early evidence from clones of L3T4 T cells (CD4+) derived from splenocytes of mice immunized with either WNV *Sarafend*, Kunjin virus or MVEV showed that they were highly cross-reactive (Uren et al., 1987) and responded to heterologous virus stimulation with a similar levels of proliferation and cytokine release as in response to the homologous virus. This indicated that some of the viral peptide determinants presented by MHC class II are conserved between these viruses. Subsequent studies showed that synthetic peptides selected from the E protein sequences of WNV, JEV and dengue virus (Kutubuddin et al., 1991) or from that of MVEV (Mathews et al., 1991) based on predictive algorithms, elicited virus-specific as well as flavivirus cross-reactive proliferative responses of virus-immune splenocytes. WNV cross-reactivity of human JEV-immune CD4+ T cells has also been observed and, comparable to that found in mice, extends to distantly related flaviviruses (Aihara et al., 1998). The strong cross-reactivity in the CD4+ T-cell response between members of the JEV serocomplex, which includes WNV (Kuno et al., 1998), could be one of the critical immunological correlates for the cross-protection noted between these viruses following live virus infection with one of the members of the serocomplex (for instance, Goverdhan et al., 1992; Monath, 2002; Tesh et al., 2002).

Assays for in vitro CD4+ T-cell proliferative responses were optimized and applied to the mapping of the antigenic determinant(s) recognized by WNV-immune $CD4^+$ T cells in the context of a suite of mouse MHC class II alleles by Kulkarni et al. (1991a–c, 1992), using a panel of vaccinia virus recombinants encoding together the entire Kunjin virus polyprotein. The MHC class II molecules of most mouse haplotypes presented determinants derived from the NH_2-terminal one-third of the polyprotein, which encodes the viral structural (C, prM and E) and the non-structural NS1 proteins. This was not unexpected, given that $CD4^+$ T cells provide "help" to B cells responsible for the humoral antiviral immune response, which is primarily directed against the E and, to a lesser degree, the NS1 proteins in flavivirus infections. The precise peptide determinants recognized by WNV-immune $CD4^+$ T cells in the context of different mouse and human MHC class II alleles remain to be identified. $CD4^+$ T-cell clones from human PBL donors immunized with an inactivated JEV vaccine also exhibited cross-reactivity with WNV and predominantly recognized peptides derived from the E protein (Aihara et al., 1998). Intriguingly, these $CD4^+$ T-cell clones exhibited cytolytic function, in vitro, the relevance of which during live virus infection is uncertain; the phenomenon has been discussed in an earlier review from our group (Müllbacher et al., 2003).

2.2 The $CD8^+$ T-Cell Response

More is known in regard to the $CD8^+$ T-cell response against WNV, in vitro, than the $CD4^+$ T-cell response, in part due to the availability of a sensitive and robust in vitro assay with functional read-out, the ^{51}Cr release assay. The first detailed analysis of conditions required for the generation of flavivirus-immune secondary, in vitro stimulated, Tc cells was by Kesson et al. (1988) using the WNV *Sarafend* model. One consistent observation made in this and subsequent studies was the high cytolytic activity of secondary flavivirus-immune Tc cells against uninfected target cells following in vitro re-stimulation of primed splenocytes with virus-infected stimulator cells (Lobigs et al., 1996). Interestingly, this high anti-self cytolytic activity was much reduced or not apparent when virus-derived peptides were used to modify stimulator cells (Regner et al., 2001c; Müllbacher et al., 2003). We have suggested previously (Lobigs et al., 1996) that the up-regulation of MHC class I on flavivirus-infected cells is one possible explanation for this phenomenon, given that the associated increased presentation of self peptides via MHC class I may exceed the activation threshold of self-reactive Tc cells. Analysis of the mechanism for flavivirus-mediated up-regulation of MHC class I revealed that WNV and other flavivirus infections increase the supply of peptides into the lumen of the endoplasmic reticulum, thereby augmenting the rate of assembly and level of cell surface expression of the class I restriction elements (Müllbacher and Lobigs, 1995; Momburg et al., 2001; Lobigs et al., 2003a).

Mapping of the peptide determinants recognized by WNV-immune Tc cells, in the context of MHC class I of various haplotypes, was first undertaken with the use of vaccinia virus recombinants encoding polyprotein regions from Kunjin virus (Hill et al., 1992, 1993) or MVEV (Lobigs et al., 1994, 1997). The dominant peptide determinants for most class I MHC alleles mapped to the non-structural region of the polyprotein with the NS3 protein as the major source. A similar non-random distribution of immunodominant Tc cell determinants on the viral polyprotein has been noted for other flaviviruses (for instance, van der Most et al., 2002; Rothman, 2003; Kumar et al., 2004). Given that all flaviviral proteins are synthesized as part of a single polyprotein in equimolar ratio, it is uncertain why this should be the case. One explanation is that the polyprotein region, which is the source of most Tc cell determinants, is subject to rapid proteolytic degradation, or forms part of short-lived alternative cleavage products. Only in the H-2^b haplotype was a significant WNV (Kunjin strain) peptide determinant mapped one-third of the polyprotein to the NH_2-terminal, in addition to another determinant found in the NS4B protein (Hill et al., 1992). This has recently been confirmed by identification of peptides from these regions of WNV (North-American isolates) with the strongest response directed against a D^b-restricted peptide derived from the NS4B protein: SSVWNATTA (single letter amino acid code; Brien et al., 2007; Purtha et al., 2007). Unprecedented CD8+ T-cell cross-reactivity between the flaviviruses at large and not only between closely related viruses became apparent from this research yet again. Most surprisingly, this cross-reactivity could not simply be explained by amino acid sequence conservation, but was dependent on homologous location of the determinants on the polyprotein of the different flaviviruses (Regner et al., 2001a). This led us to propose that highly conserved structural imprints of the NS3 protein on peptide determinants recognized by the class I restriction element, but not primary amino acid sequence of the peptides, can account for cross-recognition in the Tc cell response between distantly related members of the flavivirus genus (Regner et al., 2001a).

CD8+ T cells express their effector function via two distinct pathways: one by the release of cytokines such as interferon-γ (IFN-γ), a potent immune regulator, and the other by cellular cytotoxicity either via the death ligand/death receptor (e.g. FasL/Fas) pathway or alternatively via the exocytosis pathway mediated by perforin and granzymes (see below). $CD8^+$ T cells generated as a result of WNV *Sarafend* infection are able to lyse target cells in vitro via either of the cytolytic pathways (our unpublished observations), which corroborates the in vivo phenotypes observed when mice with deficiencies of one or the other or both cytolytic pathways are used as infection models for WNV (see below). Little is known in regard to IFN-γ release by WNV-immune $CD8^+$ T cells; however, it has been observed that MVE-immune Tc cells, when stimulated in vitro with virus-infected stimulator cells, express cytolytic effector function in the absence of IFN-γ production (Regner et al., 2001b). It appears that the generation of cytolytic activity is independent of the presence of IFN-γ during priming in vivo or boosting in vitro with WNV *Sarafend* (Wang et al., 2006).

3 Role of $CD4^+$ T Cells in Recovery from WNV Infection

$CD4^+$ T lymphocytes provide helper function to B cells to orchestrate the humoral immune response (Bishop and Hostager, 2001; Mills and Cambier, 2003) and, in some instances, to $CD8^+$ T cells for elicitation of sustained Tc cell effector and/or memory responses (Bevan, 2004; Castellino and Germain, 2006). They also can restrict viral replication by secreting cytokines (e.g. IFN-γ) (Ramshaw et al., 1997; Binder and Griffin, 2001) and can directly kill virus-infected cells by Fas- and granule exocytosis-mediated cytotoxic mechanisms (Heller et al., 2006). There is ample evidence for the vital role of antibody in the control of WNV infection (for instance, Camenga et al., 1974; Diamond et al., 2003a, c); hence, $CD4^+$ T-cell help, required for IgM-to-IgG antibody isotype switching, is expected to be a critical component of the adaptive immune response against the virus, given that IgM antibody has a much shorter half-life in serum than IgG antibody (2 and 6–8 days, respectively) (Vieira and Rajewsky, 1988) and the improved ability of IgG to diffuse into inflamed lesions (Hangartner et al., 2006). However, one report (Halevy et al., 1994) alludes to a comparable resistance of BALB/c nude mice (which are deficient in functional T cells) to intraperitoneal WNV infection (Israel 1952 strain) relative to immunocompetent ICR mice, suggesting a significant contribution of T-cell-independent (IgM) antibody in recovery. Viruses which display a highly organized, repetitive, antigenic structure can induce B-cell responses in the absence of T-cell help due to the ability to extensively cross-linked B-cell receptors, thereby delivering strong activation signals to B cells (Bachmann et al., 1993; Hangartner et al., 2006); the WNV surface envelope (E) protein conforms with this requirement (Nybakken et al., 2006). T-cell-independent IgM production at levels sufficient to clear extraneural virus replication with a kinetics identical to that in immunocompetent mice was also found in CD4$^+$ T-cell-depleted or T-cell-deficient (CD4$^{-/-}$ and MHC class II$^{-/-}$) C57Bl/6 mice following infection with a virulent strain (New York isolated from 2000) of WNV by the subcutaneous route (Sitati and Diamond, 2006). However, in the absence of functional CD4$^+$ T-cell responses, IgM antibody titres were not maintained; a precipitous decline of virus-reactive serum IgM occurred at 15 days post-infection in the absence of a vigorous IgG response (100–1,000-fold lower than in wild-type mice) and despite high virus titres in brain and spinal cord. All CD4$^+$ T-cell-deficient or T-cell-depleted mice succumbed to the infection, although with different kinetics: groups of CD4$^{-/-}$ mice showed an average survival time of 11 days, which was similar to the fraction (~30%) of wild-type mice that died, while CD4$^+$ T-cell-depleted and MHC class II$^{-/-}$ mice displayed a protracted time to death of up to 50 days post-infection (Sitati and Diamond, 2006). CD40 ligand-deficient mice, which are defective in a co-stimulatory membrane protein required for producing CD4$^+$ T-cell help for B-cell responses, produced a similar picture to that of CD4$^{-/-}$ mice when infected with WNV: virus clearance from extraneural tissues and CD8$^-$ T-cell priming were not affected; however, CD8$^+$ T-cell trafficking into the CNS was impaired, allowing the virus to grow to high titres in brain and spinal cord; all mice succumbed to infection in contrast to 70% survival in groups of wild-type mice (Sitati et al., 2007). Collectively, the data show that in the absence of functional CD4$^+$ T-cell responses a persistent and uniformly lethal WNV infection is established, which is cleared from extraneural sites, most likely by IgM antibody (Diamond et al., 2003c) and CD8$^+$ T cells (Wang et al., 2003b; Shrestha and Diamond, 2004), but not from the CNS. Sitati and Diamond (2006) also showed that CD4$^+$ T cells are important for sustaining the CD8$^+$ T-cell response against WNV, and that a marked decline in WNV-immune CD8$^+$ T cells at 15 days post-infection accounts, at least in part, for the inability of CD4$^+$ T-cell-depleted and MHC class II$^{-/-}$ mice to clear the virus from the CNS. It is not clear if the reported resistance of nude (athymic) mice against WNV infection (Halevy et al., 1994), which contrasts with the uniform susceptibility of mice lacking functional CD4$^+$ T cells (Sitati and Diamond, 2006), is reflected in complete elimination of the virus or whether delayed mortality would occur, as has been described for infections of nude mice with JEV (Lad et al., 1993); alternatively, WNV isolated with different virulence properties (Beasley et al., 2002) may have been used in the two studies.

4 Role of CD8⁺ T Cells in Recovery from WNV Infection

The role of $CD8^+$ T cells in recovery from WNV infection has been investigated in two mouse models of WNV encephalitis: 6-week-old C57Bl/6 mice infected with 10^3 PFU of the *Sarafend* strain by intravenous injection (Wang et al., 2003b) and 8–10-week-old C57Bl/6 mice infected with 10^2 PFU of the New York 2000 strain by the subcutaneous route into the foot pad (Shrestha and Diamond, 2004). In both models mortality was approximately 30% with a mean time to death of 10–12 days. Growth in extraneural tissues was more readily detectable in infections with the New York 2000 WNV strain, which produced viraemia in the first 4 days of infection and a viral burden in spleen on days 4 and 6, post-infection, while the *Sarafend* strain failed to give measurable viraemia but produced low virus titres in spleen and lymph nodes on days 4 and 5 post-infection in about half of the infected animals. Virus entry into the brain occurred slightly faster in infections with the more virulent New York 2000 strain (at 4–6 days, post-infection) (Diamond et al., 2003a) than infections with the *Sarafend* strain (at 6–8 days post-infection) (Wang et al., 2003b) and the frequency of CNS invasion was lower for the latter.

$CD8^+$ T cells were essential in the control of WNV infection in both mouse models. Depletion of $CD8^+$ T cells resulted in increased mortality and 100–1,000-fold greater viral burden in the brains of WNV *Sarafend*-infected mice (Wang et al., 2003b). A similar outcome was observed in congenic $CD8^{-/-}$ mice and mice defective in MHC class I-restricted antigen presentation due to deletion of β2-microglobulin ($\beta2\text{-m}^{-/-}$) or that of the classical MHC class I restriction elements ($\text{MHC-Ia}^{-/-}$). In $\beta2\text{-m}^{-/-}$ mice infected with WNV *Sarafend*, earlier CNS invasion and greatly increased virus load in the brain and mortality were found relative to wild-type mice. WNV infection of $CD8^{-/-}$ mice with the New York 2000 strain gave 80–90% mortality with virus persistence in spleen for ≥11 days and up to 1,000-fold increased viral load in the CNS with concomitant increased number of infected neurons and cytopathology (Shrestha and Diamond, 2004). Interestingly, the small number of $CD8^{-/-}$ and $\text{MHC-Ia}^{-/-}$ mice that survived infection with WNV (New York 2000 strain) showed residual virus in the brain even at 35 days, post-infection (Shrestha and Diamond, 2004). Finally, transfer of WNV-primed $CD8^+$ T cells also conferred disease amelioration to WNV-challenged recipient mice (Shrestha et al., 2006a; Wang et al., 2006). Accordingly, for both lineage I and lineage II WNV infections, $CD8^+$ T cells are essential (1) for the elimination of virus from extraneural tissues to reduce the incidence and delay the kinetics of virus entry into the CNS and (2) for suppression of virus replication and clearance of virus-infected cells from the CNS by immune effector mechanisms, which allow recovery of the animals.

5 Effector Mechanisms of WNV-Immune T Cells: Cytokines and Cytolytic Pathways

The effector functions of antiviral T cells can be divided into those mediated by soluble molecules (cytokines) and those that are mediated via direct contact between effector and antigen-presenting target cell. $CD4^+$ T cells produce, depending on their polarization during activation, predominantly Th1 cytokines (IFN-γ, IL-2) or Th2 cytokines (IL-4, IL-5, IL-13), or Th17 (IL-17), of which the former are considered beneficial effector cytokines to intracellular pathogens such as viruses (Boehm et al., 1997). Effector $CD8^+$ T cells and $\gamma\delta$ T cells, as well as natural killer cells produce IFN-γ and TNF-α in response to cognate antigen. Furthermore, $CD8^+$ T cells destroy cells that they recognize as virally infected (or transformed) via two distinct mechanisms: (1) ligation of the death receptor Fas (CD95) on the target cell by FasL (CD178) on the T cell or (2) via exocytosis of specialized granules that contain cytotoxic proteins, the pore-forming protein perforin and the pro-apoptotic serine proteinases, granzymes amongst others (Trapani and Smyth, 2002).

5.1 Cytokines

Historically, there is little evidence for a direct antiviral effect of IFN-γ, i.e. through inhibition of virus replication in an infected cell, although IFN-γ has recently been reported to inhibit growth of WNV in bone marrow-derived dendritic cells, in vitro (Shrestha et al., 2006b). The main effects contributing to a successful antiviral immune response are indirect. IFN-γ promotes differentiation of $CD4^+$ T cells into Th1-type cytokine producing cells, thereby ensuring production of immunoglobulin isotypes most beneficial in antiviral immune responses, activation of macrophages and natural killer cells as well as cytolytic function in

$CD8^+$ T cells and antigen presentation via the MHC class I and II pathways (Boehm et al., 1997).

IFN-γ-deficient mice succumb to infection with the virulent, North-American lineage I (Wang et al., 2003a; Shrestha et al., 2006b), but not the less virulent Kunjin or lineage II *Sarafend* (Wang et al., 2006) WNV strains. Interestingly, the main IFN-γ-producing cell type mediating this cytokine's protective effect are γδ T cells (Wang et al., 2003a; Shrestha et al., 2006b). They proliferate early in response to WNV infection (Wang et al., 2003a), and limit WNV replication in peripheral lymphoid organs, apparently delaying CNS invasion by the virus (Shrestha et al., 2006b).

$CD8^+$ T cells do not seem to employ IFN-γ against WNV, since transfer of WNV-primed IFN-γ-deficient $CD8^+$ T cells efficiently controls lineage I (Shrestha et al., 2006a) and lineage II (Wang et al., 2006) WNV infections. IFN-γ also does not seem to play a major role in the induction of anti-WNV $CD8^+$ T cells, as the cytolytic potential as well as the precursor frequency of antiviral $CD8^+$ T cells are the same in IFN-$\gamma^{-/-}$ as in wild-type mice (Wang et al., 2006). A marginally improved survival was observed when WNV-primed, IFN-γ-sufficient $CD4^+$ cells were adoptively transferred into challenged recipients, compared to IFN-$\gamma^{-/-}$ $CD4^+$ cells, although the data did not reach statistical significance (Sitati and Diamond, 2006).

TNF-α is a second pleiotropic cytokine that may be secreted by effector T cells. It activates neurovascular endothelium and attracts neutrophils (King et al., 2003), can cause cell death in susceptible cells and is a potent regulator of endothelial permeability (Tracey and Cerami, 1994).While the immunological role of TNF-α in WNV infection remains unclear, TNF-mediated inflammation has been shown to trigger, in response to Toll-like receptor-3 activation, increased permeability of the blood-brain barrier, resulting in CNS invasion by the virus (Wang et al., 2004a).

5.2 Cytotoxicity

FasL-mediated cytotoxicity contributes to the control of the New York 2000 (Shrestha and Diamond, 2007) but not *Sarafend* (Wang et al., 2004b) strains of WNV. Lack of FasL (*gld* mice) increased the viral burden and delayed clearance from the CNS compared with wild-type mice infected with the former strain. Interestingly, Fas expression is augmented on WNV-infected neurons (Shrestha and Diamond, 2007), suggesting they may become more vulnerable to attack by this pathway.

The cytotoxic granules of effector CD8+ T cells contain a variety of proteins, but the pore-forming protein, perforin, is considered essential in mediating target cell death by granule exocytosis (although human CD8+ cells also contain another pore-forming protein, granulysin) (Krensky and Clayberger, 2005). Perforin is able to form lytic pores in target cell membranes at high concentrations in vitro, leading to osmotic failure and necrotic death. However, its in vivo function is likely to deliver pro-apoptotic enzymes such as granzymes into the target cell cytosol (Voskoboinik et al., 2006). No significant difference in susceptibility to infection with WNV *Sarafend* was observed between wild-type and perforin-deficient mice (Wang et al., 2004b). In contrast, infection of mice with the New York 2000 isolate demonstrated a significant effect of perforin in recovery from the infection (Shrestha et al., 2006a). Perforin-deficiency led to an increase in mortality, despite similar kinetics and magnitude of viraemia as well as similar kinetics of CNS invasion relative to wild-type mice, although viral load in the brain was increased, and increased neuronal infection evident. Furthermore, perforin deficiency led to viral persistence in those mice that survived. Provided they were perforin sufficient, primed anti-WNV CD8+ T cells could, when transferred into challenged recipients, reduce peripheral but not CNS viral load.

The discrepancy between two strains of WNV as well as a second JEV serocomplex flavivirus, MVEV (Licon Luna et al., 2002), in the role of CD8+ T-cell effector molecules (perforin, FasL, IFN-γ) in recovery from infection (Table 1) suggests that extrapolation on the importance of specific immune mechanisms even between closely related viruses must be considered with caution (Müllbacher et al., 2004). The complex interplay between beneficial and immunopathological effects of cytolytic effector molecules may not be revealed in sufficient detail when only survival and viral load are investigated. Thus, one contributing factor to the substantial difference in disease outcome between infections of perforin$^{-/-}$, *gld* and IFN-$\gamma^{-/-}$ mice vs. wild-type mice with North-American, Kunjin or *Sarafend* strains of WNV may be their different virulence properties (Beasley et al., 2002): whereas CD8+ T-cell cytotoxicity may lead to net immunopathology in the less virulent *Sarafend* and Kunjin strains, the greater virulence of the North-American WNV isolates may render removal of infected cells in the CNS obligatory, despite the destructive nature of Tc cell-mediated cytotoxicity.

The role of granzymes have only been investigated in the mouse model of infection with WNV *Sarafend*, where low-dose infection of granzyme A × B$^{-/-}$ and perforin × granzymes A × B$^{-/-}$ mice increased

Table 1. Contribution of $CD4^+$ and $CD8^+$ T-cell immunity to recovery from infection in mouse models of WNV and that of a closely related flavivirus

Knock-out mouse or treatment	Impact of cell-mediated immunity on pathogenesis		
	WNV (lineage I)	WNV (lineage II)	MVEV
$CD4^+$ T cell deficient or depleted	↑ (Sitati and Diamond, 2006)	nt	nt
MHC class II deficient	↑ (Sitati and Diamond, 2006)	nt	nt
CD40 deficient	↑ (Sitati et al., 2007)	nt	nt
$CD8^+$ T cell deficient or depleted	↑ (Shrestha and Diamond, 2004)	↑ (Wang et al., 2003b)	nt
β2m deficient	nt	↑ (Wang et al., 2003b)	↓ (Licon Luna and Lobigs, unpublished)
Perforin deficient	↑ (Shrestha et al., 2006a)	No effect (Wang et al., 2004b)	No effect (Licon Luna et al., 2002)
Granzyme A × B deficient	Nt	[a] (Wang et al., 2004b)	No effect (Licon Luna et al., 2002)
Perforin × granzyme A × B deficient	Nt	[a] (Wang et al., 2004b)	No effect (Licon Luna et al., 2002)
Fas ligand deficient (*gld*)	↑ (Shrestha and Diamond, 2007)	No effect (Wang et al., 2004b)	↓ (Licon Luna et al., 2002)
Perforin × Fas ligand (*gld*) deficient	Nt	[a] (Wang et al., 2004b)	↓ (Licon Luna et al., 2002)
Caspase 3 deficient	↓ (Samuel et al., 2007)	Nt	nt
IFN-γ deficient	↑ (Shrestha et al., 2006b)	No effect[b] (Wang et al., 2006) or ↓ (King and Kesson, 2003)	↑ (Lobigs et al., 2003a)
CCR5 deficient	↑ (Glass et al., 2005)	nt	nt
CXCL10 deficient	↑ (Klein et al., 2005)	nt	nt

(↑) increased, (↓) decreased disease severity, *nt* not tested
[a] Increased mortality accompanied by later time to death
[b] Earlier virus entry into the CNS

mortality, but with longer average survival times (Wang et al., 2004b). This suggests that without granzyme A and/or B mice less efficiently control WNV infection, and that the granzymes ultimately mediate immunopathological effects. Interestingly, granzyme B is able to inhibit WNV growth in a human cell line by >10-fold (Fig. 1), raising the possibility that the protective effect of granzyme B observed in vivo may be, at least partially, due to a direct influence on viral replication, rather than to a T-cell-mediated clearance of infected cells. Given that neurons are a predominant target of WNV, non-lytic viral clearance would be particularly beneficial.

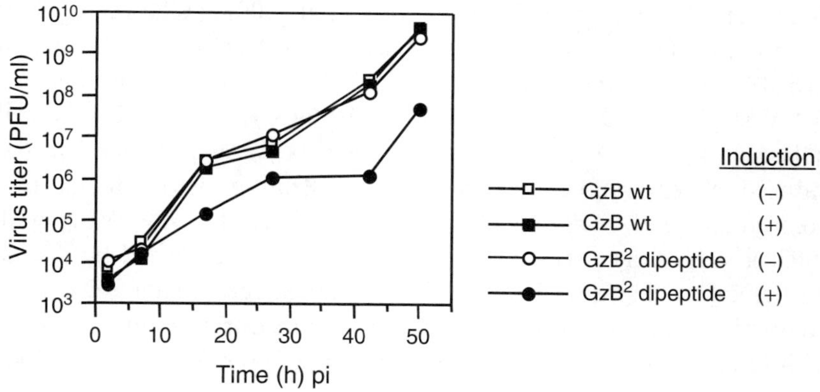

Figure 1. Intracellular granzymes B inhibits WNV growth. For stable, tetracycline-inducible expression of inactive (GzB wt; *squares*) or active, NH_2-terminal dipeptide-deleted (Smyth et al., 1995) granzymes B (GzB Δ dipeptide; *circles*) in 293 cells, mouse granzymes B cDNA with or without a NH_2-terminal dipeptide was cloned into vector pcDNA5/FRT/TO (Invitrogen, Carlsbad, CA) and transfected into Flp-In T-REx-293 cells (Invitrogen, Carlsbad, CA) according to the manufacturer's instructions. Cells were induced with tetracycline (+) for 6 h, or left untreated (−), followed by infection with WNV *Sarafend* at a multiplicity of 1 PFU/cell for 1 h. Cells were washed and growth medium with or without tetracycline (1 µg/ml final concentration) was added. Growth samples were titrated by plaque assay on Vero cells (Wang et al., 2003b).

6 T-Cell Trafficking into the CNS

Lymphocyte trafficking into the CNS is governed by a multi-step process in specialized post-capillary venules, the high-endothelial venules (Engelhardt and Ransohoff, 2005). Contact with the vascular endothelium leads to initial slowing down of the leucocytes, followed by rolling of the cell along the venule wall. This deceleration allows sensing of chemotactic factors, which, when present, trigger activation of adhesion molecules on the leucocyte surface and diapedesis through the inter-endothelial cell junctions or through the endothelial cells themselves. The uninflamed brain is an immune-privileged site, meaning that access to leucocyte and inflammatory processes are more tightly regulated than in other organs. However, while naïve T cells do not circulate through the CNS, activated T cells do traffic even to the non-inflamed CNS (Hickey et al., 1991).

In WNV infection of mice, significant leucocyte numbers have only been shown to enter the brain at the time of, or after, viral entry to the CNS, following virus replication in extraneural tissues for 4–6 days (Diamond et al., 2003b; Wang et al., 2003b). CNS infiltrating leucocytes during WNV infection of mice were predominantly T cells, NK cells and macrophages (Liu et al., 1989; Wang et al., 2003b; Glass et al., 2005; Klein et al., 2005), although the precise composition of the infiltrate has varied between studies and virus strains. Mouse infection with lineage I WNV strains led to a similar infiltration by $CD8^+$ and $CD4^+$ T cells (Glass et al., 2005; Klein et al., 2005), whereas that with WNV *Sarafend* induced a remarkable bias in T-cell infiltration, with almost total predominance of $CD8^+$ over $CD4^+$ T cells (Liu et al., 1989; Wang et al., 2003b), suggesting that these T-cell types may be governed by distinct trafficking mechanisms. Limited clinical data suggest that a $CD8^+$ T-cell predominance in the brain infiltrate also occurs during human infection with lineage I WNV strains (Sampson et al., 2000; Kelley et al., 2003; Omalu et al., 2003).

A wide range of chemokines are induced in the WNV-infected mouse brain of which CXCL10, CCL12 and CXCL12 have been detected early enough to play a role in initiating T-cell trafficking into the CNS (Klein et al., 2005; Garcia-Tapia et al., 2007). Two chemokine/chemokine-receptor pairs have been shown to exert an important, albeit not exclusive, influence on T-cell migration and outcome of viral infection with lineage I WNV strains. Remarkably, WNV-infected neurons are able to produce, in vivo and in vitro, the chemokine CXCL10 (IP-10), a chemokine attracting monocytes and T cells (Dufour et al., 2002). Although infiltrating macrophages, glial cells and some astrocytes and endothelial cells also produce CXCL10 after WNV invasion (Cheeran et al., 2005), neurons appear to be the predominant and earliest producers (Klein et al., 2005), suggesting that infected neurons may play a significant role in the establishment of an immune reaction in the CNS. CXCL10-deficient mice infected with WNV experienced slower T-cell infiltration of the CNS, and higher viral titres; infiltration of $CD8^+$ T cells bearing the receptor for CXCL10 (CXCR3) was reduced to a large extent, but not exclusively, suggesting that CXCL10 is not the sole chemoattractant in this system (Klein et al., 2005; Garcia-Tapia et al., 2007). Whereas peripheral virus titres were independent of CXCL10, CXCL10-deficient mice had higher viral loads in the brain and higher mortality than wild-type mice (Klein et al., 2005). Interestingly, CXCR3-deficiency also led to reduced $CD8^+$ T-cell infiltration and higher viral loads in dengue virus-infected brains (Hsieh et al., 2006) and was suggested to decrease resistance in tick-borne encephalitis virus infection (Lepej et al., 2007).

The second chemokine/chemokine-receptor pair involved in recruitment of WNV-immune T cells into the CNS is the CCL5-CCR5 pair. CCR5-deficient animals revealed reduced T-cell infiltration of the brain associated with increased mortality following WNV infection The maximal effect of this reduction was seen at the later stages of the infection that coincide with viral clearance from the brain by CCR5-sufficient, but not CCR5-deficient mice (Glass et al., 2005). As noted, other chemokines are likely to play additional roles in immune cell recruitment, as in these studies, deficiency in a particular chemokine or chemokine receptor only led to a relative difference in T-cell infiltration.

Clearly, much still remains to be resolved in this area. This is particularly highlighted in the recent discovery of a role for CD40, a molecule hitherto implicated in the development of B and T-cell responses, in $CD8^+$ T-cell trafficking into the WNV-infected brain (Sitati et al., 2007), as egress of $CD8^+$ T cells from the perivascular space into the parenchyma appears to be blocked in $CD40^{-/-}$ mice. Consistent with the role of $CD8^+$ T cells in recovery, this reduced CNS infiltration was associated with increased viral load and mortality.

7 T-Cell-Mediated CNS Pathology in WNV Infection

Neurons are the primary target cells in WNV infection of the CNS of mice (for instance, Eldadah and Nathanson, 1967; Diamond et al., 2003a; Shrestha et al., 2003; Garcia-Tapia et al., 2007) and humans (for instance, Sampson et al., 2000; Sejvar and Marfin, 2006). The virus is cytopathic in neuronal cell cultures, inducing cell death by apoptosis (Parquet et al., 2001; Yang et al., 2002; Shrestha et al., 2003; Diniz et al., 2006). It is the role of the immune system to control viral infections of the CNS, but the response may be pathological and life-threatening, if it results in destruction of neuronal tissue, given its limited capacity for renewal. The relative contributions of virus-induced and immune-mediated neuronal injury in WNV infection of the CNS cannot be clearly defined, other than that a correlation between severity of signs of neuronal injury, number of infected foci and number of inflammatory $CD45^+$ cells in the vicinity of infected neurons exists (Shrestha et al., 2003; Wang et al., 2006). Despite the morphological changes and damage to neuronal tissue as a consequence of WNV infection of the CNS, human patients with WNV meningitis or encephalitis can have good long-term outcomes (Sejvar and Marfin, 2006). Similarly, a significant number of mice showing signs of WNV encephalitis survive and can clear the infection (Diamond et al., 2003a; Wang et al., 2003b), in some cases even after development of hind limb paralysis (Wang et al., 2006). As discussed above, WNV-reactive $CD8^+$ T cells with cytolytic effector function traffic into and clonally expand in the CNS parenchyma and are critical for recovery. The $CD8^+$ T cells kill WNV-infected neurons mainly by cytolytic effector pathways, in contrast to other viral infections of the CNS for which non-cytolytic mechanisms of virus clearance from neurons involving antibody and IFN-γ have been observed (Griffin, 2003). Accumulation of activated Tc cells in the brain and lysis of infected neurons is dependent on antigen presentation via MHC class I; functionally intact neurons suppress expression of the restriction elements and cell adhesion molecules. However, in the event of cell damage (for instance that associated with a viral infection) MHC class I-restricted antigen presentation is induced and dysfunctional neurons become targets for lysis by Tc cells (Neumann et al., 1995, 2002; Wekerle, 2002). The observation that the protective effect of WNV-primed $CD8^+$ T-cell transfer against WNV CNS infection is lost in recipient mice lacking expression of classical MHC class I molecules (Shrestha et al., 2006a) supports the requirement for MHC class I-restricted antigen presentation on neurons for Tc cell lysis. The immune response may be potentiated by up-regulation of MHC class I expression as a result of flavivirus infection (Liu et al., 1988; Lobigs et al., 1996, 2003a; King and Kesson, 2003).

Immune-mediated destruction of infected neurons is key to resolution of WNV infection; however, there is evidence that the immune response in the brain can produces pathology greater than the minimal damage required for recovery. Groups of mice defective in MHC class I antigen presentation ($\beta 2\text{-m}^{-/-}$) or lacking IFN-γ showed a slight reduction in mortality and survived longer relative to wild-type mice, when infected with a high dose (10^8 PFU) of the WNV *Sarafend* strain, a condition, which results in rapid virus entry into the brain (Wang et al., 2003b, 2006). Intriguingly, in the absence of a key apoptotic regulator, caspase 3, mice acquired increased resistance to lethal infection with WNV, New York 2000 strain (Samuel et al., 2007). This phenotype was the result of reduced neuronal damage, despite comparable virus spread into and burden in the CNS relative to infected wild-type mice. WNV infection induced caspase 3 activation in neurons, thus the reduced apoptosis in the CNS of infected caspase $3^{-/-}$ mice could have been due to reduced virus- and/or Tc cell-mediated cell death. Taken together, the data from the mouse models of WNV encephalitis indicate that treatments targeting death or inflammatory pathways could ameliorate disease outcome.

References

Aihara H, Takasaki T, Matsutani T, Suzuki R, Kurane I (1998) Establishment and characterization of Japanese encephalitis virus-specific, human CD4(+) T-cell clones: flavivirus cross-reactivity, protein recognition, and cytotoxic activity. J Virol 72:8032–8036
Bachmann MF, Rohrer UH, Kundig TM, Burki K, Hengartner H, Zinkernagel RM (1993) The influence of antigen organization on B cell responsiveness. Science 262:1448–1451
Beasley DW, Li L, Suderman MT, Barrett AD (2002) Mouse neuroinvasive phenotype of West Nile virus strains varies depending upon virus genotype. Virology 296:17–23
Bevan MJ (2004) Helping the CD8(+) T-cell response. Nat Rev Immunol 4:595–602
Binder GK, Griffin DE (2001) Interferon-gamma-mediated site-specific clearance of alphavirus from CNS neurons. Science 293:303–306
Bishop GA, Hostager BS (2001) B lymphocyte activation by contact-mediated interactions with T lymphocytes. Curr Opin Immunol 13:278–285
Blanden RV (1974) T-cell response to viral and bacterial infections. Transplant Rev 19:56–88
Boehm U, Klamp T, Groot M, Howard JC (1997) Cellular responses to interferon-gamma. Annu Rev Immunol 15:749–795
Brien JD, Uhrlaub JL, Nikolich-Zugich J (2007) Protective capacity and epitope specificity of CD8(+) T cells responding to lethal West Nile virus infection. Eur J Immunol 37:1855–1863
Camenga DL, Nathanson N, Cole GA (1974) Cyclophosphamide-potentiated West Nile viral encephalitis: relative influence of cellular and humoral factors. J Infect Dis 130:634–641
Carding SR, Egan PJ (2002) Gammadelta T cells: functional plasticity and heterogeneity. Nat Rev Immunol 2:336–345
Castellino F, Germain RN (2006) Cooperation between CD4+ and CD8+ T cells: when, where, and how. Annu Rev Immunol 24:519–540
Cheeran MC, Hu S, Sheng WS, Rashid A, Peterson PK, Lokensgard JR (2005) Differential responses of human brain cells to West Nile virus infection. J Neurovirol 11:512–524
Diamond MS, Shrestha B, Marri A, Mahan D, Engle M (2003a) B cells and antibody play critical roles in the immediate defense of disseminated infection by West Nile encephalitis virus. J Virol 77:2578–2586
Diamond MS, Shrestha B, Mehlhop E, Sitati E, Engle M (2003b) Innate and adaptive immune responses determine protection against disseminated infection by West Nile encephalitis virus. Viral Immunol 16:259–278
Diamond MS, Sitati EM, Friend LD, Higgs S, Shrestha B, Engle M (2003c) A critical role for induced IgM in the protection against West Nile virus infection. J Exp Med 198:1853–1862
Diniz JA, Da Rosa AP, Guzman H, Xu F, Xiao SY, Popov VL, Vasconcelos, PF, Tesh RB (2006) West Nile virus infection of primary mouse neuronal and neuroglial cells: the role of astrocytes in chronic infection. Am J Trop Med Hyg 75:691–696
Doherty PC, Zinkernagel RM (1974) T-cell-mediated immunopathology in viral infections. Transplant Rev 19:89–120
Dufour JH, Dziejman M, Liu MT, Leung JH, Lane TE, Luster AD (2002) IFN-gamma-inducible protein 10 (IP-10; CXCL10)-deficient mice reveal a role for IP-10 in effector T cell generation and trafficking. J Immunol 168:3195–3204
Eldadah AH, Nathanson N (1967) Pathogenesis of West Nile Virus encephalitis in mice and rats. II. Virus multiplication, evolution of immunofluorescence, and development of histological lesions in the brain. Am J Epidemiol 86:776–790
Engelhardt B, Ransohoff RM (2005) The ins and outs of T-lymphocyte trafficking to the CNS: anatomical sites and molecular mechanisms. Trends Immunol 26:485–495
Fenner F (1949) Mouse-pox (infectious ectromelia of mice): a review. J Immunol 63:341–373
Fenner F (1982) Mousepox. In: Foster HL, Small JD, Fox JG (eds) The mouse in biomedical research, vol 2. Academic Press, New York, pp 209–230
Fulginiti VA, Kempe CH, Hathaway WE, Pearlman DS, Sieber OF, Eller JJ, Joyner JJ, Robinson A (1968) Progressive vaccinia in immunologically defiecient individuals. In:

Bergsma D (ed) Immunologic deficiency diseases in man, vol 4. The National Foundation-March of Dimes, New York, pp. 129–144

Gajdosova E, Mayer V, Oravec C (1980) Cross-reactive killer T lymphocytes in a flavivirus infection. Acta Virol 24:291–293

Gajdosova E, Oravec C, Mayer V (1981) Cell-mediated immunity in flavivirus infections. I. Induction of cytotoxic T lymphocytes in mice by an attenuated virus from the tick-borne encephalitis complex and its group-reactive character. Acta Virol 25:10–18

Garcia-Tapia D, Hassett DE, MitchellJr WJ, Johnson GC, Kleiboeker SB (2007) West Nile virus encephalitis: sequential histopathological and immunological events in a murine model of infection. J Neurovirol 13:130–138

Glass WG, Lim JK, Cholera R, Pletnev AG, Gao JL, Murphy PM (2005) Chemokine receptor CCR5 promotes leukocyte trafficking to the brain and survival in West Nile virus infection. J Exp Med 202:1087–1098

Goverdhan MK, Kulkarni AB, Gupta AK, Tupe CD, Rodrigues JJ (1992) Two-way cross-protection between West Nile and Japanese encephalitis viruses in bonnet macaques. Acta Virol 36:277–283

Griffin DE (2003) Immune responses to RNA-virus infections of the CNS. Nat Rev Immunol 3:493–502

Halevy M, Akov Y, Ben-Nathan D, Kobiler D, Lachmi B, Lustig S (1994) Loss of active neuroinvasiveness in attenuated strains of West Nile virus: pathogenicity in immunocompetent and SCID mice. Arch Virol 137:355–370

Hangartner L, Zinkernagel RM, Hengartner H (2006) Antiviral antibody responses: the two extremes of a wide spectrum. Nat Rev Immunol 6:231–243

Heller KN, Gurer C, Munz C (2006) Virus-specific CD4+ T cells: ready for direct attack. J Exp Med 203:805–808

Hickey WF, Hsu BL, Kimura H (1991) T-lymphocyte entry into the central nervous system. J Neurosci Res 28:254–260

Hill AB, Müllbacher A, Parrish C, Coia G, Westaway EG, Blanden RV (1992) Broad cross-reactivity with marked fine specificity in the cytotoxic T cell response to flaviviruses. J Gen Virol 73:1115–1123

Hill AB, Lobigs M, Blanden RV, Kulkarni A, Müllbacher A (1993) The cellular immune response to flaviviruses. In: Thomas DB (ed) Viruses and the cellular immune response. Marcel Dekker Inc., New York, pp 363–428

Hsieh MF, Lai SL, Chen JP, Sung JM, Lin YL, Wu-Hsieh BA, Gerard C, Luster A, Liao F (2006) Both CXCR3 and CXCL10/IFN-inducible protein 10 are required for resistance to primary infection by dengue virus. J Immunol 177:1855–1863

Kelley TW, Prayson RA, Ruiz AI, Isada CM, Gordon SM (2003) The neuropathology of West Nile virus meningoencephalitis. A report of two cases and review of the literature. Am J Clin Pathol 119:749–753

Kesson AM, Blanden RV, Müllbacher A (1987) The primary in vivo murine cytotoxic T cell response to the flavivirus, West Nile. J Gen Virol 68:2001–2006

Kesson AM, Blanden RV, Müllbacher A (1988) The secondary in vitro murine cytotoxic T cell response to the flavivirus, West Nile. Immunol Cell Biol 66:23–32

King NJ, Kesson AM (2003) Interaction of flaviviruses with cells of the vertebrate host and decoy of the immune response. Immunol Cell Biol 81:207–216

King NJ, Shrestha B, Kesson AM (2003) Immune modulation by flaviviruses. Adv Virus Res 60:121–155

Klein RS, Lin E, Zhang B, Luster AD, Tollett J, Samuel MA, Engle M, Diamond MS (2005) Neuronal CXCL10 directs CD8+ T-cell recruitment and control of West Nile virus encephalitis. J Virol 79:11457–11466

Krensky AM, Clayberger C (2005) Granulysin: a novel host defense molecule. Am J Transplant 5:1789–1792

Kulkarni AB, Müllbacher A, Blanden RV (1991a) Effect of high ligand concentration on West Nile virus-specific T cell proliferation. Immunol Cell Biol 69:27–38

Kulkarni AB, Müllbacher A, Blanden RV (1991b). Functional analysis of macrophages, B cells and splenic dendritic cells as antigen-presenting cells in West Nile virus-specific murine T lymphocyte proliferation. Immunol Cell Biol 69:71–80

Kulkarni AB, Müllbacher A, Blanden RV (1991c) In vitro T-cell proliferative response to the flavivirus, West Nile. Viral Immunol 4:73–82

Kulkarni AB, Müllbacher A, Parrish CR, Westaway EG, Coia G, Blanden RV (1992) Analysis of murine major histocompatibility complex class II-restricted T-cell responses to the flavivirus Kunjin by using vaccinia virus expression. J Virol 66:3583–3592

Kumar P, Sulochana P, Nirmala G, Haridattatreya M, Satchidanandam V (2004) Conserved amino acids 193–324 of non-structural protein 3 are a dominant source of peptide determinants for CD4+ and CD8+ T cells in a healthy Japanese encephalitis virus-endemic cohort. J Gen Virol 85:1131–1143

Kuno G, Chang GJ, Tsuchiya KR, Karabatsos N, Cropp CB (1998) Phylogeny of the genus flavivirus. J Virol 72:73–83

Kutubuddin M, Kolaskar AS, Galande S, Gore MM, Ghosh SN, Banerjee K (1991) Recognition of helper T cell epitopes in envelope (E) glycoprotein of Japanese encephalitis, West Nile and dengue viruses. Mol Immunol 28:149–154

Lad VJ, Gupta AK, Goverdhan MK, Ayachit VL, Rodrigues JJ, Hungund LV (1993) Susceptibility of BL6 nude (congenitally athymic) mice to Japanese encephalitis virus by the peripheral route. Acta Virol 37:232–240

Lehmann-Grube F (1982) Lymphocytic choriomeningitis virus. In: Foster HL, Small JD, Fox JG (eds) The mouse in biomedical research, vol 2. Academic Press, New York, pp 231–266

Lepej SZ, Misic-Majerus L, Jeren T, Rode OD, Remenar A, Sporec V, Vince A (2007) Chemokines CXCL10 and CXCL11 in the cerebrospinal fluid of patients with tick-borne encephalitis. Acta Neurol Scand 115:109–114

Licon Luna RM, Lee E, Müllbacher A, Blanden RV, Langman R, Lobigs M (2002) Lack of both Fas ligand and perforin protects from flavivirus-mediated encephalitis in mice. J Virol 76:3202–3211

Liu Y, King N, Kesson A, Blanden RV, Müllbacher A (1988) West Nile virus infection modulates the expression of class I and class II MHC antigens on astrocytes in vitro. Ann NY Acad Sci 540:483–485

Liu Y, Blanden RV, Müllbacher A (1989) Identification of cytolytic lymphocytes in West Nile virus-infected murine central nervous system. J Gen Virol 70:565–573

Lobigs M, Arthur CE, Müllbacher A, Blanden RV (1994) The flavivirus nonstructural protein, NS3, is a dominant source of cytotoxic T cell peptide determinants. Virology 202:195–201

Lobigs M, Blanden RV, Müllbacher A (1996) Flavivirus-induced up-regulation of MHC class I antigens; implications for the induction of CD8+ T-cell-mediated autoimmunity. Immunol Rev 152:5–19

Lobigs M, Müllbacher A, Pavy M (1997) The CD8+ cytotoxic T cell response to flavivirus infection. Arbovirus Res Aust 7:160–165

Lobigs M, Müllbacher A, Regner M (2003a) MHC class I up-regulation by flaviviruses: Immune interaction with unknown advantage to host or pathogen. Immunol Cell Biol 81:217–223

Lobigs M, Müllbacher A, Wang Y, Pavy M, Lee E (2003b) Role of type I and type II interferon responses in recovery from infection with an encephalitic flavivirus. J Gen Virol 84:567–572

Mathews JH, Allan JE, Roehrig JT, Brubaker JR, Uren MF, Hunt AR (1991) T-helper cell and associated antibody response to synthetic peptides of the E glycoprotein of Murray Valley encephalitis virus. J Virol 65:5141–5148

Mills DM, Cambier JC (2003) B lymphocyte activation during cognate interactions with CD4+ T lymphocytes: molecular dynamics and immunologic consequences. Semin Immunol 15:325–329

Momburg F, Müllbacher A, Lobigs M (2001) Modulation of transporter associated with antigen processing (TAP)- mediated peptide import into the endoplasmic reticulum by flavivirus infection. J Virol 75:5663–5671

Monath TP (2002) Editorial: Jennerian vaccination against West Nile virus. Am J Trop Med Hyg 66:113–114

Müllbacher A, Lobigs M (1995) Up-regulation of MHC class I by flavivirus-induced peptide translocation into the endoplasmic reticulum. Immunity 3:207–214

Müllbacher A, Lobigs M, Lee E (2003) Immunobiology of mosquito-borne encephalitic flaviviruses. Adv Virus Res 60:87–120

Müllbacher A, Regner M, Wang Y, Lee E, Lobigs M, Simon M (2004) Can we really learn from model pathogens? Trends Immunol 25:524–528

Neumann H, Cavalie A, Jenne DE, Wekerle H (1995) Induction of MHC class I genes in neurons. Science 269:549–552

Neumann H, Medana IM, Bauer J, Lassmann H (2002) Cytotoxic T lymphocytes in autoimmune and degenerative CNS diseases. Trends Neurosci 25:313–319

Nybakken GE, Nelson CA, Chen BR, Diamond MS, Fremont DH (2006) Crystal structure of the West Nile virus envelope glycoprotein. J Virol 80:11467–11474

Omalu BI, Shakir AA, Wang G, Lipkin WI, Wiley CA (2003) Fatal fulminant pan-meningopolioencephalitis due to West Nile virus. Brain Pathol 13:465–472

Parquet MC, Kumatori A, Hasebe F, Morita K, Igarashi A (2001) West Nile virus-induced bax-dependent apoptosis. FEBS Lett 500:17–24

Parrish CR, Coia G, Hill A, Müllbacher A, Westaway EG, Blanden RV (1991) Preliminary analysis of murine cytotoxic T cell responses to the proteins of the flavivirus Kunjin using vaccinia virus expression. J Gen Virol 72:1645–1653

Purtha WE, Myers N, Mitaksov V, Sitati E, Connolly J, Fremont DH, Hansen TH, Diamond MS (2007) Antigen-specific cytotoxic T lymphocytes protect against lethal West Nile virus encephalitis. Eur J Immunol 37:1845–1854

Ramshaw IA, Ramsay AJ, Karupiah G, Rolph MS, Mahalingam S, Ruby JC (1997) Cytokines and immunity to viral infections. Immunol Rev 159:119–135

Regner M, Lobigs M, Blanden RV, Milburn P, Müllbacher A (2001a). Antiviral cytotoxic T cells cross-reactively recognize disparate peptide determinants from related viruses but ignore more similar self- and foreign determinants. J Immunol 166:3820–3828

Regner M, Lobigs M, Blanden RV, Müllbacher A (2001b) Effector cytolotic function but not IFN-gamma production in cytotoxic T cells triggered by virus-infected target cells in vitro. Scand J Immunol 54:366–374

Regner M, Müllbacher A, Blanden RV, Lobigs M (2001c) Immunogenicity of two peptide determinants in the cytolytic T cell response to flavivirus infection: inverse correlation between peptide affinity for MHC class I and T cell precursor frequency. Viral Immunol 14:135–149

Rothman AL (2003) Immunology and immunopathogenesis of dengue disease. Adv Virus Res 60:397–419

Rudolph MG, Stanfield RL, Wilson IA (2006) How TCRs bind MHCs, peptides, and coreceptors. Annu Rev Immunol 24:419–466

Sampson BA, Ambrosi C, Charlot A, Reiber K, Veress JF, Armbrustmacher V (2000) The pathology of human West Nile virus infection. Hum Pathol 31:527–531

Samuel MA, Morrey JD, Diamond MS (2007) Caspase 3-dependent cell death of neurons contributes to the pathogenesis of West Nile virus encephalitis. J Virol 81:2614–2623

Scherret JH, Poidinger M, Mackenzie JS, Broom AK, Deubel V, Lipkin WI, Briese T, Gould EA, Hall RA (2001) The relationships between West Nile and Kunjin viruses. Emerg Infect Dis 7:697–705

Sejvar JJ, Marfin AA (2006) Manifestations of West Nile neuroinvasive disease. Rev Med Virol 16:209–224

Shrestha B, Diamond MS (2004) Role of CD8+ T cells in control of West Nile virus infection. J Virol 78:8312–8321

Shrestha B, Diamond MS (2007) Fas Ligand interactions contribute to CD8+ T cell-mediated control of West Nile virus infection in the central nervous system. J Virol 81:11749–11757

Shrestha B, Gottlieb D, Diamond MS (2003) Infection and injury of neurons by West Nile encephalitis virus. J Virol 77:13203–13213

Shrestha B, Samuel MA, Diamond MS (2006a) CD8+ T cells require perforin to clear West Nile virus from infected neurons. J Virol 80:119–129

Shrestha B, Wang T, Samuel MA, Whitby K, Craft J, Fikrig E, Diamond MS (2006b) Gamma interferon plays a crucial early antiviral role in protection against West Nile virus infection. J Virol 80:5338–5348

Sitati EM, Diamond MS (2006) CD4+ T-cell responses are required for clearance of West Nile virus from the central nervous system. J Virol 80:12060–12069

Sitati E, McCandless EE, Klein RS, Diamond MS (2007) CD40-CD40 ligand interactions promote trafficking of CD8+ T cells into the brain and protection against West Nile virus encephalitis. J Virol 81:9801–9811

Smyth MJ, McGuire MJ, Thia KY (1995) Expression of recombinant human granzyme B. A processing and activation role for dipeptidyl peptidase I. J Immunol 154:6299–6305

Tesh RB, Travassos da Rosa AP, Guzman H, Araujo TP, Xiao SY (2002) Immunization with heterologous flaviviruses protective against fatal West Nile encephalitis. Emerg Infect Dis 8:245–251

Tracey KJ, Cerami A (1994) Tumor necrosis factor: a pleiotropic cytokine and herapeutic target. Annu Rev Med 45:491–503

Trapani JA, Smyth MJ (2002) Functional significance of the perforin/granzyme cell death pathway. Nat Rev Immunol 2:735–747

Uren MF, Doherty PC, Allan JE (1987) Flavivirus-specific murine L3T4 + T cell clones: induction, characterization and cross-reactivity. J Gen Virol 68:2655–2663

van der Most RG, Harrington LE, Giuggio V, Mahar PL, Ahmed R (2002) Yellow fever virus 17D envelope and NS3 proteins are major targets of the antiviral T cell response in mice. Virology 296:117–124

Vieira P, Rajewsky K (1988) The half-lives of serum immunoglobulins in adult mice. Eur J Immunol 18:313–316

Voskoboinik I, Smyth MJ, Trapani JA (2006) Perforin-mediated target-cell death and immune homeostasis. Nat Rev Immunol 6:940–952

Wang T, Scully E, Yin Z, Kim JH, Wang S, Yan J, Mamula M, Anderson JF, Craft J, Fikrig E (2003a) IFN-gamma-producing gammadeltaT cells help control murine West Nile virus infection. J Immunol 171:2524–2531

Wang Y, Lobigs M, Lee E, Müllbacher A (2003b) CD8+ T cells mediate recovery and immunopathology in West Nile virus encephalitis. J Virol 77:13323–13334

Wang T, Town T, Alexopoulou L, Anderson JF, Fikrig E, Flavell RA (2004a) Toll-like receptor 3 mediates West Nile virus entry into the brain causing lethal encephalitis. Nat Med 10:1366–1373

Wang Y, Lobigs M, Lee E, Müllbacher A (2004b) Exocytosis and Fas mediated cytolytic mechanisms exert protection from West Nile virus induced encephalitis in mice. Immunol Cell Biol 82:170–173

Wang Y, Lobigs M, Lee E, Koskinen A, Müllbacher A (2006) CD8(+) T cell-mediated immune responses in West Nile virus (Sarafend strain) encephalitis are independent of gamma interferon. J Gen Virol 87:3599–3609

Wekerle H (2002) Immune protection of the brain – efficient and delicate. J Infect Dis 186 Suppl 2:S140–144

Yang JS, Ramanathan MP, Muthumani K, Choo AY, Jin SH, Yu QC, Hwang DS, Choo DK, Lee MD, Dang K, Tang W, Kim JJ, Weiner DB (2002) Induction of inflammation by West Nile virus capsid through the caspase-9 apoptotic pathway. Emerg Infect Dis 8:1379–1384

14. Enhanced Antigen Processing or Immune Evasion? West Nile Virus and the Induction of Immune Recognition Molecules

NICHOLAS J.C. KING, ARIANE DAVISON, DANIEL R. GETTS, DAVID PING LU, MEGHANN TEAGUE GETTS, AMANDA YEUNG, JAMES K. PETERSON, AND ALISON M. KESSON

Abstract

The neurotropic flavivirus, West Nile virus, is of widespread health concern as an emerging arthropod-borne virus. The paradoxical induction of increased immune regulatory molecule expression by WNV is of considerable interest, as it may explain the characteristic immunopathology associated with infection in both humans and laboratory animals. Many viruses with specific immune evasion mechanisms against adaptive T-cell responses reduce expression of immune recognition molecules with the result that high affinity cellular immune responses are ultimately generated, which eradicate virus-infected cells efficiently. Based on our work, however, because WNV and other flaviviruses promote expression of high levels of MHC and adhesion molecules, we hypothesise that low affinity anti-viral T cells are generated and decoyed away from subsets of cells that express low levels of MHC molecules, in which virus replicates most efficiently. This strategy may extend the period of viraemia in virus-adapted reservoir hosts, thus increasing the likelihood of virus transmission to the mosquito vector. In incidental mammalian hosts such as humans and horses, the generation of large numbers of low affinity T cells may have immunopathological consequences.

Keywords

West Nile virus, MHC, ICAM-1, Anti-viral immunity, Flavivirus, Interferon, Cell cycle, Dendritic cells, Langerhans cells, Embryo, Trophectoderm, Trophoblast, Placenta, CNS, Virus encephalitis

1 Introduction

West Nile virus (WNV), a member of the *Flaviviridae*, genus flavivirus, was originally isolated in 1937 from the West Nile province of Uganda (Smithburn et al., 1940). It is a member of the Japanese encephalitic serogroup, the most prominent members of which include St Louis encephalitis (North America), Murray Valley encephalitis, Kunjin (Australia), Japanese encephalitis (Australasia) and WNV (Africa, the Middle East, Europe, India and North America). WNV is divided into lineage I and II, with a possible lineage III proposed (Bakonyi et al., 2005). The highly pathogenic, lineage I WNV isolates are probably the most studied at present, but all three lineages are pathogenic, with Kunjin, an Australian subtype of relatively low pathogenicity, most closely related to the highly pathogenic North American lineage I WNV, both genomically (Scherret et al., 2001) and immunologically, as evidenced by the effective cross-reactive WNV immunity generated by it (Hall et al., 2003). The mechanism of the disparity in virulence between closely related strains is both a topic of intense investigation and a focus of hope for vaccine development.

With its emergence in New York in the late 1990s and subsequent spread throughout North America, WNV is now one of the most prevalent of the flaviviruses. It essentially has a worldwide distribution, and while significantly less pathogenic than Japanese encephalitis, it is nevertheless a subject of major public health concern (Briese et al., 1999; Lanciotti et al., 1999; Roehrig et al., 2002). The clinical syndromes associated with WNV infection include an undifferentiated febrile illness or meningoencephalitis, with significant morbidity, perhaps most notably exemplified by a persisting acute flaccid paralysis, and some 10% mortality (Hayes et al., 2005). Although many elements of the innate and adaptive immune response have been demonstrated to play a role in the eradication of WNV (Diamond et al., 2003a, b; Wang et al., 2003a, b, 2006; Shrestha and Diamond, 2004; Mehlhop and Diamond, 2006; Shrestha et al., 2006a, b), experimental evidence also indicates that disease is strikingly immunopatholological in nature (Andrews et al., 1999; Wang et al., 2003b, 2004; Arjona et al., 2007; Getts et al., 2007; King et al., 2007) and it has quickly become clear that WNV has a complex immunological relationship with the vertebrate host that varies with virus strain, route of infection, size of inoculum and host genetics.

2 Cell Surface Molecule Upregulation by West Nile Virus

Viruses that interact with the host adaptive immune system in some way to promote survival within infected cells, usually either decrease cell surface major histocompatibility complex antigen (MHC) expression or interfere with antigen processing and presentation, to evade the immune response. Viruses that do this include, for example, adenoviruses (Pääbo et al., 1986; Chatterjee-Kishore et al., 2000), vaccinia (Koszinowski and Ertl, 1975) and herpesviruses (Yewdell and Bennink, 1999). By contrast, infection of vertebrate cells in vitro with WNV is accompanied by a two- to tenfold increase in the expression variously, of MHC-I, MHC-II, intercellular adhesion molecule-1 (ICAM-1) (CD54), vascular cell adhesion molecule-1 (VCAM-1) (CD106), E-selectin (CD62E) and P-selectin (CD62P), most of which are members of the immunoglobulin gene superfamily. All are of major significance in cell-cell recognition and interaction during adaptive immune responses. The kinetics of these changes are shown in Fig. 1 for human umbilical vein endothelial cells. Increases in these molecules occur with varying kinetics within 30 min to 24 h, on a wide variety of cells isolated from mice, rats and humans (King and Kesson, 1988; King et al., 1989; Liu et al., 1989; Argall et al., 1991; Bao et al., 1992; Shen et al., 1995a, b, 1997). The increase in MHC-I on fibroblasts is first detectable by 6 h and continues for at least 96 h, and occurs in response to other flaviviruses such as Murray Valley encephalitis, Kunjin, Japanese encephalitis, dengue and yellow fever (17D) viruses (King and Kesson, 1988; Shen et al., 1995b; Lobigs et al., 1996). Flaviviruses, however, do not upregulate all immunoglobulin superfamily members [NCAM (CD56) and PECAM (CD31) are not upregulated], nor necessarily immunoregulatory molecules of functionally related families, such as the integrins, VLA-4 (CD49d/CD29), LFA-1 (CD11a/CD18), Mac-1 (CD11b/CD18), and CD11c, or the hyaluronan receptor, CD44 (King, unpublished observations).

2.1 West Nile Virus-Induced Cell Surface MHC Expression Is Functional

Increased MHC-I expression increases the efficiency of both induction and execution of cell mediated immune responses (O'Neill and Blanden, 1979; King et al., 1985) and the increased MHC-I expression induced by WNV infection is accompanied by a parallel increase in the susceptibility to both allo-immune (i.e. MHC-I + self-peptide-specific) and WNV-specific (i.e. MHC-I + WNV-peptide-specific)

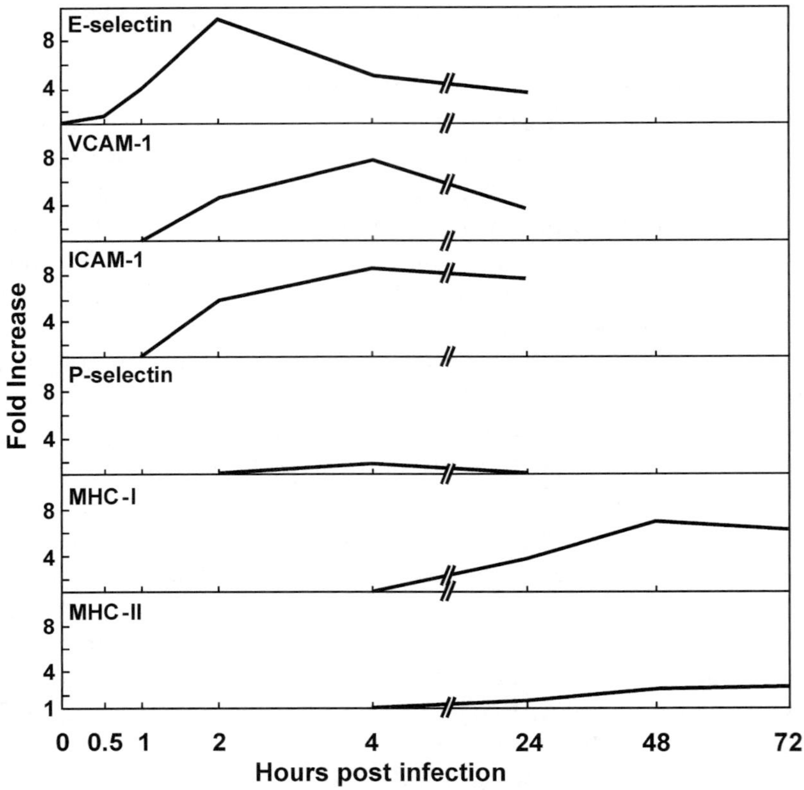

Figure 1. Kinetics of cell surface molecule upregulation on human umbilical vein endothelial cells. The graphs show the fold increase (*abscissa*) in each cell surface molecule, as determined by flow cytometry at each time point after infection with WNV (*ordinate*). Increases were determined by comparison of infected versus uninfected fluorescence values at each time point. Values start at 1.

cytotoxic T-cell (CTL) lysis (King and Kesson, 1988; Douglas et al., 1994; Kesson and King, 2001; Mullbacher and Lobigs, 1995), as well as a decrease in the susceptibility to NK cell lysis (Momburg et al., 2001). Also, WNV-induced MHC-II expression on astrocytes is recognised efficiently by MHC-restricted T-cell clones (Liu et al., 1989). Thus, WNV-induced MHC increases are functional and WNV-specific T-cell cytotoxicity indicates that virus antigen is not limiting (King and Kesson, 1988; Liu et al., 1989; King et al., 1993; Kesson and King, 2001). In the case of MHC-I, both processing and association of viral peptides with MHC-I correspond with increased MHC-I expression. This is supported by work showing an increase in supply of flavivirus peptides to the endoplasmic reticulum (Mullbacher and Lobigs, 1995), but we have also shown increased expression of LMP-2 and TAP-1, molecules crucially involved in processing and loading peptide onto MHC-I (Arnold et al., 2004).

2.2 Type I Interferon and the Cell Cycle in the Responses to West Nile Virus

The flavivirus-induced upregulation of MHC-I is partially independent of endogenous type I interferon secretion and is thus virus-specific (King and Kesson, 1988; Cheng et al., 2004a). However, the interferon-independent, WNV-induced upregulation of MHC-I is heterogeneous, i.e. when more than 80% of cells in culture are infected with WNV in the presence of neutralising polyclonal anti-type I interferon-α/β antibody (anti-IFN-α/β), only 60–65% of the population respond with an increase in MHC-I expression. Investigation of the cell cycle position at the time of infection with WNV showed that cells in G_0 were significantly more susceptible to interferon-independent, WNV-induced MHC-I upregulation than cells in G_1. Thus, cells infected in G_0 increase MHC-I in response to WNV infection five- to sixfold more than do cells infected in other phases of the cell cycle. This results in a tenfold difference in the susceptibility to WNV-specific CTL lysis (Douglas et al., 1994). On the other hand, perhaps not surprisingly, cells in G_1 produce at least tenfold more WNV than cells in G_0 (Kesson et al., 2002). WNV replication is very sensitive to type I interferon; cells support considerably more replication in its absence (Cheng et al., 2004a) and increased replication in IFN type I receptor gene knock-out (IFNAR$^{-/-}$) mice leads to accelerated mortality (Lobigs et al., 2003; Samuel and Diamond, 2005). Infection induces the expression of interferon-dependent anti-viral genes, which in turn are subverted to some extent by WNV (Perelygin et al., 2002; Liu et al., 2005, 2006; Keller et al., 2006; Samuel et al., 2006). However, both G_0 and G_1 cultures produce similar amounts of type I interferon on a cell-for-cell basis (Douglas et al., 1994).

2.3 West Nile Virus-Induced Increase in ICAM-1 Expression

Although human embryonic fibroblasts express cell surface ICAM-1 constitutively at high levels, WNV infection of these cells in G_0 further markedly upregulates ICAM-1 expression, while cells infected in G_1 show no increase in expression. The WNV-induced upregulation of ICAM-1 on cells infected in G_0 is regulated by two distinct mechanisms. There is an early dramatic increase by 2 h after infection, which is interferon-independent, and a later, weaker increase by 24 h, which is augmented by endogenous type I interferon secretion. The cell cycle status is of vital importance in determining the cellular response to virus and cytokine stimulation and hence the expression of cell surface recognition molecules. Moreover, there is a quantal threshold for the amount of infecting WNV that will induce ICAM-1 upregulation on G_0 cells, which is not seen with MHC-I increases. Thus, the induction of both MHC-I and ICAM-1 after WNV infection is cell cycle-dependent, with cells in G_0 demonstrating significantly higher levels of MHC-I and ICAM-1 expression than cells in other stages of the cell cycle (Shen et al., 1995b).

2.4 Increased Gene Transcription of WNV-Induced Immune Molecules Is Interferon Independent

Experiments with WNV-infected cells in which cellular DNA transcription was blocked by actinomycin D (controlled by monitoring ^{3}H-uridine uptake) resulted in a failure of the cells to increase surface MHC-I expression (King and Kesson, 1988). We subsequently demonstrated that the WNV-induced increased expression of MHC-I and ICAM-1 is associated with an increase in transcription of *MHC-1* and *ICAM-1* genes (Kesson and King, 2001; Kesson et al., 2002). The increase in mRNA is not solely a result of autocrine stimulation by WNV-induced type I interferon production, since treatment with anti-IFN-α/β antibody is still associated with a significant increase in MHC-I expression. Furthermore, upregulation of MHC occurs in placental trophoblast cells, which do not produce type I IFN (King et al., 1989) and in cells from IFNAR$^{-/-}$ mice, which cannot respond to endogenous type I IFN (Cheng et al., 2004a). Interestingly, although the WNV-induced increase in MHC-I protein expression on the surface of cycling (G_1) cells is significantly reduced by anti-IFN-α/β, the increase in corresponding gene expression is not, indicating that irrespective of cell cycle status, WNV-induced gene changes are interferon-independent (Douglas et al., 1994; Arnold et al., 2004). Similarly, increases in *TAP-1* and *LMP-2* gene expression in WNV-infected human skin fibroblasts are not decreased by treatment with anti-IFN-α/β (Arnold et al., 2004).

Thus, WNV evidently directs the genetic upregulation of functional cell surface immune recognition molecules, leading to demonstrably increased susceptibility to virus-specific CTL lysis and paradoxically appearing to increase the efficiency of its own eradication.

3 Intracellular Responses to West Nile Virus

3.1 Increased Immune Molecule Expression Is Mediated by NF-κB

While it has been proposed that flavivirus-induced upregulation of MHC-I antigen expression is driven by an increase in peptide supply to the ER (Mullbacher and Lobigs, 1995), this does not explain how flaviviruses upregulate the variety of cell surface recognition molecules which are not influenced by peptide supply. As mentioned, WNV upregulation of MHC-I on the cell surface requires host gene transcription (King and Kesson, 1988) and we have shown that WNV-infection is associated with increased levels of MHC-I and ICAM-1 mRNA (Kesson and King, 2001; Kesson et al., 2002). This evidence suggests a possible common mechanism driving WNV-induced transcription of the genes for these molecules, all of which, including TAP-1 and LMP-2, have sites for the transcription factor, nuclear factor kappa-B (NF-κB), in their promoter regions (Baeuerle and Henkel, 1994; Wright et al., 1995). NF-κB is held in an inactive form in the cell cytoplasm bound to its inhibitory subunit, IκB (Baeuerle and Baltimore, 1988). When IκB is phosphorylated, NF-κB is dissociated from it and becomes activated and translocates to the cell nucleus, binds its cognate sequence and initiates target gene transcription. WNV infection is associated with activation of NF-κB, producing a p65/p50 heterodimer as well as a p50/p50 homodimer (Kesson and King, 2001). The functional activation of NF-κB was demonstrated in a transactivation assay using RAWa4 LTR cells, a cell line stably transfected with a plasmid consisting of the HIV long terminal repeat which contains two NF-κB sites linked to a luciferase reporter gene. WNV infection of these cells demonstrated a significant increase in luciferase activity, thus confirming that NF-κB is activated by WNV (Kesson and King, 2001). Furthermore, blocking NF-κB activation using the protein kinase C inhibitor, 1-(5-isoquinolinesulfonyl)-2-methylpiperazine (H-7) and the serine protease inhibitor, *N*-benzoyl-l-tosyl-l-phenylalanne (BTEE) inhibits MHC-I upregulation in a dose-dependent manner.

3.2 NF-κB Activation Is Independent of IFN and TNF

In vitro, in the context of virus infection, cytokines produced by infected cells, most notably type I IFN and tumour necrosis factor (TNF), can also upregulate MHC-I expression. However, treatment of cells with exogenous type I IFN did not activate NF-κB. Thus, upregulation of MHC-I occurs by at least two mechanisms, viz., a WNV-induced, NF-κB-dependent, interferon-independent pathway and a NF-κB-independent, interferon-dependent pathway (Cheng et al., 2004a). Despite demonstrable TNF production by WNV-infected cells (Arnold et al., 2004; Cheng et al., 2004b), WNV-induced upregulation of MHC-I is independent of the actions of endogenously produced TNF; WNV upregulates cell surface MHC-I expression in cultures where TNF has been neutralised by antibody, as well as in cells from TNF gene knock out ($TNF^{-/-}$) mice (Cheng et al., 2004b). Furthermore, the kinetics of WNV-induced NF-κB activation and the subunits induced are different from those induced by TNF.

3.3 Involvement of Interferon Regulatory Factor-1 In WNV Responses

The promoters of both mouse *MHC-I* and type I *IFN* genes contain the interferon regulatory factor-1 (IRF-1) consensus binding site. Previous studies have found that the role of IRF-1 in the upregulation of *MHC-I* and type I *IFN* gene expression varies with stimulus and cell type. Our preliminary data shows that IRF-1-deficient mouse embryo fibroblasts ($IRF\text{-}1^{-/-}$ MEF) were more resistant to WNV infection than wild type (WT) cells, but paradoxically, more MHC-I was upregulated by WNV infection in the $IRF\text{-}1^{-/-}$ cells. This was not due to differing efficiency of upregulation of MHC-I by type I IFN, since MHC-I induction was similar on uninfected WT MEF and $IRF\text{-}1^{-/-}$ MEF that were treated with exogenous IFN-β. However, significantly more type I IFN was induced by WNV infection of $IRF\text{-}1^{-/-}$ MEF than in WT MEF. Furthermore, IRF-1 deficiency did not have a significant effect on WNV-induced NF-κB activation. Thus, IRF-1 appears to suppress type I IFN production by WT MEF in response to WNV infection (Kesson and King, unpublished observations). The normal suppression of this enhanced anti-viral state may protect cells, either in the short or long term: although markedly resistant to neurotropic virus infection, transgenic mice expressing type I IFN show significant CNS neurodegeneration in the absence of infection (Akwa et al., 1998; Campbell et al., 1999). This highlights the balance required between controlled antiviral action and an unrestrained response, which may have significant cytopathic effects.

4 Decoy Hypothesis

4.1 Cell Surface Molecule Concentration and the Immunological Synapse

MHC-I genes encode polymorphic membrane glycoproteins that present both self and foreign peptides to $CD8^+$ T cells. Although the efficiency of interaction between an immune T cell and the (infected) target cell is directly related to the concentration of MHC expressed on the target cell surface, several other molecules in the immunological synapse modulate the avidity of interaction, including ICAM-1 (Kuhlman et al., 1991), CD8 or CD4 (Shimonkevitz et al., 1985). It would appear paradoxical that a virus-driven response would enhance the efficiency of antiviral T-cell immunity, unless it were part of the virus survival strategy.

Our group has hypothesised that the increase in adhesion molecules, both allelic (MHC-I and MHC-II) and non-allelic (ICAM-1, VCAM-1 and E-selectin), increases the avidity of interaction between WNV-infected cells and WNV-specific T cells (King and Kesson, 2003). This is supported by the correlation of increased cellular surface expression on target cells with increased susceptibility to allo-specific and WNV-specific T-cell lysis. This is important because the cell-mediated immune response is a critical element both in the eradication of WNV (Kesson et al., 1987, 1988; Shrestha and Diamond, 2004; Shrestha et al., 2006a), and the causation of associated immunopathology (Wang et al., 2003b).

4.2 Recruitment of Low Affinity T Cells

Virus-specific CTL populations in a polyclonal response can be generated with a wide range of affinities. This range includes affinities for allogeneic MHC and heterologous viral antigens (Nahill and Welsh, 1993; Selin et al., 1994). Recruitment in the draining lymph node (DLN), should be enhanced by upregulation of the co-stimulatory molecules CD80 and CD86, and the high levels of WNV-induced MHC-I and MHC-II on antigen presenting cells (APC). This occurs on Langerhans cells (LCs) in vivo with intradermal WNV infection (Johnston et al., 1996, 2000; Byrne et al., 2001). Recruitment would therefore include T cells with affinities normally below the threshold for recognition of APC (Shimonkevitz et al., 1985; Goldstein and Mescher, 1987; Kuhlman et al., 1991). A wider affinity range than usual would include low affinity clones which, in addition to recognising MHC + virus peptide, will also be self-reactive. Since peptide specificity is not necessarily required for MHC-I-specific CTL to recognise MHC-I (Mullbacher et al., 1991), such CTL could conceivably kill uninfected cells with high MHC-I (and ICAM-1) expression, as well as WNV-infected cells.

Increased MHC expression on normal cells would be driven to some extent by the secretion of type I IFN in the local vicinity of infected cells, but later this increase would be considerably further potentiated by type II IFN (IFN-γ), when low affinity T cells secreted this cytokine upon recognition of their highly expressed cognate antigen. Initially, recognition, mainly, of infected cells by CTL would occur, but subsequently, uninfected cells would increasingly be recognised by CTL as these cells increase their

MHC and adhesion molecule expression. In an in vitro parallel for this, lysis by WNV-specific CTL of both WNV-infected and uninfected, IFN-γ-treated fibroblasts is significantly greater than lysis of uninfected, untreated fibroblasts by these same CTL (Kesson et al., 2002).

Accessory molecules, such as CD8, would further stabilise the interaction between the T αβ cell receptor (TcR) and MHC-I on the target cell, particularly in low affinity interactions. This would be less important when MHC-I expression was very high, since the avidity of multiple TcR-MHC-I interactions can overcome the need for CD8 stabilisation. Indeed, blocking CD8 monoclonal antibody studies show that the avidity of interaction between WNV-specific CTL and WNV-infected or IFN-γ-treated fibroblast targets (both high MHC-I-expressing cells) is high and functionally similar, whereas the avidity between these same CTL (i.e. with the same low affinity) and mock-infected, mock-treated (i.e. low MHC-I-expressing) fibroblasts is much lower. This low avidity interaction is markedly inhibited by anti-CD8 antibody, because of the increased contribution to avidity by CD8 on the CTL recognising MHC-I at low concentrations on target cells (Fig. 2). (Shimonkevitz

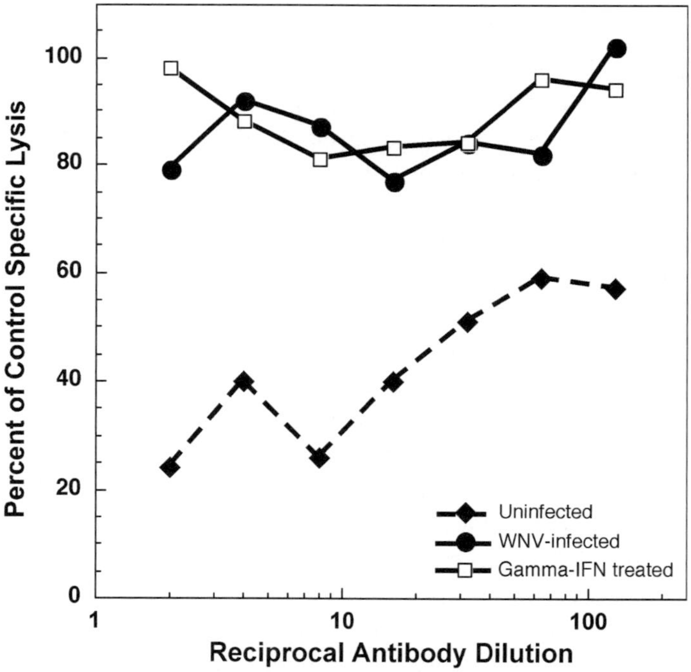

Figure 2. Avidity of interaction of secondary WNV-specific CTL. Blocking of CTL cytotoxicity in the presence of anti-CD8 monoclonal blocking antibody on ^{51}Cr-labelled mouse embryo fibroblasts that were either WNV-infected, IFN-γ-treated (500 IU for 48 h) or mock-infected, mock-treated. Values are expressed as the per cent of control lysis for each CTL target in the absence of anti-CD8 monoclonal blocking antibody.

et al., 1985; Springer, 1990; Kuhlman et al., 1991). Ongoing low affinity, high avidity interactions would support the induction of CTL clones with a wide range of affinities. Moreover, the generation of WNV quasispecies would undoubtedly broaden the range of possible reactive CTL clones (Jerzak et al., 2005; Ciota et al., 2007).

4.3 Involvement of Cell Cycle: Divide and Conquer

Low affinity CTL recognition of WNV-infected cells would be more likely to occur on cells in G_0, since these cells express very high levels of MHC-I and ICAM-1. WNV-infected cells in G_1 do not increase MHC-I to the same extent and do not increase ICAM-1 at all (Shen et al., 1995b). Although MHC-I expression increases in response to type I IFN to a relatively greater extent in G_1 than G_0 cells, WNV-infected G_0 cells still express significantly higher levels of MHC-I. Coupled to this, IFN-γ induction of MHC-I and ICAM-1 expression is significantly higher on G_0 than G_1 cells (Douglas et al., 1994). Thus, it is more likely that G_1 cells would escape detection by low affinity WNV-specific CTL, thus enabling WNV to replicate for a longer period. The majority of cells in G_0 would decoy the immune system to produce low affinity cells, which would not be of sufficient affinity to recognise WNV-infected G_1 cells. Since even high numbers of low affinity CTL clear virus poorly, compared to high affinity CTL (Derby et al., 2001), this would provide a delay in immune clearance and allow further successful virus replication. As previously mentioned, G_1 (i.e. cycling) cells support WNV replication at more than tenfold the levels seen in G_0 cells. With longer periods of greater virus output, this would lengthen the period of viraemia and facilitate transmission to the arthropod vector and the next host.

4.4 Disease Resolution or Progression to Autoimmune Disease?

In time, expansion of *high* affinity clones would eradicate low MHC-I-expressing, WNV-infected cells to clear the virus and resolve the infection. However, continued recognition of uninfected cells expressing high levels of MHC-I and ICAM-1 could contribute to the disease phenotype. This would exacerbate and prolong encephalitic illness. Consistent with this idea, damage caused by lympocytic choriomeningitis virus (LCMV), a neurotropic virus that promotes an immunopathological immune response, can be significantly reduced by restricting the range of TcR specificities available for eradicating the virus (Doherty et al.,

1994). Moreover, absence of IFN-γ, while at some inoculating doses of WNV clearly results in increased mortality (Shrestha et al., 2006b), at other doses it results in increased survival with protective immunity (Shrestha, 2002). The contribution of IFN-γ to disease can be observed in a lethal intranasal model of WNV infection, where wild type mice develop severe limbic seizures. In contrast, IFN-$\gamma^{-/-}$ mice do not develop seizures (Getts et al., 2007). Neurons may increase their MHC-I in vivo (Shrestha et al., 2006a), although they do not appear to increase MHC-II or ICAM-1 expression (King, unpublished observations). However, we have found that microglia upregulate these molecules significantly during WNV infection. The destruction of these crucial support cells, in addition to any neurons with upregulated MHC-I, would undoubtedly increase morbidity. In addition, microglia may be involved in seizure formation by virtue of the range of neuroactive inflammatory soluble factors that they produce (Getts et al., 2007, 2008). The fact that microglia do not evidently become infected further emphasises the likely susceptibility of uninfected cells to immunopathological damage, raising the possibility of a continuing autoimmune response (Bao et al., 1992; Murray and McMichael, 1992; King et al., 2003).

5 Mathematical Modelling of WNV Immune Responses

A striking finding in experimental animal models is the ragged dose–response curve that occurs with peripheral inoculation of WNV in an immune competent mouse. While very high doses inevitably kill an entire cohort of mice and very low doses will result in complete survival, there is a very wide range of inoculating doses in which the survival outcomes vary widely and are unpredictable (King et al., 2007). It is possible that low doses of virus produce an exuberant immune response, which results in immunopathological death, whereas high doses result in a response that is blunted and/or delayed, and thus insufficient to control virus replication (Wang et al., 2003b). Recently, using mathematical modelling techniques, we have attempted to replicate this "unpredictable" outcome as a computer model. This model is undergoing continual refinement, but is currently restricted to modelling the effector arm of the immune response to WNV to generate testable hypotheses that probe for crucial factors causing immunopathology.

Briefly, taking account of various elements, such as cell cycle and infection status, MHC-I and ICAM-1 expression, antiviral CTL affinity, etc., as defined by our experimental data above, we have used a series of variables that represent cell categories and factors that act on these and sought to model the lifetimes of a population of cells in three-dimensional space (i.e. representing a "host"), as they are acted upon by adjacent CTL and soluble factors produced by their interaction. A series of predicted probabilities interact with these variables and if, for example, IFN-γ upregulates MHC-I, the value for the avidity between the CTL and the cell of interest is recomputed using the relevant factor relating to IFN-γ upregulation. Using the five viral peptides commonly recognised by WNV-specific CTL on infected cells, the model generates a series of CTL that recognise each virus peptide (in the context of MHC-I) with decreasing affinities, but with varying avidities and accordingly, may lyse such cells. Varying the start amount of virus, we have run a series of iterations on a series of artificial host populations to determine "survival" at a particular dose of virus.

The advantage of the model is that very many host populations can be run to achieve the kind of statistics that are difficult to obtain in a laboratory for in vivo experiments. Clearly, in addition to being restricted to a small (but well-studied) part of the cellular immune response to WNV, this model has a number of limitations. Although it uses values based on laboratory data, (which may themselves be statistically variable), it obviously ignores unknown elements, and must inevitably estimate some values by logical or informed guesswork. Notwithstanding these deficiencies, the output of this model (Fig. 3) closely approximates the uneven mortality curves seen in experimental WNV infection (King et al., 2007). A technical report at http://www.ces.clemson.edu/~petersj/CurrentPapers/SurvivalPaper.pdf (Peterson et al., 2007) describes this model in detail.

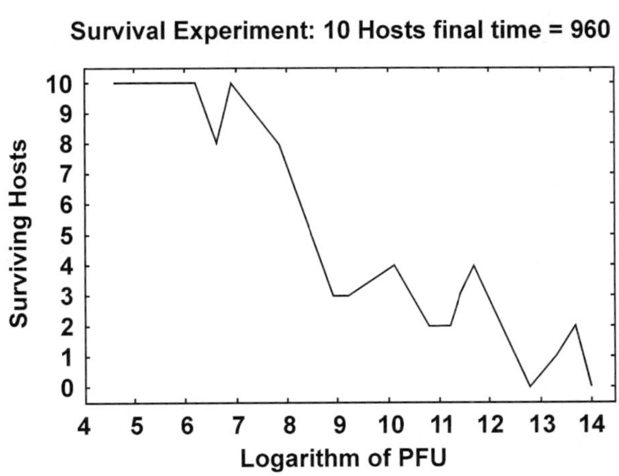

Figure 3. Survival experiment in silico. The graph shows the number of surviving artificial hosts (*ordinate*) after infection with various doses of WNV (*abscissa*). At each dose of WNV, ten artificial hosts were used. The final number of iterations was 960. The graph shows a similar ragged dose–response as seen in vivo at various doses of WNV (King et al., 2007).

6 Modelling Impact of WNV-Induced MHC Increases In Vivo: Embryonic Infection

6.1 Immunological Silence and the Developing Embryo

One scenario in which the induction of MHC class I could have drastic effects is during implantation and development of the embryo. In mammals, histo-incompatible tissue engraftment occurs in the contact between mother and embryo during gestation. The mechanisms for avoidance of rejection by the maternal immune system, enabling the survival of the semi-allogeneic fetus are multifarious. Generalised specific anti-paternal response suppression (Tafuri et al., 1995) and non-specific generation of maternal $CD4^{+}$, $CD25^{+}$ regulatory T (T_{reg}) cells (Aluvihare et al., 2004) reduce anti-paternal allogeneic adaptive T-cell responses, and modulate chronic and acutely acquired diseases in the process. Local immune modulation maintains pregnancy through the midterm, substantially underpinned, inter alia, by trophoblast expression of indoleamine 2,3 dioxygenase (IDO) (Munn et al., 1998), which suppresses T-cell activation, FasL, which ablates Fas-expressing T cells (Makrigiannakis et al., 2001) and HLA-G (in humans), which inhibits NK cell killing (Rouas-Freiss et al., 1997). During early embryonic development and implantation, the lack of allogeneic MHC antigen expression on the paternally-derived trophoblast cells is thought to contribute to the immunological silence of the embryo (Clark, 1991; Sargent, 1993), as their presence would make the embryo visible to the maternal immune system, which under normal circumstances would reject tissue expressing paternal MHC. Most of these parallel and cooperative mechanisms of immune inhibition, particularly local mechanisms, are not specific for paternal antigens. This leaves open the higher evolutionary question of how the maternal cellular immune system deals with viral infection involving the embryo. Immunological "ignorance" of the infected embryo by the mother carries the risk of continued maternal infection and possible death as a result, thus reducing species renewal by the loss of a reproductively active member. On the other hand, if the embryo becomes infected, does the mother break tolerance in order to eradicate the virus, at the expense of embryo survival, and if so, how does immune recognition of virus-infected placental and embryonic cells occur efficiently in the absence or low expression of MHC and adhesion molecules?

How WNV infection of the embryonic tissue (placenta and/or embryo) itself may affect the progression of pregnancy is therefore an interesting and potentially important question. The fact that WNV infection upregulates MHC-I and other important immune molecules arguably puts the embryo at risk of outright immunological rejection. Indeed, transplacental spread of a maternal infection with WNV and Japanese encephalitis has been documented in humans and mice and may certainly result in abortion, especially early in pregnancy (Chaturvedi et al., 1980; Mathur et al., 1982, 1986; Miura et al., 1982; Sugamata and Miura, 1982; Anonymous, 2002, 2003; Alpert et al., 2003). WNV infection of pre-implantation blastocysts from day 3.5 pregnant mice induces the expression of MHC class I on trophectoderm (King et al., 1989), which can be lysed by both paternal MHC-I-restricted and WNV-specific $CD8^+$ T cells (King et al., 1993). Importantly, primary trophoblast giant cells are unresponsive to type I or II IFN, indicating that the WNV infection alone is sufficient for MHC class I induction (King et al., 1989). Later in pregnancy, secondary trophoblast giant cells express MHC-I molecules de novo in response to both type I and II interferons (Zuckermann and Head, 1986; Drake et al., 1987; King et al., 1987) and by mid-term, IFN-γ induces ICAM-I expression on placental trophoblast cells, enhancing their susceptibility to lysis by paternal (i.e. allo)-specific CTL (Tian and King, 1994).

6.2 Model of Embryonic Infection

In systemic virus infection models, it is difficult to dissociate embryo-specific responses leading to abortion from systemic antiviral responses that could lead to embryonic abortion due to, for example, high systemic levels of inflammatory cytokines, even in the absence of embryonic infection (Haddad et al., 1997). Our group has developed a model in which the blastocyst is infected with WNV and transferred into a pseudopregnant mother. Here, WNV infection of trophectoderm on pre-implantation blastocysts induces de novo cell surface expression of ICAM-I and VCAM-I, both of which are normally absent or undetectable on the pre-implantation embryonic trophoblast (Lu et al., 2002). On day 9 of pregnancy in these mice, WNV antigen is readily detectable in the placenta, but not in the embryo itself (Fig. 4a), which is morphologically normal. ICAM-I remains detectable as well, potentially making these embryos vulnerable to maternal immune system-mediated resorption. WNV can still be detected in placental tissues isolated from the recipients of WNV-infected embryos by day 14 of pregnancy (Fig. 4b); however, in a significant percentage of these mice, resorption of all embryos occurs. Interestingly, examination of the DLN or spleens showed no changes in major cell subsets compared to non-aborting WNV-infected

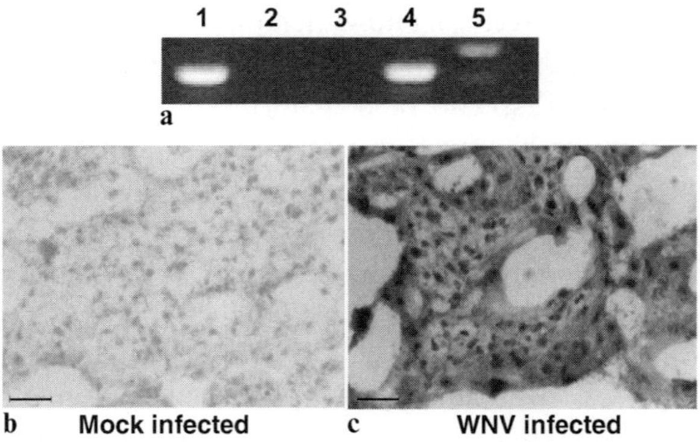

Figure 4. RT-PCR and immunohistochemistry from placental tissue at day 14 post-coitum. (**a**) RT-PCR of embryonic and placental tissue isolated at gestation day 9 and analysed for WNV E-protein transcripts. Blastocysts were mock-infected or WNV-infected before transfer on day 3 of pseudopregnancy and harvest at day 9. *Lane 1*, WNV-infected (placenta); *lane 2*, mock-infected (placenta); *lane 3*, WNV-infected (embryo); *lane 4*, WNV-infected (brain-positive control); *lane 5*, standard ladder. (**b**, **c**) Immunohistochemistry of placenta labelled with immunoperoxidase-labelled monoclonal antibody, 4G4, for WNV NS-1 protein (generous gift of Roy Hall, University of Queensland), and 3,3-diaminobenzidine tetrahydrochloride substrate, with haematoxylin as a counterstain. Blastocysts were mock-infected (**b**) or WNV-infected (**c**) before transfer on day 3 of pseudopregnancy and harvested at day 14. Fetuses were morphologically normal. Scale bar: 20 μm.

embryos and mock-infected embryos. This suggests that abortion is mediated by mechanisms responding to WNV infection within the placenta, rather than more centrally, in the DLN or the spleen.

The data from this model arise from an outbred mouse strain system. Whether resorptions occur in a syngeneic system, in which the maternal immune system would be capable of recognising WNV antigen only in the context of maternal (syngeneic) MHC and not in the context of paternal (allogeneic) MHC, remains to be seen.

6.3 Exogenous Infection in Pregnancy

In parallel, to investigate the effect of limited exogenous WNV infection, we have infected pregnant mice with WNV via intra-vaginal inoculation at mid-gestation. These tissues are drained by the same DLN as the uterus. In this model too, a significant proportion of allogeneic matings result in resorption by day 5 post-infection. In contrast, WNV-infected syngeneic matings do not and pups are born normally. However, vaginal infection does not result in placental infection. Cell subsets in placentae, including dendritic cells (DCs), are not detectably altered as a result of vaginal WNV infection. In contrast to the "direct" model using infected blastocysts, this suggests that the DLN may generate anti-paternal (allogeneic) MHC responses to the embryo. This raises the possibility that anti-viral responses set up by local infection may abrogate local tolerance, enabling the embryo to be recognised by the maternal immune system and thus aborted. If this process generated anti-paternal memory it might interfere permanently with the generation of paternal tolerance in subsequent pregnancies, causing recurrent immune-mediated abortions.

7 Initiation of Immunity: Impact on Ag Presentation

7.1 Dendritic Cells and Virus Infection

Clearly, one way in which alteration in cell surface molecules could modulate anti-viral immune responses would be during the initiation of the adaptive immune response via interaction with the most crucial cell type in this process, the DC.

DCs are distributed sparsely throughout most tissues of the body. Despite a frequency of only 1–2%, they most efficiently present antigen to initiate protective immune responses against viral infections. Mice injected with DC elicit protective immune responses against Sendai virus, Moloney leukaemia virus (Kast et al., 1988), HSV (Hengel et al., 1987), influenza virus (Nonacs et al., 1992), and clear LCMV infection from the ovaries and the brain (Ludewig et al., 1998). On the other hand, viruses can directly or indirectly modulate DCs and thus interfere with the immune response. Since DC are the primary antigen-presenting cells for inducing CTL responses, and $CD4^+$ T cells provide help for long term $CD8^+$ CTL activity, interference with DC and $CD4^+$ cell types constitute obvious avenues by which viruses can disable the immune response (Klagge and Schneider-Schaulies, 1999).

DCs also occur in the skin and mucosal surfaces. These epithelia provide a primary barrier that protects the body against damage, infection and invasion. In the epidermis, LCs, a subtype of DC form an extensive network of sentinel cells within it (Romani et al., 1985). Because of their anatomical distribution and migratory behaviour, LC were originally thought to be the primary antigen-presenting cell of the skin, particularly as the majority of pathogens entering via the skin are most likely to come into contact with these cells first. LCs phagocytose antigen, become mature and migrate to the DLN. However, recent work suggests that viral antigen presentation to T cells is principally executed by lymph node-resident CD8α^+ DC and dermal dendritic cells (DDCs) which migrate to the draining lymph node (Allan et al., 2003; Zhao et al., 2003), thus relegating LC to a putative role in tolerance (Albert et al., 2001; Steinman et al., 2003; Waithman et al., 2007). As an anti-viral strategy, the separation of first line LC from initiation of the immune response may help the vertebrate host to separate and isolate the infected periphery in the initial stages of infection from the central task of generating an effective immune response. On the other hand, it leaves open the possibility of virus manipulation of this process (Piguet and Steinman, 2007).

Several other populations of DC have been clearly defined in murine systems; however, the roles of many subsets are still unclear. Added to this, DC subsets exhibit considerable functional plasticity, depending on their location, the nature of the pathogen and route of inoculation (Heath et al., 2004), with both "lymphoid" and "myeloid" DC precursors evidently able to give rise to all DC populations (Naik et al., 2006).

It is not known which specific DC subsets take part in WNV infection in vivo. In vitro, human plasmacytoid DC (pDC), not surprisingly produce significant amounts IFN-α in response to WNV, compared to conventional monocyte-derived DC. This may occur via TLR signalling in the absence of viral replication (Silva et al., 2007). This is of interest, since pDC too, may differentiate into conventional DC during viral infection (Zuniga et al., 2004), providing a potential pathway for WNV to modulate induction of immune responses.

7.2 The Langerhans Cell Response to West Nile Virus Infection

Past studies in our laboratory have investigated the response of DC and LC to epithelial inoculation with WNV at a variety of epithelial sites (Johnston et al., 1996, 2000; Byrne et al., 2001; Burke et al., 2004). Following injection of free WNV intradermally into the skin of the ear in mice, epidermal LC density decreases within 24 h by some 25% at the site of infection, compared to mock-infection or comparable infection with Semliki Forest virus, a mosquito-borne alphavirus. This reduction correlates inversely with increases in LC numbers in the local DLN (Johnston et al., 2000). Interestingly, LC numbers in the epidermis return to normal levels by 48 h, while after SFV infection, measurable decreases in epidermal LC numbers are only detectable at 48 h. During infection, LC require IL-1β but not TNF signalling to migrate to the DLN, in contrast to the evident requirement for both cytokines when contact sensitisers are used (Cumberbatch et al., 1997). Thus, LC migration occurs as effectively in WNV-infected $TNF^{-/-}$ mice as in the wild type strain (Byrne et al., 2001). Importantly, in these studies LC migration was abrogated if UV-inactivated WNV was inoculated, instead of live replicating WNV.

LCs migrate out of the skin in response to a wide variety of inflammatory stimuli or danger signals, such as TNF and IL-1. This occurs through the simultaneous downregulation of E-cadherin and CCR6 and upregulation of CCR7, causing detachment of LC from keratinocytes, and migration via dermal lymphatics to the DLN, respectively (Schwarzenberger and Udey, 1996; Jakob et al., 2001). LC are derived from local dermal monocyte precursors during steady-state conditions (Ginhoux et al., 2006; Iijima et al., 2007). However, during vigorous immune responses involving migration of large numbers of LC, monocytes are recruited from the peripheral circulation into dermal compartments, where they migrate into the epidermis and differentiate into LC.

In vivo infection with WNV and SFV both significantly increase expression of MHC-II, CD54 and CD80 on LC, but in WNV, this occurs with accelerated kinetics, between 8 and 32 h earlier than with SFV, with initial upregulation clearly occurring prior to local infiltration of leucocytes. Moreover, increased MHC-II expression occurs in vitro on LC only from WNV-infected and not SFV-infected epidermal suspensions, supporting the notion that upregulation is WNV-specific, rather than the result of cytokine release from infiltrating leucocytes (Johnston et al., 1996). WNV antigen is detectable by flow cytometry in epidermal LC within 16–24 h following intradermal inoculation or in vitro exposure. It is not known if this represents active infection of the LC, or the uptake of large numbers of viral particles.

Thus, skin infection by WNV is associated with the rapid acquisition of a mature phenotype by LC and an accelerated migration to the DLN. It remains to be determined if WNV infection of LC, and other DC, is productive, or if the virus affects their functional properties. Moreover, the functional nature of LC leaving the dermis/epidermis from the site of infection, compared with those remaining is still unknown. Similarly, whether infectious virus or viral protein/peptide is processed and/or transferred via LC to DDC and/or $CD8\alpha^{+}$ DC in the DLN, remains to be demonstrated. This is important because if the major function of LC is to produce and maintain peripheral tolerance, the accelerated kinetics of LC maturation and migration may signal a delay, diversion or some other modulation of the initiation of the cellular immune response.

7.3 Novel Skin Model

To date, our studies and those of others modelling pathogenic skin infection (Allan et al., 2003; Belz et al., 2004; Iezzi et al., 2006) have involved the direct deposition of a large bolus of free pathogen intradermally, resulting in immediate spread to the DLN (Itano et al., 2003; Belz et al., 2004; Iezzi et al., 2006). Such a model potentially bypasses early steps in the local detection of virus infection and initiation of responses by innate immune elements, including possible local DC responses. Nowhere are local responses clearer than in vaginal infection by WNV in a mouse model. Infection of the vaginal epithelium induces the accumulation of DC under the basement membrane only in patches of demonstrable infection and results in better survival of infected animals compared to other routes of inoculation (Burke et al., 2004). Thus, in free virus injection, the kinetics of cellular, viral and immune responses may be different from true transmission by the bite of an infected mosquito, where local deposition precedes systemic spread, symptomatology and evidence of an adaptive immune response by many days (reviewed in King et al., 2007). Taken together with evidence that factors in mosquito saliva retard the local response to WNV (Schneider et al., 2006), this argues strongly that local responses are crucial in controlling the range of outcomes subtended by the initial interaction with the immune system.

To explore these issues further, we have developed a novel model of skin infection. In our model, syngeneic mouse embryonic fibroblasts infected in vitro with WNV are inoculated intradermally. By confining infection within local cells, this model retards the flow of free virus to the DLN for at least 24 h and has a similar delaying effect on leucocyte migration to the DLN (Fig. 5). In contrast to the free-virus model, this appears to involve a local DDC response, in which these cells cluster tightly around the focus of infected MEF within the dermis, similar to the sub-basement membrane DC clustering seen in vaginal WNV infection (Burke et al., 2004), as well as migrating to the DLN. In this model, LC migrate to the DLN with similar kinetics; however, blood-derived $CD11b^{hi}$ DC and lymph node-resident $CD8\alpha^{+}$ DC were primarily responsible for the initiation of WNV-specific T-cell immunity in the DLN following intradermal skin infection (Davison and King, submitted). Skin-derived DC (i.e. DDC and LC) accounted for <15% of total DC infiltrating the DLN. This suggests that neither LC nor DDC are directly presenting antigen in the DLN to initiate immune responses, although they may be involved in antigen transfer to the DC that do (Carbone et al., 2004). It has been proposed that the infiltration of LC

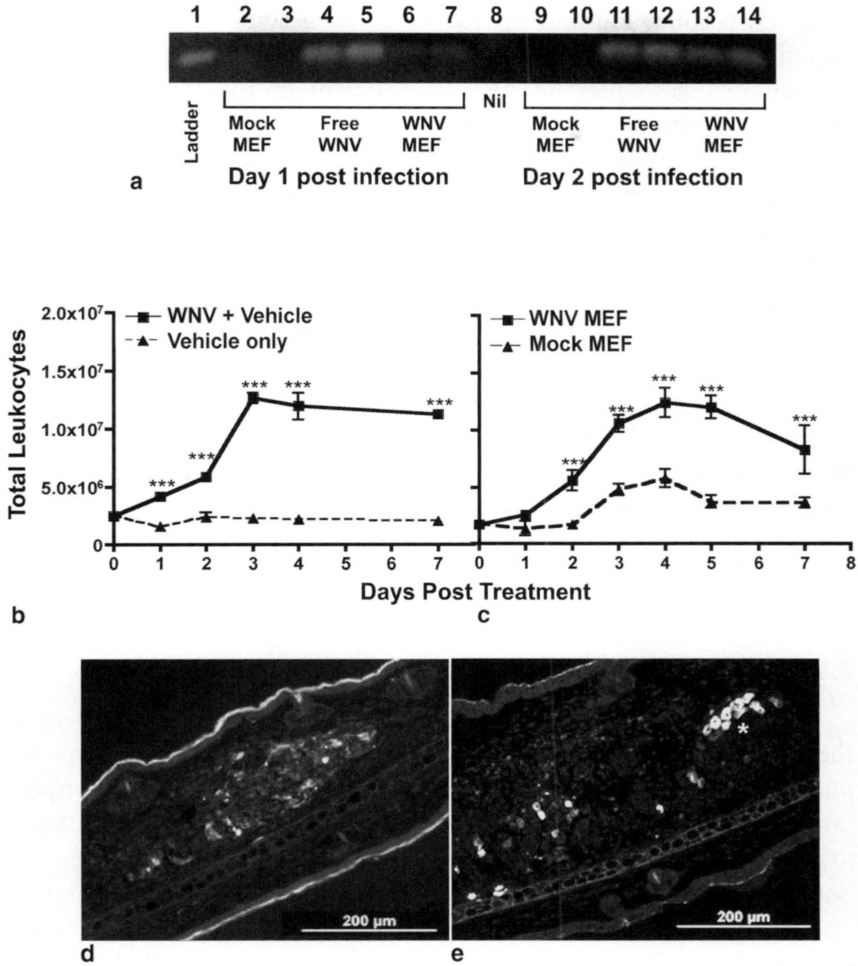

Figure 5. Kinetics of WNV infection in comparative skin infection models. (**a**) Representative RT-PCR from mice analysed for WNV E-protein transcripts in draining ALN at day 1 and day 2 post-infection. *Lane 1*, standard ladder; *lanes 2 and 3*, day 1 mock-infected MEF (mouse 1 and 2); *lanes 4 and 5*, free WNV (mouse 3 and 4); *lanes 6 and 7*, WNV$^+$ MEF (mouse 5 and 6); *lane 8*, empty; *lanes 9 and 10*, day 2 mock-infected MEF (mouse 7 and 8); *lanes 11 and 12*, free WNV (mouse 9 and 10); *lanes 13 and 14*, WNV$^+$ MEF (mouse 11 and 12). Each lane represents mRNA from the ALN of one mouse. Two mice were used for each treatment group and time point. Total number of leucocytes in ALN of mice injected intradermally in the ear with free virus compared to vehicle only (**b**) or WNV$^+$ MEF compared to mock-infected MEF (**c**) at indicated time points post-infection. Means ± SE are shown. $***P < 0.001$. At least four mice were used for each treatment and time point in three separate experiments. Cryosections (7 µm) of ear inoculated with WNV$^+$ MEF, labelled with anti-NS-1-FITC antibody, at day 1 post-infection (**d**), showing infected MEF adjacent to the ear cartilage, and at day 5 post-infection, showing few remaining infected MEF (**e**). Local muscle fibres became infected, as indicated by a *white asterisk*. The *stratum corneum* of the epidermis is autofluorescent.

in the DLN following skin-derived inoculation represents the second of two distinct waves of migrating skin-derived DC. If these waves occur sequentially, this "permits" the development of full fledged immune responses (Itano et al., 2003). However, in our models of WNV skin infection, LC migration always precedes DDC migration. It is possible therefore that LC may alter the local immune response in some way by migrating into the DLN as mature LC prior to the DDC. It is also likely that local infection stimulates the recruitment of monocyte precursors from the blood to contribute to the pool of DC interacting with the primary site of infection and migrating on to the DLN, and/or the recruitment of these precursors to immigrate into the epidermis to become resident LC to replenish the loss of emigrant LC from the epidermis. In the former scenario, particularly, this may promote anti-WNV immune responses by enhancing antigen transfer to lymph node-resident $CD8\alpha^+$ DC by cell-cell interactions in the lymph node, consistent with the paradoxical increase in MHC and adhesion molecules.

7.4 Dendritic Cells in the CNS

The migration of monocytes to the CNS is an important event in the pathogenesis of WNV encephalitis. As previously shown, macrophages derived from a circulating monocyte precursor, play an important role in driving both viral clearance and pathology during WNV infection (Ben-Nathan et al., 1996; Arjona et al., 2007). However, the possible role of DC arising from blood monocyte precursors migrating into the WNV-infected brain remains unknown. While DC have been shown to be critical in driving immune responses in other CNS diseases, especially in experimental autoimmune encephalitis and Theiler's encephalomyelitis virus infection (McMahon et al., 2005), they have not been investigated in the WNV-infected brain. Using an intranasal inoculation model, where WNV is limited to neurons in the brain (Getts et al., 2007, 2008a,b; Wacher et al., 2007), we found a small percentage (2–5%) of the leucocytes expressing CD11c in the WNV-infected brain at day 7 post-infection. Detailed phenotyping of these DC showed >75% of expressed CD11b. Since normally the brain is essentially devoid of these cells, this argues that they were derived from the periphery. As we have shown that $CD45^{int}$ microglia are derived from $Ly6C^+$ inflammatory monocytes during WNV infection (Getts et al., 2008b), we hypothesised that the $CD11b^+/CD11c^+$ DC subsets in the CNS may also be derived from inflammatory monocytes. Indeed, co-expression of Ly6C on cFMS-EGFP⁺ CD11b⁺ DC in WNV-infected chimeric mice support this hypothesis (Fig. 6). While the presence of DC in the WNV-infected brain suggests a role for antigen presentation within the CNS, these DC were not capable of inducing anti-WNV memory CD4⁺ T-cell proliferation, in contrast to infiltrating macrophages ($CD45^{hi}/CD11c^{-}$) and

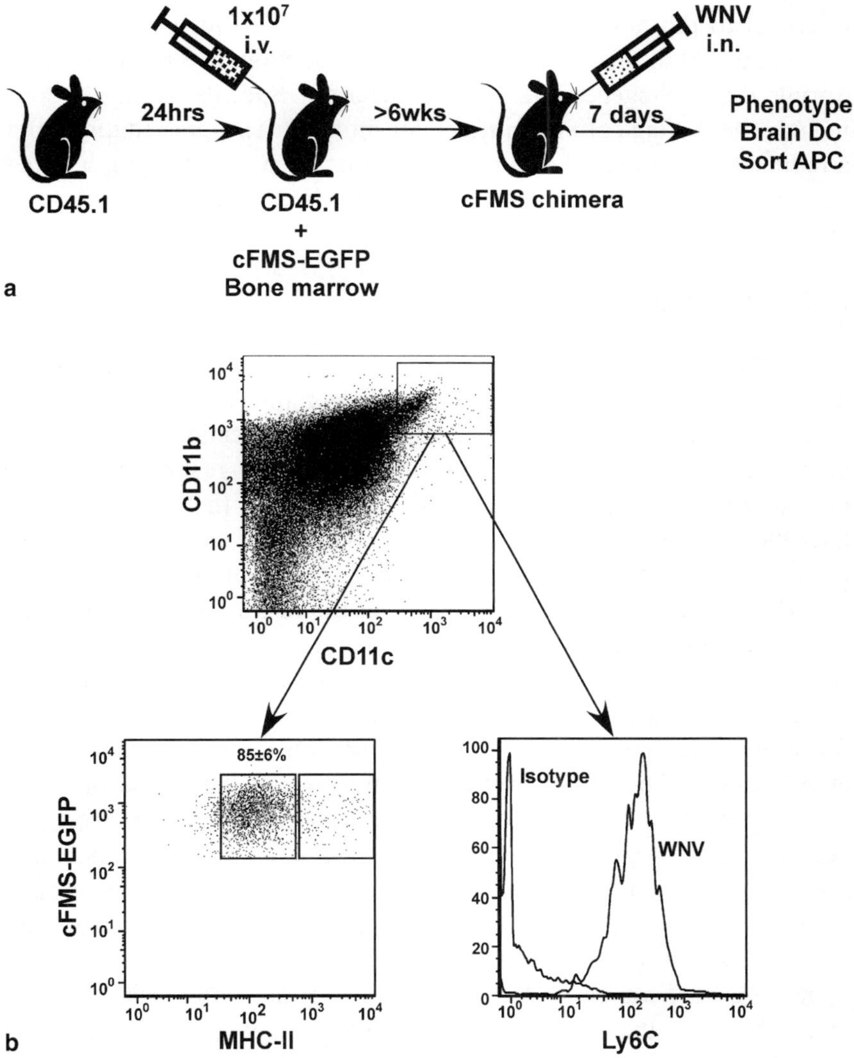

Figure 6. Dendritic cell isolation from WNV-infected cfms chimeric brain. (**a**) CD45.1 mice are irradiated and 24 h later are reconstituted with bone marrow from a CD45.2 cFMS-EGFP transgenic mouse. Six weeks later, mice are infected with WNV intranasally and their brains harvested on day 7 post-infection. (**b**) Leucocytes ($CD45^{+}$) are isolated from the brains of infected mice and analysed and sorted for $EGFP^{+}$, $C11c^{+}$ and $CD11b^{+}$ cells. The majority of these cells are $MHC\text{-}II^{int}$ (*lower left panel*) and express Ly6C (*lower right panel*).

microglia ($CD45^{lo\text{-}int}/CD11c^{-}$). Investigation into the level of MHC-II on the cell surface of these infiltrating DC clearly indicated that they mostly expressed intermediate levels. This suggests that the lack of proliferation induced by DC is a by-product of their immature phenotype. However, adoptive transfer of mature bone marrow-derived DC loaded with WNV failed to confer protection although they evidently migrated to the brain. Thus, in the brain of WNV-infected mice, DC may either fail to present antigen or have a function as yet to be defined during the pathogenesis of WNV infection. Speculatively, this may include the carriage of viral antigen for more efficient presentation after a requisite maturational step in the draining deep cervical lymph nodes outside the brain, (Getts and King, unpublished observations).

8 Conclusions

WNV paradoxically induces increased cell surface expression of molecules that are recognised by immune cells, appearing, on the face of it, to enhance the mechanisms that lead to its own eradication. Experimental evidence suggests, however, that these antiviral responses include populations of low affinity T cells that cross-react with "self" MHC. We hypothesise that these T cells are preferentially decoyed into a high avidity interactions, initially with WNV-infected cells that express high levels of MHC following infection by WNV in the resting phase (G_0) of the cell cycle. In this way, WNV-infected cycling (G_1) cells, which by comparison, produce tenfold more virus, but do not increase MHC and adhesion molecule expression to any marked degree, escape detection for an extended period, until high affinity clones are eventually generated to destroy them. This would probably extend the period of viraemia, improving the chances of the virus transmission to an arthropod vector. Immunological adaptation to such a strategy by normal avian reservoir hosts could explain the typically prolonged high viraemic phase in the absence of disease, otherwise lethal in unaccustomed avian species. Such a strategy may also explain the characteristic immunopathological features of WNV infection in incidental hosts such as man. Moreover, this process might be the basis in genetically susceptible individuals of virus-triggered autoimmune disease later in life.

Most self-limiting viruses that interfere with recognition by the cellular immune response do so by reducing the expression of the molecules involved, leading to the generation of high affinity interactions that efficiently eradicate virus-infected cells. The effects of MHC *upregulation* by WNV in vivo remain to be clearly demonstrated. We anticipate that in silico modelling should elucidate experimental directions that could be profitably explored in this regard. As such, we have generated a range of in vivo systems that attempt to model processes most likely to be influenced by WNV-induced MHC upregulation, including embryo implantation and antigen presentation at various stages of the immune response. We are also trying to understand the mechanisms used by WNV at the cellular level to accomplish this, to determine whether this is more likely to be an evolved strategy for virus survival or a paradoxical by-product of virus replication resulting in a justifiable sacrifice in the interests of transmission at the host level. Current experiments are directed at the further elucidation of these issues to test our hypotheses more rigorously.

References

Akwa Y, Hassett DE, Eloranta ML, Sandberg K, Masliah E, Powell H, Whitton JL, Bloom FE, Campbell IL (1998) Transgenic expression of IFN-alpha in the central nervous system of mice protects against lethal neurotropic viral infection but induces inflammation and neurodegeneration. J Immunol 161:5016–5026

Albert ML, Jegathesan M, Darnell RB (2001) Dendritic cell maturation is required for the cross-tolerization of CD8+ T cells. Nat Immunol 2:1010–1017

Allan RS, Smith CM, Belz GT, van Lint AL, Wakim LM, Heath WR, Carbone FR (2003) Epidermal viral immunity induced by CD8alpha+ dendritic cells but not by Langerhans cells. Science 301:1925–1928

Alpert SG, Fergerson J, Noel LP (2003) Intrauterine West Nile virus: ocular and systemic findings. Am J Ophthalmol 136:733–735

Aluvihare VR, Kallikourdis M, Betz AG (2004) Regulatory T cells mediate maternal tolerance to the fetus. Nat Immunol 5:266–271

Andrews DM, Matthews VB, Sammels LM, Carrello AC, McMinn PC (1999) The severity of murray valley encephalitis in mice is linked to neutrophil infiltration and inducible nitric oxide synthase activity in the central nervous system. J Virol 73:8781–8790

Argall KG, Armati PJ, King NJC, Douglas MW (1991) The effects of West Nile virus on major histocompatibility complex class I and II molecule expression by Lewis rat Schwann cells in vitro. J Neuroimmunol 35:273–284

Arjona A, Foellmer HG, Town T, Leng L, McDonald C, Wang T., Wong SJ, Montgomery RR, Fikrig E, Bucala R (2007) Abrogation of macrophage migration inhibitory factor decreases West Nile virus lethality by limiting viral neuroinvasion. J Clin Invest 117:3059–3066

Arnold SJ, Osvath SR, Hall RA, King NJC, Sedger LM (2004) Regulation of antigen processing and presentation molecules in West Nile virus-infected human skin fibroblasts. Virology 324:286–296

Baeuerle PA, Baltimore D (1988) I kappa B: a specific inhibitor of the NF-kappa B transcription factor. Science 242:540–546

Baeuerle PA, Henkel T (1994) Function and activation of NF-kappa B in the immune system. Annu Rev Immunol 12:141–179

Bakonyi T, Hubalek Z, Rudolf I, Nowotny N (2005) Novel flavivirus or new lineage of West Nile virus, central Europe. Emerg Infect Dis 11:225–231

Bao S, King NJ, Dos Remedios CG (1992) Flavivirus induces MHC antigen on human myoblasts: a model of autoimmune myositis? Muscle Nerve 15:1271–1277

Belz GT, Smith CM, Kleinert L, Reading P, Brooks A, Shortman K, Carbone FR, Heath WR (2004) Distinct migrating and nonmigrating dendritic cell populations are involved in

MHC class I-restricted antigen presentation after lung infection with virus. Proc Natl Acad Sci USA 101:8670–8675
Ben-Nathan D, Huitinga I, Lustig S, van Rooijen N, Kobiler D (1996) West Nile virus neuroinvasion and encephalitis induced by macrophage depletion in mice. Arch Virol 141:459–469
Briese T, Jia XY, Huang C, Grady LJ, Lipkin WI (1999) Identification of a Kunjin/West Nile-like flavivirus in brains of patients with New York encephalitis. Lancet 354:1261–1262
Burke SA, Wen L, King NJ (2004) Routes of inoculation and the immune response to a resolving genital flavivirus infection in a novel murine model. Immunol Cell Biol 82:174–183
Byrne SN, Halliday GM, Johnston LJ, King NJC (2001) Interleukin-1beta but not tumor necrosis factor is involved in West Nile virus-induced Langerhans cell migration from the skin in C57BL/6 mice. J Invest Dermatol 117:702–709
Campbell IL, Krucker T, Steffensen S, Akwa Y, Powell HC, Lane T, Carr DJ, Gold LH, Henriksen SJ, Siggins GR (1999) Structural and functional neuropathology in transgenic mice with CNS expression of IFN-alpha. Brain Res 835:46–61
Carbone FR, Belz GT, Heath WR (2004) Transfer of antigen between migrating and lymph node-resident DCs in peripheral T-cell tolerance and immunity. Trends Immunol 25:655–658
Centers for Disease Control and Prevention (CDC) (2002) Intrauterine West Nile virus infection – New York, 2002. MMWR Morb Mortal Weekly Rep 51:1135–1136
Centers for Disease Control and Prevention (CDC) (2003) Intrauterine West Nile virus infection – New York, 2002. JAMA 289:295–296
Chatterjee-Kishore M, van Den Akker F, Stark GR (2000) Adenovirus E1A down-regulates LMP2 transcription by interfering with the binding of stat1 to IRF1. J Biol Chem 275:20406–20411
Chaturvedi UC, Mathur A, Chandra A, Das SK, Tandon HO, Singh UK (1980) Transplacental infection with Japanese encephalitis virus. J Infect Dis 141:712–715
Cheng Y, King NJ, Kesson AM (2004a) Major histocompatibility complex class I (MHC-I) induction by West Nile virus: involvement of 2 signaling pathways in MHC-I up-regulation. J Infect Dis 189:658–668
Cheng Y, King NJ, Kesson AM (2004b) The role of tumor necrosis factor in modulating responses of murine embryo fibroblasts by flavivirus, West Nile. Virology 329:361–370
Ciota AT, Ngo KA, Lovelace AO, Payne AF, Zhou Y, Shi PY, Kramer LD (2007) Role of the mutant spectrum in adaptation and replication of West Nile virus. J Gen Virol 88:865–874
Clark DA (1991) Controversies in reproductive immunology. Crit Rev Immunol 11:215–247
Cumberbatch M, Dearman RJ, Kimber I (1997) Langerhans cells require signals from both tumour necrosis factor-alpha and interleukin-1 beta for migration. Immunology 92:388–395
Derby MA, Alexander-Miller MA, Tse R, Berzofsky JA (2001) High-avidity CTL exploit two complementary mechanisms to provide better protection against viral infection than low-avidity CTL. J Immunol 166:1690–1697
Diamond MS, Shrestha B, Marri A, Mahan D, Engle M (2003a) B cells and antibody play critical roles in the immediate defense of disseminated infection by West Nile encephalitis virus. J Virol 77:2578–2586
Diamond MS, Sitati EM, Friend LD, Higgs S, Shrestha B, Engle M (2003b) A critical role for induced IgM in the protection against West Nile virus infection. J Exp Med 198:1853–1862
Doherty PC, Hou S, Evans CF, Whitton JL, Oldstone MB, Blackman MA (1994) Limiting the available T cell receptor repertoire modifies acute lymphocytic choriomeningitis virus-induced immunopathology. J Neuroimmunol 51:147–152
Douglas MW, Kesson AM, King NJ (1994) CTL recognition of west Nile virus-infected fibroblasts is cell cycle dependent and is associated with virus-induced increases in class I MHC antigen expression. Immunology 82:561–570

Drake BL, King NJ, Maxwell LE, Rodger JC (1987) Class I major histocompatibility complex antigen expression on early murine trophoblast and its induction by lymphokines in vitro. J Reprod Immunol 10:319–328

Getts DR, Matsumoto I, Muller M, Getts MT, Radford J, Shrestha B, Campbell IL, King NJ (2007) Role of IFN-gamma in an experimental murine model of West Nile virus-induced seizures. J Neurochem 103:1019–1030

Getts DR, Balcar V, Matsumoto I, Müller M, King NJC (2008a) Viruses and the immune system, their roles in seizure cascade development. J Neurochem 104:1167–1176

Getts DR, Terry RL, Teague Getts M, Müller M, Rana S, Shrestha B, Radford J, Van Rooijen N, Campbell IL, King NJC (2008b) $Ly6C^{+}$ "inflammatory monocytes" are microglial precursors recruited in a pathogenic manner in West Nile Virus Encephalitis. J Exp Med 205:2319–2337

Ginhoux F, Tacke F, Angeli V, Bogunovic M, Loubeau M, Dai XM, Stanley ER, Randolph GJ, Merad M (2006) Langerhans cells arise from monocytes in vivo. Nat Immunol 7:265–273

Goldstein SAN, Mescher MF (1987) Cytotoxic T cell activation by class I protein on cell-size artificial membranes: antigen density and Lyt-2/3 function. J Immunol 138:2034–2043

Haddad EK, Duclos AJ, Antecka E, Lapp WS, Baines MG (1997) Role of interferon-gamma in the priming of decidual macrophages for nitric oxide production and early pregnancy loss. Cell Immunol 181:68–75

Hall RA, Nisbet DJ, Pham KB, Pyke AT, Smith GA, Khromykh AA (2003) DNA vaccine coding for the full-length infectious Kunjin virus RNA protects mice against the New York strain of West Nile virus. Proc Natl Acad Sci USA 100:10460–10464

Hayes EB, Sejvar JJ, Zaki SR, Lanciotti RS, Bode AV, Campbell GL (2005) Virology, pathology, and clinical manifestations of West Nile virus disease. Emerg Infect Dis 11:1174–1179

Heath WR, Belz GT, Behrens GM, Smith CM, Forehan SP, Parish IA, Davey GM, Wilson NS., Carbone FR, Villadangos JA (2004) Cross-presentation, dendritic cell subsets, and the generation of immunity to cellular antigens. Immunol Rev 199:9–26

Hengel H, Lindner M, Wagner H, Heeg K (1987) Frequency of herpes simplex virus-specific murine cytotoxic T lymphocyte precursors in mitogen- and antigen-driven primary in vitro T cell responses. J Immunol 139:4196–4202

Iezzi G, Frohlich A, Ernst B, Ampenberger F, Saeland S, Glaichenhaus N, Kopf M (2006) Lymph node resident rather than skin-derived dendritic cells initiate specific T cell responses after Leishmania major infection. J Immunol 177:1250–1256

Iijima N, Linehan MM, Saeland S, Iwasaki A (2007) Vaginal epithelial dendritic cells renew from bone marrow precursors. Proc Natl Acad Sci USA 104:19061–19066

Itano AA, McSorley SJ, Reinhardt RL, Ehst BD, Ingulli E, Rudensky AY, Jenkins MK (2003) Distinct dendritic cell populations sequentially present antigen to CD4 T cells and stimulate different aspects of cell-mediated immunity. Immunity 19:47–57

Jakob T, Ring J, Udey MC (2001) Multistep navigation of Langerhans/dendritic cells in and out of the skin. J Allergy Clin Immunol 108:688–696

Jerzak G, Bernard KA, Kramer LD, Ebel GD (2005) Genetic variation in West Nile virus from naturally infected mosquitoes and birds suggests quasispecies structure and strong purifying selection. J Gen Virol 86:2175–2183

Johnston LJ, Halliday GM, King NJ (1996) Phenotypic changes in Langerhans' cells after infection with arboviruses: a role in the immune response to epidermally acquired viral infection? J Virol 70:4761–4766

Johnston LJ., Halliday GM, King NJ (2000) Langerhans cells migrate to local lymph nodes following cutaneous infection with an arbovirus. J Invest Dermatol 114:560–568

Kast WM, Boog CJ, Roep BO, Voordouw AC, Melief CJ (1988) Failure or success in the restoration of virus-specific cytotoxic T lymphocyte response defects by dendritic cells. J Immunol 140:3186–3193

Keller BC, Fredericksen BL, Samuel MA, Mock RE, Mason PW, Diamond MS, Gale M, Jr (2006) Resistance to alpha/beta interferon is a determinant of West Nile virus replication fitness and virulence. J Virol 80:9424–9434

Kesson AM, King NJ (2001) Transcriptional regulation of major histocompatibility complex class I by flavivirus West Nile is dependent on NF-kappaB activation. J Infect Dis 184:947–954

Kesson AM, Blanden RV, Müllbacher A (1987) The primary in vivo murine cytotoxic T cell response to the flavivirus, West Nile. J Gen Virol 68:2001–2006

Kesson AM, Blanden RV, Müllbacher A (1988) The secondary in vitro murine cytotoxic T cell response to the flavivirus, West Nile. Immunol Cell Biol 66:23–32

Kesson AM, Cheng Y, King NJ (2002) Regulation of immune recognition molecules by flavivirus, West Nile. Viral Immunol 15:273–283

King NJ, Kesson AM (1988) Interferon-independent increases in class I major histocompatibility complex antigen expression follow flavivirus infection. J Gen Virol 69(Pt 10):2535–2543

King NJ, Kesson AM (2003) Interaction of flaviviruses with cells of the vertebrate host and decoy of the immune response. Immunol Cell Biol 81:207–216

King NJC, Sinickas VG, Blanden RV (1985) H-2K and H-2D antigens are independently regulated in mouse embryo fibroblasts. Exp Clin Immunogenet 2:206–214

King NJ, Drake BL, Maxwell LE, Rodger JC (1987) Class I major histocompatibility complex antigen expression on early murine trophoblast and its induction by lymphokines in vitro. II. The role of gamma interferon in the responses of primary and secondary giant cells. J Reprod Immunol 12:13–21

King NJ, Maxwell LE, Kesson AM (1989) Induction of class I major histocompatibility complex antigen expression by West Nile virus on gamma interferon-refractory early murine trophoblast cells. Proc Natl Acad Sci USA 86:911–915

King NJ, Mullbacher A, Tian L, Rodger JC, Lidbury B, Hla RT (1993) West Nile virus infection induces susceptibility of in vitro outgrown murine blastocysts to specific lysis by paternally directed allo-immune and virus-immune cytotoxic T cells. J Reprod Immunol 23:131–144

King NJ, Shrestha B, Kesson AM (2003) Immune modulation by flaviviruses. Adv Virus Res 60:121–155

King NJC, Getts DR, Getts MT, Rana S, Shrestha B, Kesson AM (2007) Immunopathology of flavivirus infections. Immunol Cell Biol 85:33–42

Klagge I M, Schneider-Schaulies S (1999) Virus interactions with dendritic cells. J Gen Virol 80(Pt 4):823–833

Koszinowski U, Ertl H (1975) Altered serological and cellular reactivity to H-2 antigens after target cell infection with vaccinia virus. Nature 257:596–597

Kuhlman P, Moy VT, Lollo BA, Brian AA (1991) The accessory function of murine intercellular adhesion molecule-1 in T lymphocyte activation. Contributions of adhesion and co-activation. J Immunol 146:1773–1782

Lanciotti RS, Roehrig JT, Deubel V, Smith J, Parker M, Steele K, Crise B, Volpe KE, Crabtree MB, Scherret JH et-al. (1999) Origin of the West Nile virus responsible for an outbreak of encephalitis in the north-eastern United States. Science 286:2333–2337

Liu Y, King N, Kesson A, Blanden RV, Mullbacher A (1989) Flavivirus infection up-regulates the expression of class I and class II major histocompatibility antigens on and enhances T cell recognition of astrocytes in vitro. J Neuroimmunol 21:157–168

Liu WJ, Wang XJ, Mokhonov VV, Shi PY, Randall R, Khromykh AA (2005) Inhibition of interferon signaling by the New York 99 strain and Kunjin subtype of West Nile virus involves blockage of STAT1 and STAT2 activation by nonstructural proteins. J Virol 79:1934–1942

Liu WJ, Wang XJ, Clark DC, Lobigs M, Hall RA, Khromykh AA (2006) A single amino acid substitution in the West Nile virus nonstructural protein NS2A disables its ability to inhibit alpha/beta interferon induction and attenuates virus virulence in mice. J Virol 80:2396–2404

Lobigs M, Blanden RV, Mullbacher A (1996) Flavivirus-induced up-regulation of MHC class I antigens; implications for the induction of CD8+ T-cell-mediated autoimmunity. Immunol Rev 152:5–19

Lobigs M, Mullbacher A, Wang Y, Pavy M, Lee E (2003) Role of type I and type II interferon responses in recovery from infection with an encephalitic flavivirus. J Gen Virol 84:567–572
Lu DP, Tian L, O'Neill C, King NJC (2002) Regulation of cellular adhesion molecule expression in murine oocytes, peri-implantation and post-implantation embryos. Cell Res 12:373–383
Ludewig B, Ehl S, Karrer U, Odermatt B, Hengartner H, Zinkernagel RM (1998) Dendritic cells efficiently induce protective antiviral immunity. J Virol 72:3812–3818
Makrigiannakis A, Zoumakis E, Kalantaridou S, Coutifaris C, Margioris AN, Coukos G, Rice KC, Gravanis A, Chrousos GP (2001) Corticotropin-releasing hormone promotes blastocyst implantation and early maternal tolerance. Nat Immunol 2:1018–1024
Mathur A, Arora KL, Chaturvedi UC (1982) Transplacental Japanese encephalitis virus (JEV) infection in mice during consecutive pregnancies. J Gen Virol 59:213–217
Mathur A, Arora KL, Rawat S, Chaturvedi UC (1986) Japanese encephalitis virus latency following congenital infection in mice. J Gen Virol 67(Pt 5):945–947
McMahon EJ, Bailey SL, Castenada CV, Waldner H, Miller SD (2005) Epitope spreading initiates in the CNS in two mouse models of multiple sclerosis. Nat Med 11:335–339
Mehlhop E, Diamond MS (2006) Protective immune responses against West Nile virus are primed by distinct complement activation pathways. J Exp Med 203:1371–1381
Miura T, Sugamata M, Ogata T, Matsuda R (1982) Japanese encephalitis virus infection in fetal mice at different stages of pregnancy. II. Resistance to Japanese encephalitis virus infection. Acta Virol 26:283–287
Momburg F, Mullbacher A, Lobigs M (2001) Modulation of transporter associated with antigen processing (TAP)-mediated peptide import into the endoplasmic reticulum by flavivirus infection. J Virol 75:5663–5671
Mullbacher A, Lobigs M (1995) Up-regulation of MHC class I by flavivirus-induced peptide translocation into the endoplasmic reticulum. Immunity 3:207–214
Mullbacher A, Hill AB, Blanden RV, Cowden WB, King NJ, Hla RT (1991) Alloreactive cytotoxic T cells recognize MHC class I antigen without peptide specificity. J Immunol 147:1765–1772
Munn DH, Zhou M, Attwood JT, Bondarev I, Conway SJ, Marshall B, Brown C, Mellor AL (1998) Prevention of allogeneic fetal rejection by tryptophan catabolism. Science 281:1191–1193
Murray N, McMichael A (1992) Antigen presentation in virus infection. Curr Opin Immunol 4:401–407
Nahill SR, Welsh RM (1993) High frequency of cross-reactive cytotoxic T lymphocytes elicited during the virus-induced polyclonal cytotoxic T lymphocyte response. J Exp Med 177:317–327
Naik SH, Metcalf D, van Nieuwenhuijze A, Wicks I, Wu L, O'Keeffe M, Shortman K (2006) Intrasplenic steady-state dendritic cell precursors that are distinct from monocytes. Nat Immunol 7:663–671
Nonacs R, Humborg C, Tam JP, Steinman RM (1992) Mechanisms of mouse spleen dendritic cell function in the generation of influenza-specific, cytolytic T lymphocytes. J Exp Med 176:519–529
O'Neill HC, Blanden RV (1979) Quantitative differences in the expression of parentally-derived H-2 antigens in F1 hybrid mice affect T-cell responses. J Exp Med 149:724–731
Pääbo S, Nilsson T, Peterson PA (1986) Adenoviruses of subgenera B, C, D, and E modulate cell-surface expression of major histocompatibility complex class I antigens. Proc Natl Acad Sci USA 83:9665–9669
Perelygin AA, Scherbik SV, Zhulin IB, Stockman BM, Li Y, Brinton MA (2002) Positional cloning of the murine flavivirus resistance gene. Proc Natl Acad Sci USA 99:9322–9327
Peterson JK, King NJC, Kesson AM (2007) West Nile Virus Survival Curve Models. Available at: http://www.ces.clemson.edu/~petersj/CurrentPapers/SurvivalPaper.pdf
Piguet V, Steinman RM (2007) The interaction of HIV with dendritic cells: outcomes and pathways. Trends Immunol 28:503–510
Roehrig JT, Layton M, Smith P, Campbell GL, Nasci R, Lanciotti RS (2002) The emergence of West Nile virus in North America: ecology, epidemiology, and surveillance. Curr Top Microbiol Immunol 267:223–240

Romani N, Stingl G, Tschachler E, Witmer MD, Steinman RM, Shevach EM, Schuler G (1985) The Thy-1-bearing cell of murine epidermis. A distinctive leukocyte perhaps related to natural killer cells. J Exp Med 161:1368–1383

Rouas-Freiss N, Goncalves RM, Menier C, Dausset J, Carosella ED (1997) Direct evidence to support the role of HLA-G in protecting the fetus from maternal uterine natural killer cytolysis. Proc Natl Acad Sci USA 94:11520–11525

Samuel MA, Diamond MS (2005) Alpha/beta interferon protects against lethal West Nile virus infection by restricting cellular tropism and enhancing neuronal survival. J Virol 79:13350–13361

Samuel MA, Whitby K, Keller BC, Marri A, Barchet W, Williams BR, Silverman RH, Gale M, Jr, Diamond MS (2006) PKR and RNase L contribute to protection against lethal West Nile Virus infection by controlling early viral spread in the periphery and replication in neurons. J Virol 80:7009–7019

Sargent IL (1993) Maternal and fetal immune responses during pregnancy. Exp Clin Immunogenet 10:85–102

Scherret JH, Poidinger M, Mackenzie JS, Broom AK, Deubel V, Lipkin WI, Briese T, Gould EA, Hall RA (2001) The relationships between West Nile and Kunjin viruses. Emerg Infect Dis 7:697–705

Schneider BS, Soong L, Girard YA, Campbell G, Mason P, Higgs S (2006) Potentiation of West Nile encephalitis by mosquito feeding. Viral Immunol 19:74–82

Schwarzenberger K, Udey MC (1996) Contact allergens and epidermal proinflammatory cytokines modulate Langerhans cell E-cadherin expression in situ. J Invest Dermatol 106:553–558

Selin LK, Nahill SR, Welsh RM (1994) Cross-reactivities in memory cytotoxic T lymphocyte recognition of heterologous viruses. J Exp Med 179:1933–1943

Shen J, Devery JM, King NJ (1995a) Adherence status regulates the primary cellular activation responses to the flavivirus West Nile. Immunology 84:254–264

Shen J, Devery JM, King NJ (1995b) Early induction of interferon-independent virus-specific ICAM-1 (CD54) expression by flavivirus in quiescent but not proliferating fibroblasts–implications for virus-host interactions. Virology 208:437–449

Shen J, T-To SS, Schrieber L, King NJ (1997) Early E-selectin, VCAM-1, ICAM-1, and late major histocompatibility complex antigen induction on human endothelial cells by flavivirus and comodulation of adhesion molecule expression by immune cytokines. J Virol 71:9323–9332

Shimonkevitz R, Luescher B, Cerottini JC, MacDonald HR (1985) Clonal analysis of cytolytic T lymphocyte-mediated lysis of target cells with inducible antigen expression: correlation between antigen density and requirement for Lyt-2/3 function. J Immunol 135:892–899

Shrestha B (2002) A study of pathogenesis of West Nile virus encephalitis in the adult murine model, Ph.D. Thesis, The University of Sydney, Sydney

Shrestha B, Diamond MS (2004) Role of CD8+ T cells in control of West Nile virus infection. J Virol 78:8312–8321

Shrestha B, Samuel MA, Diamond MS (2006a) CD8+ T cells require perforin to clear West Nile virus from infected neurons. J Virol 80:119–129

Shrestha B, Wang T, Samuel MA, Whitby K, Craft J, Fikrig E, Diamond MS (2006b) Gamma interferon plays a crucial early antiviral role in protection against West Nile virus infection. J Virol 80:5338–5348

Silva MC, Guerrero-Plat A, Gilfoy FD, Garofalo RP, Mason PW (2007) Differential activation of human monocyte-derived and plasmacytoid dendritic cells by West Nile virus generated in different host cells. J Virol

Smithburn KC, Hughes TP, Burke AW, Paul JH (1940) A neurotropic virus isolated from the blood of a native of Uganda. Am J Trop Med 20:471–492

Springer TA (1990) Adhesion receptors of the immune system. Nature 346:425–434

Steinman RM, Hawiger D, Nussenzweig MC (2003) Tolerogenic dendritic cells. Annu Rev Immunol 21:685–711

Sugamata M, Miura T (1982) Japanese encephalitis virus infection in fetal mice at different stages of pregnancy I. Stillbirth. Acta Virol 26:279–282

Tafuri A, Alferink J, Moller P, Hammerling GJ, Arnold B (1995) T cell awareness of paternal alloantigens during pregnancy. Science 270:630–633

Tian L, King NJ (1994) Interferon gamma induces intercellular adhesion molecule-1 on murine midterm trophoblast and enhances susceptibility to specific lysis by paternally directed allo-immune cytotoxic T cells. Biol Reprod 51:1164–1172

Wacher C, Muller M, Hofer MJ, Getts DR, Zabaras R, Ousman SS, Terenzi F, Sen GC, King NJC, Campbell IL (2007) Coordinated regulation and widespread cellular expression of interferon-stimulated genes (ISG) ISG-49, ISG-54, and ISG-56 in the central nervous system after infection with distinct viruses. J Virol 81:860–871

Waithman J, Allan RS, Kosaka H, Azukizawa H, Shortman K, Lutz MB, Heath WR, Carbone FR, Belz GT (2007) Skin-derived dendritic cells can mediate deletional tolerance of class I-restricted self-reactive T cells. J Immunol 179:4535–4541

Wang T, Scully E, Yin Z, Kim JH, Wang S, Yan J, Mamula M, Anderson JF, Craft J, Fikrig E (2003a) IFN-gamma-producing gamma delta T cells help control murine West Nile virus infection. J Immunol 171:2524–2531

Wang Y, Lobigs M, Lee E, Mullbacher A (2003b) CD8+ T cells mediate recovery and immunopathology in West Nile virus encephalitis. J Virol 77:13323–13334

Wang T, Town T, Alexopoulou L, Anderson JF, Fikrig E, Flavell RA (2004) Toll-like receptor 3 mediates West Nile virus entry into the brain causing lethal encephalitis. Nat Med 10:1366–1373

Wang T, Gao Y, Scully E, Davis CT, Anderson JF, Welte T, Ledizet M, Koski R, Madri JA, Barrett A, et-al. (2006) Gamma delta T cells facilitate adaptive immunity against West Nile virus infection in mice. J Immunol 177:1825–1832

Wright KL, White LC, Kelly A, Beck S, Trowsdale J, Ting JP (1995) Coordinate regulation of the human TAP1 and LMP2 genes from a shared bidirectional promoter. J Exp Med 181:1459–1471

Yewdell JW, Bennink JR (1999) Mechanisms of viral interference with MHC class I antigen processing and presentation. Annu Rev Cell Dev Biol 15:579–606

Zhao X, Deak E, Soderberg K, Linehan M, Spezzano D, Zhu J, Knipe DM, Iwasaki A (2003) Vaginal submucosal dendritic cells, but not Langerhans cells, induce protective Th1 responses to herpes simplex virus-2. J Exp Med 197:153–162

Zuckermann FA, Head JR (1986) Expression of MHC antigens on murine trophoblast and their modulation by interferon. J Immunol 137:846–853

Zuniga EI, McGavern DB, Pruneda-Paz JL, Teng C, Oldstone MB (2004) Bone marrow plasmacytoid dendritic cells can differentiate into myeloid dendritic cells upon virus infection. Nat Immunol 5:1227–1234

15. Chemokines and Clearance of West Nile Virus Infection

ROBYN S. KLEIN

Abstract

West Nile virus (WNV) is a neurotropic flavivirus that can lead to fatal neuroinvasive infection. Recent studies indicate that innate and adaptive immune responses are critical for clearing infection in the periphery and preventing viral dissemination into the central nervous system (CNS). Clearance of WNV within the CNS compartment, however, specifically requires the infiltration of virus-specific T cells. Chemokines are chemoattractant molecules expressed early in the course of WNV infection that participate in leukocyte migration and activation both in the periphery and during CNS dissemination. As WNV leads to predominantly Th1 antiviral immune responses, chemokines that attract these cells might be particularly important in disease progression especially with regard to neuroinvasive disease. In this review, we focus on the role of chemokine members of the CXC or CC subfamilies. Among the CXC chemokines, the non-ELR group comprised of CXCL9–11, attract Th1 cells through the interaction with their receptor, CXCR3. Among the CC subfamily, Th1-associated chemokines include CCL3–5. These chemokines attract cells through an interaction with their receptor, CCR5. Secondary lymphoid and CNS tissue levels of all of these chemokines are elevated during primary infection with WNV, while homeostatic chemokines, including CXCL12 are decreased at the CNS microvasculature during WNV encephalitis. Among the proinflammatory chemokines, CXCL10 is dominantly expressed by WNV-infected neurons. Thus, chemokine gradients tightly regulate T-cell entry into and within the CNS. In the future, manipulation of chemokine receptor activation might be used to control the trafficking of virus-specific T cells and improve viral clearance within the CNS in patients with WNV encephalitis

Keywords

Leukocyte trafficking, innate immunity, chemoattractants, central nervous system, viral clearance

1 Introduction

Leukocyte movement is an essential component of cell-mediated immunity; cells need to move from peripheral to lymphoid tissues and back to carry out their effector functions during immune surveillance and in response to pathogen invasion. These immune cells migrate in response to a network of chemoattractants that include classes of bioactive lipids, microbial and complement products, cytokines, and a large family of small, secreted proteins called chemokines (Kim, 2005). Chemokines are a superfamily of over 50 chemoattractant cytokines with their target cell specificity conferred by pertussis toxin (PTX) sensitive, $G_{\alpha i}$-coupled seven transmembrane glycoprotein chemokine receptors. Along with their small size (8–10 kDa), chemokines share 20–70% amino acid identity and the presence of four cysteine residues in conserved locations at their amino-termini that are both essential for their three-dimensional structure and allow categorization into four groups based on the spacing of the first two cysteine residues. Within these categories, two major families have been extensively characterized: the CXC chemokines, which attract neutrophils or lymphocytes, contain an amino acid between the first two cysteine residues while the CC chemokines, which attract monocytes, eosinophils, basophils, and lymphocytes with variable selectivity, do not. Besides the CC and CXC chemokines, there are two minor groups, CX_3C and C chemokines, which contain one and two members, respectively. CXC chemokines are subdivided into those with a motif of gluamate-leucine-arginine (ELR) prior to the CXC motif, which primarily attract neutrophils, and those without this motif, which primarily attract lymphocytes. The nomenclature of chemokines and their receptors was recently simplified to refer to the subfamily designation (C, CC, CXC, CX_3C), whether the molecule is a ligand (L) or receptor (R), and its order of discovery.

Chemokines are also categorized according to their pattern of expression. Among CC and CXC chemokines are a group of homeostatic chemokines, which are constitutively expressed by secondary lymphoid tissues and orchestrate the trafficking of leukocytes during immune surveillance (Goetzl et al., 2004). Additional chemokines are proinflammatory in that they are induced by inflamed tissues during immune responses to pathogen invasion and promote the movement of leukocytes during infectious diseases. In general, chemokines bind their receptors in a promiscuous fashion with most ligands binding multiple receptors and many receptors binding multiple ligands. Chemokines and their receptors are expressed by leukocytes and nonleukocytes, including endothelial cells and stromal cells. In the CNS, chemokines and their receptors are expressed by both neurons and glia, depending on the cell type, its activation and the extent of its differentiation (Klein et al., 2005b). Chemokines also bind heparan sulfate proteoglycans, which may participate in chemokine receptor signaling through concentration of ligand at cellular surfaces (Handel et al., 2005).

The role of cell-mediated immunity in the clearance of flavivirus infections is well appreciated (King et al., 2007). CD8$^+$ cytotoxic T cells directly lyse virally infected cells, produce additional antiviral cytokines, including TNF-α and IFN-γ, and also express chemokines (Melchjorsen et al., 2003). CD4$^+$ T cells, which participate in cross-presentation and other helper T-cell responses, also produce IFN-γ and chemokines. Chemokines are now recognized to participate in host antiviral responses not only through recruitment of virus-specific lymphocytes but by direct activation of leukocytes, leading to additional cytokine production and release of granula contents that kill virally infected cells. In certain circumstances, the expression of proinflammatory chemokines may also induce immunopathology within virally infected tissues beyond the cytopathic damage induced by viruses themselves.

In this chapter, we will discuss the roles of chemokines in antiviral defenses within lymphoid compartments, emphasizing those involved in the clearance of West Nile virus (WNV) in the periphery and within the CNS. WNV is divided into lineage 1 and 2 strains, which are isolated by geographic distribution with less virulent lineage 2 strains only found in sub-Saharan Africa and Madagascar (Hayes et al., 2005). Most of the data concerning chemokines and WNV infection has been gathered from experiments utilizing murine models infected with lethal and nonlethal lineage 1 strains. Additional studies examining chemokine expression in human patients infected with other flaviviruses will also be discussed.

2 Chemokines and Viral Infections

There are more than 30 virally encoded chemokines and chemokine receptors, whose function in derailing antiviral immune responses highlights the critical roles of chemokines in antiviral defenses (Vischer et al., 2006). Chemokines participate in every phase of antiviral immune responses (Esche et al., 2005; Piqueras et al., 2006) (Table 1). Levels of expression of homeostatic chemokines that direct the trafficking of T (CCL21) and B (CXCL13) lymphocytes into secondary lymphoid tissues are transiently down-regulated during primary viral infections (Mueller et al., 2007) whereas levels of CCL19, which recruits dendritic cells

Table 1. Chemokines involved in innate and adaptive responses to viral infections

Chemokine	Chemokine receptor	Target cell
Innate immune responses		
CCL7 (MCP-7)	CCR1, -2, -5	PMN, Mo, iDC
CL4 (MIP-1β)	CCR5	NK
CL20 (MIP-3α)	CCR6	mT, B, iDC
XCL1 (GRO-1/NAP-3/MSGA-α)	CXCR1, -2	PMN
XCL2 (MIP-2α/GRO-b/GRO2-)	CXCR2	PMN
XCL8 (IL-8)	CXCR1, -2	aT, NK
XCL14 (BRAK)	?	Mo
XCL16	CXCR6	aT
Effector memory responses		
CCL2 (MCP-1)	CCR2	Mo, aT, NK, iDC
CCL3 (MIP-1α)	CCR1, -5	Mo, aT, NK, iDC
CCL4 (MIP-1β)	CCR5	Mo, aT, NK, iDC
CCL5 (RANTES)	CCR1, -3, -5	Mo, aT, NK, iDC
CXCL9 (Mig)	CXCR3	aT, NK
CXCL10 (IP-10)	CXCR3	aT, NK
CXCL11 (I-TAC)	CXCR3, CXCR7	aT, NK
Primary immune responses		
CCL19 (MIP-3β)	CCR7	T, mDC
CCL22 (MDC/STCP-1)	CCR4	aT, iDC
CXCL12 (SDF-1)	CXCR4, CXCR7	All leukocytes
CXCL13 (BLC)	CXCR5	B

Abbreviations: PMN polymorphonuclear (neutrophils), *aT* activated T cell, *NK* natural killer cells, *Mo* monocytes, *iDC* immature dendritic cells, *mDC* mature dendritic cells, *T* resting T cell, *mT* memory T cell, *B* B cell.

(DCs) into lymph nodes (LNs), are increased (Choi et al., 2003). The proinflammatory chemokines expressed by virus-infected tissues are those that recruit NK and activated T cells (CCL2–5, CCL8, CXCL9, CXCL10) via the chemokine receptors CXCR3, CCR1, and CCR5, which are up-regulated by T cells after antigen encounter and TCR signaling (Glass and Lane, 2003; Sarkar et al., 2003; Trifilo et al., 2003). In general, production of CC chemokines appears to dominate over CXC chemokines and the pattern of CC expression during viral infections appears to differ depending on the type of virus and target cell. For example, while influenza and RSV both infect lung tissues, only RSV-infected lungs express CCL3 and CCL5 (Haeberle et al., 2001). In contrast, CXC chemokines such as CXCL10 are expressed by a variety of cells infected during many viral infections including those caused by encephalitic flaviviruses (Klein et al., 2005a; Lepej et al., 2007; Silva et al., 2007; Winter et al., 2004). CXCL10 and its sister chemokines CXCL9 and CXCL11 are non-ELR chemokines that all bind the receptor CXCR3. As this receptor is up-regulated in all activated T cells, the non-ELR chemokines are essential for mounting an antiviral defense

(Liu et al., 2001). In this section we will discuss the patterns of chemokine expression that occur in secondary lymphoid tissues during infection with WNV and how these contribute to antiviral defenses in the periphery.

2.1 Chemokines Participate in Innate and Adaptive Immune Responses During WNV Infection

Following subcutaneous inoculation, WNV undergoes limited replication in the skin (Garcia-Tapia et al., 2007; Schneider et al., 2007), infecting monocyte-derived DCs (mDCs) that reside within the dermis. In vitro studies indicate that mDCs produce moderate amounts of IFN-α and CXCL8 in response to WNV grown in mosquito cells (Silva et al., 2007). IFN-α induces the expression of PKR, which inhibits viral translation (Gilfoy and Mason, 2007) and 2 ,5 -olgoadenylate synthetase, which activates latent RNase L to degrade viral and cellular RNAs, inhibiting viral replication and spread to adjacent cells (Kajaste-Rudnitski et al., 2006; Lucas et al., 2003; Mashimo et al., 2002; Perelygin et al., 2002; Samuel et al., 2006b; Scherbik et al., 2006). CXCL8 expression is downstream of toll-like receptors (TLRs), pattern recognition receptors (PRRs) that are directly activated by proteins and nucleic acids related to pathogen invasion (Sabroe et al., 2005). CXCL8 binds the receptors CXCR1 and CXCR2, which are present on neutrophils, and has been associated with neutrophilic meningitis in about half of the patients who present with WNV neuroinvasive disease (Jeha et al., 2003). It is unclear whether CXCL8 expression occurs in the periphery and plays any role in viral clearance within secondary lymphoid tissues. WNV-infected mDCs then migrate from the epidermis to local LNs (Byrne et al., 2001), presumably via up-regulation of CCR7 (MartIn-Fontecha et al., 2003), and participate in antigen presentation and induce adaptive immune responses, including the activation and expansion of B and T cells (Diamond et al., 2003a, b). The activation of lymphocyte effector functions are critical in the defense against severe WNV infection as humoral immune responses control viral load in the periphery and prevent dissemination whereas T-cell responses are more required for the clearance of WNV specifically within the CNS.

The delivery of infectious WNV LNs may lead to infection of other leukocytes including plasmacytoid DCs (pDCs) and monocytes (Silva et al., 2007). Following pathogen-mediated up-regulation of IFN, pDCs differentiate into mature DCs capable of activating and modulating T-cell responses via chemokine-mediated recruitment of leukocytes (Piqueras et al., 2006). WNV grown in mammalian cells has been shown to induce high levels of expression of IFN-α and CXCL10 within pDCs in vitro. Type I IFN also activates NK cells, which produce IFN-γ, leading to up-regulation of CXCL9 and CXCL10, IFN-inducible chemokines that are expressed by many cell types during viral infections (Proost et al., 2004). Although these chemokines have been shown to recruit effector CD8 T cells, recent studies suggest they may play essential roles in counter-regulatory immune responses through the recruitment of CD4 and CD8 regulatory T cells. CXCL10 has been observed to participate in regulatory T-cell recruitment in the context of several inflammatory diseases including atherogenesis (Heller et al., 2006), inflammatory bowel disease (Kristensen et al., 2006) and experimental autoimmune encephalomyelitis (EAE) (Muller et al., 2007). Consistent with this, CD8 T cells derived from WNV-infected mice with targeted deletion of CXCL10 or its receptor CXCR3 express significantly higher levels of IFN-γ after ex vivo stimulation with an immunodominant D^b-restricted NS4B peptide (Purtha et al., 2007) than those derived from WNV-infected wild-type mice (Chan and Klein, unpublished results). In addition, while mice with targeted deletion of CXCL10 or CXCR3 are unable to clear WNV within the CNS compartment (see below), viral loads in the periphery are similar to their wild-type counterparts (Klein et al., 2005a; Zhang et al., 2008), suggesting these molecules are dispensable for WNV clearance outside the CNS.

CCL3–5 are additional chemokines that are invariably up-regulated with many types of viral infections including WNV (Shirato et al., 2004). Although studies indicate that these chemokines recruit CCR1- and/or CCR5-expressing effector mononuclear and dendritic cells within secondary lymphoid tissues after viral infections, mice with targeted deletion of CCR5 do not display altered ability to clear virus within the spleen, although they do exhibit impaired antiviral responses within the CNS (Glass et al., 2005) (see below). In contrast, targeted deletion of CCR1 does not affect the clearance of WNV at any site of infection (Glass et al., 2005). In general, WNV-mediated up-regulation of chemokines in peripheral tissues and within the CNS occurs predominantly with neuroinvasive, lineage 1 strains (Glass et al., 2005; Klein et al., 2005a; Shirato et al., 2004). Nonlethal, lineage 1 strains do not invade the CNS nor replicate there, and only minimally replicate in peripheral organs. Nevertheless, these strains lead to patterns of proinflammatory chemokine within secondary lymphoid tissues similar to those observed with neuroinvasive, lineage 1 strains (Table 2). An exception is CXCL14, a proinflammatory chemokine that recruits and, in the presence of prostaglandin-E2, activates immature DCs, monocytes, and NK cells

Table 2. Chemokines expressed during West Nile virus infection

Strain (all lineage 1)	Tissue site	Chemokine	Reference
NY99-6922	Spleen	CCL3–5, CXCL10	Shirato et al. (2004)
	Brain	CCL3–5, CXCL10	
Eg101	Spleen	CCL3–5, CXCL10, CXCL14	
	Brain	CCL3–5, CXCL10	
NY2000-3000.0259	Brain	CCL2–5, CCL7, CCL19, CCL21, CXCL9–12, CXCL14	Klein et al. (2005a)
NY99-35262	Brain	CCL2–5, CXCL1/2, CXCL9, –10	Glass et al. (2005)
NY99-35262-11	Primary microglia	CCL1–5, CXCL8, CXCL10	Cheeran et al. (2005)
	Primary astrocytes	CCL2, CCL5, CXCL10	
MO2002	Brain	CCL2, CCL12, CXCL 9, –10	Garcia-Tapia et al. (2007)
TX2002	Primary pDC	CXCL10	Silva et al. (2007)
	Primary mDC	CXCL10	

(Shirato et al., 2004). CXCL14 expression within secondary lymphoid tissues of mice infected with nonlethal lineage 1 WNV strain was increased to a greater extent than other chemokines (Shirato et al., 2004). CXCL14 is constitutively expressed at high levels in many normal tissues, where its cellular source is thought to be fibroblasts. Although CXCL14 had been shown to act as an antiangiogenic factor during tumorigenesis, little is known regarding the physiologic function of CXCL14 and its receptor has yet to be identified.

2.2 WNV-Mediated Regulation of Chemokine Expression

Many chemokine genes contain common as well as promoter-specific regulatory elements that are variably induced during viral infections (Fig. 1). This explains the largely overlapping yet distinctive patterns of expression observed during different viral infections. Although few studies have directly examined WNV-mediated pathways leading to chemokine expression in distinct cell types, several studies examining WNV activation of IFN pathways suggest it triggers predictable regulators of chemokine expression. Replication of WNV within cells is recognized through endosomal nucleic acid sensors, TLR-3 and -7, and dsRNA sensors, retinoic acid-inducible gene I (RIG-I) and melanoma-differentiation-associated gene 5 (MDA-5) (Kawai and Akira, 2007; Yoneyama et al., 2004). These pathogen recognition receptors (PRR) then activate transcription factors, such as interferon regulatory factors 3 and 7 (IRF-3 and IRF-7), leading

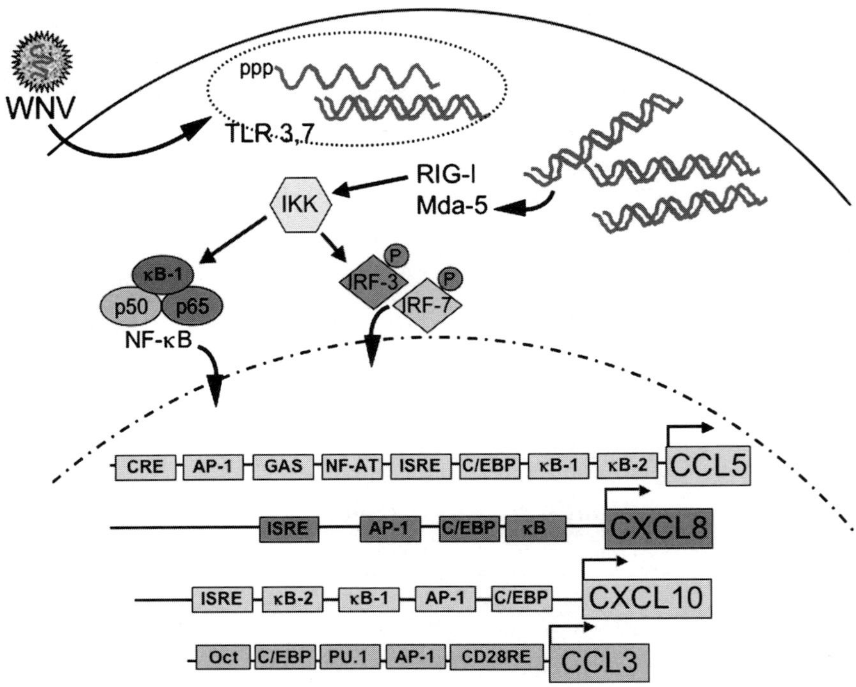

Figure 1. Viral RNA sensing pathways lead to chemokine expression during WNV infection. Replication of WNV within infected cells leads to generation of 5 -triphosphate viral RNA moieties that trigger the activation of several PRRs including endosomally based toll-like receptors (TLR)-3 and -7 and cytoplasmic dsRNA sensors, retinoic acid-inducible gene I (RIG-I) and melanoma-differentiation-associated gene 5 (MDA-5). RIG-1/MDA5-5 activates the inhibitor of IκB kinase (IKK), which leads to both the activation of the transcription factors interferon response factor (IRF)-3 and -7 and the dissociation of κB-1 from NF-κB. These transcription factors, in conjunction with those induced by cytokine-mediated pathways, migrate into the nucleus (*dotted line*) and lead to expression of proinflammatory chemokines CCL5, CXCL8, CXCL10, and CCL3. *Abbreviations: Cre* cyclic adenosine monophosphate response element, *AP-1* activator protein-1, *GAS* IFN-γ activation site, *NF-AT* nuclear factor of activated T cells, *ISRE* IFN-stimulated response element, *C/EBP* CCAAT enhancer-binding proteins, *κB-1*, *-2* NF-κB binding sites, *CD28RE* CD28 response element. (*See Color Plates*)

to the expression of IFNs and IFN-stimulated genes (ISG). Each of these PRRs demonstrates specificity for different RNA virus families with RIG-I, MDA5, and TLR-3 essential for IFN responses in response to flavivirus infections (Kato et al., 2006). Following infection, IRF-3 and IRF-7 are serine/threonine phosphorylated, which allows them to homo- or heterodimerize. IRF-3 is also downstream of the kinases TBK1 and IKKε (Silva et al., 2007), which, when activated induce phosphorylation of IκB by the IκB kinase, disassociating it from the proinflammatory transcription factor NF-κB. IRF-3/7 dimers and NF-κB migrate into the nucleus and synergistically promote the transcription of chemokines such as CCL5, CXCL8, and CXCL10 via stimulation of IFN-stimulated response elements (ISREs) and distal κB sites (Fig. 1) (Brat et al., 2005; Casola et al., 2001; Cheng et al., 1998; Leung et al., 2004; Lu et al., 2002; Servant et al., 2003).

Up-regulation of monocytes chemoattractants, such as CCL5 (Shirato et al., 2004), within WNV-infected lymphoid tissues may promote the recruitment of WNV-permissive cells, such as monocytes, which then relocate to the brain during normal physiologic turnover of perivascular macrophages (Lassmann et al., 1993), thus supporting viral spread within the CNS. In addition, CCL5 has been shown to prevent apoptosis of macrophages in the context of infections with pulmonary viruses (Tyner et al., 2005). CCL5 prosurvival effects may therefore directly promote the delivery of WNV-infected macrophages to the CNS.

2.3 Chemokines Shape Immune Responses to WNV Infection Within the CNS

Although in vitro studies indicate that glial cells may be infected with WNV (Cheeran et al., 2005), in vivo examination of WNV-infected CNS tissues detects viral antigen primarily within neurons. Neuronal targets of WNV include cortical, midbrain, cerebellar, and spinal cord neurons and, in rodent models with peripheral routes of inoculation, WNV dissemination occurs in a caudal to rostral direction beginning at about 4 days postinfection (Fratkin et al., 2004; Hunsperger and Roehrig, 2006; Samuel et al., 2006a; Shrestha et al., 2003). As WNV neuroinvasive disease occurs predominantly within immunocompromised hosts, whether due to age, genetic defect, hematologic malignancy or pharmacologic immunosuppression, intact cell-mediated immunity is required not only to prevent CNS invasion but also to clear the virus within the CNS compartment. In murine models, $CD8^+$ T cells traffic into WNV-infected CNS shortly after local viral replication, at about 5 days postinfection. Although initial studies suggested that $CD8^+$ T cells may both clear virus and induce neuropathology due to injury of WNV-infected neurons (Wang et al., 2003), studies using targeted deletion of T-cell chemoattractants to thwart leukocyte trafficking indicate that the CNS entry of virus-specific $CD8^+$ T cells is essential for clearance of WNV within the CNS.

Initial studies examining patterns of proinflammatory chemokine expression within the CNS of WNV-infected mice demonstrated significant up-regulation of all mononuclear chemoattractants (Glass et al., 2005; Klein et al., 2005a). Most of these chemokines peaked at about 1 week postinfection, coinciding with the timing of infiltrating mononuclear cells. Although the specific cellular sources of chemokines CCL2–5 have not been determined, in many CNS viral infections, glial cells initiate inflammation in response to cytokines released by antigen-specific T cells that enter the CNS in search of virally infected cells. In culture WNV-infected microglia, however, have been shown to express TNF-α, IL-6, CXCL8, CXCL10, CCL2, CCL4, and CCL5, as assessed by ELISA (Cheeran et al., 2005). Given that macrophages support WNV replication, it is possible that perivascular microglia which are similarly derived from myeloid lineage cells, may become infected with WNV and express proinflammatory responses that recruit virus-specific T cells.

CCL3–5 all bind the chemokine receptor CCR5, which is strongly up-regulated within the brain after WNV infection (Glass et al., 2005, 2006; Klein et al., 2005a; Shirato et al., 2004). CCR5, which is expressed by activated T cells and macrophages, not only plays an essential role in controlling infection with intracellular pathogens but also protects against postinfection inflammatory responses through participation in counter-regulatory T-cell responses. WNV-infected mice with CCR5 deletion had decreased trafficking of T cells, NK cells and macrophages, higher CNS viral loads and enhanced mortality (Glass et al., 2005). These studies did not examine the regional expression of CCR5 chemoattractants or the CNS compartmental trafficking patterns of CCR5-expressing leukocytes.

Increased expression of CXCL10, however, can be detected shortly after neuronal infection, in a pattern consistent with the caudal to rostral spread of WNV (Klein et al., 2005a). Thus, CXCL10 can be detected in the cerebellum by day 5 postinfection whereas cortical expression lags behind, with increased levels occurring at about 1 week postinfection (Klein et al., 2005a). This regional heterogeneity in CXCL10 expression is due to differential up-regulation of the chemokine by WNV-infected cortical versus cerebellar granule cell neurons and translates into increased trafficking of CXCR3-expressing T cells into the hindbrain versus the forebrain (Zhang et al., 2008). Loss of CXCL10 or CXCR3 via targeted deletion or antibody administration leads to decreased recruitment of WNV-specific $CD8^+$ T cells, increased viral loads and enhanced mortality. Thus, WNV-infected neurons directly induce the recruitment of virus-specific T cells for the purpose of viral clearance. The differential expression of CXCL10 by WNV-infected hindbrain versus forebrain neurons may reflect differences in the expression of viral sensing proteins. Given that CXC chemokines are believed to evolutionarily precede CC chemokines, it is logical that viral sensing pathways that regulate chemokine expression within neurons might also exhibit differential expression within evolutionarily older versus newer brain regions.

The CNS is a rich source of constitutive chemokines that are normally expressed by secondary lymphoid tissues where they function in regulating immune surveillance and primary immune responses. Many of these chemokines, including CXCL12 and CCL19, are detected within endothelial cells of the CNS microvasculature, where they are postulated to regulate the trafficking of leukocytes into the CNS parenchyma during neuroinflammatory diseases (Alt et al., 2002; Columba-Cabezas et al., 2003; McCandless et al., 2006; Pashenkov et al., 2003; Stumm et al., 2002). These hypotheses, however, are contrary to the inflammatory lesion observed during viral infections of the CNS, as most infiltrating leukocytes are prevented from entering the CNS parenchyma and instead are localized to perivascular spaces (Bouffard et al., 2004). The molecular basis for the formation of the observed perivascular infiltrate and its role in CNS immune responses is unknown. Although few studies have examined the CNS expression of constitutive chemokines during viral encephalitis, recent studies indicate that polarized expression of the chemokine CXCL12 at the microvasculature is responsible for this localization and regulates the parenchymal entry of infiltrating leukocytes during WNV encephalitis (McCandless et al., 2006, 2008).

CXCL12 is normally expressed by high endothelial venules of LNs and mediates the homing and localization of mononuclear cells to lymphoid compartments (Okada et al., 2002; Scimone et al., 2004). The release of effector lymphocytes from secondary lymphoid tissues is accomplished via alterations in the expression of several chemokine receptors, including the CXCL12 receptor, CXCR4 (Nombela-Arrieta et al., 2007; Yopp et al., 2004, 2005). Thus, the role of CXCL12 in retaining leukocytes within the perivascular spaces of the CNS microvasculature is consistent with its role in the periphery. In a recent study, mice with WNV encephalitis exhibited decreased levels of microvasculature CXCL12 expression when compared with uninfected counterparts, suggesting a mechanism for promoting leukocyte entry for viral clearance (Klein et al., 2005a). Consistent with this, pharmacological antagonism of CXCL12 signaling was observed to increase the numbers of parenchymal T cells, (McCandless et al., 2006) and, during WNV CNS infection, this improved viral clearance, decreased immunopathology and enhanced survival (McCandless et al., 2008). These data provide evidence that the parenchymal location of virus-specific T cells is essential for them to effectively clear virus and suggest that chemokine receptor antagonists may provide an immunotherapeutic approach to improve outcome during viral encephalitis. In addition, they suggest that, contrary to the role of CXCL12 within secondary lymphoid tissues, the movement of leukocytes into and out of CNS perivascular spaces relies on regulation of ligand expression rather than of its receptor.

In general, the regulation of CXCL12 expression is incompletely understood, however several studies have demonstrated that interactions between members of the TNF ligand/receptor superfamilies, including TNF-α/TNFR and CD40/CD154, up-regulate CXCL12 in various cell types (Jung et al., 2006; Nanki et al., 2000) while growth factors, such as granulocyte colony-stimulating factor (G-CSF) inhibit CXCL12 mRNA expression, especially within the bone marrow (Semerad et al., 2005). CD40 and TNFR-1 (p55) expression in the CNS is required for the development of CNS autoimmunity (Becher et al., 2001; Gimenez et al., 2004, 2006), and activation of CD40 within the CNS microvasculature is associated with the retention of myelin-specific T cells (Howard et al., 1999; Howard and Miller, 2001). Mice with targeted deletion of either CD40 or TNFR-1 exhibit decreased trafficking of $CD8^+$ T cells into the CNS and loss of viral control during WNV encephalitis (Sitati et al., 2007) (and unpublished results). Further studies are needed to determine whether these molecules exert their T-cell trafficking effects via regulation of CXCL12 expression.

3 Chemokine Studies in Human Cases of Flavivirus Encephalitis

WNV is related to ~70 other members of the *Flaviviridae* family, which includes several other clinically relevant viruses that cause encephalitis including St. Louis encephalitis virus (SLE), Japanese encephalitis virus (JEV), Murray Valley encephalitis virus (MVE) (Mukhopadhyay et al., 2005), and tick-borne encephalitis virus (TBEV) (Kindberg et al., 2008). WNV, SLE, JEV, and MVE comprise the neurotropic Japanese encephalitis (JE) serogroup. Studies of patients with meningitis and encephalitis due to flaviviral infections indicate roles for neutrophil and lymphocyte chemoattractants, which have been detected within the sera and cerebrospinal fluid (CSF) during clinical disease. JEV, for example, induces human peripheral blood monocytes to secrete a chemotactic cytokine, human macrophage-derived factor (hMDF), which causes chemotaxis of neutrophils (Singh et al., 2000).

Elevated levels of hMDF, CXCL8, and CCL5 may be detected within the sera of patients with JEV encephalitis and, in the case of CCL5, they were the highest in nonsurvivors (Shirato et al., 2004; Winter et al., 2004). In CSF studies, elevated levels of CXCL8 and CCL5 and CXCL10, -11 were detected in patients with encephalitides due to JEV and TBEV, respectively, with CXCL8 levels highest in nonsurvivors (Lepej et al., 2007; Winter et al., 2004). This inverse relationship between CSF CXCL8 and clinical outcome may relate to CXCL8-mediated BBB disruption through the induction of matrix metalloproteinases (Li et al., 2005). More curious, is the finding that patients with JEV encephalitis and high level of serum CCL5 were less likely to survive their infection (Winter et al., 2004). CCL5 binds the receptor CCR5 which, as with all chemokine receptor activation, leads to receptor desensitization and internalization (Oppermann, 2004). As discussed above, loss of CCR5 in murine models is associated with enhanced mortality due to WNV encephalitis (Glass et al., 2005). Studies of the relationship between severity of WNV infection and carriage of the *CCR5δ32* allele, a 32-bp deletion within the coding sequence of CCR5 that results in a complete loss of function of the receptor in homozygous individuals (Liu et al., 1996; Samson et al., 1996) revealed that homozygosity for the allele correlates with increased risk for symptomatic disease and fatal outcome (Glass et al., 2006; Lim et al., 2008). Thus, increased serum levels of CCL5 could possibly contribute to the loss of CCR5 function in the periphery, leading to decreased CNS leukocyte trafficking and viral control. Use of anti-CCL5 antibodies in severe cases with high serum CCL5 might be a therapeutic option to allow CCR5-expressing leukocytes to traffic into the CNS for the purpose of viral clearance.

4 Conclusions

Chemokines were initially discovered by immunologists and named for their critical functions in leukocyte movement during immune surveillance and inflammation. More recently, they have been recognized for their important roles in a variety of immunological responses to pathogen invasion including leukocyte activation and antiviral T-cell responses. As chemokine promoter systems are targeted by transcription factors activated by viral nucleic acids, chemokine expression is not only a significant component of early immune recognition of viral infection but is also an essential bridge between innate and adaptive immune responses in the periphery. The high level of expression of proinflammatory, T-cell chemoattractants with the CNS is essential for promoting the parenchymal entry of virus-specific T cells to clear virus during WNV encephalitis. However, the discovery that the CNS constitutively expresses relatively high levels of secondary lymphoid chemokines suggests they participate in the stringent regulation of leukocyte entry that maintains immune privilege at this site, which, in the case of viral encephalitis may negatively impact on the kinetics of viral clearance. The balance between excessive and effective leukocyte entry may thus rely on appropriate chemokine expression patterns within WNV-infected tissues.

Acknowledgements This work was supported by NIH/NINDS NS045607 and NS052632 and by a grant from the Midwest Regional Center for Excellence in Emerging Infectious Diseases.

References

Alt C, Laschinger M, Engelhardt B (2002) Functional expression of the lymphoid chemokines CCL19 (ELC) and CCL 21 (SLC) at the blood–brain barrier suggests their involvement in G-protein-dependent lymphocyte recruitment into the central nervous system during experimental autoimmune encephalomyelitis. Eur J Immunol 32:2133–2144

Becher B, Durell BG, Miga AV, Hickey WF, Noelle RJ (2001) The clinical course of experimental autoimmune encephalomyelitis and inflammation is controlled by the expression of CD40 within the central nervous system. J Exp Med 193:967–974

Bouffard JP, Riudavets MA, Holman R, Rushing EJ (2004) Neuropathology of the brain and spinal cord in human West Nile virus infection. Clin Neuropathol 23:59–61

Brat DJ, Bellail AC, Van Meir EG (2005) The role of interleukin-8 and its receptors in gliomagenesis and tumoral angiogenesis. Neuro Oncol 7:122–133

Byrne SN, Halliday GM, Johnston LJ, King NJ (2001) Interleukin-1beta but not tumor necrosis factor is involved in West Nile virus-induced Langerhans cell migration from the skin in C57BL/6 mice. J Invest Dermatol 117:702–709

Casola A, Garofalo RP, Haeberle H, Elliott TF, Lin R, Jamaluddin M, Brasier AR (2001) Multiple cis regulatory elements control RANTES promoter activity in alveolar epithelial cells infected with respiratory syncytial virus. J Virol 75:6428–6439

Cheeran MC, Hu S, Sheng WS, Rashid A, Peterson PK, Lokensgard JR (2005) Differential responses of human brain cells to West Nile virus infection. J Neurovirol 11:512–524

Cheng G, Nazar AS, Shin HS, Vanguri P, Shin ML (1998) IP-10 gene transcription by virus in astrocytes requires cooperation of ISRE with adjacent kappaB site but not IRF-1 or viral transcription. J Interferon Cytokine Res 18:987–997

Choi YK, Fallert BA, Murphey-Corb MA, Reinhart TA (2003) Simian immunodeficiency virus dramatically alters expression of homeostatic chemokines and dendritic cell markers during infection in vivo. Blood 101:1684–1691

Columba-Cabezas S, Serafini B, Ambrosini E, Aloisi F (2003) Lymphoid chemokines CCL19 and CCL21 are expressed in the central nervous system during experimental autoimmune encephalomyelitis: implications for the maintenance of chronic neuroinflammation. Brain Pathol 13:38–51

Diamond MS, Shrestha B, Marri A, Mahan D, Engle M (2003a) B cells and antibody play critical roles in the immediate defense of disseminated infection by West Nile encephalitis virus. J Virol 77:2578–2586

Diamond MS, Shrestha B, Mehlhop E, Sitati E, Engle M (2003b) Innate and adaptive immune responses determine protection against disseminated infection by West Nile encephalitis virus. Viral Immunol 16:259–278

Esche C, Stellato C, Beck LA (2005) Chemokines: key players in innate and adaptive immunity. J Invest Dermatol 125:615–628

Fratkin JD, Leis AA, Stokic DS, Slavinski SA, Geiss RW (2004) Spinal cord neuropathology in human West Nile virus infection. Arch Pathol Lab Med 128:533–537

Garcia-Tapia D, Hassett DE, Mitchell WJ Jr, Johnson GC, Kleiboeker SB (2007) West Nile virus encephalitis: sequential histopathological and immunological events in a murine model of infection. J Neurovirol 13:130–138

Gilfoy FD, Mason PW (2007) West Nile virus-induced interferon production is mediated by the double-stranded RNA-dependent protein kinase PKR. J Virol 81:11148–11158

Gimenez MA, Sim JE, Russell JH (2004) TNFR1-dependent VCAM-1 expression by astrocytes exposes the CNS to destructive inflammation. J Neuroimmunol 151:116–125

Gimenez MA, Sim J, Archambault AS, Klein RS, Russell JH (2006) A tumor necrosis factor receptor 1-dependent conversation between central nervous system-specific T cells and the central nervous system is required for inflammatory infiltration of the spinal cord. Am J Pathol 168:1200–1209

Glass WG, Lane TE (2003) Functional expression of chemokine receptor CCR5 on CD4(+) T cells during virus-induced central nervous system disease. J Virol 77:191–198

Glass WG, Lim JK, Cholera R, Pletnev AG, Gao JL, Murphy PM (2005) Chemokine receptor CCR5 promotes leukocyte trafficking to the brain and survival in West Nile virus infection. J Exp Med 202:1087–1098

Glass WG, McDermott DH, Lim JK, Lekhong S, Yu SF, Frank WA, Pape J, Cheshier RC, Murphy PM (2006) CCR5 deficiency increases risk of symptomatic West Nile virus infection. J Exp Med 203:35–40

Goetzl EJ, Wang W, McGiffert C, Huang MC, Graler MH (2004) Sphingosine 1-phosphate and its G protein-coupled receptors constitute a multifunctional immunoregulatory system. J Cell Biochem 92:1104–1114

Haeberle HA, Kuziel WA, Dieterich HJ, Casola A, Gatalica Z, Garofalo RP (2001) Inducible expression of inflammatory chemokines in respiratory syncytial virus-infected mice: role of MIP-1alpha in lung pathology. J Virol 75:878–890

Handel TM, Johnson Z, Crown SE, Lau EK, Proudfoot AE (2005) Regulation of protein function by glycosaminoglycans – as exemplified by chemokines. Annu Rev Biochem 74:385–410

Hayes EB, Sejvar JJ, Zaki SR, Lanciotti RS, Bode AV, Campbell GL (2005) Virology, pathology, and clinical manifestations of West Nile virus disease. Emerg Infect Dis 11:1174–1179

Heller EA, Liu E, Tager AM, Yuan Q, Lin AY, Ahluwalia N, Jones K, Koehn SL, Lok VM, Aikawa E, et-al. (2006) Chemokine CXCL10 promotes atherogenesis by modulating the local balance of effector and regulatory T cells. Circulation 113:2301–2312

Howard LM, Miller SD (2001) Autoimmune intervention by CD154 blockade prevents T cell retention and effector function in the target organ. J Immunol 166:1547–1553

Howard LM, Miga AJ, Vanderlugt CL, Dal Canto MC, Laman JD, Noelle RJ, Miller SD (1999) Mechanisms of immunotherapeutic intervention by anti-CD40L (CD154) antibody in an animal model of multiple sclerosis. J Clin Invest 103:281–290

Hunsperger EA, Roehrig JT (2006) Temporal analyses of the neuropathogenesis of a West Nile virus infection in mice. J Neurovirol 12:129–139

Jeha LE, Sila CA, Lederman RJ, Prayson RA, Isada CM, Gordon SM (2003) West Nile virus infection: a new acute paralytic illness. Neurology 61:55–59

Jung Y, Wang J, Schneider A, Sun YX, Koh-Paige AJ, Osman NI, McCauley LK, Taichman RS (2006) Regulation of SDF-1 (CXCL12) production by osteoblasts; a possible mechanism for stem cell homing. Bone 38:497–508

Kajaste-Rudnitski A, Mashimo T, Frenkiel MP, Guenet JL, Lucas M, Despres P (2006) The 2 ,5 -oligoadenylate synthetase 1b is a potent inhibitor of West Nile virus replication inside infected cells. J Biol Chem 281:4624–4637

Kato H, Takeuchi O, Sato S, Yoneyama M, Yamamoto M, Matsui K, Uematsu S, Jung A, Kawai T, Ishii KJ, et-al. (2006) Differential roles of MDA5 and RIG-I helicases in the recognition of RNA viruses. Nature 441:101–105

Kawai T, Akira S (2007) Antiviral signaling through pattern recognition receptors. J Biochem 141:137–145

Kim CH (2005) The greater chemotactic network for lymphocyte trafficking: chemokines and beyond. Curr Opin Hematol 12:298–304

Kindberg E, Mickiene A, Ax C, Akerlind B, Vene S, Lindquist L, Lundkvist A, Svensson L (2008) A deletion in the chemokine receptor 5 (CCR5) gene is associated with tickborne encephalitis. J Infect Dis 197:266–269

King NJ, Getts DR, Getts MT, Rana S, Shrestha B, Kesson AM (2007) Immunopathology of flavivirus infections. Immunol Cell Biol 85:33–42

Klein RS, Lin E, Zhang B, Luster AD, Tollett J, Samuel MA, Engle M, Diamond MS (2005a) Neuronal CXCL10 directs CD8+ T-cell recruitment and control of West Nile virus encephalitis. J Virol 79:11457–11466

Klein RS, Rubin JB, Luster AD (2005b) Chemokines and central nervous system physiology. In: Schwiebert LM (ed) Chemokines, chemokine receptors and disease, current topics in membranes, vol 55. Elsevier, San Diego, CA, pp 159–187

Kristensen NN, Brudzewsky D, Gad M, Claesson MH (2006) Chemokines involved in protection from colitis by CD4+CD25+ regulatory T cells. Inflamm Bowel Dis 12:612–618

Lassmann H, Schmied M, Vass K, Hickey WF (1993) Bone marrow derived elements and resident microglia in brain inflammation. Glia 7:19–24

Lepej SZ, Misic-Majerus L, Jeren T, Rode OD, Remenar A, Sporec V, Vince A (2007) Chemokines CXCL10 and CXCL11 in the cerebrospinal fluid of patients with tick-borne encephalitis. Acta Neurol Scand 115:109–114

Leung TH, Hoffmann A, Baltimore D (2004) One nucleotide in a kappaB site can determine cofactor specificity for NF-kappaB dimers. Cell 118:453–464

Li A, Varney ML, Valasek J, Godfrey M, Dave BJ, Singh RK (2005) Autocrine role of interleukin-8 in induction of endothelial cell proliferation, survival, migration and MMP-2 production and angiogenesis. Angiogenesis 8:63–71

Lim JK, Louie CY, Glaser C, Jean C, Johnson B, Johnson H, McDermott DH, Murphy PM (2008) Genetic deficiency of chemokine receptor CCR5 is a strong risk factor for symptomatic West Nile virus infection: a meta-analysis of 4 cohorts in the US epidemic. J Infect Dis 197(2): 262-5

Liu R, Paxton WA, Choe S, Ceradini D, Martin SR, Horuk R, MacDonald ME, Stuhlmann H, Koup RA, Landau NR (1996) Homozygous defect in HIV-1 coreceptor accounts for resistance of some multiply-exposed individuals to HIV-1 infection. Cell 86:367–377

Liu MT, Chen BP, Oertel P, Buchmeier MJ, Hamilton TA, Armstrong DA, Lane TE (2001) The CXC chemokines IP-10 and Mig are essential in host defense following infection with a neurotropic coronavirus. Adv Exp Med Biol 494:323–327

Lu R, Moore PA, Pitha PM (2002) Stimulation of IRF-7 gene expression by tumor necrosis factor alpha: requirement for NFkappa B transcription factor and gene accessibility. J Biol Chem 277:16592–16598

Lucas M, Mashimo T, Frenkiel MP, Simon-Chazottes D, Montagutelli X, Ceccaldi PE, Guenet JL, Despres P (2003) Infection of mouse neurones by West Nile virus is modulated by the interferon-inducible 2 -5 oligoadenylate synthetase 1b protein. Immunol Cell Biol 81:230–236

MartIn-Fontecha A, Sebastiani S, Hopken UE, Uguccioni M, Lipp M, Lanzavecchia A, Sallusto F (2003) Regulation of dendritic cell migration to the draining lymph node: impact on T lymphocyte traffic and priming. J Exp Med 198:615–621

Mashimo T, Lucas M, Simon-Chazottes D, Frenkiel MP, Montagutelli X, Ceccaldi PE, Deubel V, Guenet JL, Despres P (2002) A nonsense mutation in the gene encoding 2 -5 -oligoadenylate synthetase/L1 isoform is associated with West Nile virus susceptibility in laboratory mice. Proc Natl Acad Sci U S A 99:11311–11316

McCandless EE, Wang Q, Woerner BM, Harper JM, Klein RS (2006) CXCL12 limits inflammation by localizing mononuclear infiltrates to the perivascular space during experimental autoimmune encephalomyelitis. J Immunol 177:8053–8064

McCandless EE, Piccio L, Woerner BM, Schmidt RE, Rubin JB, Cross AH, Klein RS (2008) Pathologic expression of CXCL12 at the blood–brain barrier correlates with severity of multiple sclerosis. Am J Pathol 172(3): 799–808

McCandless EE, Zhang B, Diamond MS, Klein RS (2008) CXCR4 antagonism increases T cell trafficking in the central nervous system and improves survival from West Nile virus encephalitis. PNAS 105(32):11270–5

Melchjorsen J, Sorensen LN, Paludan SR (2003) Expression and function of chemokines during viral infections: from molecular mechanisms to in vivo function. J Leukoc Biol 74:331–343

Mueller SN, Hosiawa-Meagher KA, Konieczny BT, Sullivan BM, Bachmann MF, Locksley RM, Ahmed R, Matloubian M (2007) Regulation of homeostatic chemokine expression and cell trafficking during immune responses. Science 317:670–674

Mukhopadhyay S, Kuhn RJ, Rossmann MG (2005) A structural perspective of the flavivirus life cycle. Nat Rev Microbiol 3:13–22

Muller M, Carter SL, Hofer MJ, Manders P, Getts DR, Getts MT, Dreykluft A, Lu B, Gerard C, King NJ, Campbell IL (2007) CXCR3 signaling reduces the severity of experimental autoimmune encephalomyelitis by controlling the parenchymal distribution of effector and regulatory T cells in the central nervous system. J Immunol 179:2774–2786

Nanki T, Hayashida K, El-Gabalawy HS, Suson S, Shi K, Girschick HJ, Yavuz S, Lipsky PE (2000) Stromal cell-derived factor-1-CXC chemokine receptor 4 interactions play a central role in CD4+ T cell accumulation in rheumatoid arthritis synovium. J Immunol 165:6590–6598

Nombela-Arrieta C, Mempel TR, Soriano SF, Mazo I, Wymann MP, Hirsch E, Martinez AC, Fukui Y, von Andrian UH, Stein JV (2007) A central role for DOCK2 during interstitial lymphocyte motility and sphingosine-1-phosphate-mediated egress. J Exp Med 204:497–510

Okada T, Ngo VN, Ekland EH, Forster R, Lipp M, Littman DR, Cyster JG (2002) Chemokine requirements for B cell entry to lymph nodes and Peyer's patches. J Exp Med 196:65–75

Oppermann M (2004) Chemokine receptor CCR5: insights into structure, function, and regulation. Cell Signal 16:1201–1210

Pashenkov M, Soderstrom M, Link H (2003) Secondary lymphoid organ chemokines are elevated in the cerebrospinal fluid during central nervous system inflammation. J Neuroimmunol 135:154–160

Perelygin AA, Scherbik SV, Zhulin IB, Stockman BM, Li Y, Brinton MA (2002) Positional cloning of the murine flavivirus resistance gene. Proc Natl Acad Sci U S A 99:9322–9327

Piqueras B, Connolly J, Freitas H, Palucka AK, Banchereau J (2006) Upon viral exposure, myeloid and plasmacytoid dendritic cells produce 3 waves of distinct chemokines to recruit immune effectors. Blood 107:2613–2618

Proost P, Verpoest S, Van de Borne K, Schutyser E, Struyf S, Put W, Ronsse I, Grillet B, Opdenakker G, Van Damme J (2004) Synergistic induction of CXCL9 and CXCL11 by Toll-like receptor ligands and interferon-gamma in fibroblasts correlates with elevated levels of CXCR3 ligands in septic arthritis synovial fluids. J Leukoc Biol 75:777–784

Purtha WE, Myers N, Mitaksov V, Sitati E, Connolly J, Fremont DH, Hansen TH, Diamond MS (2007) Antigen-specific cytotoxic T lymphocytes protect against lethal West Nile virus encephalitis. Eur J Immunol 37:1845–1854

Sabroe I, Jones EC, Whyte MK, Dower SK (2005) Regulation of human neutrophil chemokine receptor expression and function by activation of Toll-like receptors 2 and 4. Immunology 115:90–98

Samson M, Libert F, Doranz BJ, Rucker J, Liesnard C, Farber CM, Saragosti S, Lapoumeroulie C, Cognaux J, Forceille C, et-al. (1996) Resistance to HIV-1 infection in caucasian individuals bearing mutant alleles of the CCR-5 chemokine receptor gene. Nature 382:722–725

Samuel MA, Morrey JD, Diamond MS (2007) Caspase-3 dependent cell death of neurons contributes to the pathogenesis of West Nile virus encephalitis. J Virol 81(6):2614–23
Samuel MA, Whitby K, Keller BC, Marri A, Barchet W, Williams BR, Silverman RH, Gale M Jr, Diamond MS (2006b) PKR and RNase L contribute to protection against lethal West Nile Virus infection by controlling early viral spread in the periphery and replication in neurons. J Virol 80:7009–7019
Sarkar S, Kalia V, Murphey-Corb M, Montelaro RC, Reinhart TA (2003) Expression of IFN-gamma induced CXCR3 agonist chemokines and compartmentalization of CXCR3+ cells in the periphery and lymph nodes of rhesus macaques during simian immunodeficiency virus infection and acquired immunodeficiency syndrome. J Med Primatol 32:247–264
Scherbik SV, Paranjape JM, Stockman BM, Silverman RH, Brinton MA (2006) RNase L plays a role in the antiviral response to West Nile virus. J Virol 80:2987–2999
Schneider BS, McGee CE, Jordan JM, Stevenson HL, Soong L, Higgs S (2007) Prior exposure to uninfected mosquitoes enhances mortality in naturally-transmitted west nile virus infection. PLoS One 2:e1171
Scimone ML, Felbinger TW, Mazo IB, Stein JV, Von Andrian UH, Weninger W (2004) CXCL12 mediates CCR7-independent homing of central memory cells, but not naive T cells, in peripheral lymph nodes. J Exp Med 199:1113–1120
Semerad CL, Christopher MJ, Liu F, Short B, Simmons PJ, Winkler I, Levesque JP, Chappel J, Ross FP, Link DC (2005) G-CSF potently inhibits osteoblast activity and CXCL12 mRNA expression in the bone marrow. Blood 106:3020–3027
Servant MJ, Grandvaux N, tenOever BR, Duguay D, Lin R, Hiscott J (2003) Identification of the minimal phosphoacceptor site required for in vivo activation of interferon regulatory factor 3 in response to virus and double-stranded RNA. J Biol Chem 278:9441–9447
Shirato K, Kimura T, Mizutani T, Kariwa H, Takashima I (2004) Different chemokine expression in lethal and non-lethal murine West Nile virus infection. J Med Virol 74:507–513
Shrestha B, Gottlieb D, Diamond MS (2003) Infection and injury of neurons by West Nile encephalitis virus. J Virol 77:13203–13213
Silva MC, Guerrero-Plata A, Gilfoy FD, Garofalo RP, Mason PW (2007) Differential activation of human monocyte-derived and plasmacytoid dendritic cells by West Nile virus generated in different host cells. J Virol 81:13640–13648
Singh A, Kulshreshtha R, Mathur A (2000) An enzyme immunoassay for detection of Japanese encephalitis virus-induced chemotactic cytokine. J Biosci 25:47–55
Sitati E, McCandless EE, Klein RS, Diamond MS (2007) CD40-CD40 ligand interactions promote trafficking of CD8+ T Cells into the brain and protection against West Nile virus encephalitis. J Virol 81:9801–9811
Stumm RK, Rummel J, Junker V, Culmsee C, Pfeiffer M, Krieglstein J, Hollt V, Schulz S (2002) A dual role for the SDF-1/CXCR4 chemokine receptor system in adult brain: isoform-selective regulation of SDF-1 expression modulates CXCR4-dependent neuronal plasticity and cerebral leukocyte recruitment after focal ischemia. J Neurosci 22:5865–5878
Trifilo MJ, Bergmann CC, Kuziel WA, Lane TE (2003) CC chemokine ligand 3 (CCL3) regulates CD8(+)-T-cell effector function and migration following viral infection. J Virol 77:4004–4014
Tyner JW, Uchida O, Kajiwara N, Kim EY, Patel AC, O'Sullivan MP, Walter MJ, Schwendener RA, Cook DN, Danoff TM, Holtzman MJ (2005) CCL5-CCR5 interaction provides antiapoptotic signals for macrophage survival during viral infection. Nat Med 11:1180–1187
Vischer HF, Hulshof JW, de Esch IJ, Smit MJ, Leurs R (2006) Virus-encoded G-protein-coupled receptors: constitutively active (dys)regulators of cell function and their potential as drug target. Ernst Schering Found Symp Proc 2:187–209
Wang Y, Lobigs M, Lee E, Mullbacher A (2003) CD8+ T cells mediate recovery and immunopathology in West Nile virus encephalitis. J Virol 77:13323–13334

Winter PM, Dung NM, Loan HT, Kneen R, Wills B, Thu le T, House D, White NJ, Farrar JJ, Hart CA, Solomon T (2004) Proinflammatory cytokines and chemokines in humans with Japanese encephalitis. J Infect Dis 190:1618–1626

Yoneyama M, Kikuchi M, Natsukawa T, Shinobu N, Imaizumi T, Miyagishi M, Taira K, Akira S, Fujita T (2004) The RNA helicase RIG-I has an essential function in double-stranded RNA-induced innate antiviral responses. Nat Immunol 5:730–737

Yopp AC, Fu S, Honig SM, Randolph GJ, Ding Y, Krieger NR, Bromberg JS (2004) FTY720-enhanced T cell homing is dependent on CCR2, CCR5, CCR7, and CXCR4: evidence for distinct chemokine compartments. J Immunol 173:855–865

Yopp AC, Ochando JC, Mao M, Ledgerwood L, Ding Y, Bromberg JS (2005) Sphingosine 1-phosphate receptors regulate chemokine-driven transendothelial migration of lymph node but not splenic T cells. J Immunol 175:2913–2924

Zhang B, Chan YK, Lu B, Diamond MS, Klein RS (2008) CXCR3 mediates region-specific antiviral T cell trafficking within the central nervous system during West Nile virus encephalitis. J Immunol 180(4):2641–9

16. Persistence of West Nile Virus Infection in Vertebrates

ROBERT B. TESH AND SHU-YUAN XIAO

Abstract

This chapter reviews the experimental evidence for persistence of West Nile virus (WNV) in vertebrates. Results of studies in monkeys, hamsters, and birds all indicate that WNV can produce an asymptomatic persistent infection in immune competent vertebrates. In the case of hamsters, the most thoroughly studied model, the infected animals develop a persistent renal infection with chronic shedding of WNV in their urine. Similar reports with a diverse group of other flaviviruses suggest that persistent infection is not a rare phenomenon and that it may be a general characteristic of viruses in this genus.

Keywords

West Nile virus, flaviviruses, persistent infection, arbovirus, viral pathogenesis, renal tropism

1 Introduction

In an immune competent vertebrate host, infection with an arthropod-borne flavivirus generally produces a transient viremia followed by a rapid immune response and, if the animal survives, viral clearance (Monath and Tsai, 2002). In the case of West Nile virus (WNV), the initial infection results in a brief period of viremia, lasting about 3–6 days depending on the animal species (Xiao et al., 2001; Komar et al., 2003; Ratterree et al., 2004; Samuel and Diamond, 2006). The initial WNV infection initiates both innate and adaptive immune responses (Chambers and Diamond, 2003; Samuel and Diamond, 2006), which are usually sufficient to eliminate infectious virus from the blood during the first week of infection. Despite the absence of viremia and the presence of specific humoral antibodies (i.e., IgM), some vertebrates go on to develop neuroinvasive disease during the second week. But it has generally been assumed that if the host survives WNV infection, with or without evidence of neuroinvasive disease, then the virus has been controlled and probably eliminated. However, reports of persistent flavivirus infection *in vitro* and *in vivo* have raised questions as to whether the acute infection paradigm described above is the complete story.

Most studies of persistent flavivirus infection have been done *in vitro*, using cell culture methods (Dinitz et al., 2006). The phenomenon of flavivirus persistence in vertebrate cells in culture has usually been attributed to genetic changes (mutations) in the infecting virus, resulting in alterations of its phenotype or from defective viral particles or defective RNAs. It is not the intent of this chapter to review WNV persistence *in vitro*; instead the focus will be on the evidence for and the mechanism of WNV persistence in vivo.

2 WNV Persistent Infection in Nonhuman Primates

The first well-documented report of WNV persistence *in vivo* was published by Pogodina et al. (1983). These workers infected 61 rhesus monkeys with several different WNV strains by intracerebral (IC) or subcutaneous inoculation. Seven of the 52 (13.5%) IC infected monkeys developed fatal encephalitis; the remaining animals survived, although some of the surviving monkeys showed clinical signs of encephalitis but subsequently recovered. These surviving animals were sacrificed at various times after infection (up to 5½ months) and representative tissues were cultured for virus. WNV was recovered from tissues of some of the animals up to 167 days after infection, principally from brain, lymph nodes, spleen, and kidneys. At the time of death, the monkeys did not appear clinically ill and many had detectable levels of WNV-neutralizing and hemagglutination-inhibiting (HI) antibodies.

Pogodina et al. (1983) also observed that the phenotype of the virus changed during persistent infection. During the first 2 months after infection, WNV could be readily detected and recovered by IC inoculation of newborn mice or by plaque assay in chick embryo fibroblasts. After about 60 days, virus in the persistently infected monkeys no longer killed newborn mice and irregularly produced plaques in chick fibroblasts. However, the virus could still be detected by co-cultivation of trypsinized monkey organ cells on an indicator culture of pig embryo kidney cells. Although viral cytopathic effect was rarely noted, subsequent examination of the indicator cells by immunofluorescence or HI test demonstrated the presence of WNV antigen (Pogodina et al., 1983). No genetic studies were done with the virus isolates from the persistently infected macaques, as these techniques were not widely available at the time. Nonetheless, the results of this experiment indicated that WNV had persisted in some of the experimentally infected monkeys for up to 167 days and that the phenotype of the persistent virus had changed.

3 WNV Persistent Infection in Hamsters

Based on the findings of Pogodina et al. (1983), we decided to further investigate the phenomenon of WNV persistent infection, using a previously described hamster model of West Nile encephalitis (Xiao et al., 2001). All experiments were done in adult female golden hamsters (*Mesocricetus auratus*). In the initial studies, an infecting dose of 10^4 tissue culture infectious dose$_{50}$ units of a 1999 New York strain of WNV (NY385-99) was used (Xiao et al., 2001).

Following intraperitoneal (IP) inoculation of this WNV strain, hamsters develop a moderate viremia which lasts about 6 days (Fig. 1). IgM and HI antibodies start to appear about 5 days postinfection (PI) and rapidly increase in titer. Neutralizing antibodies can be detected in the animals' sera between the eighth and tenth day. WNV generally cannot be cultured from the hamsters' blood after the sixth day (Xiao et al., 2001).

During the first week of WNV infection, the hamsters show no signs of clinical illness; but starting from about day 7 or 8, the animals become lethargic, anorexic, and irritable. During the second week of

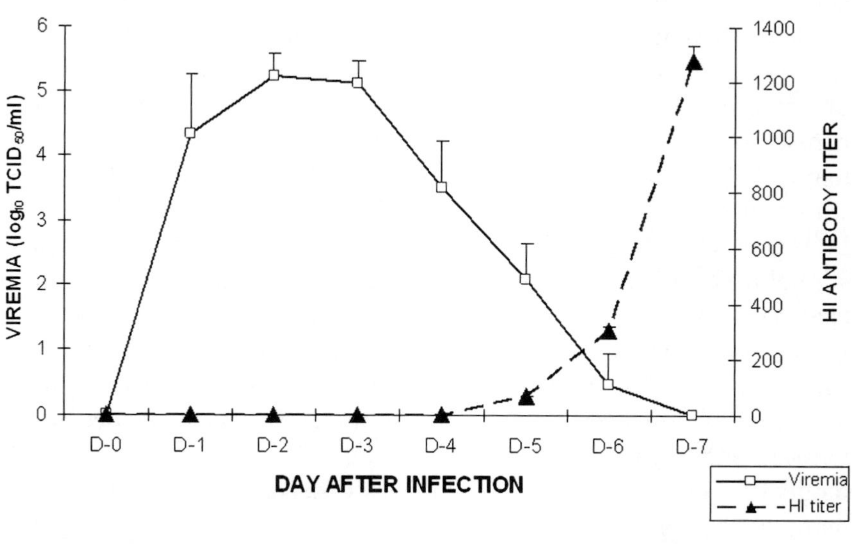

Figure 1. Daily mean level of viremia (standard deviation) and hemagglutination-inhibition (HI) antibody titers in hamsters following infection with WNV.

infection, many of the hamsters develop symptoms of encephalitis such as somnolence, muscle weakness and paralysis, loss of balance, and coma (Xiao et al., 2001; Morrey et al., 2004). During this period, 40–50% of the infected hamsters die. However, by the third week, the remaining animals begin to recover, gain weight and become active again.

WNV can be readily recovered by direct culture of homogenized brain and renal tissue taken at necropsy from animals dying during the second week (Xiao et al., 2001; Tonry et al., 2005). WNV can also be recovered by direct culture from the brain tissue of some of the surviving convalescent hamsters for up to 2 months after their initial infection with strain NY385-99, despite persistent high levels of HI and neutralizing antibodies in their sera (Xiao et al., 2001). We have been unable to isolate the virus by direct culture of brain or other tissues from the convalescent hamsters beyond the second month. In this regard, our results are in agreement with those reported earlier by Pogodina et al. (1983) in nonhuman primates.

Within 2 or 3 days after the initial WNV infection, infectious virus also appears in the hamsters' urine. The virus can be readily isolated by direct culture of freshly voided urine in Vero or BHK cells. The amount of infectious virus in the urine generally ranges from 10^2 to 10^4 plaque forming units (PFU) per milliliter. The viruria persists in the convalescent hamsters for many months, despite the absence of viremia and continuing high levels of WNV-neutralizing antibodies in their sera (Tesh et al., 2005). Some of the convalescent hamsters appear to clear their infection after several months, as culture of their urine eventually becomes negative; but other animals continue to shed infectious WNV in their urine for as long as 8 months after the infection (the longest period tested).

No significant or consistent histopathologic changes could be observed in the lungs, liver, spleen, or brain of these chronically viruric hamsters (Tesh et al., 2005). Starting from about week 20 after infection, many of the convalescent hamsters show dilatation of the renal tubules, with atrophy and flattening of the tubular epithelia, mostly in the cortical region. However, it is uncertain whether these changes are associated with chronic WNV infection, or are simply the result of aging. Similar changes (progressive glomerulonephropathy and tubular atrophy) have been reported to occur spontaneously in older female hamsters (Van Marck et al., 1978; Perry and Barthold, 2001). Immunohistochemical (IHC) staining of liver, spleen, lungs, brain, and urinary bladder of the viruric hamsters was negative for specific WNV antigen. In contrast, moderate to strong antigen staining was observed in kidney tissues of many of the convalescent viruric hamsters. WNV antigen staining was mostly in the interstitial area between the tubules in the renal papillae (Tesh et al., 2005). Reverse-transcription polymerase chain reaction (RT-PCR) performed on extracts from fragments of the IHC-positive hamster kidneys also confirmed the presence of WNV RNA.

Despite these indications of persistent WNV renal infection, repeated attempts to culture the virus directly from homogenates of the infected kidneys were unsuccessful. Consequently, we attempted the co-cultivation technique described earlier by Pogadina et al. (1983) and others (Shope et al., 1972). Trypsinized cells from fragments of fresh kidneys from the convalescent hamsters were inoculated onto monolayer cultures of Vero cells (Tesh et al., 2005). By this method, we were usually able to recover infectious WNV from kidneys of viruric convalescent hamsters. Viral CPE can be observed in the indicator Vero cells within 7–10 days after inoculation of the trypsinized hamster kidney cells. Once again, we were able to confirm the earlier Russian work (Pogodina et al., 1983).

4 Serial Passage of WNV in Hamster Urine

Following our initial finding that many hamsters inoculated with WNV develop a chronic renal infection with persistent viruria (Tonry et al., 2005; Tesh et al., 2005), we passaged infectious urine from a single infected hamster three additional times in hamsters, to determine if genotypic and/or phenotypic changes would occur in the virus with serial renal passage. Figure 2 summarizes the procedure followed. Initially ten hamsters were inoculated IP with 10^4 PFU of the parent WNV strain NY389-99. Fifty-four days later, infectious urine from one of the surviving experimentally infected hamsters (H9317B) was collected and inoculated IP into a second hamster (designated T-35639). Fifteen days later, urine from hamster T-35639 was collected, and a portion was cultured to confirm that WNV was present. A diluted sample of infectious urine from hamster T-35639 was then inoculated IP into a third hamster (H8535). A urine sample from hamster H8535 was collected 136 days later and was subsequently inoculated IP into a fourth hamster (H8912). Infectious urine was collected 195 days later from hamster H8912. In this manner, WNV strain was serially passaged through the kidneys of four different infected hamsters over a period of 400 days (Fig. 2). After each passage, some of the infectious hamster urine was cultured in Vero cells to confirm the presence of WNV and to prepare Vero stocks of virus from each passage for subsequent genetic and phenotypic analyses.

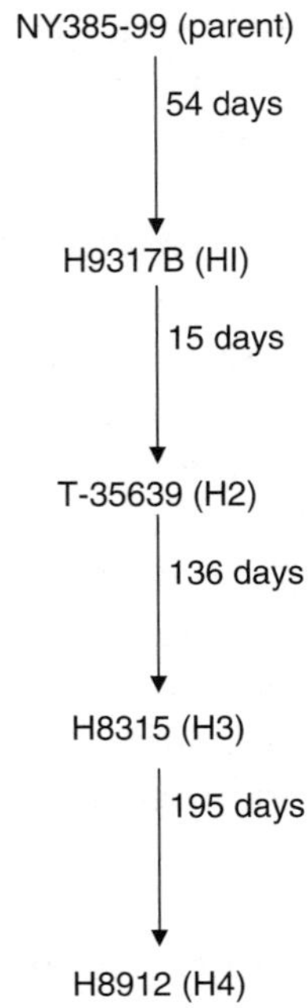

Figure 2. Flow chart of the serial passage of WNV strain NY385-99 through hamster urine to obtain the virus isolates used for experimental infections and sequence comparisons.

Despite their persistent WNV renal infection and chronic viruria, the experimentally infected hamsters had a normal antibody response initially. Table 1 shows the results of serial antibody determinations and urine cultures done on three persistently infected hamsters. When tested 7 days after infection, the animals had already developed IgM and HI antibodies. At day 18 after infection, the IgM antibody reading (OD_{405}) was lower but the HI titer was higher, reflecting a rising IgG antibody level. By day 95 after infection, the IgM ELISA was negative, but high levels of HI and neutralizing antibodies persisted for the duration of the experiment (8 months). Although only two dates are shown in Table 1 after the third week, multiple urine samples collected from these hamsters at various times between the third and eighth months consistently yielded WNV upon culture. When the three hamsters studied (numbers H8535, H8537, and H8538) were euthanized and necropsied at 242, 247, and 243 days after infection, respectively, WNV was cultured directly from urine aspirated from their bladders and by co-cultivation of trypsinized fragments of renal tissue (Tesh et al., 2005). The point to be emphasized here is that except for their persistent renal infection, the pattern of viremia observed in these hamsters after the initial WNV infection and their subsequent antibody responses otherwise conformed

Table 1. Results of West Nile virus antibody determinations [IgM ELISA, hemagglutination-inhibition (HI), and plaque reduction neutralization tests (PRNT)] and of urine cultures on three persistently infected hamsters at various times after infection

Day after WNV infection	Hamster number		
	H8535	H8537	H8538
Day 7			
IgM ELISA	2.50[a]	2.07	1.82
HI	1:20[b]	1:40	1:20
PRNT	NT	NT	NT
Urine culture	NT	NT	NT
Day 18			
IgM ELISA	1.23	1.57	0.71
HI	1:640	1:640	1:640
PRNT	NT	NT	NT
Urine culture	NT	NT	NT
Day 95			
IgM ELISA	0.13	0.12	0.12
HI	1:2,560	1:2,560	1:2,560
PRNT	NT	NT	NT
Urine culture	Positive	Positive	Positive
Day 242–247			
IgM ELISA	0.15	0.12	0.12
HI	1:2,560	1:2,560	1:640
PRNT	1:2,560[b]	1:2,560	1:2,560
Urine culture	Positive	Positive	Positive

NT not tested.
[a]Expressed as optical density at 405 nm at a 1:40 screening dilution.
[b]Highest positive serum dilution.

to the expected vertebrate host response to an acute arthropod-borne flavivirus infection (Monath and Tsai, 2002).

5 Genetic Changes Associated with Persistent Infection

To investigate a possible genetic basis for the phenotypic changes observed in the hamster-passaged viruses, full-length sequence was obtained from the parent WNV strain 385-99, several passage 1 viruses (Ding et al, 2005), including H9317B, passage 3 virus H8535, and passage 4 virus H8912 (Table 2) (Wu, et al., in press). When compared to the parent virus, there was sequence divergence ranging from 0.045 to 0.091% in the passage 1 progenies (Ding et al, 2005). In the first hamster passage (H9317B), two changes occurred in the E protein coding region; the remaining changes were in the nonstructural (NS) protein genes and the 3′-noncoding

Table 2. Nucleotide changes observed during serial hamster (urine) passages of West Nile virus strain YN385-99

	Virus strains				
Nucleotides	NY385-99	H9317B	H8535	H8912	Genome region
296	A	A	A	G	C[a]
315	C	C	C	T	C
348	T	T	T	C	C
1027	G	A	A	A	E[a]
1163	G	G	G	A	E[a]
1465	C	T	T	T	E[a]
1787	T	T	T	C	E[a]
1793	G	G	G	A	E[a]
2828	A	A	A	G	NS1[a]
3017	T	C	C	C	NS1[a]
3592	G	G	G	A	NS2A[a]
3708	G	G	A	G	NS2A
4297	G	G	G	A	NS2B[a]
4332	C	C	C	T	NS2B
4515	G	A	A	A	NS2B[a]
6405	T	C	C	C	NS3
7855	A	A	A	G	NS5[a]
7893	C	C	C	G	NS5
8235	T	C	C	C	NS5
8373	T	T	T	C	NS5
8849	A	G	G	G	NS5[a]
8856	T	T	T	C	NS5
9663	C	C	C	T	NS5
9972	C	C	C	T	NS5
10472	T	T	T	C	3′-NCR
10545	C	T	T	T	3′-NCR
10655	C	T	T	T	3′-NCR
10777	T	T	C	C	3′-NCR
10822	C	C	T		3′-NCR

Bold: changes in all urine isolates.
Note: shaded cells indicate de novo nucleotide changes from a previous passage Wu et al. (in press).
[a]Resulting in amino acid substitution.

region (NCR). Interestingly, by the third hamster passage H8535, only three additional changes had occurred (Table 2), suggesting that after passage through the kidney, the virus remained relatively stable genetically for two generations. These findings suggest that some of the nucleotide changes may be associated with a persistent renal infection phenotype and with loss of neurotropism (Ding et al., 2005). This must be confirmed by experiments using infectious clones of the parent virus and introducing individual or a combination of the nucleotide changes identified.

A question remains as to the nature of these nucleotide variations. They may represent *de novo* mutations of the parent strain, or they could be the result of "selection" from a pool of existing quasispecies in the parent virus stock. Most likely, both mechanisms play a role in this process. In the latter scenario, the kidney simply served as a "tissue sieve," which filtered and "amplified" the viruses possessing higher renal tropism but lacking neurotropism. If this were the case, then the sequences obtained from the urine isolates came from a genotype that was already present in the parental pool and which became dominant through the tissue selection process. Work is underway to directly sequence multiple plaques of viruses recovered from both brain and kidney tissues of single infected hamsters during the acute infection phase. More recent work (Wu et al. in press) with the hamster passage 3 (H8535) and passage 4 (H8912) viruses revealed that *de novo* mutations do occur, as shown by the many additional nucleotide changes observed in the passage 4 virus H8912 (Table 2) (Wu et al. in press). Interestingly, some of these mutations had been described by others in less virulent WNV strains (Davis et al., 2004; Granwehr et al., 2004).

6 Phenotypic Changes Associated with Persistent Infection

Experiments were also done to compare the pathogenesis (phenotype) of several of the hamster-passaged urine isolates of WNV with the parent strain. Four groups of adult hamsters were inoculated IP with the same dose (10^4 PFU) of each of the following WNV isolates: group 1 (n = 19 animals) received the parent virus NY385-99; group 2 (n = 10) received the second hamster passage (HP) urine isolate T35639; group 3 (n = 10) received the third HP isolate H8535; and group 4 (n = 10) received the fourth HP urine isolate H8912 (Fig. 2).

Figure 3 shows the mean level and duration of viremia in a sample of five hamsters from each of the four groups. Although the mean daily levels of viremia on days 3, 4, and 5 were slightly higher in group 1 than in group 4, overall the pattern of viremia and its duration were similar among the four groups, suggesting that persistent infection and renal passage did not significantly alter the ability of the virus to replicate or to produce viremia in hamsters. However, serial renal passage did appear to reduce neurovirulence of the virus. Seven of 19 (36.8%) hamsters inoculated with the parent virus (group 1) developed severe encephalitis during the second week and had to be euthanized. This is consistent with the death rates reported with WNV strain NY385-99 in hamsters before (Xiao et al., 2001; Morrey et al., 2004). In contrast, none of the 30 hamsters in groups 2, 3, or 4 became clinically ill or died, despite a comparable viremia. Some of the hamsters in each group were euthanized and

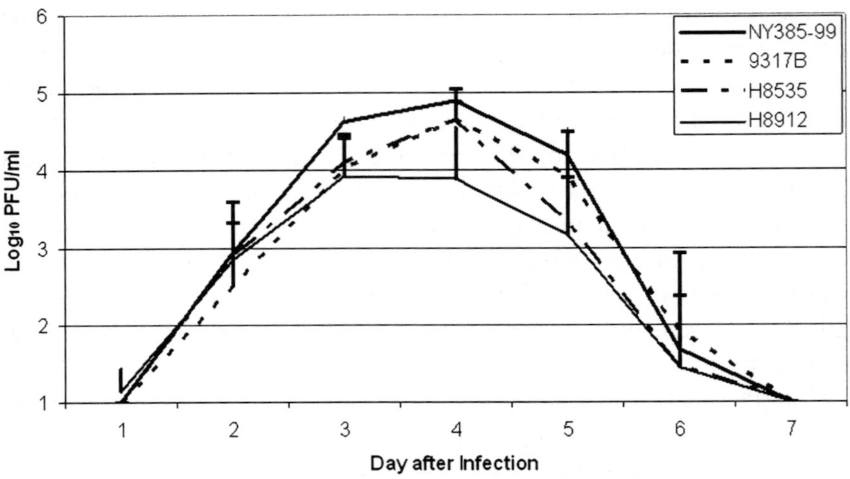

Figure. 3. Mean level of viremia in hamsters inoculated with various urine passages of WNV strain NY385-99.

Table 3. Virus load in blood, brain, and kidney of hamsters following infection with WNV

	Days postinfection	WNV titer			Comments
		Blood	Brain	Kidney	
Group 1 (NY 385-99)					
H1091	9	0[a]	4.3	4.9	Encephalitis[b]
H1092	10	0	3.9	3.2	Encephalitis
H1094	12	0	0	3.7	Asymptomatic
H1096	9	0	5.2	3.8	Encephalitis
H1097	11	0	0	4.7	Asymptomatic
Group 2 (T-35639)					
H1084	12	0	0	4.4	Asymptomatic
H1087	10	0	0	3.6	Asymptomatic
H1089	11	0	0	2.5	Asymptomatic
H1090	9	0	0	2.0	Asymptomatic
Group 3 (H-8535)					
H1062	10	0	0	3.9	Asymptomatic
H1063	11	0	0	2.9	Asymptomatic
H1065	12	0	0	3.2	Asymptomatic
Group 4 (H-8912)					
H1051	10	0	0	3.4	Asymptomatic
H1053	10	0	0	3.8	Asymptomatic
H1054	11	0	0	4.3	Asymptomatic
H1057	12	0	0	3.3	Asymptomatic

[a]WNV titer = $\log_{10}$ PFU/ml of blood or tissue homogenate. 0 = <0.7 PFU/ml.
[b]Animal had symptoms of severe clinical encephalitis when euthanized.

necropsied during the second week of infection to determine the distribution and level of virus in their blood, brain, and kidney (Table 3). None of the animals had detectable viremia, when they work euthanized. WNV was cultured from brain homogenates of three animals from group 1 (H1091, H1092, and H1096) that had clinical encephalitis, but it was not detected in the brain of any of the others. However, WNV was cultured from kidney homogenates of all of the hamsters sampled between the ninth and twelfth days after infection. While the sample sizes were small, these results suggest that although the three WNV strains isolated from hamster urine were less neurotropic than the parent virus; they retained their renal tropism.

Table 4 shows the pattern and duration of viruria in the surviving hamsters in the four groups. Urine was collected from the animals at

Table 4. Duration of viuria in hamsters following experimental infection with various West Nile virus strains

Animal number	Days after initial WNV infection				
	24	56	72	95	128
Group 1 (NY385-99)					
H1100	P	P	NS	NS	N
H1099	P	P	P	N	P
H1098	N	N	N	N	N
H1093	P	P	NS	N	N
Group 2 (T35639)					
H1088	P	P	P	N	N
H1086	P	P	P	NS	P
H1085	P	P	NS	NS	NS
H1083	P	NS	NS	P	P
H1082	P	P	P	P	P
H1081	P	N	P	P	N
Group 3 (H8535)					
H1080	P	NS	NS	NS	P
H1079	P	NS	P	P	NS
H1078	P	NS	NS	P	N
H1077	P	P	P	P	N
H1076	P	NS	NS	P	P
H1066	NS	NS	NS	P	NS
H1064	P	P	NS	NS	P
Group 4 (H8912)					
H1060	P	P	P	P	P
H1059	NS	P	NS	NS	P
H1058	P	P	NS	P	P
H1056	P	P	P	NS	N
H1055	P	P	NS	P	N
H1052	N	N	P	NS	P

P urine culture positive, yielding WNV; *N* urine culture negative; *NS* no urine sample obtained from hamster.

intervals of 24, 56, 72, 95, and 128 days after infection. To collect these samples, each hamster was held over a sterile plastic Petri dish and manual pressure was applied over the suprapubic area, inducing the animal to urinate (Tesh et al., 2005). The freshly voided urine was aspirated, diluted 1:10 in phosphate-buffered saline containing 25% fetal bovine serum, and then 200 μl of the diluted sample was inoculated into a flask culture of Vero cells. Although it was not possible to obtain a urine specimen from every animal on a given day, the pattern seen with most of the hamsters was chronic viruria. Some of the hamsters were viruric on each day sampled, while others (i.e., H1099, H1081, H1052) appeared to shed WNV intermittently. Previous studies (Tesh et al., 2005) found that infectious hamster urine contains about 10^2–10^4 PFU of WNV per milliliter. Most of the hamsters shown in Table 4 were still shedding WNV in their urine when tested 128 days after the initial infection, indicating persistent renal infection.

7 Evidence for Persistent Infection with Other Flaviviruses

Review of the literature suggests that other viruses in the genus *Flavivirus* also produce persistent infection in their vertebrate hosts and that this phenomenon is not limited to WNV. We recently demonstrated that hamsters experimentally infected with St. Louis encephalitis virus (SLEV) also develop persistent infection with chronic viruria (Siirin et al., 2007). Although the animals remained asymptomatic and developed high levels of specific neutralizing antibodies following infection, SLEV could be recovered by co-cultivation of tissues from kidney, lung, and brain of some of the hamsters for up to 185 days after initial infection. No specific histopathologic changes were observed in these tissues; but SLEV antigen was demonstrated by IHC in the interstitium and tubular epithelium of the renal cortex and in a few large neurons in the cerebral cortex. The pattern of antigen distribution in the kidneys of the SLEV persistently infected hamsters was similar to that observed in the hamsters chronically infected with WNV. Nucleotide sequencing of the envelope genes of the parent SLEV strain used to infect the hamsters and of 17 SLEV isolates from urine and tissues of the persistently infected animals indicated that all of the tissue and urine isolates had two common nucleotide mutations at the 1425 and 1430 positions, which caused amino acid changes at the 154 and 156 positions and which were different from the parent SLEV strain (Siirin et al., 2007). The analogy with the genetic changes noted in urine isolates from the WNV infected was striking.

Another example of persistent flavivirus infection is the Modoc virus (MODV), which naturally infects deer mice (*Peromyscus maniculatus*) in the western United States and Canada. The available information suggests that MODV is not arthropod-transmitted, but instead is maintained in deer mouse populations by horizontal and possibly vertical transmission (Johnson, 1970; Davis and Hardy, 1974; Davis et al., 1974). Laboratory studies have demonstrated that both deer mice and hamsters develop a persistent MODV infection and chronic viruria (Johnson, 1970; Davis and Hardy, 1974; Davis et al., 1974). Hamsters experimentally infected with MODV developed a brief viremia lasting 2–6 days, followed by the development of HI and neutralizing antibodies by day 7 after infection. Despite a brisk antibody response, MODV was chronically shed in the hamsters' urine for at least 12 weeks after infection. During this chronic phase, virus also could be isolated by co-cultivation of tissues from the kidneys, spleen, and lungs of the experimentally infected hamsters for up to 32 weeks after infection. Similar results were obtained with naturally and experimentally infected deer mice (Johnson, 1970; Davis et al., 1974).

A similar outcome was reported by Pogodina et al. (1981a, b) in experiments with rhesus monkeys infected with the Far Eastern subtype of tick-borne encephalitis virus (TBEV). TBEV could be isolated by co-cultivation from brain, liver, spleen, lymph nodes, and kidneys of rhesus macaques inoculated subcutaneously or intracerebrally with the virus, for up to 2 years after infection. Furthermore, the TBEV strains recovered from tissues of the persistently infected rhesus macaques lost their pathogenicity for mice and, in some cases, the ability to produce CPE in vertebrate cell cultures (Pogodina et al., 1981b). As observed with WNV, SLEV, and MODV (see above), TBEV persisted in certain tissues despite high levels of specific neutralizing antibodies in the convalescent monkeys' sera.

Other less well-documented reports indicate that persistent infection of the bat salivary glands occurs with Rio Bravo virus (Constatine and Woodall, 1964; Baer and Woodall, 1966); chronic renal infection and viuria with Omsk hemorrhagic fever virus in muskrats (Kharitonova and Leonov, 1985), and persistent Japanese encephalitis virus infection occurs in mice and in humans (Mathur et al., 1986; Sharms et al., 1991; Ravi. et al., 1993).

8 Summary and Speculation

Collectively, results of the aforementioned animal studies suggest that persistent infection with WNV is not uncommon. The fact that persistent infection seems to occur with a biologically and genetically diverse group of flaviviruses (WNV, SLEV, TBEV, MODV, and Rio Bravo virus) (Kuno et al., 1998) and in a variety of mammalian species also argues that it is not an isolated phenomenon. Persistent infection may be a principal mechanism by which some of these viruses are maintained in nature. In the case of Modoc and Rio Bravo viruses, two flaviviruses, which are not thought to have an arthropod vector, persistent infection with chronic shedding of the viruses in urine and saliva, respectively, would seem to be the probable modes of transmission. Persistent renal infection could also be one method by which WNV is maintained in nature.

Recent reports from investigators in California and New York (Reisen et al., 2006; Dawson et al., 2007) of sporadic crow deaths, during the winter season when adult mosquito activity is reduced or absent, has been interpreted as evidence of lateral (horizontal) transmission among these birds in their roosts. Crows experimentally infected with WNV shed large amounts of infectious virus in their cloacal excretions, and bird-to-bird transmission has been demonstrated among caged crows (Komar et al., 2003; Kipp et al., 2006). In birds, the cloaca is a common chamber into which the bladder, hindgut, and genital ducts all empty. Based on the results of animal studies described in this review, it seems plausible that some infected surviving crows (and possibly other less susceptible avian species) could develop a persistent renal infection and chronically shed the virus in their cloacal excretions. This material, in turn, could serve as a source of WNV to other birds, thereby eliminating the necessity of an infected mosquito to complete the virus transmission cycle from one vertebrate host to another.

There is experimental evidence that birds develop chronic WNV renal infection. Reisen et al. (2006) reported that 28 (34%) of 82 experimentally infected birds, representing six different avian species, had one or more organs positive for WNV RNA when tested by RT-PCR at necropsy >6 weeks after their initial infection. Infectious WNV was recovered by culture in Vero cells from four of six RT-PCR-positive house finches (*Carpodacus mexicanus*) after one blind passage of their organ extracts through mosquito cell (C6/36) cultures (Reisen et al., 2006). The WNV RNA-positive organs were spleen, kidney, and lung. Although cloacal cultures were not taken from these birds, the results are nonetheless remarkably similar to those described above with experimentally infected hamsters and monkeys. It has generally been assumed that once a vertebrate host has been infected with WNV and has developed humoral antibodies and recovered, the animal has eliminated the virus and is no longer infectious. Perhaps this paradigm should be reexamined.

One could begin by testing cloacal swabs from birds and urine from convalescing mammals, respectively, for infectious WNV.

The implication of persistent WNV infection for humans is less clear. There have been a few reports of renal involvement or persistent WNV infection in humans (Guarner et al., 2004; Penn et al., 2006; Armach et al., 2007). Most of the reported cases have been in immunocompromised patients who died, and a complete necropsy was done. But in immunosuppressed people, as in immunosuppressed hamsters (Mateo et al., 2006), the pathology and distribution of WNV antigen in the tissues is diffuse and quite different from the pattern seen in an immunocompetent host. As most humans infected with WNV survive, we really do not know whether some of these individuals had renal involvement and/or persistent infection. In view of the evidence from the aforementioned monkey and hamster studies, further investigation of this possibility may lead us to reevaluate our current concepts about the interaction of WNV and other flaviviruses with their vertebrate hosts.

Acknowledgements This work was supported by contracts NO1-AI25489 and NO1-AI30027 from the National Institutes of Health.

References

Armach HB, Wang G, Omalu BI, Tesh RB, Gyare KA, Chute DJ, Smith RD, Dulai P, Vinters HV, Kleinschmidt-McMasters BK, Wiley CA (2007) Systemic distribution of West Nile virus during encephalitis: postmortem immunohistochemical study of six cases. J Comp Int Soc Neuropath 17:354–362

Baer GM and Woodall DF (1966) Bat salivary gland virus carrier state in a naturally infected Mexican free tail bat. Am J Trop Med Hyg 15:769–771

Chambers TJ and Diamond MS (2003) Pathogenesis of flavivirus encephalitis. Adv Virus Res 60:273–342

Constatine DG and Woodall DF (1964) Latent infection of Rio Bravo virus in salivary glands of bats. Public Health Rep 79:1033–1039

Davis JW and Hardy JL (1974) Characterization of persistent Modoc viral infection in Syrian hamsters. Infect Immun 10:328–334

Davis JW, Hardy JL, Reeves WC (1974) Modoc viral infections in the deer mouse *Peromyscus maniculatus*. Infect Immun 10:1362–1369

Davis CT, Beasley DW, Guzman H, Siirin M, Parsons RE, Tesh RB, Barrett AD (2004) Emergence of attenuated West Nile virus variants in Texas, 2003. Virology 330:342–350

Dawson JR, Stone WB, Ebel GD, Young DS, Galinski DS, Pansabene JP, Franke MA, Eidson M , Kramer LD (2007) Crow deaths caused by West Nile virus during winter. Emerg Infect Dis 13:1912–1914

Ding X, Wu T, Duan T, Siirin M, Guzman H, Yang Z, Tesh RB and Xiao SY (2005) Nucleotide and amino acid changes in West Nile virus strains exhibiting renal tropism in hamsters. Am J Trop Med Hyg 73:803–807

Dinitz JAP, Travassos da Rosa APA, Guzman H, Xu F, Xiao S-Y, Popov VL, Vasconcelos PFC, Tesh RB (2006) West Nile virus infection of primary mouse neuronal and neuroglial cells: the role of astrocytes in chronic infection. Am J Trop Med Hyg 75:691–696

Granwehr BP, Li L, Davis CT, Beasley DW, Barrett AD (2004) Characterization of a West Nile virus isolate from a human on the Gulf Coast of Texas. J Clin Microbiol 42: 5375–5377

Guarner J, Shieh WJ, Hunter S, Paddock CD, Morken T, Cambell GL, Marfin AA, Zaki SR (2004) Clinicopathological study and laboratory diagnosis of 23 cases with West Nile encephalitis. Hum Pathol 35:983–990

Johnson HN (1970) Long-term persistence of Modoc virus in hamster kidney cells: *in vivo* and *in vitro* demonstration. Am J Trop Med Hyg 19:537–539

Kharitonova NN and Leonov YA (1985) Omsk hemorrhagic fever: ecology of the agent and epizootiology (in Russian). Amerind, New Delhi

Kipp AM, Lehman JA, Bowen RA, Fox PA, Stephens MR, Klenk K, Komar N and Bunning ML (2006) West Nile virus quantification in feces of experimentally infected American and fish crows. Am J Trop Med Hyg 75:688–690

Komar N, Langevin S, Hinten S, Nemeth N, Edward E, Hettler D, Davis B, Bowen R, Bunning M (2003) Experimental infection of North American birds with the New York 1999 strain of West Nile virus. Emerg Infect Dis 9:311–322

Kuno G, Chang GJJ, Tsuchiya KR, Karabatsos N , Cropp CB (1998) Phylogeny of the genus Flavivirus. J Virol 72:73–83

Mateo R, Xiao S-Y, Guzman H, Lei H, Travassos da Rosa APA, Tesh RB (2006) Effects of immunosuppression on West Nile virus infection in hamsters. Am J Trop Med Hyg 75:356–362

Mathur A, Arora KL, Rawat S, Chaturvedi UC (1986) Persistence, latency and reactivation of Japanese encephalitis in mice. J Gen Virol 67:381–385

Monath TP and Tsai TF (2002) Flaviviruses. In: Richman DD, Whitley RJ , Haydan FG (eds)Clinical virology, 2nd edn. ASM Press, Washington, pp 1097–1151

Morrey JD, Day CW, Julander JG, Olson AL, Sidwell RW, Cheney CD, Blatt LM (2004) Modeling hamsters for evaluating West Nile virus therapies. Antiviral Res 63:41–50

Penn RG, Guarner J, Sejvar JJ, Hartman H, McComb RD, Nevins DL, Bhatnagar J , Zaki SR (2006) Persistent neuroinvasive West Nile virus infection in an immunocomprised patient. Clin Infect Dis 42:680–683

Perry DH and Barthold SW (2001) Pathology of laboratory rodents and rabbits, 2nd edn. Iowa State Press, Ames, IA, p 168–196

Pogodina VV, Frolova MP, Malenko GV, Fokina GI, Levina LS, Mamonenko LL, Koreshkova GV, Ralf NM (1981a) Persistence of tick-borne encephalitis virus in monkeys. 1. Features of experimental infection. Acta Virol 25:337–343

Pogodina VV, Levina LS, Fokina GI, Koreshkova GV, Melenko GV, Bochkova NG, Rzhakhova OE (1981b) Persistence of tick-borne encephalitis virus in monkeys. 3. Phenotypes of the persisting virus. Acta Virol 25:352–360

Pogodina VV, Frolova MP, Malenko GV, Fokina GI, Koreshkova GV, Kiseleva LL, Bochkova NG , Ralph NM (1983) Study of West Nile virus persistence in monkeys. Arch Virol 75:71–86

Ratterree MS, Gutierrez RA, Travassos da Rosa APA, Dille JB, Beasley DW, Bohm RP, Desai SM, Didier PJ, Bikenmeyer LG, Dawson GL, Leary TP, Schochetmann G, Phillippi-Falkenstein, Arroyo J, Barrett ADT , Tesh RB (2004) Experimental infection of rhesus monkeys with West Nile virus: level and duration of viremia and kenetics of the antibody response. J Infect Dis 189:669–676

Ravi V, Desai AS, Shenov PK, Satishchandra P, Cahdramuki A, Gouri-Devi M (1993) Persistence of Japanese encephalitis virus in the human nervous system. J Med Virol 40:326–329

Reisen WK, Fang Y, Lothrop HD, Martinez VM, Wilson J, O'Connor P, Carney R, Cahoon-Young B, Shafii M, Brault AC (2006) Overwintering of West Nile virus in southern California. J Med Entomol 43:344–355

Samuel MA and Diamond MS (2006) Pathogenesis of West Nile virus infection: a balance between virulence, innate and adaptive immunity, and viral evolution. J Virol 80:9349–9360

Sharms S, Mathur A, Prakash V, Kulshreshtha R, Kumar R, Chaturvedi US (1991) Japanese encephalitis virus latency in peripheral blood lymphocytes and recurrence of infection in children. Clin Exp Immunol 85:85–89

Shope TC, Klein-Robbenhaar J, Miller H (1972) Fatal encephalitis due to *Herpesvirus hominus*: use of intact brain cells for isolation of virus. J Infect Dis 125:542–544

Siirin MT, Duan T, Lei H, Guzman H, Travassos da Rosa APA, Watts DM, Xiao S-Y, Tesh RB (2007) Chronic St. Louis encephalitis virus infection in the golden hamster (*Mesocricetus auratus*). Am J Trop Med Hyg 76:299–306

Tesh RB, Siirin M, Guzman H, Travassos da Rosa APA, Wu X, Duan T, Lei H, Nunes MR and Xiao S-Y (2005) Persistent West Nile virus infection in the golden hamster: studies on its mechanism and possible implications for other flavivirus infections. J Infect Dis 192:287–295

Tonry JH, Xiao S-Y, Sirrin M, Chen H, Travassos da Rosa APA , Tesh RB (2005) Persistent shedding of West Nile virus in urine of experimentally infected hamsters. Am J Trop Med Hyg 73:320–324

Van Marck EAE, Jacob W, Deelder AM, Gigase PII (1978) Spontaneous glomerular basement membrane changes in the golden Syrian hamster (*Mesocricetus auratus*): a light and electron microscope study. Lab Anim 12:207–211

Wu X, Lu L, Guzman, H, Tesh RB, and Xiao S-Y (2008) Persistent infection and associated nucleotide changes of West Nile virus serially passaged in hamsters. *J Gen Virol* (in press)

Xiao S-Y, Guzman H, Zhang H, Travassos da Rosa APA and Tesh RB (2001) West Nile virus infection in the golden hamster (*Mesocricetus auratus*): a model for West Nile encephalitis. Emerg Infect Dis 7:714–721

17. West Nile Virus Infection of the Central Nervous System

Melanie A. Samuel and Michael S. Diamond

Abstract

West Nile virus (WNV) infects thousands of humans annually and causes a spectrum of disease ranging from an acute febrile illness to paralysis and lethal encephalitis. This chapter will review our current understanding of the interactions between WNV pathognesis and the host response in the central nervous system that modulate disease outcome.

Keywords

Encephalitis, neuron, flavivirus, disease, immune response

1 Introduction

West Nile virus (WNV) has emerged as a significant agent of neurological infection and is now the most common cause of epidemic viral encephalitis in the United States. WNV is an arthropod-borne virus that cycles between mosquitoes and birds, but can also infect and cause disease in other vertebrate species (Hayes et al., 2005a; van der Meulen et al., 2005). In humans, WNV induces distinct clinical syndromes, ranging from an asymptomatic infection to a febrile illness that can progress to lethal neurological disease (Ceausu et al., 1997; Petersen and Marfin, 2002; Sejvar et al., 2003). Outbreaks of WNV fever and neuroinvasive disease have occurred more frequently since the 1990s in endemic regions throughout the world, including Africa, the Middle East, Europe, and Central and Western Asia (Dauphin et al., 2004). Following its emergence in the United States in 1999, WNV rapidly disseminated across North America and has been reported in Central and South America (Deardorff et al., 2006; Komar and Clark, 2006; Lanciotti et al., 1999). Annual epidemics of human WNV infection now occur in North America and have resulted in the largest outbreak of WNV encephalitis reported to date (Granwehr et al., 2004; Solomon, 2004).

Between 1999 and 2007, approximately 27,000 human WNV cases were diagnosed in the United States, although human blood supply analyses estimate millions of undiagnosed cases (Busch, 2006). Seroprevalence studies suggest that while the majority of human WNV infections are subclinical, 20–30% of infected individuals develop flu-like symptoms termed WNV fever (Petersen and Marfin, 2002; Watson et al., 2004). Approximately 1 in 150 develop neuroinvasive disease, with a resulting mortality rate of approximately 5–10% (Petersen and Roehrig, 2001; Weaver and Barrett, 2004). The elderly and immunocompromised are the most susceptible to WNV neurological infection, with a 20-fold increased risk of severe disease and death in those over age 50 (Chowers et al., 2001; Hayes et al., 2005a; Huhn et al., 2005; Nash et al., 2001). WNV neurological disease develops less frequently in children, though from 2002 to 2004 more than 300 pediatric cases were reported (Hayes and O'Leary, 2004). Other species, including birds and horses, are also susceptible to central nervous system (CNS) infection (van der Meulen et al., 2005). The *Corvidae* family of birds are especially vulnerable, with American crow populations in some US states reduced up to 45% (LaDeau et al., 2007).

The increase in neurological disease burden following the emergence of WNV in the Western Hemisphere has fostered extensive research. Experimental studies have provided key insights into the mechanisms of WNV CNS infection and the molecular determinates of neuroinvasive disease (Samuel and Diamond, 2006). This chapter will focus on the pathogenesis of WNV in the nervous system and host immune responses to CNS infection.

2 Clinical Features of WNV Neuroinvasive Disease

The clinical spectrum of WNV neuroinvasive disease is extensive (Table 1) and includes symptoms such as severe headache, cognitive dysfunction, muscle weakness or paralysis, seizures, and tremors that develop within 1–2 weeks after viral exposure (Campbell et al., 2002). WNV infection of the nervous system commonly presents as encephalitis, meningitis, or poliomyelitis-like acute flaccid paralysis. Overlapping disease features and mixed clinical criteria complicate the division between disease states, but in general 60% of individuals with WNV neurological disease develop encephalitis, while 40% are diagnosed with meningitis (Bode et al., 2006; Brilla et al., 2004; Nash et al., 2001). Subgroups within both categories develop acute flaccid paralysis,

Table 1. Clinical spectrum of WNV neuroinvasive disease

Symptom/Sign	Frequency (%)	References
Fever	85–100	Brilla et al. (2004), Klein et al. (2002), Nash et al. (2001), Sayao et al. (2004) and Sejvar et al. (2003)
Headache	50–100	Brilla et al. (2004), Chowers et al. (2001), Klee et al. (2004), Klein et al. (2002), Nash et al. (2001), Sayao et al. (2004) and Sejvar et al. (2003)
Generalized weakness/fatigue	20–82	Brilla et al. (2004), Klee et al. (2004), Klein et al. (2002), Nash et al. (2001), Sayao et al. (2004) and Sejvar et al. (2003)
Chills/rigors	50–85	Brilla et al. (2004) and Sejvar et al. (2003)
Nausea and vomiting	31–80	Brilla et al. (2004), Chowers et al. (2001), Klein et al. (2002), Nash et al. (2001), Sayao et al. (2004) and Sejvar et al. (2003)
Myalgia	10–80	Brilla et al. (2004), Chowers et al. (2001), Nash et al. (2001), Sayao et al. (2004) and Sejvar et al. (2003)
Neck pain or stiffness	19–80	Chowers et al. (2001), Klein et al. (2002), Nash et al. (2001) and Sejvar et al. (2003)
Parkinsonism (rigidity bradykinesia)	40–75	Sejvar et al. (2003)
Rash	19–57	Brilla et al. (2004), Chowers et al. (2001), Klein et al. (2002), Nash et al. (2001) and Sayao et al. (2004)
Uncoordinated gait/ataxia	57	Sayao et al. (2004)
Confusion/delirium	39–56	Brilla et al. (2004) and Chowers et al. (2001)
Altered mental status	28–48	Brilla et al. (2004), Chowers et al. (2001), Klee et al. (2004), Klein et al. (2002), Nash et al. (2001) and Sayao et al. (2004)
Diarrhea	14–43	Chowers et al. (2001), Klein et al. (2002), Nash et al. (2001) and Sayao et al. (2004)
Slurred speech	8–43	Nash et al. (2001) and Sayao et al. (2004)
Tremor	12–43	Nash et al. (2001) and Sayao et al. (2004)
Back pain	38–40	Sejvar et al. (2003)
Postural instability (imbalance)	0–50	Sejvar et al. (2003)
Somnolence (marked sleepiness)	38	Klein et al. (2002)
Myoclonus (involuntary twitching)	14–38	Sayao et al. (2004) and Sejvar et al. (2003)
Babinski sign	0–38	Sejvar et al. (2003)
Arthralgia (joint pains)	15–30	Klee et al. (2004) and Nash et al. (2001)
Facial weakness	28	Sayao et al. (2004)
Acute flaccid paralysis	14–27	Burton et al. (2004), Nash et al. (2001), Sayao et al. (2004) and Sejvar et al (2003)
Cervical pain	14	Klein et al. (2002)
Dizziness	14	Brilla et al. (2004)

(continued)

Table 1. (continued)

Symptom/Sign	Frequency (%)	References
Photophobia (sensitivity to light)	9–14	Klein et al. (2002) and Nash et al. (2001)
Coma	0–16	Burton et al. (2004) and Chowers et al. (2001)
Seizures	3–9	Brilla et al. (2004), Klein et al. (2002) and Nash et al. (2001)
Focal sensory change	7	Nash et al. (2001)
Abdominal pain	7	Nash et al. (2001)
Pharyngitis	5	Nash et al. (2001)
Lymphadenopathy	2–4	Chowers et al. (2001) and Klein et al. (2002)
Conjunctivitis	3	Klein et al. (2002)

with 10–50% of patients experiencing asymmetric muscle weakness (Davis et al., 2006b; Debiasi and Tyler, 2006). Ocular complications have been increasingly documented, and include altered vision associated with chorioretinitis, retinal hemorrhages, or optic neuritis (Bakri and Kaiser, 2004). Although rare, additional neurologic features can also develop, including demyelinating neuropathies (Omalu et al., 2003; Sejvar et al., 2005).

2.1 WNV Meningitis

Meningitis is defined as inflammation of the meninges, the membranes that cover and protect the surface of the CNS. In human and rodent histological studies, WNV antigen is excluded from both meningeal and endotheial cells (Guarner et al., 2004), suggesting that WNV meningitis does not involve direct infection of the meninges, but may occur secondary to parenchymal or parameningeal involvement. In patients diagnosed with WNV infection, WNV meningitis is defined by symptoms including headache, fever, stiff neck, or photophobia and cerebrospinal fluid (CSF) pleocytosis in the absence of muscle weakness or altered mental status. Most patients with WNV meningitis are hospitalized and recover relatively quickly, although some may experience persistent fatigue, weakness, and memory problems (Sejvar, 2007).

2.2 WNV Encephalitis

WNV encephalitis resulting from acute infection of the brain parenchyma occurs most commonly in older patients and is frequently associated with meningitis (19–57%) (Bode et al., 2006; Davis et al., 2006b). In 2006, the mean age for patients with neuroinvasive WNV disease was 58 years, and in the initial 1999 WNV outbreak 88% of individuals with neurological disease were older than 50 years (2007; Nash et al., 2001). WNV targets multiple and diverse regions in the CNS (see below), including the brain stem, hippocampus, cerebral cortex, cerebellum, and spinal cord (Fratkin et al., 2004; Kleinschmidt-DeMasters et al., 2004; Omalu et al., 2003). The relative involvement of distinct CNS areas in individual patients results in a spectrum of clinical features, ranging from mild disorientation to severe disease and death (Hayes et al., 2005b). The most common symptoms (>38%) of WNV encephalitis include fatigue or weakness, and altered mental status (Davis et al., 2006b). Individuals with WNV encephalitis also more frequently had memory problems, focal motor abnormalities, and speech disorders, whereas seizures, tremors, and limb weakness were observed equally in both encephalitis and meningitis patient groups (Bode et al., 2006).

In contrast to other encephalitides, neuromuscular manifestations are a common feature of WNV neuroinvasive disease. Parkinsonian movement disorders, including bradykinesia and rigidity, occur in up to two-thirds of reported cases, and cranial palsies involving the seventh cranial nerve are common (Saad et al., 2005; Sejvar et al., 2003). Dizziness and difficulty swallowing or speaking may result from brainstem and cranial nerve involvement (Davis et al., 2006b; Nash et al., 2001). Though rare in nonflavivirus encephalitis, myoclonic muscle twitching occurred in 33% of WNV infected patients in one study, while tremors have been reported in anywhere from 12% to 100% of WNV neuroinvasive disease cases (Nash et al., 2001; Sejvar et al., 2003). Seizures are less common (1–4%), whereas gait and balance abnormalities indicative of cerebellar involvement occur more frequently (Davis et al., 2006b). In addition to these physical abnormalities, WNV neuroinvasive disease patients often experience significant and sustained cognitive defects. Though comprehensive neurophysiological studies have been limited, patients are often characterized as confused or with altered consciousness (Brilla et al., 2004; Burton et al., 2004; Chowers et al., 2001).

2.3 Acute Flaccid Paralysis

WNV can infect and injure motor neurons in the anterior horn of the spinal cord, resulting in a poliomyelitis-like flaccid paralysis that is characterized by rapid onset (24–48 h after symptoms) and progression of asymmetric muscle weakness (Davis et al., 2006b). Unlike meningitis and encephalitis, humans of all age and immune status groups are at risk for developing acute flaccid paralysis (Debiasi and Tyler, 2006). Acute flaccid paralysis can occur alone or in conjunction with meningitis and/or encephalitis and affects up to 50% individuals with neuroinvasive WNV disease (Tyler, 2004). Although a WNV-associated Guillain–Barré-like syndrome causing sensory and motor defects has been reported (Ahmed et al., 2000), the majority (84%) of flaccid paralysis cases are associated with motor defects alone (Sejvar, 2006). Muscle weakness occurs most often in the lower extremities, but can also involve multiple limbs and result in bladder and bowel dysfunction (Debiasi and Tyler, 2006; Leis et al., 2003). Acute respiratory distress due to respiratory muscle motor neuron injury is the most severe manifestation of WNV-associated paralysis and occurs in up to one-third of affected patients (Sejvar, 2006, 2007).

2.4 Laboratory Findings and Diagnosis

WNV neurological infection can result in nonspecific clinical laboratory findings, including reduced lymphocyte and platelet counts in the blood and elevated lymphocyte numbers in the CSF (~200 cells/mm^3) (Davis et al., 2006b; Hayes et al., 2005b). However, definitive diagnosis requires detection of WNV-specific antibodies, nucleic acid, or infectious virus in the blood or CSF. Computed tomography (CT) scans generally do not show brain abnormalities (Nash et al., 2001; Sejvar et al., 2003). However, electroencephalographic abnormalities have been reported frequently (~50–80%) and commonly reflect slower activity of the anterior or temporal regions of the brain (Gandelman-Marton et al., 2003; Sejvar et al., 2003; Weiss et al., 2001). Magnetic resonance imaging (MRI) has also variably detected abnormalities in 2.5–70% of WNV neuroinvasive disease cases (Ali et al., 2005; Brilla et al., 2004; Petropoulou et al., 2005). When present, signal variations are most commonly identified in the basal ganglia, brain stem, and thalamus, though cortical, cerebellar, and spinal cord findings have been documented (Ali et al., 2005; Petropoulou et al., 2005; Sejvar et al., 2003). To better assess the neurological imaging characteristics of human WNV infection, an MRI registry has been established (Robertson and Sejvar, 2003).

2.5 Prognosis and Outcome

Between 1999 and 2007, 10,615 cases of acute WNV neuroinvasive disease in the United States were reported with 966 deaths (9% mortality rate), though in elderly individuals this frequency increases significantly (up to 35% mortality rate). While most patients survive acute infection, long-term neurological sequelae are common (>50%) and represent a significant source of morbidity (Klee et al., 2004; Sejvar, 2006, 2007). Patients with WNV neurological disease require longer hospital stays, experience more complications, and have prolonged recovery times compared to those with other WNV diseases (Sejvar, 2007). Movement disorders, such as tremors or parkinsonism, may resolve within weeks (Robinson et al., 2003) but can persist for months and were present in ~20% of patients for up to 1.5 years in recent outbreaks (Carson et al., 2006). Patients with WNV encephalitis frequently require placement in assisted-living facilities after hospitalization (75%) as well as intensive physical, speech, or occupational therapy (Bode et al., 2006; Patnaik et al., 2006). While many patients experience full cognitive recovery, persistent systems of fatigue, muscle weakness, headache, and memory problems are self-reported by 20–50% of patients up to one year after WNV infection (Gottfried et al., 2005; Klee et al., 2004; Ou and Ratard, 2005). Overall, patients with acute flaccid paralysis have the worst prognosis (Debiasi and Tyler, 2006; Sejvar et al., 2005). Recovery is generally correlated with preservation of >50% of motor units within the muscle, and significant persistent weakness often occurs (Cao et al., 2005).

2.6 Potential Therapies

At present, no specific therapy for WNV peripheral or neurological infection is available. Given the importance of neuronal death in the pathogenesis of WNV encephalitis (see below), pharmacologic strategies that target cell death pathways have the potential to mitigate disease (Samuel et al., 2007a). A key challenge will be to develop therapeutics that effectively cross the blood–brain barrier (BBB) and target infected neuronal populations.

3 Pathogenesis of WNV in the CNS

3.1 Neuropathology

WNV primarily replicates within neurons and is largely absent from endothelial cells, astrocytes, other glial cells, or cells lining the meninges (Armah et al., 2007; Guarner et al., 2004). Histological studies in human WNV encephalitis cases and experimentally infected rodents demonstrate focal WNV antigen staining predominantly in neurons of

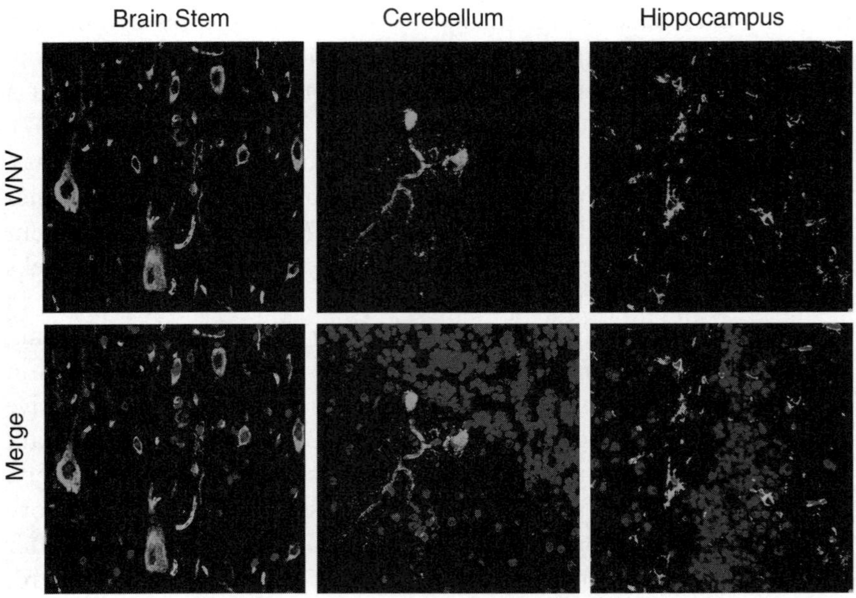

Figure 1. WNV antigen staining in infected CNS tissues. Mice were inoculated intradermally with WNV and immunohistochemistry was performed on brains at day 10 after infection for WNV antigen (*green*) and nuclear staining (*blue*). WNV was detected at high levels in neurons throughout the brain stem and hippocampus and in the Purkinje layer of the cerebellum. (*See Color Plates*)

the brainstem, hippocampus, and Purkinje layer of the cerebellum (Fig. 1) (Diamond et al., 2003a; Eldadah and Nathanson, 1967; Fratkin et al., 2004; Omalu et al., 2003; Shrestha et al., 2003; Xiao et al., 2001). Within the spinal cord, WNV antigen is most frequently detected in anterior horn motor neurons, though posterior horn sensory neurons may also be infected (Guarner et al., 2004). In severe cases of immunosuppression or insufficiency, fulminant infection may occur with WNV infection in neurons throughout the brain, suggesting that host factors restrict WNV tropism within the CNS (see below) (Guarner et al., 2004; Omalu et al., 2003; Samuel and Diamond, 2005). WNV infection induces neuronal injury and loss that may appear as apoptosis, or more rarely as necrosis and neuronophagia (Bouffard et al., 2004; Guarner et al., 2004). Perivascular cuffing, mononuclear cell infiltration, and glial nodules are also hallmarks of fatal WNV CNS infection. In one recent human study, inflammation composed primarily of B and T cells was noted in all cases (20/20) in the brainstem and spinal cord and in ~60% of cases in the cerebellum and cortex (Guarner et al., 2004).

3.2 Mechanisms of CNS Injury and Disease

Within the CNS, WNV infection may induce injury either directly through replication within neurons or indirectly through bystander or immune-mediated mechanisms. Studies to date suggest that immune-independent cell death due to viral infection may be the primary disease pathway of WNV encephalitis. Cell death generally occurs by two biologically distinct processes, apoptosis and necrosis. While apoptosis is a highly regulated process characterized by a defined sequence of biochemical and morphologic changes (Majno and Joris, 1995), necrosis involves cytoplasmic swelling and plasma membrane rupture (Griffin and Hardwick, 1999). In vitro, WNV is cytolytic in primary neurons and induces hallmarks of the apoptotic pathway, including caspase-3 activation, chromatin condensation, and exposure of phosphatidylserine on the outer leaflet of the plasma membrane (Parquet et al., 2001; Samuel et al., 2007a; Shrestha et al., 2003). Recent evidence suggests that host cell death pathways directly promote the pathological manifestations of WNV encephalitis in vivo. Neurons in the CNS of mice infected with WNV stain positively for TUNEL, which labels free ends of DNA generated during apoptosis (Samuel et al., 2007a; Shrestha et al., 2003). Moreover, in mice, a genetic deficiency of caspase-3 significantly improved the survival of WNV-infected neurons in vitro and in vivo (Samuel et al., 2007a).

While apoptosis appears to be a general pathway by which WNV induces neuronal death, apoptotic programs may vary in different neuronal subtypes. A genetic deficiency of caspase-3 did not protect all CNS neurons equivalently, indicating that additional caspases or other proteases may have dominant functions within diverse neuronal populations. In addition, different neurons show distinct survival kinetics after WNV infection in vitro: CNS-derived neurons largely die by day 4 after WNV infection whereas peripheral neurons survive for up to 2–4 weeks (Hunsperger and Roehrig, 2005; Samuel and Diamond, 2005; Samuel et al., 2007a; Shrestha et al., 2003). Thus, the molecular mechanisms by which WNV induces injury and death of distinct neuronal subtypes warrant further study.

Viral determinants also directly affect the cellular outcome of WNV infection. UV-inactivated WNV fails to induce cell death, indicating that viral replication is required (Parquet et al., 2001). Several WNV proteins have been suggested to contribute to this process: ectopic expression of NS3 protein in vitro induced apoptosis (Ramanathan et al., 2006), and expression of the WNV caspid protein either in vitro or in the striatum of mouse brains triggered apoptosis downstream of caspase-3 and caspase-9 (Yang et al., 2002). WNV replication or expression of nonstructural proteins also directly activates unfolded protein response pathways and induces the proapoptotic cyclic AMP response factor CHOP (Medigeshi et al., 2007). In addition, phosphatidylinositol 3-kinase signaling is activated by flavivirus infection and may counterbalance the induction of rapid cell death, thereby prolonging virus replication (Lee et al., 2005).

Although several studies have shown that the dominant function of the innate and adaptive immune response in the CNS is to restrict WNV pathogenesis (Glass et al., 2005; Klein et al., 2005; Mehlhop et al., 2005; Samuel and Diamond, 2005; Samuel et al., 2006; Shrestha and Diamond, 2004; Shrestha et al., 2006a; Wang et al., 2003b), in some cases neuroinflamation may contribute to injury in the CNS. Neurological WNV infection is accompanied by cytokine and chemokine modulation, lymphocyte recruitment, and microglial and astrocyte activation (Guarner et al., 2004; King et al., 2007). Glial activation can induce CNS pathology through the production of reactive oxygen species, proteases, and neurotoxic cytokines (Dheen et al., 2007; Ghoshal et al., 2007; Irani and Prow, 2007). Following intravenous infection with supra-physiologic levels of WNV (10^8 PFU), $CD8^+$ T cells contribute to immunopathology in the CNS (Wang et al., 2003b). In addition, some cytokines have been implicated in WNV neurological disease by regulating viral entry into the CNS. Migration inhibitory factor (MIF) and tumor necrosis factor (TNF)-α may facilitate WNV neuroinvasion through modulating the BBB permeability (Arjona et al., 2007; Wang et al., 2004). Though not reported to date, additional pathways may also contribute to WNV pathogenesis in the CNS, including excitotoxic injury by excessive glutamate levels. Thus, much remains to be learned about the specific roles of diverse cell death pathways and inflammatory processes in the context of WNV encephalitis.

4 WNV Entry into the CNS

Animal models have helped elucidate the mechanisms by which WNV disseminates and enters the CNS. Following peripheral inoculation, WNV is believed to infect dendritic cells in the skin, which presumably traffic to and seed the draining lymph node (Byrne et al., 2001). The resulting viremia spreads infection to peripheral lymphoid and visceral tissues, including the spleen and kidney. By approximately 6 days, WNV is cleared from peripheral organs and serum and has entered the CNS (Samuel and Diamond, 2006). The capacity of WNV to invade the CNS therefore depends on its ability to replicate and evade clearance by the peripheral immune system. Several aspects of WNV biology facilitate its ability to cause severe disease, including (1) broad tropism and likely use of multiple receptors (Chu and Ng, 2004; Davis et al., 2006a), (2) rapid induction of cell death (Parquet et al., 2001; Shrestha et al., 2003), and (3) multiple immune evasion strategies (Best et al., 2005; Chung et al., 2006; Guo et al., 2005; Lin et al., 2004, 2006; Liu et al., 2005; Munoz-Jordan et al., 2003, 2005). Phylogenic analyses have defined two genetic lineages of WNV, lineage I and II, and this genetic variation results in viral strains with distinct capacities to invade the CNS (Lanciotti et al., 2002; Scherret et al., 2002). Lineage I strains are detected worldwide and have been responsible for recent lethal human outbreaks (Dauphin et al., 2004; Mackenzie et al., 2004), whereas lineage II strains predominantly circulate in Africa and have caused occasional human infections (Jupp, 2001; Lanciotti et al., 2002). Though lineage I strains generally induce significant encephalitis and mortality in birds and mammals, isolates from both lineages can be neuroinvasive (Beasley et al., 2002; Brault et al., 2004; Diamond et al., 2003a; Langevin et al., 2005; Xiao et al., 2001).

The mechanisms by which WNV and other neurotropic flaviviruses enter and spread within the CNS are now beginning to be defined. Neurotropic viruses may utilize several routes to invade the CNS, including hematogenous spread, retrograde axonal transport from peripheral nerves, or spread via the olfactory system (Johnson, 1998). Though WNV seeding of the CNS via a hematogenous route has not been directly demonstrated, data from several studies suggest this may contribute to WNV entry. Increased viral burden in the serum of immunodeficient mice (Diamond et al., 2003a,b; Mehlhop and Diamond, 2006; Samuel and Diamond, 2005; Samuel et al., 2006; Shrestha and Diamond, 2004; Shrestha et al., 2006b; Wang et al., 2003a, 2006) and TNF-α- and MIF-mediated changes in BBB permeability (Arjona et al., 2007; Wang et al., 2004) correlate with earlier CNS entry. In addition to passive transport through the cerebrovascular endothelial cells, hematogenous spread may also occur by trafficking of infected myeloid cells to the CNS. WNV infection of myeloid cells has been documented in severely immunodeficient mice (Samuel and Diamond, 2005), suggesting that these cells could serve as a "Trojan hose" for viral CNS entry.

Retrograde axonal transport has also been suggested as a means of WNV entry into and spread within the CNS. Intranasal inoculation results in olfactory bulb infection and rostral-to-caudal viral spread, and in some peripheral inoculation models spinal cord infection precedes replication in the brain (King et al., 2007; Nir et al., 1965).

Recent data have validated this hypothesis, as WNV undergoes transneuronal spread to enter the CNS and induce acute flaccid limb paralysis (Samuel et al., 2007b). In vitro, WNV spread within axons in both anterograde and retrograde directions, and direct infection of the sciatic nerve in vivo promoted viral transport to the spinal cord, neuronal death, and acute flaccid paralysis. Transneuronal spread was blocked by nerve ligation or neutralizing antibody, indicating that early therapeutic intervention with antibodies could block CNS entry and amplification. Because acute flaccid paralysis can occur as an isolated disease in the absence of encephalitis or meningitis and patients of all ages and immune status groups are at risk (Sejvar et al., 2005), these experiments suggest that the route of WNV entry into the CNS in part determines pathological outcome. In turn, the immune status of the host may determine the entry routes available for WNV to invade the CNS.

5 CNS Immune Responses to WNV

The CNS represents a unique challenge to the immune system, as effective resolution of infection requires clearance with limited damage to neurons, a critical nonrenewing cell population (Griffin, 2003). Though CNS immune responses could in part contribute to neuronal injury through the induction of neurotoxic cytokines (see above), immune responses to WNV infection appear in most cases to be protective rather than pathologic. Studies in animal models show that both innate and adaptive immunity control WNV pathogenesis in the CNS and are required for limiting viral replication and clearing neuronal infection.

5.1 Innate Immune Responses in the CNS

Type I (IFN-α and IFN-β), type II (IFN-γ), and type III (IFN-λ) interferons (IFNs) present a first line of defense against viral infections (Pestka et al., 2004; Platanias, 2005). Cells detect and respond to RNA viruses through binding of RNA to intracellular sensors [retinoic acid-inducible gene (RIG)-I, melanoma differentiation antigen (MDA)5, and toll-like receptor (TLR)3, TLR7, or TLR8], resulting in activation of transcription factors, [interferon regulatory factors 3 and 7 (IRF-3 and IRF-7)], type I IFN production, and the induction of IFN-stimulated genes (ISGs). Type I IFN is produced in the CNS after WNV infection and is made at high levels directly by infected neurons (Daffis et al., 2007; Samuel and Diamond, 2005). Experiments indicate that type I

IFN contributes to the control of WNV infection and injury in the CNS. Mice lacking the type I IFN receptor (IFNαβR$^{-/-}$) showed decreased survival and increased viral burdens after intracranial (IC) inoculation. Treatment of primary neurons with type I IFN increased neuronal survival, suggesting IFN may directly mitigate neuronal pathogenesis. In contrast, control of WNV infection in the CNS is largely independent of IFN-γ, which is produced primarily by γδ T cells, CD8^{+} T cells, and natural killer (NK) cells. Mice deficient in IFN-γ did not demonstrate delayed viral clearance from the CNS or enhanced lethality after intracranial infection (Shrestha et al., 2006b).

Unique neuronal subtypes may differ both in the pathways involved in type I IFN induction as well as the IFN-stimulated antiviral effectors that inhibit WNV infection. Mice lacking the IFN-induced antiviral effectors PKR and RNase L demonstrated increased viral burdens in the CNS after IC inoculation (Samuel et al., 2006). WNV infection rapidly activated the PKR pathway in wild-type CNS-derived cortical neurons, and cortical neurons deficient in PKR and RNase L showed decreased IFN-induced inhibition of WNV. In contrast, this pathway was dispensable for IFN-mediated inhibition of WNV in primary peripheral sympathetic neurons derived from the superior cervical ganglia (SCG). However, SCG neurons show dramatically higher IFN-mediated inhibition of WNV compared to cortical neurons (1,000-fold versus 8-fold inhibition, respectively), suggesting that IFN-induced effector pathways differ significantly between the two neuronal types. Additional studies have shown that neurons are distinct from other cell types in their detection of and response to WNV. Unlike macrophages, primary cortical neurons do not basally express RIG-I, MDA5, ISG54, and ISG56 and require IRF3 and IRF7 for efficient type I IFN production (Daffis et al., 2007) (S. Daffis, M. Gale, and M.S. Diamond, unpublished data). Moreover, mice lacking IRF3, IRF7, or TLR3 demonstrated increased viral titers in CNS tissues following IC inoculation, suggesting that recognition and signaling molecules restrict WNV replication in the nervous system. Taken together, these results suggest that CNS neurons may possess attenuated innate immune pathways and therefore represent a uniquely permissive environment for viral replication. Additional studies will be required to determine the distinct pathways through which different neuronal subtypes maintain a baseline antiviral state.

5.2 Adaptive Immune Responses in the CNS

T-cell-mediated immunity is essential for controlling WNV infection in the CNS. In mice, CD8^{+} T cells traffic to the brain beginning on day 7 after infection, and WNV clearance from the CNS correlates with $CD8^+$ T cell recruitment. (Mehlhop et al., 2005; Shrestha and Diamond, 2004; Shrestha et al., 2006a; Wang et al., 2003b). Mice deficient in $CD8^+$ T cells, MHC class I molecules, perforin, and Fas ligand all showed WNV persistence in the brain (Shrestha and Diamond, 2004, 2007; Shrestha et al., 2006a). Moreover, primed $CD8^+$ T cells require both perforin and Fas ligand to clear WNV from infected primary neurons in vitro (Shrestha and Diamond, 2007; Shrestha et al., 2006a). $CD4^+$ T cell responses are also required in the CNS, as they sustain primary $CD8^+$ T-cell responses (Sitati and Diamond, 2006). Mice deficient in MHC class II molecules showed blunted $CD8^+$ T-cell trafficking and persistent WNV infection in the CNS for up to 45 days. $CD4^+$ T-cell help is mediated in part through CD40–CD40 ligand interactions. Interestingly, WNV infection models revealed a novel function for CD40: in addition to providing an important costimulatory signal, CD40 interactions facilitate T-cell migration across the BBB (Sitati et al., 2007).

Since the BBB restricts leukocyte entry into the CNS, it experiences limited immune surveillance in the absence of inflammation (Ransohoff et al., 2003). Thus, chemokine-dependent leukocyte recruitment is important for modulating viral encephalitis. T-cell trafficking to the brain during WNV infection is regulated in part by the chemokine receptor CCR5 and its congnate ligand CCL5. A genetic deficiency in CCR5 resulted in depressed CNS leukocyte migration and increased lethality in mice (Glass et al., 2005). Correspondingly, in humans, deletions in the CCR5 gene (CCR5Δ32) have been identified as a genetic risk factor for increased susceptibility to WNV encephalitis. (Glass et al., 2006). Rodent models have implicated neurons as active participants in the adaptive immune response against WNV. Although chemokines expression by trafficking leukocytes and resident glia following viral infection has been well documented (Asensio et al., 2001; Lane et al., 1998; Palma and Kim, 2004; Ransohoff et al., 2002), neurons unexpectedly expressed high levels of the chemokine CXCL10 after WNV infection (Klein et al., 2005). CXCL10 production recruited $CD8^+$ T cells through its cognate ligand CXCR3, and its absence reduced T-cell trafficking and enhanced mortality. In contrast to CXCL10, CXCL12 and its ligand CXCR4 help to localize lymphocytes to the perivascular space and restrict entry into the brain parenchyma during WNV infection. Administration of AMD-100, a specific CXCR4 antagonist, increased T-cell trafficking into the parenchyma, decreased WNV burden, and improved survival (McCandiess et al., 2008). Thus, multiple aspects of the innate and adaptive immune responses must function together to control WNV infection in the CNS.

6 Future Perspectives

Since its emergence in the United States in 1999, WNV has become the dominant cause of epidemic meningeoencephaltisis in humans in North America and will likely continue to induce significant morbidity and mortality. Human studies and animal models have begun to define the pathogenesis of WNV in the CNS and elucidate the mechanisms by which WNV induces clinically diverse neurological diseases. Several important questions regarding WNV infection of the CNS remain. Experiments in animal models should continue to elucidate the specific routes by which WNV enters the CNS, examine the molecular pathways that regulate injury and death of diverse neuronal subtypes, and further define the protective and potentially pathologic functions of CNS immunity. An improved understanding of WNV neuronal pathogenesis will facilitate the development of targeted strategies to prevent neurological infection and mitigate neurological diseases in affected individuals. New classes of antiviral agents may be required to meet the diverse pharmacologic challenges posed by WNV disease in the nervous system. Effective therapeutics must cross the BBB, inhibit viral infection, minimize neuronal death, and restrict immunopathology without hampering protective immune responses. Treatment strategies that meet these challenges will not only help combat the ongoing WNV epidemic but may prove useful in restricting the pathogenesis of related and unrelated encephalitic viruses that cause significant human disease.

Acknowledgements The authors thank S. Daffis, E. McCandless, and R. Klein for helpul discussions, and B Shrestha for help with the figure preparation. This work was supported by a Predoctoral Fellowship from the Howard Hughes Medical Institute (M.A.S.), the Ellison Medical Foundation Global Infectious Diseases Program (M.S.D.) and the NIH (AI53870 to M.S.D.).

References

Ahmed, S., Libman, R., Wesson, K., Ahmed, F., and Einberg, K. (2000). Guillain–Barre syndrome: an unusual presentation of West Nile virus infection. Neurology *55*, 144–146.

Ali, M., Safriel, Y., Sohi, J., Llave, A., and Weathers, S. (2005). West Nile virus infection: MR imaging findings in the nervous system. AJNR Am J Neuroradiol *26*, 289–297.

Arjona, A., Foellmer, H. G., Town, T., Leng, L., McDonald, C., Wang, T., Wong, S. J., Montgomery, R. R., Fikrig, E., and Bucala, R. (2007). Abrogation of macrophage migration inhibitory factor decreases West Nile virus lethality by limiting viral neuroinvasion. J Clin Invest *117*, 3059–3066.

Armah, H. B., Wang, G., Omalu, B. I., Tesh, R. B., Gyure, K. A., Chute, D. J., Smith, R. D., Dulai, P., Vinters, H. V., Kleinschmidt-Demasters, B. K., and Wiley, C. A. (2007). Systemic distribution of West Nile virus infection: postmortem immunohistochemical study of six cases. Brain Pathol *17*, 354–362.

Asensio, V. C., Maier, J., Milner, R., Boztug, K., Kincaid, C., Moulard, M., Phillipson, C., Lindsley, K., Krucker, T., Fox, H. S., and Campbell, I. L. (2001). Interferon-independent, human immunodeficiency virus type 1 gp120-mediated induction of CXCL10/IP-10 gene expression by astrocytes in vivo and in vitro. J Virol *75*, 7067–7077.

Bakri, S. J., and Kaiser, P. K. (2004). Ocular manifestations of West Nile virus. Curr Opin Ophthalmol *15*, 537–540.

Beasley, D. W., Li, L., Suderman, M. T., and Barrett, A. D. (2002). Mouse neuroinvasive phenotype of West Nile virus strains varies depending upon virus genotype. Virology *296*, 17–23.

Best, S. M., Morris, K. L., Shannon, J. G., Robertson, S. J., Mitzel, D. N., Park, G. S., Boer, E., Wolfinbarger, J. B., and Bloom, M. E. (2005). Inhibition of interferon-stimulated JAK-STAT signaling by a tick-borne flavivirus and identification of NS5 as an interferon antagonist. J Virol *79*, 12828–12839.

Bode, A. V., Sejvar, J. J., Pape, W. J., Campbell, G. L., and Marfin, A. A. (2006). West Nile virus disease: a descriptive study of 228 patients hospitalized in a 4-county region of Colorado in 2003. Clin Infect Dis *42*, 1234–1240.

Bouffard, J. P., Riudavets, M. A., Holman, R., and Rushing, E. J. (2004). Neuropathology of the brain and spinal cord in human West Nile virus infection. Clin Neuropathol *23*, 59–61.

Brault, A. C., Langevin, S. A., Bowen, R. A., Panella, N. A., Biggerstaff, B. J., Miller, B. R., and Nicholas, K. (2004). Differential virulence of West Nile strains for American crows. Emerg Infect Dis *10*, 2161–2168.

Brilla, R., Block, M., Geremia, G., and Wichter, M. (2004). Clinical and neuroradiologic features of 39 consecutive cases of West Nile virus meningoencephalitis. J Neurol Sci *220*, 37–40.

Burton, J. M., Kern, R. Z., Halliday, W., Mikulis, D., Brunton, J., Fearon, M., Pepperell, C., and Jaigobin, C. (2004). Neurological manifestations of West Nile virus infection. Can J Neurol Sci *31*, 185–193.

Busch, M. P. (2006). West Nile virus infections projected from blood donor screening data, United States, 2003. Emerg Infect Dis *12*, 395–402.

Byrne, S. N., Halliday, G. M., Johnston, L. J., and King, N. J. (2001). Interleukin-1beta but not tumor necrosis factor is involved in West Nile virus-induced Langerhans cell migration from the skin in C57BL/6 mice. J Invest Dermatol *117*, 702–709.

Campbell, G. L., Marfin, A. A., Lanciotti, R. S., and Gubler, D. J. (2002). West Nile virus. Lancet Infect Dis *2*, 519–529.

Cao, N. J., Ranganathan, C., Kupsky, W. J., and Li, J. (2005). Recovery and prognosticators of paralysis in West Nile virus infection. J Neurol Sci *236*, 73–80.

Carson, P. J., Konewko, P., Wold, K. S., Mariani, P., Goli, S., Bergloff, P., and Crosby, R. D. (2006). Long-term clinical and neuropsychological outcomes of West Nile virus infection. Clin Infect Dis *43*, 723–730.

Ceausu, E., Erscoiu, S., Calistru, P., Ispas, D., Dorobat, O., Homos, M., Barbulescu, C., Cojocaru, I., Simion, C. V., Cristea, C., et-al. (1997). Clinical manifestations in the West Nile virus outbreak. Rom J Virol *48*, 3–11.

Chowers, M. Y., Lang, R., Nassar, F., Ben-David, D., Giladi, M., Rubinshtein, E., Itzhaki, A., Mishal, J., Siegman-Igra, Y., Kitzes, R., et-al. (2001). Clinical characteristics of the West Nile fever outbreak, Israel, 2000. Emerg Infect Dis *7*, 675–678.

Chu, J. J., and Ng, M. L. (2004). Interaction of West Nile virus with alpha v beta 3 integrin mediates virus entry into cells. J Biol Chem *279*, 54533–54541.

Chung, K. M., Liszewski, M. K., Nybakken, G., Davis, A. E., Townsend, R. R., Fremont, D. H., Atkinson, J. P., and Diamond, M. S. (2006). West Nile virus nonstructural protein NS1 inhibits complement activation by binding the regulatory protein factor H. Proc Natl Acad Sci U S A *103*, 19111–19116.

Daffis, S., Samuel, M. A., Keller, B. C., Gale, M., Jr., and Diamond, M. S. (2007). Cell-specific IRF-3 responses protect against West Nile virus infection by interferon-dependent and -independent mechanisms. PLoS Pathog *3*, e106.

Dauphin, G., Zientara, S., Zeller, H., and Murgue, B. (2004). West Nile: worldwide current situation in animals and humans. Comp Immunol Microbiol Infect Dis *27*, 343–355.

Davis, C. W., Nguyen, H. Y., Hanna, S. L., Sanchez, M. D., Doms, R. W., and Pierson, T. C. (2006a). West Nile virus discriminates between DC-SIGN and DC-SIGNR for cellular attachment and infection. J Virol *80*, 1290–1301.

Davis, L. E., DeBiasi, R., Goade, D. E., Haaland, K. Y., Harrington, J. A., Harnar, J. B., Pergam, S. A., King, M. K., DeMasters, B. K., and Tyler, K. L. (2006b). West Nile virus neuroinvasive disease. Ann Neurol *60*, 286–300.

Deardorff, E., Estrada-Franco, J., Brault, A. C., Navarro-Lopez, R., Campomanes-Cortes, A., Paz-Ramirez, P., Solis-Hernandez, M., Ramey, W. N., Davis, C. T., Beasley, D. W., et-al. (2006). Introductions of West Nile virus strains to Mexico. Emerg Infect Dis *12*, 314–318.

Debiasi, R. L., and Tyler, K. L. (2006). West Nile virus meningoencephalitis. Nat Clin Pract Neurol *2*, 264–275.

Dheen, S. T., Kaur, C., and Ling, E. A. (2007). Microglial activation and its implications in the brain diseases. Curr Med Chem *14*, 1189–1197.

Diamond, M. S., Shrestha, B., Marri, A., Mahan, D., and Engle, M. (2003a). B cells and antibody play critical roles in the immediate defense of disseminated infection by West Nile encephalitis virus. J Virol *77*, 2578–2586.

Diamond, M. S., Sitati, E. M., Friend, L. D., Higgs, S., Shrestha, B., and Engle, M. (2003b). A critical role for induced IgM in the protection against West Nile virus infection. J Exp Med *198*, 1853–1862.

Eldadah, A. H., and Nathanson, N. (1967). Pathogenesis of West Nile virus encepahlitis in mice and rats. II. Virus multiplication, evolution of immunofluorescence, and development of histological lesions in the brain. Am J Epidemiol *86*, 776–790.

Fratkin, J. D., Leis, A. A., Stokic, D. S., Slavinski, S. A., and Geiss, R. W. (2004). Spinal cord neuropathology in human West Nile virus infection. Arch Pathol Lab Med *128*, 533–537.

Gandelman-Marton, R., Kimiagar, I., Itzhaki, A., Klein, C., Theitler, J., and Rabey, J. M. (2003). Electroencephalography findings in adult patients with West Nile virus-associated meningitis and meningoencephalitis. Clin Infect Dis *37*, 1573–1578.

Ghoshal, A., Das, S., Ghosh, S., Mishra, M. K., Sharma, V., Koli, P., Sen, E., and Basu, A. (2007). Proinflammatory mediators released by activated microglia induces neuronal death in Japanese encephalitis. Glia *55*, 483–496.

Glass, W. G., Lim, J. K., Cholera, R., Pletnev, A. G., Gao, J. L., and Murphy, P. M. (2005). Chemokine receptor CCR5 promotes leukocyte trafficking to the brain and survival in West Nile virus infection. J Exp Med *202*, 1087–1098.

Glass, W. G., McDermott, D. H., Lim, J. K., Lekhong, S., Yu, S. F., Frank, W. A., Pape, J., Cheshier, R. C., and Murphy, P. M. (2006). CCR5 deficiency increases risk of symptomatic West Nile virus infection. J Exp Med *203*, 35–40.

Gottfried, K., Quinn, R., and Jones, T. (2005). Clinical description and follow-up investigation of human West Nile virus cases. South Med J *98*, 603–606.

Granwehr, B. P., Lillibridge, K. M., Higgs, S., Mason, P. W., Aronson, J. F., Campbell, G. A., and Barrett, A. D. (2004). West Nile virus: where are we now? Lancet Infect Dis *4*, 547–556.

Griffin, D. E. (2003). Immune responses to RNA-virus infections of the CNS. Nat Rev Immunol *3*, 493–502.

Griffin, D. E., and Hardwick, J. M. (1999). Perspective: virus infections and the death of neurons. Trends Microbiol *7*, 155–160.

Guarner, J., Shieh, W. J., Hunter, S., Paddock, C. D., Morken, T., Campbell, G. L., Marfin, A. A., and Zaki, S. R. (2004). Clinicopathologic study and laboratory diagnosis of 23 cases with West Nile virus encephalomyelitis. Hum Pathol *35*, 983–990.

Guo, J. T., Hayashi, J., and Seeger, C. (2005). West Nile virus inhibits the signal transduction pathway of alpha interferon. J Virol *79*, 1343–1350.

Hayes, E. B., and O'Leary, D. R. (2004). West Nile virus infection: a pediatric perspective. Pediatrics *113*, 1375–1381.

Hayes, E. B., Komar, N., Nasci, R. S., Montgomery, S. P., O'Leary, D. R., and Campbell, G. L. (2005a). Epidemiology and transmission dynamics of West Nile virus disease. Emerg Infect Dis *11*, 1167–1173.

Hayes, E. B., Sejvar, J. J., Zaki, S. R., Lanciotti, R. S., Bode, A. V., and Campbell, G. L. (2005b). Virology, pathology, and clinical manifestations of West Nile virus disease. Emerg Infect Dis *11*, 1174–1179.

Huhn, G. D., Austin, C., Langkop, C., Kelly, K., Lucht, R., Lampman, R., Novak, R., Haramis, L., Boker, R., Smith, S., et-al. (2005). The emergence of West Nile virus during a large outbreak in Illinois in 2002. Am J Trop Med Hyg *72*, 768–776.

Hunsperger, E., and Roehrig, J. T. (2005). Characterization of West Nile viral replication and maturation in peripheral neurons in culture. J Neurovirol *11*, 11–22.

Irani, D. N., and Prow, N. A. (2007). Neuroprotective interventions targeting detrimental host immune responses protect mice from fatal alphavirus encephalitis. J Neuropathol Exp Neurol *66*, 533–544.

Johnson, R. T. (1998). Viral infections of the nervous system, 2nd edn (Philadelphia: Lippincott Williams & Wilkins).

Jupp, P. G. (2001). The ecology of West Nile virus in South Africa and the occurrence of outbreaks in humans. Ann N Y Acad Sci *951*, 143–152.

King, N. J., Getts, D. R., Getts, M. T., Rana, S., Shrestha, B., and Kesson, A. M. (2007). Immunopathology of flavivirus infections. Immunol Cell Biol *85*, 33–42.

Klee, A. L., Maidin, B., Edwin, B., Poshni, I., Mostashari, F., Fine, A., Layton, M., and Nash, D. (2004). Long-term prognosis for clinical West Nile virus infection. Emerg Infect Dis *10*, 1405–1411.

Klein, C., Kimiagar, I., Pollak, L., Gandelman-Marton, R., Itzhaki, A., Milo, R., and Rabey, J. M. (2002). Neurological features of West Nile virus infection during the 2000 outbreak in a regional hospital in Israel. J Neurol Sci *200*, 63–66.

Klein, R. S., Lin, E., Zhang, B., Luster, A. D., Tollett, J., Samuel, M. A., Engle, M., and Diamond, M. S. (2005). Neuronal CXCL10 directs CD8+ T-cell recruitment and control of West Nile virus encephalitis. J Virol *79*, 11457–11466.

Kleinschmidt-DeMasters, B. K., Marder, B. A., Levi, M. E., Laird, S. P., McNutt, J. T., Escott, E. J., Everson, G. T., and Tyler, K. L. (2004). Naturally acquired West Nile virus encephalomyelitis in transplant recipients: clinical, laboratory, diagnostic, and neuropathological features. Arch Neurol *61*, 1210–1220.

Komar, N., and Clark, G. G. (2006). West Nile virus activity in Latin America and the Caribbean. Rev Panam Salud Publica *19*, 112–117.

LaDeau, S. L., Kilpatrick, A. M., and Marra, P. P. (2007). West Nile virus emergence and large-scale declines of North American bird populations. Nature *447*, 710–713.

Lanciotti, R. S., Roehrig, J. T., Deubel, V., Smith, J., Parker, M., Steele, K., Crise, B., Volpe, K. E., Crabtree, M. B., Scherret, J. H., et-al. (1999). Origin of the West Nile virus responsible for an outbreak of encephalitis in the northeastern United States. Science *286*, 2333–2337.

Lanciotti, R. S., Ebel, G. D., Deubel, V., Kerst, A. J., Murri, S., Meyer, R., Bowen, M., McKinney, N., Morrill, W. E., Crabtree, M. B., et-al. (2002). Complete genome sequences and phylogenetic analysis of West Nile virus strains isolated from the United States, Europe, and the Middle East. Virology *298*, 96–105.

Lane, T. E., Asensio, V. C., Yu, N., Paoletti, A. D., Campbell, I. L., and Buchmeier, M. J. (1998). Dynamic regulation of alpha- and beta-chemokine expression in the central nervous system during mouse hepatitis virus-induced demyelinating disease. J Immunol *160*, 970–978.

Langevin, S. A., Brault, A. C., Panella, N. A., Bowen, R. A., and Komar, N. (2005). Variation in virulence of West Nile virus strains for house sparrows (*Passer domesticus*). Am J Trop Med Hyg *72*, 99–102.

Lee, C. J., Liao, C. L., and Lin, Y. L. (2005). Flavivirus activates phosphatidylinositol 3-kinase signaling to block caspase-dependent apoptotic cell death at the early stage of virus infection. J Virol *79*, 8388–8399.

Leis, A. A., Stokic, D. S., Webb, R. M., Slavinski, S. A., and Fratkin, J. (2003). Clinical spectrum of muscle weakness in human West Nile virus infection. Muscle Nerve *28*, 302–308.

Lin, R. J., Liao, C. L., Lin, E., and Lin, Y. L. (2004). Blocking of the alpha interferon-induced Jak-Stat signaling pathway by Japanese encephalitis virus infection. J Virol *78*, 9285–9294.

Lin, R.-J., Chang, B.-L., Yu, H.-P., Liao, C.-L., and Lin, Y.-L. (2006). Blocking of Interferon-induced Jak-Stat signaling by Japanese encephalitis virus NS5 through a protein tyrosine phosphatase-mediated mechanism. J Virol *80*, 5908–5918.

Liu, W. J., Wang, X. J., Mokhonov, V. V., Shi, P. Y., Randall, R., and Khromykh, A. A. (2005). Inhibition of interferon signaling by the New York 99 strain and Kunjin subtype of West Nile virus involves blockage of STAT1 and STAT2 activation by nonstructural proteins. J Virol *79*, 1934–1942.

Mackenzie, J. S., Gubler, D. J., and Petersen, L. R. (2004). Emerging flaviviruses: the spread and resurgence of Japanese encephalitis, West Nile and dengue viruses. Nat Med *10*, S98–109.

Majno, G., and Joris, I. (1995). Apoptosis, oncosis, and necrosis. An overview of cell death. Am J Pathol *146*, 3–15.

McCandess, E. E., Zhang, B., Diamond, M. S., and Klein, R. S. (2008) CXCR4 antagonism increases T cell trafficking in the central nervous system and improves survival from West Nile virus encephalitis. *Proc Nati Acad Sci USA* 105, 11270–11275.

Medigeshi, G. R., Lancaster, A. M., Hirsch, A. J., Briese, T., Lipkin, W. I., Defilippis, V., Fruh, K., Mason, P. W., Nikolich-Zugich, J., and Nelson, J. A. (2007). West Nile virus infection activates the unfolded protein response, leading to CHOP induction and apoptosis. J Virol *81*, 10849–10860.

Mehlhop, E., and Diamond, M. S. (2006). Protective immune responses against West Nile virus are primed by distinct complement activation pathways. J Exp Med *203*, 1371–1381.

Mehlhop, E., Whitby, K., Oliphant, T., Marri, A., Engle, M., and Diamond, M. S. (2005). Complement activation is required for induction of a protective antibody response against West Nile virus infection. J Virol *79*, 7466–7477.

Munoz-Jordan, J. L., Sanchez-Burgos, G. G., Laurent-Rolle, M., and Garcia-Sastre, A. (2003). Inhibition of interferon signaling by dengue virus. Proc Natl Acad Sci U S A *100*, 14333–14338.

Munoz-Jordan, J. L., Laurent-Rolle, M., Ashour, J., Martinez-Sobrido, L., Ashok, M., Lipkin, W. I., and Garcia-Sastre, A. (2005). Inhibition of alpha/beta interferon signaling by the NS4B protein of flaviviruses. J Virol *79*, 8004–8013.

Nash, D., Mostashari, F., Fine, A., Miller, J., O'Leary, D., Murray, K., Huang, A., Rosenberg, A., Greenberg, A., Sherman, M., et-al. (2001). The outbreak of West Nile virus infection in the New York City area in 1999. N Engl J Med *344*, 1807–1814.

Nir, Y., Beemer, A., and Goldwasser, R. A. (1965). West Nile virus infection in mice following exposure to a viral aerosol. Br J Exp Pathol *46*, 443–449.

Omalu, B. I., Shakir, A. A., Wang, G., Lipkin, W. I., and Wiley, C. A. (2003). Fatal fulminant pan-meningo-polioencephalitis due to West Nile virus. Brain Pathol *13*, 465–472.

Ou, A. C., and Ratard, R. C. (2005). One-year sequelae in patients with West Nile virus encephalitis and meningitis in Louisiana. J La State Med Soc *157*, 42–46.

Palma, J. P., and Kim, B. S. (2004). The scope and activation mechanisms of chemokine gene expression in primary astrocytes following infection with Theiler's virus. J Neuroimmunol *149*, 121–129.

Parquet, M. C., Kumatori, A., Hasebe, F., Morita, K., and Igarashi, A. (2001). West Nile virus-induced bax-dependent apoptosis. FEBS Lett *500*, 17–24.

Patnaik, J. L., Harmon, H., and Vogt, R. L. (2006). Follow-up of 2003 human West Nile virus infections, Denver, Colorado. Emerg Infect Dis *12*, 1129–1131.

Pestka, S., Krause, C. D., and Walter, M. R. (2004). Interferons, interferon-like cytokines, and their receptors. Immunol Rev *202*, 8–32.

Petersen, L. R., and Marfin, A. A. (2002). West Nile virus: a primer for the clinician. Ann Intern Med *137*, 173–179.

Petersen, L. R., and Roehrig, J. T. (2001). West Nile virus: a reemerging global pathogen. Emerg Infect Dis *7*, 611–614.

Petropoulou, K. A., Gordon, S. M., Prayson, R. A., and Ruggierri, P. M. (2005). West Nile virus meningoencephalitis: MR imaging findings. AJNR Am J Neuroradiol *26*, 1986–1995.

Platanias, L. C. (2005). Mechanisms of type-I- and type-II-interferon-mediated signalling. Nat Rev Immunol *5*, 375–386.

Ramanathan, M. P., Chambers, J. A., Pankhong, P., Chattergoon, M., Attatippaholkun, W., Dang, K., Shah, N., and Weiner, D. B. (2006). Host cell killing by the West Nile virus NS2B-NS3 proteolytic complex: NS3 alone is sufficient to recruit caspase-8-based apoptotic pathway. Virology *345*, 56–72.

Ransohoff, R. M., Wei, T., Pavelko, K. D., Lee, J. C., Murray, P. D., and Rodriguez, M. (2002). Chemokine expression in the central nervous system of mice with a viral disease resembling multiple sclerosis: roles of CD4+ and CD8+ T cells and viral persistence. J Virol *76*, 2217–2224.

Ransohoff, R. M., Kivisakk, P., and Kidd, G. (2003). Three or more routes for leukocyte migration into the central nervous system. Nat Rev Immunol *3*, 569–581.

Robertson, H. J., and Sejvar, J. J. (2003). The need for a West Nile virus MRI registry. AJNR Am J Neuroradiol *24*, 1741–1742.

Robinson, R. L., Shahida, S., Madan, N., Rao, S., and Khardori, N. (2003). Transient parkinsonism in West Nile virus encephalitis. Am J Med *115*, 252–253.

Saad, M., Youssef, S., Kirschke, D., Shubair, M., Haddadin, D., Myers, J., and Moorman, J. (2005). Acute flaccid paralysis: the spectrum of a newly recognized complication of West Nile virus infection. J Infect *51*, 120–127.

Samuel, M. A., and Diamond, M. S. (2005). Alpha/beta interferon protects against lethal West Nile virus infection by restricting cellular tropism and enhancing neuronal survival. J Virol *79*, 13350–13361.

Samuel, M. A., and Diamond, M. S. (2006). Pathogenesis of West Nile virus infection: a balance between virulence, innate, and adaptive immunity and viral evasion. J Virol *80*, 9349–9360.

Samuel, M. A., Whitby, K., Keller, B. C., Marri, A., Barchet, W., Williams, B. R. G., Silverman, R. H., Gale, M., Jr., and Diamond, M. S. (2006). PKR and RNase L contribute to protection against lethal West Nile virus infection by controlling early viral spread in the periphery and replication in neurons. J Virol *80*, 7009–7019.

Samuel, M. A., Morrey, J. D., and Diamond, M. S. (2007a). Caspase 3-dependent cell death of neurons contributes to the pathogenesis of West Nile virus encephalitis. J Virol *81*, 2614–2623.

Samuel, M. A., Wang, H., Siddharthan, V., Morrey, J. D., and Diamond, M. S. (2007b). Axonal transport mediates West Nile virus entry into the central nervous system and induces acute flaccid paralysis. Proc Natl Acad Sci U S A *104*, 17140–17145.

Sayao, A. L., Suchowersky, O., Al-Khathaami, A., Klassen, B., Katz, N. R., Sevick, R., Tilley, P., Fox, J., and Patry, D. (2004). Calgary experience with West Nile virus neurological syndrome during the late summer of 2003. Can J Neurol Sci *31*, 194–203.

Scherret, J. H., Mackenzie, J. S., Hall, R. A., Deubel, V., and Gould, E. A. (2002). Phylogeny and molecular epidemiology of West Nile and Kunjin viruses. Curr Top Microbiol Immunol *267*, 373–390.

Sejvar, J. J. (2006). West Nile virus-associated flaccid paralysis outcome. Emerg Infect Dis *12*, 514–516.

Sejvar, J. J. (2007). The long-term outcomes of human West Nile virus infection. Clin Infect Dis *44*, 1617–1624.

Sejvar, J. J., Haddad, M. B., Tierney, B. C., Campbell, G. L., Marfin, A. A., Van Gerpen, J. A., Fleischauer, A., Leis, A. A., Stokic, D. S., and Petersen, L. R. (2003). Neurologic manifestations and outcome of West Nile virus infection. JAMA *290*, 511–515.

Sejvar, J. J., Bode, A. V., Marfin, A. A., Campbell, G. L., Ewing, D., Mazowiecki, M., Pavot, P. V., Schmitt, J., Pape, J., Biggerstaff, B. J., and Petersen, L. R. (2005). West Nile virus-associated flaccid paralysis. Emerg Infect Dis *11*, 1021–1027.

Shrestha, B., and Diamond, M. S. (2004). Role of CD8+ T cells in control of West Nile virus infection. J Virol *78*, 8312–8321.

Shrestha, B., and Diamond, M. S. (2007). Fas ligand interactions contribute to CD8+ T cell-mediated control of West Nile virus infection in the central nervous system. J Virol *81*, 11749–11757.

Shrestha, B., Gottlieb, D., and Diamond, M. S. (2003). Infection and injury of neurons by West Nile encephalitis virus. J Virol *77*, 13203–13213.

Shrestha, B., Samuel, M. A., and Diamond, M. S. (2006a). CD8+ T cells require perforin to clear West Nile virus from infected neurons. J Virol *80*, 119–129.

Shrestha, B., Wang, T., Samuel, M. A., Whitby, K., Craft, J., Fikrig, E., and Diamond, M. S. (2006b). Gamma interferon plays a crucial early antiviral role in protection against West Nile virus infection. J Virol *80*, 5338–5348.

Sitati, E. M., and Diamond, M. S. (2006). CD4+ T-cell responses are required for clearance of West Nile virus from the central nervous system. J Virol *80*, 12060–12069.

Sitati, E., McCandless, E. E., Klein, R. S., and Diamond, M. S. (2007). CD40–CD40 ligand interactions promote trafficking of CD8+ T cells into the brain and protection against West Nile virus encephalitis. J Virol *81*, 9801–9811.

Solomon, T. (2004). Flavivirus encephalitis. N Engl J Med *351*, 370–378.

Tyler, K. L. (2004). West Nile virus infection in the United States. Arch Neurol *61*, 1190–1195.

van der Meulen, K. M., Pensaert, M. B., and Nauwynck, H. J. (2005). West Nile virus in the vertebrate world. Arch Virol *150*, 637–657.

Wang, T., Scully, E., Yin, Z., Kim, J. H., Wang, S., Yan, J., Mamula, M., Anderson, J. F., Craft, J., and Fikrig, E. (2003a). IFN-gamma-producing gamma delta T cells help control murine West Nile virus infection. J Immunol *171*, 2524–2531.

Wang, Y., Lobigs, M., Lee, E., and Mullbacher, A. (2003b). CD8+ T cells mediate recovery and immunopathology in West Nile virus encephalitis. J Virol *77*, 13323–13334.

Wang, T., Town, T., Alexopoulou, L., Anderson, J. F., Fikrig, E., and Flavell, R. A. (2004). Toll-like receptor 3 mediates West Nile virus entry into the brain causing lethal encephalitis. Nat Med *10*, 1366–1373.

Wang, T., Gao, Y., Scully, E., Davis, C. T., Anderson, J. F., Welte, T., Ledizet, M., Koski, R., Madri, J. A., Barrett, A. D., et al. (2006). Gamma-delta T cells facilitate adaptive immunity against West Nile virus. *J Immunol* 177, 1825–1832.

Watson, J. T., Pertel, P. E., Jones, R. C., Siston, A. M., Paul, W. S., Austin, C. C., and Gerber, S. I. (2004). Clinical characteristics and functional outcomes of West Nile Fever. Ann Intern Med *141*, 360–365.

Weaver, S. C., and Barrett, A. D. (2004). Transmission cycles, host range, evolution and emergence of arboviral disease. Nat Rev Microbiol *2*, 789–801.

Weiss, D., Carr, D., Kellachan, J., Tan, C., Phillips, M., Bresnitz, E., and Layton, M. (2001). Clinical findings of West Nile virus infection in hospitalized patients, New York and New Jersey, 2000. Emerg Infect Dis *7*, 654–658.

Xiao, S. Y., Guzman, H., Zhang, H., Travassos da Rosa, A. P., and Tesh, R. B. (2001). West Nile virus infection in the golden hamster (*Mesocricetus auratus*): a model for West Nile encephalitis. Emerg Infect Dis *7*, 714–721.

Yang, J. S., Ramanathan, M. P., Muthumani, K., Choo, A. Y., Jin, S. H., Yu, Q. C., Hwang, D. S., Choo, D. K., Lee, M. D., Dang, K., et-al. (2002). Induction of inflammation by West Nile virus capsid through the caspase-9 apoptotic pathway. Emerg Infect Dis *8*, 1379–1384.

18. The Human Antibody Response Against WNV

MARK THROSBY, JAAP GOUDSMIT, AND JOHN DE KRUIF

Abstract

Experimental evidence has shown that antibody responses to West Nile virus (WNV) are critical for protection from WNV-mediated disease. Antibody responses are also an important immune correlate of protection for the clinical evaluation of WNV vaccines. However, little direct study has been carried out on the characteristics of the human antibody response to natural WNV infection. Preliminary evidence suggests that there are important differences in the way humans and experimental animals mount humoral responses to WNV. In humans, IgM is remarkably persistent in the serum and specific IgG is slow to appear. In addition, mapping of the IgG response to the functionally relevant E-protein suggests that it directed away from critical protective epitopes and towards weakly neutralizing immunodominant epitopes. These findings have important implications for vaccine design and testing.

Keywords

antibody, flavivines repertoire human vaccine

1 Introduction

In experimental models, humoral immunity is critical to protection from WNV-mediated disease (Ben-Nathan et al., 2003; Camenga et al., 1974; Diamond et al., 2003a,b; Engle and Diamond, 2003; Gould et al., 2005; Tesh et al., 2002; Wang et al., 2001). B cell-deficient mice are more susceptible to disease (Diamond et al., 2003a; Halevy et al., 1994), and passive transfer of antibodies to these mice restores immunity (Engle and Diamond, 2003). Clinical case reports describing the use of human intravenous immunoglobulin (Ig) containing WNV antibodies in patients with neurological disease suggest some therapeutic benefit (Haley et al., 2003; Hamdan et al., 2002; Shimoni et al., 2001). Placebo-controlled clinical trials are in progress to determine the validity of these observations (http://clinicaltrials.gov.htm).

Analysis of humoral responses to WNV have mainly been carried out after experimental infection of rodents (Oliphant et al., 2005; Sanchez et al., 2005). Isolation of monoclonal antibody (mAb) panels has been used to determine the antigenic variability of WNV, and challenge studies have been performed to test vaccines for clinical development. Classically, the human antibody response to WNV infection has been studied by looking at serological responses to infection. Recently, more advanced protein engineering approaches and display technologies have been utilized to probe the human repertoire against WNV. Collectively, these studies have identified important differences in the humoral response against natural WNV infection in humans compared to experimental WNV infection in rodents. In particular, the persistence of WNV-specific IgM after infection and the immunodominance of weakly neutralizing epitopes of the domain II region of E-protein appear to be unique to the human immune response to WNV. These differences potentially have important consequences for the design and testing of preventative vaccines.

2 Natural Infections in Humans

The source of human WNV infections is almost invariably through the bite of an infected mosquito, although transmissions through other routes such as transplantation or blood transfusion have been reported (Iwamoto et al., 2003). The initial replication of WNV after inoculation is thought to occur in dendritic cells in the skin, which migrate to secondary lymphoid tissues and the replicating virus enters the circulation (Byrne et al., 2001). Viremia is transient and of low titer. Virus can typically be detected by reverse transcriptase polymerase chain reaction (RT-PCR)-based assays 4–9 days post infection (Lanciotti et al., 2000). Symptoms generally appear at the time viremia disappears, and the virus becomes detectable in the cerebral spinal fluid (CSF). The period in which WNV is detectable in the CSF is variable, but can last several weeks in some patients often correlating with the severity of symptoms. The clinical course ranges from a self-limited febrile disease to a severe neuroinvasive disease that can lead to permanent disability or death (Klee et al., 2004).

Serological surveillance data from the New York epidemic of 1999 showed that one in five people infected with WNV develop fever and 1 in 150 develop neurological symptoms (Mostashari et al., 2001). These data are supported by subsequent seroprevalence studies which suggest that most human WNV infections do not result in significant symptoms (Busch et al., 2006). Risk factors for severe disease have not been well defined apart from age. Individuals over the age of 50 have up to a 20-fold increased risk to develop severe disease. Hypertension and diabetes have also been causally associated with serious WNV disease (Granwehr et al., 2004).

3 Serology

Patients suspected of WNV infection are tested for the presence of IgM and IgG antibodies to WNV using a commercially available IgM-capture enzyme-linked immunosorbent assay (ELISA) and indirect IgG ELISA, respectively. In some cases, a diagnosis of WNV infection is confirmed by a definitive plaque reduction neutralization assay (PRNT). However, while the PRNT assay is more specific for WNV, it requires specialized laboratory facilities and is not routinely performed. For this reason, most serological data relating to human WNV infections is based on ELISA experiments alone, which give no direct information on the functional activity of detected antibodies.

The first antibody responses detected in humans after primary infection with WNV are of the decavalent, pentameric IgM isotype. Clinical case studies report that serum IgM levels appear 10–14 days after infection (Gea-Banacloche et al., 2004). This correlates well with an analysis of WNV-positive blood donations where the time between infectious mosquito bite and index donation was calculated using the measured viral load and models of virus doubling time (Prince et al., 2005). The peak IgM serum titers are reached 18 days after infection and can persist for more than 12 months. On the basis of regression analysis, IgM levels were estimated to persist above baseline on average 7–8 months post infection (Prince et al., 2005). In one study of patients with meningitis and encephalitis, specific IgM against WNV was measured 500 days after infection (Roehrig et al., 2003). The serum persistence of IgM against WNV is unusual particularly given that viremia is transient and, in general, the titer of WNV antibodies is low. However, it is consistent with reported serology to other flaviviruses such as dengue and St. Louis encephalitis viruses (Han et al., 1988; Summers et al., 1984). Somewhat surprisingly, WNV-specific IgM can appear in the CSF before detection in serum, soon after the virus is detected in the CSF (Gea-Banacloche et al., 2004). IgM is too large to diffuse across the blood–brain barrier (BBB), and therefore may be produced locally by B cells that have trafficked into the CSF (Binder and Griffin, 2003) or though inflamed vessels when the BBB integrity has been compromised

(Wang et al., 2004). In addition, consistent with the persistence of IgM in the serum, IgM can be measured in the CSF 199 days post infection (Kapoor et al., 2004).

Extended maintenance of serological responses have been proposed to result from the presence of long-lived plasma cells in the bone marrow that secrete antibody, from boosting of memory responses through interaction with antigen retained on follicular dendritic cells, or through bystander stimulation of memory B cells by heterologous vaccination or infection (Bernasconi et al., 2002). These responses are generally associated with isotype-switched and affinity-matured immunoglobulin (e.g. IgG). The short serum half-life of human IgM (2 days) requires a larger number of plasma cells to maintain serum levels in contrast to switched isotypes (e.g. IgG1 has a serum half-life ~21 days). Although there is a limited capacity in the bone marrow and spleen to accommodate plasma cells (Cassese et al., 2003; Manz and Radbruch, 2002), there may be a benefit in retaining the IgM isotype compared to switched isotypes at the cost of efficiency in maintaining serological surveillance. This could be due to specific functionality of the IgM constant region (e.g., efficient complement fixation) or detrimental functionality of a switch isotype constant region. Alternatively, viral elements could block isotype switching by interfering with T-cell help to reduce the potential effectiveness of the immune response. Longitudinal studies analyzing the neutralizing potency of serum IgM, and investigation into the nature of the persistent IgM-secreting plasma cells in humans, are required to better understand the relationship between IgM levels, WNV pathogenesis, and clinical outcome.

The serum persistence of IgM has diagnostic implications. WNV-reactive IgM in serum samples is generally regarded as confirmation of WNV infection when taken with the seasonal and clinical information. However, if IgM levels persist over 1 year, they can overlap WNV seasons of infectivity, creating the possibility of false positives. The IgM assay can be additionally confounded, as it fails to distinguish WNV infection from other closely related flaviviruses such as the Japanese or St. Louis encephalitis virus (Martin et al., 2002).

Isotype-switched isoforms of immunoglobulin generally follow the initial detection of IgM in a primary immune response. Potent virus-neutralizing antibodies are generally of the IgG isotype and reflect a process of affinity maturation (Hangartner et al., 2006). In humans, measurable IgG against WNV does not appear in the serum until after the second or third week of infection, a time during which most symptoms would have disappeared (Prince et al., 2005). In most patients, serum IgG titers are low, a finding that is also observed with live attenuated viral vaccines that have been tested in clinical trials (Monath et al., 2006). The avidity of IgG increases over time and in response to a secondary infection (Fox et al., 2006), consistent with somatic hypermutation. IgG can sometimes be found in the CSF but this is highly variable and does not necessarily follow an IgM response even when IgG is measurable in the serum (Gea-Banacloche et al., 2004).

4 Antibody Repertoire Analysis

The antigen-binding site of human Ig is derived from the translated product of two rearranged genes VH and VL coupled via constant regions. The VH gene is made up of the V–D–J regions that are chosen from pools of V (~50), D (25), and J (6). Similarly the VL gene is made up of two segments V (~70) and J (9) (V-base http://vbase.mrc-cpe.cam.ac.uk/). The rearrangement of these gene segments theoretically gives rise to more than 10 million combinations, which is certain to be an underestimation given the extra variability introduced at junctions and through processes such as receptor editing (Neuberger, 2002). The process of somatic hypermutation (SHM), which occurs in B cells specifically activated by T-cell interaction, can further modify the antigen-binding site to increase its binding affinity to cognate antigen (Di Noia and Neuberger, 2007).

Analyses of antibody repertoires generated against infectious pathogens have revealed diverse strategies to elicit protective immunity. In some cases, a single V gene combination binding to a conserved carbohydrate epitope dominates the immune response. This kind of restriction usually results from surface exposure of a low-complexity antigen, such as the repetitive capsule polysaccharide of *Haemophilus influenza* (Lucas and Reason, 1999). In cases where the pathogen displays a more complex antigenic surface, greater repertoire diversity is observed. In one study analyzing the antibody repertoire in two donors after boosting with tetanus vaccine, ~100 unique VH–VL combinations were calculated to comprise the complete response to the complex protein target tetanus toxoid (Poulsen et al., 2007).

Two studies have analyzed the human antibody repertoire against WNV using display approaches. In one approach, scFv phage libraries were constructed from B cells of donors following a symptomatic infection with WNV (Throsby et al., 2006). Blood samples were taken from convalescent patients 1, 2, and 3 months after clinical presentation. Serum IgG ELISA titers were maximum at 1 month after infection. The

ELISA titers from the different patients did not correlate with neutralization, nor did neutralization titers increase over the sampled time points. The IgG VH and VL genes were amplified from these patients and pooled in combinatorial libraries (Kramer et al., 2005). VH and VL gene representation in clones sequenced from the unselected libraries were in accordance with previously published analysis of naïve human repertoires (de Wildt et al., 1999; Poulsen et al., 2007). These libraries were screened for antibodies binding to WNV antigenic preparations that included either purified inactivated virus, virus-like particles (VLPs) consisting of prM and E-protein, or soluble recombinant E-protein. A total of 72 different unique VH genes defined at the level of HCDR3 and VH gene alignment use were identified. Sequencing of the V genes demonstrated ample diversity among the selected scFv VH-VL combinations; but in contrast to the unselected library, the selection of anti-WNV clones revealed a bias to VH1 (32% vs. 21%) and VH3 genes (39% vs. 23%) and to Vλ 1–3 (67% vs. 28%). The diversity in the repertoire is consistent with antigenic complexity displayed on the WNV surface. Structural analysis of the flavivirus envelope protein has revealed that a relatively large proportion of the envelope protein is solvent exposed (Rey et al., 1995) compared to the dense packing of spike glycoproteins such as on the vesicular stomatitis virus (VSV) or the SARS coronavirus. In addition, the structural proteins of WNV are arranged in a complex pattern on the viral surface that potentially creates further antigenic diversity (Kaufmann et al., 2006). From the unique human mAbs identified from immune libraries, ~90% bound to E-protein, a single mAb was reactive with prM, and the remaining small number was bound to unidentified targets. Heterogeneity was observed among the panel in terms of ELISA, immunoprecipitation, and western blot reactivity to E-protein, indicating that the WNV mAb binds to a diverse number of epitopes. In contrast, only nine of the mAbs in the panel demonstrated activity in a microneutralization assay against WNV. This observation was not due to affinity, as some of the highest affinity antibodies in the panel had poor or no neutralizing activity, but rather suggests that only a restricted number of the epitopes displayed by WNV are targets of neutralizing mAb.

Unfortunately, in this study the repertoire of IgM expressing B cells was not examined. It would be valuable to compare the IgM to the IgG repertoire in convalescent patients after WNV infection given the important protective role of anti-WNV IgM in animal models and its persistence in serum.

In a complementary set of experiments, two large combinatorial libraries built from naive donors were extensively screened for antibodies binding to recombinant WNV-E-protein (Gould et al., 2005). Although not formally tested, it is assumed that none of the 57 donors used to generate the library was seropositive for WNV. Only five unique scFv against WNV were isolated, of which two had neutralizing activity when converted into a bivalent IgG-like format. Although the poor recovery of WNV-specific binders from these libraries could be related to technical issues (library panning was only carried out on recombinant E-protein, which may have restricted the diversity), high-affinity scFv have been isolated against a variety of targets in previous experiments using these libraries (http://research.dcfi.harvard.edu/nfcr-ctae/research/mehta.php), including a potent neutralizing mAb against SARS coronavirus with nanomolar affinity (Sui et al., 2004). This may indicate that in the normal repertoire of individuals not infected with WNV, rearranged immunoglobulins with specificity to WNV are rare.

5 Epitope Mapping

The antigenic structure of various flaviviruses including WNV has been extensively studied. Effective development of vaccines against WNV depends on a sound knowledge of the key neutralizing and protective epitopes on WNV. However, a detailed understanding of how the human immune system responds to the presentation of these epitopes is currently lacking. In experimental WNV infections, antibodies develop against E, prM, NS1, NS3, and NS5 but only antibodies against E-protein have been identified as neutralizing and protective. E-protein mediates two important functions in flavivirus replication: virus attachment to susceptible host cells and the fusion of the viral and the cellular membranes (Mukhopadhyay et al., 2005). The only other target of protective antibodies that has been described is the NS1 protein, which is secreted from infected cells and not present on the virus surface (Chung et al., 2006; Henchal et al., 1988; Roehrig, 2003). The mechanism of protection by these NS1-specific antibodies has not been formally elucidated but is likely due to antibody-dependent cellular cytotoxicity of infected cells expressing NS1 on their surface (Chung et al., 2006).

Fine mapping of neutralizing epitopes on E-protein domains using a novel yeast display system (Oliphant et al., 2005), (escape) mutant generation (Beasley and Barrett, 2002; Sanchez et al., 2005), NMR (Wu et al., 2003), and strain comparison (Li et al., 2005) have identified several critical residues required for neutralization activity. Residues consistently identified as important for potent neutralization cluster around E-protein amino acid positions 305–312, 330–333, and 365. The recent co-crystallization of WNV E-protein domain III in complex with the potent neutralizing mAb E16 showed that these residues, located on adjacent exposed loops of the domain III, form a discontinuous epitope referred to as the domain III lateral ridge (DIII-lr) (Nybakken et al., 2005). When sequences of these loops from other flaviviruses such as Japanese encephalitis and dengue viruses are aligned, considerable variation is observed compared to the E-protein as a whole (Nybakken et al., 2005). This is consistent with the observation that E-protein DIII binding neutralizing antibodies are virus-type specific (Roehrig, 2003). Several studies have mapped cross-reactive, neutralizing mAbs to regions outside of DIII. Most of these bind in or around the fusion loop at the distal tip of domain II (DII-fl) (Crill and Chang, 2004; Goncalvez et al., 2004; Oliphant et al., 2006; Stiasny et al., 2006). Antibodies directed against this region are not as potent in vitro or protective in vivo as the DIII-binding antibodies. Recent evidence suggests that many of these epitopes are partially occluded on the surface of the infectious virus, in part explaining their lack of efficient neutralizing activity (Oliphant et al., 2006; Stiasny et al., 2006).

Mapping studies were carried out with a representative panel of human mAbs isolated from the WNV-immune libraries described above (Throsby et al., 2006). In an ELISA competition assay, 47% of WNV antibodies from the panel competed for binding with the DII-fl-binding mAb 6B6C, while in contrast only 8% competed with DIII-lr-binding mAb 7H2. Results from in vitro and in vivo functional testing of the human mAb panel were consistent with experimental systems: potently neutralizing and protective mAbs were exclusively directed to domain III of E-protein, whereas mAbs that recognized domain II had weak, if any, in vitro neutralizing activity and no in vivo protective activity (Throsby et al., 2006). CR4374 was the most potent inhibitory human anti-WNV mAb identified from the panel. Fine mapping using yeast display demonstrated that the binding activity of CR4374 was lost by mutation at residue E307, indicating that its epitope is located in the DIII-lr (Oliphant et al., 2007). However, in contrast to E16, 7H2, and other potent mAbs binding in this region, mutation at position 332 did not affect binding, suggesting it may bind a novel overlapping epitope. Taken together, the data from these experiments indicate that in humans the immunodominant epitopes on WNV are situated predominantly in E-protein domain II and generate only weak neutralizing activity, while the weakly immunogenic E-protein domain III is associated with the most potent antiviral activity.

Experimental evidence has suggested that antibodies against E-protein domain II of flaviviruses can, under certain conditions, enhance infectivity of WNV in vitro (Nybakken et al., 2005; Pierson et al., 2007). In animal models, sub-neutralizing concentrations of antibodies against a related flaviviruses resulted in lower levels of survival compared to animals that were treated with a control antibody (Hawkes and Lafferty, 1967), and the administration of nonprotective YF mAb at the time of YF or JEV virus challenge was shown to enhance neurovirulence (Gould and Buckley, 1989). In humans, a serious hemorrhagic disease is associated with previous exposure to a heterologous subtype of dengue (Cardosa, 1998; Halstead, 1988). This process called antibody-dependent enhancement (ADE) is thought to result from the internalization and productive infection of macrophages and other myeloid cells through Fc-γ receptors (Cardosa et al., 1986; Iankov et al., 2006; Takada and Kawaoka, 2003; Tirado and Yoon, 2003). In a similar way, complement fixation and viral uptake through complement receptors on macrophages have also been associated with enhanced infectivity (Cardosa et al., 1986; Tirado and Yoon, 2003). A significant proportion of non-neutralizing mAbs (~30%) isolated from the WNV-immune repertoire above bound to an epitope depending on the homodimeric form of the E-protein and were particularly active in fixing the complement (Throsby et al., 2006). Thus, the most commonly identified antibodies from the human antibody repertoire could, in theory, enhance infection via mechanisms that allow the virus to be internalized by a larger number of cell types. It is important to note that enhancement of WNV infection has also been demonstrated in vitro for the very potent anti-E-protein domain III-neutralizing antibodies at low concentration and that neutralization and enhancement activity overlap (Pierson et al., 2006, 2007); however, this finding has not been reproduced experimentally in vivo (Engle and Diamond, 2003).

Two studies, one screening the naive human B cell repertoire before infection (Gould et al., 2005) and the other looking at the B cell repertoire after infection (Throsby et al., 2006), indicate that the human immune response is skewed toward E-protein domain II of WNV, and the potent neutralizing E-protein domain III antibodies are rare in the human B cell repertoire. To address this point further, analysis of human serum samples has been carried out with gain or loss of function WNV E-protein mutants (Oliphant et al., 2007). Convalescent serum samples screened at various time points after clinical disease demonstrated a skewing toward DII epitopes compared to DIII-binding epitopes. Although substantial variation was observed, on average, only 7.3%

(range, 0.6–50.5%) of the total IgG response was directed to DIII and an even smaller fraction (1.6%) to the potent neutralizing epitope DIII-lr. These percentages are remarkably close to those derived from repertoire analysis described above (8% for DIII and 2% for DIII-lr). In contrast, 61% (range 8.8–91%) of total IgG was directed to a single epitope at the tip of domain II. Again, this is consistent with repertoire analysis, where 47% of antibodies were shown to bind in the same region (Throsby et al., 2006).

A second observation to emerge was the clear difference in the response of experimental animals and humans to WNV infections. In mice, the WNV E-protein-specific IgM appeared early after infection, peaked by 8 days, and declined rapidly. Interestingly, and in contrast to the situation in humans, a specific IgM response against DIII was already present in naïve serum and increased after infection. Pre-infection immunoreactivity to viral proteins is frequently observed (Avrameas, 1991) and is generally ascribed to natural IgM (Casali and Schettino, 1996; Ochsenbein and Zinkernagel, 2000). In mice, natural IgM is produced in a T-independent fashion by B1 cells that are defined by surface expression of CD5 and localization to the peritoneum (Baumgarth et al., 2000; Casali and Notkins, 1989). Natural antibodies are characterized by polyspecific low-affinity binding and generally encoded by germline V genes without much evidence of somatic mutation. Natural antibodies are proposed to form a first line of defense against infection (Ochsenbein and Zinkernagel, 2000). In humans, the existence of a specific lineage of B cells producing natural antibodies is controversial and, to date, has not been conclusively demonstrated. T-independent antibody responses to carbohydrate antigens are mediated by marginal zone B cells which harbor mutated V genes (Lucas and Reason, 1999) that may be prediversified (Weller et al., 2004). However, most human polyreactive immunoreactivity is ascribed to naïve B cells that have not undergone antigen-driven proliferation and maturation (Tsuiji et al., 2006).

Although there was a prompt IgM response directed toward DIII-lr in mice, IgG against this critical protective epitope was not detected until days 10–15. Sequence analysis of several strongly neutralizing murine mAb binding the DIII-lr epitope showed that their V regions were in germline configuration (Oliphant et al., 2007). Under these circumstances, a rapid protective IgG response would be expected as observed for other acute cytopathic viruses (Bachmann et al., 1997). A possible explanation is that during the explosive B-cell proliferation and expansion in germinal centers, antibody specificities directed against the multiple immunodominant DI and DII epitopes simply outcompete the more restricted DIII-lr epitope.

Although delayed, a robust murine IgG response was observed against the DIII-lr epitope, indicating that there is nothing intrinsic in its molecular structure that makes DIII-lr nonimmunogenic. Rather, the failure in humans to generate IgM and consistent IgG responses against DIII-lr may result from a hole in the human antibody germline repertoire, which makes it difficult to develop high-affinity antibodies to the DIII-lr epitope. This phenomenon has been reported in the immune response to lymphocytic choriomeningitis virus (LCMV) in mice. LCMV was shown to be an effective immunogen when transgenic mice already bearing high-affinity cognate receptors were infected, but in wild-type mice infection results in a very poor antibody response against the principal neutralizing epitope (Hangartner et al., 2003). It is speculated that the coevolution of LCMV with its host has selected a neutralizing epitope not recognized with high avidity by antibodies in the murine germine repertoire (Hangartner et al., 2006). It is unlikely that WNV has evolved in an analogous manner. Most mammals, including humans, are dead-end hosts for the virus (Hayes and Gubler, 2006). B-cell immunoglobulin receptors against DIII-lr may simply be absent in the repertoire of many individuals, or are so rare that they are outcompeted for antigen by antibodies against immunodominant epitopes. Alternatively, the conformational epitope on DIII-lr may mimic antigenic structures on human self-proteins. If this were the case, then B cells expressing high-avidity Ig receptors to these cross-reactive epitopes would be purged by negative selection during B-cell differentiation and not contribute to the humoral immune response against WNV.

6 Conclusion

Healthy individuals appear to be able to mount a robust protective response against WNV infection. Only a relatively small number of those infected display clinical signs and an even smaller number develop serious symptoms. Innate resistance mechanisms such as interferon secretion and signaling likely explain much of this protective activity. In experimental models, humoral immunity plays an important role in protection from lethal infection; however, the protective role antibodies play in humans is less clear. The studies performed to date suggest that there are several deficits in the human antibody response to natural WNV infection when compared to experimental models. In particular, the antibody response to a critical potent neutralizing epitope on

E-protein domain III of the envelope protein only develops in a minority of individuals compared to immunodominant but poorly neutralizing epitopes. Another anomalous observation is the persistence of IgM serum levels that could result from viral interference in isotype switching. Deficits in the human antibody response to WNV infection may impact protection from disease when the innate response is weakened owing to age or immunosuppression. Further analysis of the human antibody response to WNV will likely help direct development of effective and safe vaccines that can be targeted to high-risk groups such as the elderly. The current findings suggest that vaccine developmental efforts should focus on increasing the immunogenicity of E-protein domain III while maintaining the broad coverage of WNV strains and preventing escape within an individual due to the formation of quasispecies.

References

Avrameas S (1991) Natural autoantibodies: from 'horror autotoxicus' to 'gnothi seauton'. Immunol Today, 12:154–9.

Bachmann MF, Kalinke U, Althage A, Freer G, Burkhart C, Roost H, Aguet M, Hengartner H, Zinkernagel RM, Roost HP, Haag A, and Pliska V (1997) The role of antibody concentration and avidity in antiviral protection. Early high-affinity neutralizing anti-viral IgG responses without further overall improvements of affinity. Science, 276:2024–7.

Baumgarth N, Herman OC, Jager GC, Brown LE, Herzenberg LA, and Chen J (2000) B-1 and B-2 cell-derived immunoglobulin M antibodies are nonredundant components of the protective response to influenza virus infection. J Exp Med, 192:271–80.

Beasley DW, and Barrett AD (2002) Identification of neutralizing epitopes within structural domain III of the West Nile virus envelope protein. J Virol, 76:13097–100.

Ben-Nathan D, Lustig S, Tam G, Robinzon S, Segal S, Rager-Zisman B, Roehrig JT, Staudinger LA, Hunt AR, Mathews JH, and Blair CD (2003) Prophylactic and therapeutic efficacy of human intravenous immunoglobulin in treating West Nile virus infection in mice Antibody prophylaxis and therapy for flavivirus encephalitis infections. J Infect Dis, 188:5–12.

Bernasconi NL, Traggiai E, and Lanzavecchia A (2002) Maintenance of serological memory by polyclonal activation of human memory B cells. Science, 298:2199–202.

Binder GK, and Griffin DE (2003) Immune-mediated clearance of virus from the central nervous system. Microbes Infect, 5:439–48.

Busch MP, Wright DJ, Custer B, Tobler LH, Stramer SL, Kleinman SH, Prince HE, Bianco C, Foster G, Petersen LR, Nemo G, and Glynn SA (2006) West Nile virus infections projected from blood donor screening data, United States, 2003. Emerg Infect Dis, 12:395–402.

Byrne SN, Halliday GM, Johnston LJ, and King NJ (2001) Interleukin-1beta but not tumor necrosis factor is involved in West Nile virus-induced Langerhans cell migration from the skin in C57BL/6 mice. J Invest Dermatol, 117:702–9.

Camenga DL, Nathanson N, and Cole GA (1974) Cyclophosphamide-potentiated West Nile viral encephalitis: relative influence of cellular and humoral factors. J Infect Dis, 130:634–41.

Cardosa MJ (1998) Dengue vaccine design: issues and challenges. Br Med Bull, 54:395–405.

Cardosa MJ, Gordon S, Hirsch S, Springer TA, and Porterfield JS (1986) Interaction of West Nile virus with primary murine macrophages: role of cell activation and receptors for antibody and complement. J Virol, 57:952–9.

Casali P, and Notkins AL (1989) CD5 + B lymphocytes, polyreactive antibodies and the human B-cell repertoire. Immunol Today, 10:364–8.

Casali P, and Schettino EW (1996) Structure and function of natural antibodies. Curr Top Microbiol Immunol, 210:167–79.

Cassese G, Arce S, Hauser AE, Lehnert K, Moewes B, Mostarac M, Muehlinghaus G, Szyska M, Radbruch A, and Manz RA (2003) Plasma cell survival is mediated by synergistic effects of cytokines and adhesion-dependent signals. J Immunol, 171:1684–90.

Chung KM, Nybakken GE, Thompson BS, Engle MJ, Marri A, Fremont DH, and Diamond MS (2006) Antibodies against West Nile virus nonstructural protein NS1 prevent lethal infection through Fc gamma receptor-dependent and -independent mechanisms. J Virol, 80:1340–51.

Crill WD, and Chang GJ (2004) Localization and characterization of flavivirus envelope glycoprotein cross-reactive epitopes. J Virol, 78:13975–86.

de Wildt RM, Hoet RM, van Venrooij WJ, Tomlinson IM, and Winter G (1999) Analysis of heavy and light chain pairings indicates that receptor editing shapes the human antibody repertoire. J Mol Biol, 285:895–901.

Di Noia JM, and Neuberger MS (2007) Molecular mechanisms of antibody somatic hypermutation. Annu Rev Biochem, 76:1–22.

Diamond MS, Shrestha B, Marri A, Mahan D, and Engle M (2003a) B cells and antibody play critical roles in the immediate defense of disseminated infection by West Nile encephalitis virus. J Virol, 77:2578–86.

Diamond MS, Sitati EM, Friend LD, Higgs S, Shrestha B, and Engle M (2003b) A critical role for induced IgM in the protection against West Nile virus infection. J Exp Med, 198:1853–62.

Engle MJ, and Diamond MS (2003) Antibody prophylaxis and therapy against West Nile virus infection in wild-type and immunodeficient mice. J Virol, 77:12941–9.

Fox JL, Hazell SL, Tobler LH, and Busch MP (2006) Immunoglobulin G avidity in differentiation between early and late antibody responses to West Nile virus. Clin Vaccine Immunol, 13:33–6.

Gea-Banacloche J, Johnson RT, Bagic A, Butman JA, Murray PR, and Agrawal AG (2004) West Nile virus: pathogenesis and therapeutic options. Ann Intern Med, 140:545–53.

Goncalvez AP, Purcell RH, and Lai CJ (2004) Epitope determinants of a chimpanzee Fab antibody that efficiently cross-neutralizes dengue type 1 and type 2 viruses map to inside and in close proximity to fusion loop of the dengue type 2 virus envelope glycoprotein. J Virol, 78:12919–28.

Gould EA, and Buckley A (1989) Antibody-dependent enhancement of yellow fever and Japanese encephalitis virus neurovirulence. J Gen Virol, 70:1605–8.

Gould LH, Sui J, Foellmer H, Oliphant T, Wang T, Ledizet M, Murakami A, Noonan K, Lambeth C, Kar K, Anderson JF, de Silva AM, Diamond MS, Koski RA, Marasco WA, and Fikrig E (2005) Protective and therapeutic capacity of human single-chain Fv-Fc fusion proteins against West Nile virus. J Virol, 79:14606–13.

Granwehr BP, Lillibridge KM, Higgs S, Mason PW, Aronson JF, Campbell GA, and Barrett AD (2004) West Nile virus: where are we now? Lancet Infect Dis, 4:547–56.

Halevy M, Akov Y, Ben-Nathan D, Kobiler D, Lachmi B, and Lustig S (1994) Loss of active neuroinvasiveness in attenuated strains of West Nile virus: pathogenicity in immunocompetent and SCID mice. Arch Virol, 137:355–70.

Haley M, Retter AS, Fowler D, Gea-Banacloche J, and O'Grady NP (2003) The role for intravenous immunoglobulin in the treatment of West Nile virus encephalitis. Clin Infect Dis, 37:88–90.

Halstead SB (1988) Pathogenesis of dengue: challenges to molecular biology. Science, 239: 476–81.

Hamdan A, Green P, Mendelson E, Kramer MR, Pitlik S, and Weinberger M (2002) Possible benefit of intravenous immunoglobulin therapy in a lung transplant recipient with West Nile virus encephalitis. Transpl Infect Dis, 4:160–2.

Han XY, Ren QW, Xu ZY, and Tsai TF (1988) Serum and cerebrospinal fluid immunoglobulins M, A, and G in Japanese encephalitis. J Clin Microbiol, 26:976–8.

Hangartner L, Senn BM, Ledermann B, Kalinke U, Seiler P, Bucher E, Zellweger RM, Fink K, Odermatt B, Burki K, Zinkernagel RM, and Hengartner H (2003) Antiviral immune responses in gene-targeted mice expressing the immunoglobulin heavy chain of virus-neutralizing antibodies. Proc Natl Acad Sci U S A, 100:12883–8.

Hangartner L, Zinkernagel RM, and Hengartner H (2006) Antiviral antibody responses: the two extremes of a wide spectrum. Nat Rev Immunol, 6:231–43.

Hawkes RA, and Lafferty KJ (1967) The enchancement of virus infectivity by antibody. Virology, 33:250–61.

Hayes EB, and Gubler DJ (2006) WEST NILE VIRUS: epidemiology and clinical features of an emerging epidemic in the United States. Annu Rev Med, 57:181–94.

Henchal EA, Henchal LS, and Schlesinger JJ (1988) Synergistic interactions of anti-NS1 monoclonal antibodies protect passively immunized mice from lethal challenge with dengue 2 virus. J Gen Virol, 69 (Pt 8):2101–7.

Iankov ID, Pandey M, Harvey M, Griesmann GE, Federspiel MJ, and Russell SJ (2006) Immunoglobulin G antibody-mediated enhancement of measles virus infection can bypass the protective antiviral immune response. J Virol, 80:8530–40.

Iwamoto M, Jernigan DB, Guasch A, Trepka MJ, Blackmore CG, Hellinger WC, Pham SM, Zaki S, Lanciotti RS, Lance-Parker SE, DiazGranados CA, Winquist AG, Perlino CA, Wiersma S, Hillyer KL, Goodman JL, Marfin AA, Chamberland ME, and Petersen LR (2003) Transmission of West Nile virus from an organ donor to four transplant recipients. N Engl J Med, 348:2196–203.

Kapoor H, Signs K, Somsel P, Downes FP, Clark PA, and Massey JP (2004) Persistence of West Nile Virus (WNV) IgM antibodies in cerebrospinal fluid from patients with CNS disease. J Clin Virol, 31:289–91.

Kaufmann B, Nybakken GE, Chipman PR, Zhang W, Diamond MS, Fremont DH, Kuhn RJ, and Rossmann MG (2006) West Nile virus in complex with the Fab fragment of a neutralizing monoclonal antibody. Proc Natl Acad Sci U S A, 103:12400–4.

Klee AL, Maidin B, Edwin B, Poshni I, Mostashari F, Fine A, Layton M, and Nash D (2004) Long-term prognosis for clinical West Nile virus infection. Emerg Infect Dis, 10:1405–11.

Kramer RA, Marissen WE, Goudsmit J, Visser TJ, Clijsters-Van der Horst M, Bakker AQ, de Jong M, Jongeneelen M, Thijsse S, Backus HH, Rice AB, Weldon WC, Rupprecht CE, Dietzschold B, Bakker AB, and de Kruif J (2005) The human antibody repertoire specific for rabies virus glycoprotein as selected from immune libraries. Eur J Immunol, 35:2131–45.

Lanciotti RS, Kerst AJ, Nasci RS, Godsey MS, Mitchell CJ, Savage HM, Komar N, Panella NA, Allen BC, Volpe KE, Davis BS, and Roehrig JT (2000) Rapid detection of West Nile virus from human clinical specimens, field-collected mosquitoes, and avian samples by a TaqMan reverse transcriptase-PCR assay. J Clin Microbiol, 38:4066–71.

Li L, Barrett AD, and Beasley DW (2005) Differential expression of domain III neutralizing epitopes on the envelope proteins of West Nile virus strains. Virology, 335:99–105.

Lucas AH, and Reason DC (1999) Polysaccharide vaccines as probes of antibody repertoires in man. Immunol Rev, 171:89–104.

Manz RA, and Radbruch A (2002) Plasma cells for a lifetime? Eur J Immunol, 32:923–7.

Martin DA, Biggerstaff BJ, Allen B, Johnson AJ, Lanciotti RS, and Roehrig JT (2002) Use of immunoglobulin m cross-reactions in differential diagnosis of human flaviviral encephalitis infections in the United States. Clin Diagn Lab Immunol, 9:544–9.

Monath TP, Liu J, Kanesa-Thasan N, Myers GA, Nichols R, Deary A, McCarthy K, Johnson C, Ermak T, Shin S, Arroyo J, Guirakhoo F, Kennedy JS, Ennis FA, Green S, and Bedford P (2006) A live, attenuated recombinant West Nile virus vaccine. Proc Natl Acad Sci U S A, 103:6694–9.

Mostashari F, Bunning ML, Kitsutani PT, Singer DA, Nash D, Cooper MJ, Katz N, Liljebjelke KA, Biggerstaff BJ, Fine AD, Layton MC, Mullin SM, Johnson AJ, Martin DA, Hayes EB, and Campbell GL (2001) Epidemic West Nile encephalitis, New York, 1999: results of a household-based seroepidemiological survey. Lancet, 358:261–4.

Mukhopadhyay S, Kuhn RJ, and Rossmann MG (2005) A structural perspective of the flavivirus life cycle. Nat Rev Microbiol, 3:13–22.

Neuberger MS (2002) Novartis Medal Lecture. Antibodies: a paradigm for the evolution of molecular recognition. Biochem Soc Trans, 30:341–50.

Nybakken GE, Oliphant T, Johnson S, Burke S, Diamond MS, and Fremont DH (2005) Structural basis of West Nile virus neutralization by a therapeutic antibody. Nature, 437:764–9.

Ochsenbein AF, and Zinkernagel RM (2000) Natural antibodies and complement link innate and acquired immunity. Immunol Today, 21:624–30.

Oliphant T, Engle M, Nybakken GE, Doane C, Johnson S, Huang L, Gorlatov S, Mehlhop E, Marri A, Chung KM, Ebel GD, Kramer LD, Fremont DH, and Diamond MS (2005) Development of a humanized monoclonal antibody with therapeutic potential against West Nile virus. Nat Med, 11:522–30.

Oliphant T, Nybakken GE, Engle M, Xu Q, Nelson CA, Sukupolvi-Petty S, Marri A, Lachmi BE, Olshevsky U, Fremont DH, Pierson TC, and Diamond MS (2006) Antibody recognition and neutralization determinants on domains I and II of West Nile virus envelope protein. J Virol, 80:12149–59.

Oliphant T, Nybakken GE, Austin SK, Xu Q, Bramson J, Loeb M, Throsby M, Fremont DH, Pierson TC, and Diamond MS (2007) Induction of epitope-specific neutralizing antibodies against West Nile virus. J Virol, 81:11828–39.

Pierson TC, Sanchez MD, Puffer BA, Ahmed AA, Geiss BJ, Valentine LE, Altamura LA, Diamond MS, and Doms RW (2006) A rapid and quantitative assay for measuring antibody-mediated neutralization of West Nile virus infection. Virology, 346:53–65.

Pierson TC, Xu Q, Nelson S, Oliphant T, Nybakken GE, Fremont DH, and Diamond MS (2007) The stoichiometry of antibody-mediated neutralization and enhancement of West Nile virus infection. Cell Host Microbe, 1:135–45.

Poulsen TR, Meijer PJ, Jensen A, Nielsen LS, and Andersen PS (2007) Kinetic, affinity, and diversity limits of human polyclonal antibody responses against tetanus toxoid. J Immunol, 179:3841–50.

Prince HE, Tobler LH, Lape-Nixon M, Foster GA, Stramer SL, and Busch MP (2005) Development and persistence of West Nile virus-specific immunoglobulin M (IgM), IgA, and IgG in viremic blood donors. J Clin Microbiol, 43:4316–20.

Rey FA, Heinz FX, Mandl C, Kunz C, and Harrison SC (1995) The envelope glycoprotein from tick-borne encephalitis virus at 2 A resolution. Nature, 375:291–8.

Roehrig JT (2003) Antigenic structure of flavivirus proteins. Adv Virus Res, 59:141–75.

Roehrig JT, Nash D, Maldin B, Labowitz A, Martin DA, Lanciotti RS, and Campbell GL (2003) Persistence of virus-reactive serum immunoglobulin m antibody in confirmed West Nile virus encephalitis cases. Emerg Infect Dis, 9:376–9.

Sanchez MD, Pierson TC, McAllister D, Hanna SL, Puffer BA, Valentine LE, Murtadha MM, Hoxie JA, and Doms RW (2005) Characterization of neutralizing antibodies to West Nile virus. Virology, 336:70–82.

Shimoni Z, Niven MJ, Pitlick S, and Bulvik S (2001) Treatment of West Nile virus encephalitis with intravenous immunoglobulin. Emerg Infect Dis, 7:759.

Stiasny K, Kiermayr S, Holzmann H, and Heinz FX (2006) Cryptic properties of a cluster of dominant flavivirus cross-reactive antigenic sites. J Virol, 80:9557–68.

Sui J, Li W, Murakami A, Tamin A, Matthews LJ, Wong SK, Moore MJ, St Clair Tallarico A, Olurinde M, Choe H, Anderson LJ, Bellini WJ, Farzan M, and Marasco WA (2004) Potent neutralization of severe acute respiratory syndrome (SARS) coronavirus by a human mAb to S1 protein that blocks receptor association. Proc Natl Acad Sci U S A, 6:6.

Summers PL, Eckels KH, Dalrymple JM, Scott RM, and Boyd VA (1984) Antibody response to dengue-2 vaccine measured by two different radioimmunoassay methods. J Clin Microbiol, 19:651–9.

Takada A, and Kawaoka Y (2003) Antibody-dependent enhancement of viral infection: molecular mechanisms and in vivo implications. Rev Med Virol, 13:387–98.

Tesh RB, Arroyo J, Travassos Da Rosa AP, Guzman H, Xiao SY, and Monath TP (2002) Efficacy of killed virus vaccine, live attenuated chimeric virus vaccine, and passive immunization for prevention of West Nile virus encephalitis in hamster model. Emerg Infect Dis, 8:1392–7.

Throsby M, Geuijen C, Goudsmit J, Bakker AQ, Korimbocus J, Kramer RA, Clijsters-van der Horst M, de Jong M, Jongeneelen M, Thijsse S, Smit R, Visser TJ, Bijl N, Marissen WE, Loeb M, Kelvin DJ, Preiser W, ter Meulen J, and de Kruif J (2006) Isolation and characterization of human monoclonal antibodies from individuals infected with West Nile virus. J Virol, 80:6982–92.

Tirado SM, and Yoon KJ (2003) Antibody-dependent enhancement of virus infection and disease. Viral Immunol, 16:69–86.

Tsuiji M, Yurasov S, Velinzon K, Thomas S, Nussenzweig MC, and Wardemann H (2006) A checkpoint for autoreactivity in human IgM + memory B cell development. J Exp Med, 203:393–400.

Wang T, Anderson JF, Magnarelli LA, Wong SJ, Koski RA, and Fikrig E (2001) Immunization of mice against West Nile virus with recombinant envelope protein. J Immunol, 167:5273–7.

Wang T, Town T, Alexopoulou L, Anderson JF, Fikrig E, and Flavell RA (2004) Toll-like receptor 3 mediates West Nile virus entry into the brain causing lethal encephalitis. Nat Med, 10:1366–73.

Weller S, Braun MC, Tan BK, Rosenwald A, Cordier C, Conley ME, Plebani A, Kumararatne DS, Bonnet D, Tournilhac O, Tchernia G, Steiniger B, Staudt LM, Casanova JL, Reynaud CA, and Weill JC (2004) Human blood IgM "memory" B cells are circulating splenic marginal zone B cells harboring a prediversified immunoglobulin repertoire. Blood, 104:3647–54.

Wu KP, Wu CW, Tsao YP, Kuo TW, Lou YC, Lin CW, Wu SC, and Cheng JW (2003) Structural basis of a flavivirus recognized by its neutralizing antibody: solution structure of the domain III of the Japanese encephalitis virus envelope protein. J Biol Chem, 278:46007–13.

19. Antibody Protection and Therapy for West Nile Virus Infections

John T. Roehrig

Abstract

Although West Nile virus (WNV) vaccines are currently under development, an approved WNV vaccine for human use is likely years away. Because the incidence of WNV human infection is very low, it will be difficult to target the human population in need of vaccination and to assess a vaccine's economic feasibility. At present, there are also no specific therapeutic agents for flaviviral infections. It has been demonstrated, however, that prophylactic administration of anti-envelope (anti-E) protein antibody can protect animals from virus challenge. Antibody prophylaxis has been evaluated using monoclonal antibodies (MAbs), primarily in murine animal models, to determine the appropriate timing of antibody injection and epitope specificity of MAb necessary for the most efficacious prophylaxis. The protective efficacy of an E protein-specific MAb is directly related to its ability to neutralize virus infectivity. The window for successful therapeutic administration of antibody closes at about 4–5 days postinfection concomitant with viral invasion of the brain in small animal models. Efforts are now being focused on developing humanized murine or fully human MAbs with antiviral protective or therapeutic activity, for use in humans. Antiflaviviral human or humanized MAbs might be practical and cost-effective reagents for preventing or modifying the outcome of flaviviral infections.

Keywords

antibody, neutralization, therapy, flavivirus

1 Introduction

The outbreak of West Nile virus (WNV) in the USA has spurred interest in developing direct methods for prevention and control of human WNV infections (Craven and Roehrig, 2001; Petersen and

Roehrig, 2001; Petersen et al., 2002; Roehrig et al., 2002). Although equine WNV vaccines have been developed, approved, and deployed, a WNV vaccine for humans is years away. In areas of intense WNV transmission, the human infection rate is only around 2.5%, so the overall incidence of WNV human infection is still very low (Mostashari et al., 2001). The low rate of human exposure will make it difficult to target the human populations in need of vaccination and to assess the vaccine's protective capacity and economic viability.

While there are also no specific therapeutic agents for flaviviral infections, prophylactic application of antiflaviviral antibody can protect laboratory animals from virus challenge (Roehrig et al., 2001). Antibody prophylaxis has been evaluated using murine and human monoclonal antibodies (MAbs) to determine the timing of antibody application, the protein/epitope specificity, and the quantity of applied antibody necessary for successful prophylaxis (Roehrig et al., 2001). The major virion structural protein is the immunodominant envelope (E) glycoprotein. The E protein binds cellular receptors, mediates cell membrane fusion, and contains an array of epitopes that elicit virus-neutralizing and non-neutralizing antibodies (Roehrig, 2003). The protective efficacy of an E protein-specific MAb is directly related to its ability to neutralize virus infectivity. The use of murine MAbs in humans results in a human anti-mouse antibody (HAMA) response, which eventually limits the effectiveness of subsequent murine antibody treatments. To reduce the HAMA response and make these MAbs more generally useful for humans, murine MAbs can be "humanized," or human MAbs with analogous reactivities can be developed. Antiflaviviral human or humanized MAbs might be practical and cost-effective reagents for preventing or modifying flaviviral diseases.

2 The Role of the E Protein in Antiflaviviral Immunity

The structure, function, and role of the E protein in antiflaviviral immunity are reviewed in great detail elsewhere in this book. The E protein can be divided into three structural and functional domains: I, II, and III (Roehrig, 2003). The quality of the antibody elicited by each of these domains varies. Domain II elicits more cross-reactive antibodies that block virus-mediated cell membrane fusion, and to a lesser degree viral attachment (Oliphant and Diamond, 2007; Roehrig et al., 1998). The binding sites for some DII-reactive MAbs have been mapped by mutagenesis of expressed DI/DII of WNV or virus-like particles of WNV and other flaviviruses (Crill et al., 2007; Oliphant et al., 2006; Roberson et al., 2007; Sukupolvi-Petty et al., 2007; Trainor et al., 2007). These antibodies appear to bind to amino acids in or near the E protein fusion tip (amino acids 98–110). Antibodies reactive with DIII are more virus type specific; however, epitopes that elicit cross-reactive antiviral antibodies can also be identified within this domain. Antibodies specific for DIII of dengue (DEN) virus are potent neutralizers of viral infection, and appear to be the antibodies that best block viral attachment to cells (Crill and Roehrig, 2001). Additionally, WNV anti-DIII MAbs have been shown to block viral infection at a step postattachment (Oliphant and Diamond, 2007). Virus-neutralizing DIII MAbs bind to the upper and lateral surfaces of DIII (Oliphant and Diamond, 2007).

3 Using Small Animal Models of Infection to Define Antibody-Mediated Protection and Therapy

The laboratory mouse has been used successfully to investigate the ability of MAbs to protect animals from or cure them of infections with neurotropic flaviviruses (Ben-Nathan et al., 2003; Broom et al., 2000; Diamond et al., 2003; Engle and Diamond, 2003; Gould et al., 2005; Julander et al., 2005; Kreil, 2004; Kreil et al., 1997, 1998a, b; Kreil and Eibl, 1997; Morrey et al., 2004, 2006, 2007; Nybakken et al., 2005; Oliphant et al., 2005; Roehrig et al., 2001; Throsby et al., 2006). Early studies used passive antibody transfer into mice to demonstrate that the anti-E protein MAbs could protect mice from infection. Results were remarkably similar, regardless of the challenge flavivirus. Transfer of as little as 5 μg of antibody could protect mice from challenge with either St. Louis encephalitis (SLE), Murray Valley encephalitis (MVE), or Japanese encephalitis (JE) viruses (Hawkes et al., 1988; Mathews and Roehrig, 1984; Roehrig et al., 2001). Subsequent work with WNV confirmed these earlier results. Additionally, complement has been confirmed as an important component of the antiviral immune response to WNV (Mehlhop et al., 2005).

Generally, the protective capacity of these MAbs depend on their specificity, the time, and quantity of inoculation (Fig. 1). The MAbs that define critical virus neutralization epitopes in DIII are the most efficient at protecting or curing mice from virus infection. A single 2 mg dose of anti-WNV DIII MAb E16 cured 90% of WNV-infected C57BL/6J mice at 5 days postinfection (Oliphant et al., 2005).

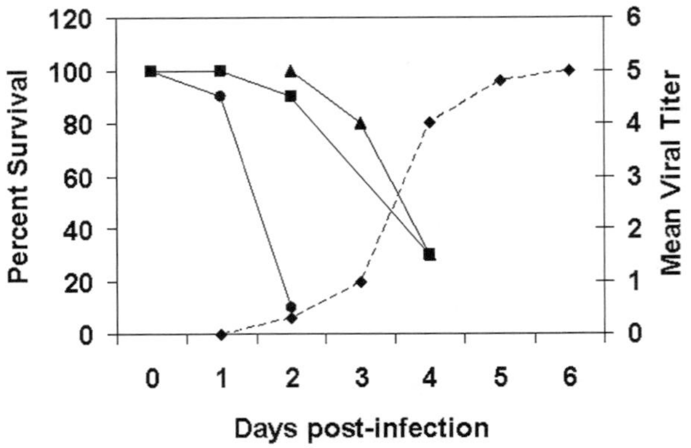

Figure 1. Ability of anti-E-1c mab to cure SLE virus-infected mice as a function of viral titer in the brain. Twenty-one-day-old mice ($n = 10$) were intravenously administered purified anti-E-1c Mab 24 h prior to or 1, 2, 3, 4, 5, or 6 days after i.p. challenge with 100 i.p. LD_{50} of SLE virus strain MSI-7. Amounts of transferred antibody were 1 μg (*filled circle*), 20 μg (*filled square*), or 100 μg (*filled triangle*). Mice were observed daily for clinical signs of viral encephalitis for 15 days. Percent protection was calculated as the number of mice without symptoms divided by the total number of mice challenged. Mean viral titers ($\log_{10}$ PFU/ml) in brain (*dashed line*) were determined by plaque assay. (Adapted from Mathews and Roehrig, 1984).

4 Previous Use of Human Antibody for Prophylaxis and Therapy for Flaviviral Infections

Applications of antibody, either prophylactically or therapeutically, continue to be evaluated as interventional approaches for a number of viral infections (Sawyer, 2000). There are, however, no specific approved therapeutic agents for flaviviral encephalitis. Anti-tick-borne encephalitis (TBE) virus immune serum has been used in Austria for nonimmunized humans following a possible tick-bite exposure to TBE virus. In an attempt to determine the efficacy of this practice, a telephone survey was conducted in Vienna (Kunz et al., 1981). The incidence of clinical TBE for unvaccinated persons in TBE virus endemic areas was approximately 1:1,000. Administration of TBE immunoglobulin after a tick bite reduced this disease incidence to 1:2,500, thus showing an apparent rate of protection of approximately 60%. While this analysis was not scientifically rigorous, the low incidence of human TBE virus infection makes the use of more standard epidemiological approaches to answer this question difficult. Some concerns have been raised as to whether this treatment may result in an "antibody-enhanced" infection with the demonstration of more severe clinical symptoms, similar to that proposed with DEN viruses (Kluger et al., 1995; Waldvogel et al., 1996). In general, however, if such "enhanced disease" does occur with TBE virus infections, it is of very low frequency (Arras et al., 1996).

Three case reports using intravenous immunoglobulin (IVIG) for treatment of WNV-infected individuals have been documented (Agrawal and Petersen, 2003; Haley et al., 2003; Hamdan et al., 2002; Shimoni et al., 2001). The first example was a comatose WNV-infected woman with chronic lymphocytic leukemia (Shimoni et al., 2001). She recovered after treatment with IVIG of a presumed anti-WNV antibody titer of 1:1,600. The second patient was a lung transplant recipient who recovered from WNV encephalitis following administration of IVIG (Hamdan et al., 2002). Six other patients had variable outcomes (Agrawal and Petersen, 2003; Haley et al., 2003). One problem with this approach is that the level of anti-WNV antibody in IVIG preparations is dependent on the incidence of WNV infection (Planitzer et al., 2007). Low incidence of WNV disease – which is currently the case for the USA – results in the collection of specimens with low anti-WNV antibody titers resulting in a poor IVIG product.

5 MAbs as Human Therapeutics

Successful immunotherapy for WNV infection will depend on three factors: (1) the time postinfection the antibody is administered, (2) the titer of the immunoglobulin, and (3) the protective biological activity (protective capacity) of the transferred immunoglobulin. Commercial value of this approach will depend upon the cost of production of the MAb and the number of doses needed. To make this approach most efficacious and cost effective, substituting virus-reactive, protective human or humanized MAbs for IVIG is appropriate. Unlike IVIG, whose concentration of virus-reactive antibody may vary from lot to lot, MAbs are well-defined proteins of known specificity, concentration, and protective capacity in animal models of virus challenge (Reichert and Dewitz, 2006; Zhu et al., 2006).

This approach is currently being investigated for WNV (Gould et al., 2005; Oliphant et al., 2005; Throsby et al., 2006). A similar approach has been applied successfully for neurotropic alphaviral infections (Hunt et al., 2006). A WNV-reactive, protective murine MAb, E16, that reacts with four discontinuous segments of DIII has been humanized, hE16, and is currently in clinical trials. When delivered as a single dose of 4 μg at 2 days postinfection, this humanized MAb cured 60% of WNV-infected mice (Oliphant et al., 2005). A 100 μg dose cured 90% of WNV-infected mice. The therapeutic potential of hE16 has also been evaluated in hamsters. A single 100 mg/kg peripheral dose of hE16 cured 80% of WNV-challenged hamsters, whereas 37% of placebo-inoculated hamsters survived this challenge (Morrey et al., 2006). As with the mouse model, this temporal limit of protection correlated with the entry of virus into the brain of the infected animals. Eighty-eight percent of hamsters survived WNV challenge if the hE16 was injected directly into the brain. Therapeutic efficacy could be extended to 6 days postinfection if antibody was delivered by convection-enhanced delivery (Morrey et al., 2007).

6 Conclusion

More work is needed to fully assess the efficacy of antibody in prophylaxis and therapy of flaviviral encephalitis. Follow-up studies in nonhuman primates would be informative, even though these viruses do not cause severe symptoms in primates. It is clear from the work with small animal models, however, that the efficiency of passive antibody protection depends upon (1) the dose of antibody transferred, (2) the timing of the transfer, and (3) the biological characteristics of transferred antibody (neutralizing activity, epitope specificity, etc.). For an antibody to be effective therapeutically, it must be administered as early in infection as possible, preferably prior to the time that the virus enters the central nervous system. Because WNV-infected individuals are symptomatic at the time of their clinical presentation, serologic or virologic confirmation of WNV infection prior to antibody administration may not be possible. Because of the low disease incidence of flaviviral encephalitis, antibody prophylaxis/therapy will require an artificial source of protective antibodies. To reduce the HAMA response and make MAbs more generally useful for humans, "humanized" murine MAbs or fully human MAbs with analogous reactivities are being developed. On the basis of the discussions here, administration of antiflaviviral human or humanized MAbs may be a practical and cost-effective alternative to vaccines for preventing or modifying flaviviral diseases.

References

Agrawal AG, Petersen LR (2003) Human immunoglobulin as a treatment for West Nile virus infection. J Infect Dis 188:1–4

Arras C, Fescharek R, Gregersen JP (1996) Do specific hyperimmunoglobulins aggravate clinical course of tick-borne encephalitis? Lancet 347:1331

Ben-Nathan D, Lustig S, Tam G, Robinzon S, Segal S, Rager-Zisman B (2003) Prophylactic and therapeutic efficacy of human intravenous immunoglobulin in treating West Nile virus infection in mice. J Infect Dis 188:5–12

Broom AK, Wallace MJ, Mackenzie JS, Smith DW, Hall RA (2000) Immunisation with gamma globulin to Murray Valley encephalitis virus and with an inactivated Japanese encephalitis virus vaccine as prophylaxis against Australian encephalitis: evaluation in a mouse model. J Med Virol 61:259–265

Craven RB, Roehrig JT (2001) West Nile virus. JAMA 286:651–653

Crill WD, Roehrig JT (2001) Monoclonal antibodies that bind to domain III of dengue virus E glycoprotein are the most efficient blockers of virus adsorption to Vero cells. J Virol 75:7769–7773

Crill WD, Trainor NB, Chang GJ (2007) A detailed mutagenesis study of flavivirus cross-reactive epitopes using West Nile virus-like particles. J Gen Virol 88:1169–1174

Diamond MS, Sitati EM, Friend LD, Higgs S, Shrestha B, Engle M (2003) A critical role for induced IgM in the protection against West Nile virus infection. J Exp Med 198:1853–1862

Engle MJ, Diamond MS (2003) Antibody prophylaxis and therapy against West Nile virus infection in wild-type and immunodeficient mice. J Virol 77:12941–12949

Gould LH, Sui J, Foellmer H, Oliphant T, Wang T, Ledizet M, Murakami A, Noonan K, Lambeth C, Kar Ket al, (2005) Protective and therapeutic capacity of human single-chain Fv-Fc fusion proteins against West Nile virus. J Virol 79:14606–14613

Haley M, Retter AS, Fowler D, Gea-Banacloche J, O'Grady NP (2003) The role for intravenous immunoglobulin in the treatment of West Nile virus encephalitis. Clin Infect Dis 37:e88–e90

Hamdan A, Green P, Mendelson E, Kramer MR, Pitlik S, Weinberger M (2002) Possible benefit of intravenous immunoglobulin therapy in a lung transplant recipient with West Nile virus encephalitis. Transpl Infect Dis 4:160–162

Hawkes RA, Roehrig JT, Hunt AR, Moore GA (1988) Antigenic structure of the Murray Valley encephalitis virus E glycoprotein. J Gen Virol 69(Pt 5):1105–1109

Hunt AR, Frederickson S, Hinkel C, Bowdish KS, Roehrig JT (2006) A humanized murine monoclonal antibody protects mice either before or after challenge with virulent Venezuelan equine encephalomyelitis virus. J Gen Virol 87:2467–2476

Julander JG, Winger QA, Olsen AL, Day CW, Sidwell RW, Morre JD (2005) Treatment of West Nile virus-infected mice with reactive immunoglobulin reduces fetal titers and increases dam survival. Antiviral Res 65:79–85

Kluger G, Schottler A, Waldvogel K, Nadal D, Hinrichs W, Wundisch GF, Laub MC (1995) Tickborne encephalitis despite specific immunoglobulin prophylaxis. Lancet 346:1502

Kreil TR (2004) West Nile virus: recent experience with the model virus approach. Dev Biol (Basel) 118:101–105

Kreil TR, Eibl MM (1997) Pre- and postexposure protection by passive immunoglobulin but no enhancement of infection with a flavivirus in a mouse model. J Virol 71:2921–2927

Kreil TR, Burger I, Bachmann M, Fraiss S, Eibl MM (1997) Antibodies protect mice against challenge with tick-borne encephalitis virus (TBEV)-infected macrophages. Clin Exp Immunol 110:358–361

Kreil TR, Burger I, Attakpah E, Olas K, Eibl MM (1998a) Passive immunization reduces immunity that results from simultaneous active immunization against tick-borne encephalitis virus in a mouse model. Vaccine 16:955–959

Kreil TR, Maier E, Fraiss S, Eibl MM (1998b) Neutralizing antibodies protect against lethal flavivirus challenge but allow for the development of active humoral immunity to a nonstructural virus protein. J Virol 72:3076–3081

Kunz C, Hofmann H, Kundi M, Mayer K (1981) [Efficacy of specific immunoglobulin against TBE (author's transl)]. Wien Klin Wochenschr 93:665–667

Mathews JH, Roehrig JT (1984) Elucidation of the topography and determination of the protective epitopes on the E glycoprotein of Saint Louis encephalitis virus by passive transfer with monoclonal antibodies. J Immunol 132:1533–1537

Mehlhop E, Whitby K, Oliphant T, Marri A, Engle M, Diamond MS (2005) Complement activation is required for induction of a protective antibody response against West Nile virus infection. J Virol 79:7466–7477

Morrey JD, Day CW, Julander JG, Olsen AL, Sidwell RW, Cheney CD, Blatt LM (2004) Modeling hamsters for evaluating West Nile virus therapies. Antiviral Res 63:41–50
Morrey JD, Siddharthan V, Olsen AL, Roper GY, Wang H, Baldwin TJ, Koenig S, Johnson S, Nordstrom JL, Diamond MS (2006) Humanized monoclonal antibody against West Nile virus envelope protein administered after neuronal infection protects against lethal encephalitis in Hamsters. J Infect Dis 194:1300–1308
Morrey JD, Siddharthan V, Olsen AL, Wang H, Julander JG, Hall JO, Li H, Nordstrom JL, Koenig S, Johnson S,et al. (2007) Defining limits of treatment with humanized neutralizing monoclonal antibody for West Nile virus neurological infection in a hamster model. Antimicrob Agents Chemother 51:2396–2402
Mostashari F, Bunning ML, Kitsutani PT, Singer DA, Nash D, Cooper MJ, Katz N, Liljebjelke KA, Biggerstaff BJ, Fine AD,et-al. (2001) Epidemic West Nile encephalitis, New York, 1999: results of a household-based seroepidemiological survey. Lancet 358:261–264
Nybakken GE, Oliphant T, Johnson S, Burke S, Diamond MS, Fremont DH (2005) Structural basis of West Nile virus neutralization by a therapeutic antibody. Nature 437:764–769
Oliphant T, Diamond MS (2007) The molecular basis of antibody-mediated neutralization of West Nile virus. Expert Opin Biol Ther 7:885–892
Oliphant T, Engle M, Nybakken GE, Doane C, Johnson S, Huang L, Gorlatov S, Mehlhop E, Marri A, Chung KM,et al. (2005) Development of a humanized monoclonal antibody with therapeutic potential against West Nile virus. Nat Med 11:522–530
Oliphant T, Nybakken GE, Engle M, Xu Q, Nelson CA, Sukupolvi-Petty S, Marri A, Lachmi BE, Olshevsky U, Fremont DH,et al. (2006) Antibody recognition and neutralization determinants on domains I and II of West Nile virus envelope protein. J Virol 80:12149–12159
Petersen LR, Roehrig JT (2001) West Nile virus: a reemerging global pathogen. Emerg Infect Dis 7:611–614
Petersen LR, Roehrig JT, Hughes JM (2002) West Nile virus encephalitis. N Engl J Med 347:1225–1226
Planitzer CB, Modrof J, Kreil TR (2007) West Nile virus neutralization by US plasma-derived immunoglobulin products. J Infect Dis 196:435–440
Reichert JM, Dewitz MC (2006) Anti-infective monoclonal antibodies: perils and promise of development. Nat Rev Drug Discov 5:191–195
Roberson JA, Crill WD, Chang GJ (2007) Differentiation of West Nile and St. Louis encephalitis virus infections using cross-reactivity reduced noninfectious virus-like particles. J Clin Microbiol 45:3167–3174
Roehrig JT (2003) Antigenic structure of flavivirus proteins. Adv Virus Res 59:141–175
Roehrig JT, Bolin RA, Kelly RG (1998) Monoclonal antibody mapping of the envelope glycoprotein of the dengue 2 virus, Jamaica. Virology 246:317–328
Roehrig JT, Staudinger LA, Hunt AR, Mathews JH, Blair CD (2001) Antibody prophylaxis and therapy for flavivirus encephalitis infections. Ann NY Acad Sci 951:286–297
Roehrig JT, Layton M, Smith P, Campbell GL, Nasci R, Lanciotti RS (2002) The emergence of West Nile virus in North America: ecology, epidemiology, and surveillance. Curr Top Microbiol Immunol 267:223–240
Sawyer L (2000) Antibodies for prevention and treatment of viral diseases. Antiviral Res 47:57–77
Shimoni Z, Niven MJ, Pitlick S, Bulvik S (2001) Treatment of West Nile virus encephalitis with intravenous immunoglobulin. Emerg Infect Dis 7:759
Sukupolvi-Petty S, Austin SK, Purtha WE, Oliphant T, Nybakken GE, Schlesinger JJ, Roehrig JT, Gromowski GD, Barrett AD, Fremont DH,et-al. (2007) Type- and subcomplex-specific neutralizing antibodies against domain III of dengue virus type 2 envelope protein recognize adjacent epitopes. J Virol 81:12816–12826
Throsby M, Geuijen C, Goudsmit J, Bakker AQ, Korimbocus J, Kramer RA, Clijsters-van der Horst M, de Jong M, Jongeneelen M, Thijsse S, et al. (2006) Isolation and characterization of human monoclonal antibodies from individuals infected with West Nile virus. J Virol 80:6982–6992

Trainor NB, Crill WD, Roberson JA, Chang GJ (2007) Mutation analysis of the fusion domain region of St.Louis encephalitis virus envelope protein. Virology 360:398–406

Waldvogel K, Bossart W, Huisman T, Boltshauser E, Nadal D (1996) Severe tick-borne encephalitis following passive immunization. Eur J Pediatr 155:775–779

Zhu Z, Dimitrov AS, Chakraborti S, Dimitrova D, Xiao X, Broder CC, Dimitrov DS (2006) Development of human monoclonal antibodies against diseases caused by emerging and biodefense-related viruses. Expert Rev Anti Infect Ther 4:57–66

20. Vaccine Development Against West Nile Virus

Alexander A. Khromykh, David C. Chang and Roy A. Hall

Abstract

Since the late 1990s, disease outbreaks associated with the mosquito-borne West Nile virus (WNV) have become a public heath concern of global significance. Outbreaks of fever and encephalitis in Eastern Europe, the Middle East and North America have resulted in almost 25,000 cases and nearly 1,500 deaths. Since its introduction into North America almost a decade ago, the virus has not only persisted, but spread throughout the USA and regions of Canada, Mexico and countries of Central and South America. In 2006, the virus caused more than 4,000 cases and 100 deaths, with an estimated 750,000 undiagnosed infections. This continued spread of the virus and sustained incidence of WNV infection indicate that the virus will remain a serious public health problem for the foreseeable future. Justifiably, considerable effort has been spent in developing a suitable vaccine and a number of experimental vaccines have shown promise in pre-clinical and clinical trials. However, the standard issues of vaccine efficacy and safety are further complicated by the fact that the population most at risk of severe and fatal WNV infection, and hence the likely recipients of vaccination, are the elderly and immunosuppressed.

In a previous article (Hall and Khromykh, 2004) we extensively reviewed the status of WNV vaccines and readers are directed to this reference for details of earlier developments in the field. In this chapter we review recent developments of safe and efficacious new-generation WNV vaccines, with specific emphasis on recombinant strategies to produce infectious, clone-derived, live, attenuated preparations and non-infectious, self-replicating vaccines.

Keywords

West Nile virus, glavivirus, vaccine, inactivated, recombinant subunit, viral-vector, live attenuated, chimeric, nucleic acid-based vaccine

1 Background

WNV is a mosquito-borne virus of the genus *Flavivirus.* The genus also includes globally relevant pathogens such as the dengue viruses (DENV), yellow fever virus (YFV) and Japanese encephalitis virus (JEV) (Heinz et al., 2000). The majority of WNV infections are asymptomatic – in approximately 20% of infections a febrile influenza-like illness (West Nile fever) is observed with symptoms of fever, headache, body aches, myalgia and anorexia. In about 5–10% of cases, a neuroinvasive form of the disease is observed, manifesting as encephalitis, meningitis or flaccid paralysis. WNV neurological disease is fatal in approximately 10% of these patients, as documented by the World Health Organization (WHO). Between 1999 and 2007, approximately 27,000 cases of WNV disease have occurred in the USA, resulting in more than 1,000 fatalities. Clinical data from the USA outbreak also reveal that there is much higher incidence of severe disease (causing encephalitis or death) in people older than 60 years.

WNV strains comprise two genetic lineages. Lineage I include highly pathogenic strains associated with outbreaks of neurological disease in North America, Eastern Europe and the Middle East. Lineage II WNV isolates are generally restricted to Africa and the Middle East, and are usually associated with febrile illness rather than neurological disease (Lanciotti et al., 1999). A comparatively benign subtype of WNV known as Kunjin virus (KUNV) belongs to lineage I and occurs in Australasia (Hall et al., 2002).

WNV appears to initially replicate in Langerhans cells at the site of dermal infection (Johnston et al., 2000). Infected cells then migrate to regional lymph nodes where further replication seeds viraemia and dissemination of the virus to peripheral organs. Infection of the central nervous system (CNS) then occurs if the viraemia is sufficiently high. Appearance of disease symptoms is dependent on the race between viral replication and the host immune response. Innate and adaptive immune responses are required for protection against WNV disease (Diamond et al., 2003) with type I (α/β) and type II (γ) interferons (IFNs) involved in controlling peripheral dissemination of infection. Neutralizing antibody is also crucial to recovery and long-term immunity, while cytotoxic $CD8^+$ T-cell responses are responsible for viral clearance (Diamond et al., 2003).

No vaccines have yet been approved for use against WNV in humans, and no effective therapies are currently available against the disease. Several experimental vaccines are in various stages of pre-clinical or clinical evaluation (Hall and Khromykh, 2004). The Chimerivax-WN02 vaccine, a live, attenuated, recombinant virus developed by Acambis is currently the most advanced in clinical trials (Monath et al., 2006), while inactivated, DNA and viral-vectored recombinant vaccines have been approved for veterinary use (Hall and Khromykh, 2004).

2 Inactivated Vaccines

In 2000, the Kimron Veterinary Institute in Israel developed a mouse brain-derived, formalin-inactivated WNV vaccine to protect domestic geese flocks. Two intramuscular doses of the preparation that was derived from a 1998 Israeli WNV isolate were sufficient to protect birds from lethal viral challenge (Malkinson et al., 2001). The vaccine was further refined by culturing the virus in the human PER.C6™ cell line developed by the Dutch biotechnology company Crucell and was approved for veterinary use in 2004 (Samina et al., 2007). A similar vaccine was developed by Crucell for use in humans and recently underwent phase I trials in Belgium. Two doses of the alum-adjuvanted formulation of the whole inactivated vaccine were administered intramuscularly to 47 healthy subjects enrolled into a randomized, double-blind, placebo-controlled, dose-escalation study. Administration of the vaccine was systemically and locally well tolerated with headache and reactions at the injection site as the most common adverse events reported.

In 2001 a formalin-inactivated vaccine "West Nile Innovator" derived form the North American strain of WNV was developed by Fort Dodge Veterinary and approved for veterinary use (Ng et al., 2003). Several million horses were immunized with this preparation in the USA and Canada, considerably reducing equine morbidity and mortality from WNV infection. Fort Dodge has now replaced the inactivated vaccine with a DNA-based vaccine (see below).

3 Recombinant Subunit Vaccines

Several groups have used a recombinant subunit approach to develop a vaccine against WNV. Lieberman et al. (2007) stably expressed the WNV structural proteins (prM plus the amino terminal 80% of E) in *Drosophila* S2 cells. After immunoaffinity purification of the secreted truncated E protein, the preparation was shown to be immunogenic in mice when administered with a saponin-based adjuvant containing Quil A (*Quillaja saponaria Molina*) and cholesterol components, inducing high levels of antibody and cell-mediated immunity as measured by lymphocyte proliferation and production of both Th1 (IFN-γ) and Th2

(IL-4 and IL-10) cytokines. Immunization of golden hamsters with this preparation induced high titres of WNV-neutralizing antibodies and animals were protected from lethal WNV challenge for up to 12 months after vaccination (Watts et al., 2007).

Domain III (DIII) of the West Nile E protein is known to contain major neutralizing epitopes and, as such, has been assessed by several groups as a candidate subunit vaccine. The immunoglobulin-like DIII can be independently expressed and folded to a functional fragment which is capable of eliciting neutralizing antibodies (Chu et al., 2007). Mice vaccinated with three doses of WNV-DIII protein (100 μg) expressed and purified from *E. coli* along with the CpG adjuvant induced neutralizing antibodies and protection from WNV challenge (Chu et al., 2007). Similar results have recently been reported by Martina et al. (2007). These findings demonstrate that immunization with the DIII fragment, in the absence of full-length E or virus particles, induces protective immunity to WNV infection. The efficiency of the DIII vaccine was further improved when expressed as a recombinant fusion protein with a modified version of bacterial flagellin (STF2Δ) (McDonald et al., 2007). The fused STF2Δ aided the delivery of DIII to antigen-presenting cells by linking it to the ligand of Toll-like receptor (TLR)-5. Mice vaccinated with WNV-DIII-STF2Δ developed protective immunity with a significantly lower dosage (2.5 μg delivered twice) and without the need of an adjuvant.

4 Viral-Vectored WNV Vaccines

4.1 Canarypox Vector

A recombinant vaccine was licensed in 2004 for equine use (Recombitek equine WNV vaccine; Merial vaccine, Athens, GA), consisting of a canarypox virus vector (ALVAC) expressing the prM and E genes derived from the 1999 New York isolate of WNV, and an adjuvant (Minke et al., 2004). Horses vaccinated with two dosages at 5-week intervals developed neutralizing antibodies, and were protected against WNV challenge by infected mosquitoes at 2 weeks and up to 1 year post-vaccination. Similar protection results were also observed in a more intrusive intrathecal short-term challenge (Siger et al., 2006). In a separate study where a single dose was used, all vaccinated horses were also protected; however, no antibodies were detected pre-challenge (Siger et al., 2004). Cats and dogs were also protected against a mosquito WNV challenge after vaccination with two doses of $10^{7.5}$TCID$_{50}$ and

$10^{5.6}$TCID$_{50}$, respectively (Karaca et al., 2005). A comparative study between Recombitek and two commercially available vaccines for equine use, (1) West Nile-Innovator; Fort Dodge, Fort Dodge, IA (formalin-inactivated) and (2) PreveNile; Intervet, De Soto, KS (live flavivirus chimera), concluded that while very low levels of neutralizing antibodies were induced from these vaccines, the response was adequate to provide protection against WNV challenge of the CNS in horses (Seino et al., 2007). Further study on the long-term efficacy and cellular immunity (CTLs) needs to be carried out for the Recombitek equine WNV vaccine.

4.2 Measles Virus Vector

A vector derived from the live attenuated Schwarz strain of the measles virus (MV) was shown to strongly and stably express genes encoding proteins from heterologous viruses and to induce specific humoral and cellular immune responses to these proteins in vivo (Brandler and Tangy, 2007). Proof of concept of the vector as a vaccine delivery system for flaviviruses was demonstrated by constructing a recombinant measles vector that encoded the secreted form of the E protein of an Israeli 1998 WNV strain (Despres et al., 2005). The recombinant virus (MVSchw–sEWNV) efficiently expressed WNV E, which was secreted from infected cells. Mice immunized with a single, low dose of the recombinant virus induced high levels of WNV-neutralizing antibodies that provided protection against WNV challenge by 8 days post-immunization, with immunity persisting for up to 6 months. Protection against WNV challenge was also reported after a single immunization of squirrel monkeys [unpublished data cited in Brandler and Tangy (2007)]. The potential problem of pre-existing immunity to measles restricting replication of recombinant MV constructs was previously addressed by immunizing MV-immune mice and macaques with an MV–HIV recombinant virus (Lorin et al., 2004). After vaccination with two doses of the recombinant virus, antibody responses to HIV were induced, indicating the MV–WNV may be effective for immunizing MV-immune adult humans (Brandler and Tangy, 2007).

4.3 Lentiviral Vector

A lentiviral-vectored vaccine expressing a secreted form of the WNV E protein (sEWNV) was assessed in mice for immunogenicity and protective efficacy against lethal WNV challenge (Iglesias et al.,

2006). Immunization with a single dose, containing as little as 50 ng of p24 antigen in the TRIP/sEWNV vector particles, protected mice against WNV challenge by 7 days post-vaccination. The authors acknowledged that retroviral vectors, requiring transduction of cells in vivo, pose serious safety issues in regard to their use for human prophylaxis. However, the minute vaccine doses required for rapid induction of protective immunity and the broad host cell range of the VSV-G envelope pseudotyped vector particles suggest that the vaccine could have significant veterinary potential for a range of vertebrate species at risk of WNV disease, including horses, domestic poultry and zoo animals.

4.4 Equine Herpesvirus Vector

A recombinant equine herpesvirus type 1 (EHV-1) vaccine expressing WNV prM and E proteins was assessed for immunogenicity in horses. Five of six horses seroconverted to WNV after three vaccinations at one monthly intervals, with high titres of WNV E-specific antibodies detected for more than 2 months after vaccination (Rosas et al., 2007). Of interest, 4/6 animals developed low to moderate 50% plaque reduction WNV-neutralizing titres after a single dose; however, titres were not boosted by further vaccination and fell to undetectable levels in all animals 66 days after the initial dose.

5 Live Attenuated Vaccines

Despite the development of the highly successful live attenuated 17D vaccine against YFV, no other live flavivirus vaccines have been approved for use in the wider community. However, there are many reasons for choosing a live attenuated vaccine strategy over inactivated or recombinant subunit preparations. Because they simulate natural infection, live vaccines generally provide a stronger and long-lived immune response from a single immunization, stimulating innate, humoral and cell-mediated responses from a single dose (Monath and Heinz, 1996). Nevertheless, there are many crucial safety issues associated with live vaccines. Of primary importance is the remote possibility that the vaccine virus will revert to virulence. Evidence that candidate vaccines are phenotypically and genetically stable, after serial passage in vitro and in vivo, is therefore essential for the demonstration of vaccine safety. To ensure that the risk of reversion to the virulent phenotype is minimal, it is also highly desirable to include multiple mutations that have different mechanisms of attenuation. Since those most at risk from severe WNV disease are the elderly and the immunosuppressed (Monath et al., 2001), it is also crucial to demonstrate that the vaccine is attenuated in immunodeficient animal models. The release of live vaccines derived from a vector-borne virus into the community also carries the possibility that the vaccine virus may reach sufficient viraemic titre in the vaccinee to infect blood feeding arthropods and enter natural transmission cycles. There is also a risk that live vaccines may be contaminated with adventitious viruses that are inadvertently introduced into the virus culture during the manufacturing process.

Several live vaccines have been developed against WNV and are in various stages of pre-clinical and clinical evaluation. These include naturally attenuated WNV isolates, and mutants attenuated in the laboratory by serial passage or neutralization escape selection [reviewed in Hall and Khromykh (2004)]. In this section we focus on attenuated WNV vaccines prepared from infectious DNA clones and recombinant viruses derived from chimerization of WNV with attenuated heterologous flaviviruses.

5.1 Attenuated WNV Isolates

Attenuated lineage II WNV isolates have been used for the development of WNV vaccine candidates. The most advanced of these is an attenuated cDNA clone of the lineage II WNV strain, WN 956 (Yamshchikov et al., 2001). Virus recovered from the cDNA was highly attenuated in mice, and i.m. immunization induced strong humoral immunity and provided protection against challenge with the virulent NY99 strain of WNV (Yamshchikov et al., 2004). Borisevich et al. (2006) also created chimeric viruses by exchanging the structural genes between infectious clones of the lineage I NY99 strain and the lineage II attenuated W956. Chimeric viruses carrying the NY99 structural genes in the lineage II WNV backbone remained attenuated for neuroinvasiveness in mice but showed superior immunogenicity compared to the parental WN 956 virus.

5.2 Kunjin Virus

KUNV is an Australian subtype of WNV that is antigenically and genetically closely related to lineage I West Nile strains (Hayes, 1988; Heinz et al., 2000). Although KUNV shares 98–99% amino acid homology with the virulent North American strain WNV NY99, it is relatively benign. Few, mostly mild human cases have been attributed to KUNV infection with no reported fatalities. Indeed, studies by Beasley et al. (2002) in a 3–4-week-old mouse model indicate that KUNV is ~10,000-fold less neuroinvasive than WNV NY99. The low level of virulence observed for KUNV and its antigenic similarity to WNV NY99, make this virus a naturally attenuated vaccine candidate against more pathogenic WNV strains.

Studies in our laboratories have shown that infection of mice with KUN virus, containing a mutation in the NS1 protein (^{250}Pro to Leu), which further attenuates the virus (Hall et al., 1999), produced no disease but induced WNV-neutralizing antibodies and protected animals from subsequent lethal challenge with NY99 strain of WNV [see Table 2 in Hall et al. (2003)]. Our recent studies with cell-culture-adapted KUN virus replicons identified a mutation in NS2A protein (^{30}Ala to Pro), which when introduced into an infectious clone resulted in substantially reduced ability of the virus to inhibit IFN responses and significantly attenuated virus virulence in mice (Liu et al., 2005, 2006). Importantly, the immunization with NS2A A30P-mutated KUN virus provided complete protection from a lethal challenge with NY99 strain. KUN NS2A A30P mutant and wild-type KUN viruses were also examined for attenuation and protective efficacy in American crows by Brault et al. (unpublished data). Seven American crows (*Corvus brachyrhynchos*) were inoculated with 750 PFU of wild-type KUNV and six crows were inoculated with 900 PFU of the NS2A A30P mutant virus. There was no mortality or signs of illness associated with infection in any of the inoculated birds with very little virus detected in blood between days 3 and 5 after inoculation (peak titres of 2.8 $\log_{10}$ for wild-type Kunjin virus and less than 2.5 $\log_{10}$ for the mutant virus). In contrast, NY99 challenge resulted in high viraemia with titres reaching 8–10 $\log_{10}$ PFU/mL and all birds were dead by 5 days after inoculation. All wild-type and NS2A A30P mutant KUN virus-immunized birds developed similar neutralizing antibody titres to both KUNV and WNV. When the KUNV-immune crows were challenged with 1,500 LD_{50} of WNV NY99, no morbidity or mortality was observed, and no viraemia detected.

Taken together, these results indicate that KUNV provides a safe option for an efficient live vaccine against WNV. Although wild-type KUNV is already substantially naturally attenuated, the inclusion of mutations in the NS1 and/or NS2A proteins further reduces the virulence of KUNV.

6 Chimeric Viruses

Chimeric flavivirus vaccines are based on the principle of exchanging the two structural viral genes (prM and E) between two different flaviviruses (Fig. 1) (Bray and Lai, 1991; Lai and Monath, 2003). This exchange was shown to produce viable viral chimeras and has been extensively used to develop vaccines against a number of pathogenic flaviviruses including WNV. In general, attenuated viral strains are used as the genomic backbone with donor prM–E genes derived from more pathogenic viruses for which vaccines are developed. YFV 17D, attenuated DEN4 and DEN2 viruses and, more recently, KUN virus have been used as the recipient backbones for developing chimeric vaccines. The chimeric viruses based on YFV-17D and on DENV backbones have produced the most advanced vaccine candidates with several undergoing clinical trials. A number of comprehensive reviews on chimeric flavivirus vaccines have been recently published (Hall and Khromykh, 2007; Jones et al., 2005; Lai and Monath, 2003) and readers are referred to these articles for more detail. In this section we will briefly describe the most recent advances in development of chimeric vaccines for WNV (Table1).

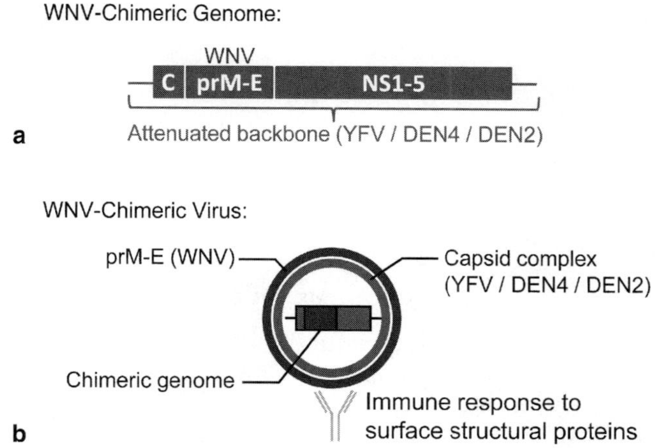

Figure 1. Chimeric live attenuated vaccines. (**a**) The genes coding for the structural proteins prM and E of West Nile virus (WNV) are used to replace the corresponding genes in the attenuated host virus backbone. Host viruses used to develop WNV chimeras include attenuated strains of YFV, DEN4 and DEN2. (**b**) The chimeric virus consists of the WNV prM and E proteins and the nucleocapsid of the host virus with the encapsidated chimeric genome. The surface structural proteins of the virus elicit a specific immune response to WNV. (*See Color Plates*)

Table 1. Summary of candidate WNV vaccines

Vaccine	Formulation	Pre-clinical or clinical trials	References
ChimeriVax-WNV (Acambis)	Live attenuated YFV/WNV chimera	Mice, hamsters, monkeys, birds Phase I and II clinical trials	Monath et al. (2001), Tesh et al. (2002), Langevin et al. (2003), Arroyo et al. (2004) and Monath et al. (2006)
WN/DEN4 (NIH)	Live attenuated DENV4/WNV chimera	Mice, monkeys, geese Phase I clinical trial	Pletnev et al. (2002, 2003, 2006)
D2/WN (CDC)	Live attenuated DENV2/WNV chimera	Mice	Huang et al. (2005)
pKUN1 (Replikun, University of Queensland)	DNA coding for live attenuated KUNV	Mice	Hall et al. (2003)
PER.C6 WNV vaccine (Crucell/Kimron)	Inactivated whole virus produced by PER.C6° technology	Mice, geesePhase I clinical trial	Malkinson et al. (2001) and Samina et al. (2007)
pCBWN (CDC)	DNA vaccine coding for WNV prM–E genes	Mice, horses, birds Phase I Clinical trial	Davis et al. (2001) and Turell et al. (2003)
WN preM-E/LAMP (John Hopkins)	DNA vaccine coding for WNV prM–E-LAMP chimera	Mice	Anwar et al. (2005)
Subunit WNV vaccine (Hawaii Biotech)	Recombinant WNV E and NS1 expressed in *Drosophila* S2 cells	Mice, hamsters	Lieberman et al. (2003) and Watts et al. (2007)
pCMVWN (University of Kansas)	Infectious DNA coding for attenuated lineage II strain B956	Mice	Seregin et al. (2006)
pCMVWN ΔC (University of Kansas)	DNA coding for capsid-deleted genome of lineage II strain WN 956	Mice	Seregin et al. (2006)
pKUNdC/C Vaccine (University of Queensland)	DNA coding for capsid-deleted lineage I strain Kunjin RNA packaged into virus-like particles	Mice, horses	Chang et al. (2008)
PIV vaccine (UTMB)	Virus-like particles containing capsid-deleted WNV genomic RNA	Mice	Mason et al. (2006)
TRIP/sEWNV (Institut Pasteur)	Lentivirus vector expressing secreted form of WNV E protein	Mice	Iglesias et al. (2006)

(continued)

Table 1. (continued)

Vaccine	Formulation	Pre-clinical or clinical trials	References
Recombitek equine WNV vaccine (Merial)	Canarypox vector expressing WNV prM–E proteins	Mice, horses, cats, dogs	Minke et al. (2004), Siger et al. (2004, 2006), Karaca et al. (2005) and Seino et al. (2007)
MVSchw–sEWNV (Institut Pasteur)	Measles vector expressing secreted form of WNV E protein	Mice	Despres et al. (2005)
Equine herpes virus-prM–E (Cornell University)	Equine herpes virus vector, expressing prM–E proteins	Horses	Rosas et al. (2007)
WNV-DIII (National University of Singapore)	E Domain III with CpG adjuvant	Mice	Chu et al. (2007)
WNV-DIII-STF2Δ (VaxInnate)	E Domain III fused to bacterial flagellin sequences	Mice	McDonald et al. (2007)

6.1 ChimeriVax-WN

The ChimeriVax-WN vaccine was produced by Acambis by replacing the surface glycoprotein pre-membrane and envelope gene E in an infectious cDNA clone of the 17D yellow fever vaccine strain (YFV 17D) (Rice et al., 1989) with corresponding genes of the wild-type WNV NY99 strain (Lanciotti et al., 1999) such that a YF–WN recombinant chimera was formed. The design of this vaccine is based on the ChimeriVax vaccine technology platform developed at St Louis University (Chambers et al., 1999), and the original strategy to produce chimeric flaviviruses (Bray and Lai, 1991). A similar strategy has been used to develop ChimeriVax vaccines to Japanese encephalitis and Dengue viruses (Jones, 2004; Monath et al., 2002).

On the basis of their finding with the ChimeriVax-JE vaccine, additional mutations were introduced into the ChimeriVax-WN vaccine at residues 107 (L→F), 316 (A→V) and 440 (K→R) of the E gene to further attenuate the virus and reduce the risk of reversion to a more virulent phenotype (Arroyo et al., 2004). These mutations were stable and were retained even after repeated serial passage in culture. Although this modified virus was less neurovirulent than the original chimera in mice, it was also less immunogenic, with a tenfold reduction in neutralizing antibody response (Arroyo et al., 2004). Nevertheless, subcutaneous inoculation into monkeys induced satisfactory levels of neutralizing antibody, with serum $PRNT_{50}$ values ranging from 80 to 640 (Arroyo et al., 2004).

Between 2003 and 2005 a phase I clinical trial of ChimeriVax-WN was conducted in healthy volunteers (Monath et al., 2006). The vaccine was well tolerated with minimal adverse effects reported. WNV-neutralizing antibodies were induced in all 45 recipients (mean titre of 1,280) receiving a single dose of 10^3–10^5 PFU of virus and persisted for 12 months in all but two of the vaccines. IFN-γ ELISPOT assays also revealed that $CD8^+$ T-cell responses to WNV envelope-specific peptides were induced in 43/45 recipients between 2 and 4 weeks after vaccination. Earlier and stronger T-cell responses appeared to correlate with reduced viraemia after vaccination (Monath et al., 2006). This finding suggests that levels of viraemia should be carefully monitored in future trials on elderly individuals who may fail to mount a strong cellular response to the vaccine virus.

Additional studies indicate that ChimeriVax-WN does not replicate in the major mosquito vectors of WNV (Johnson et al., 2003) nor in avian hosts (Langevin et al., 2003), suggesting that inadvertent release will not result in uncontrolled amplification of the virus in natural transmission cycles.

6.2 WNV/DEN4 Chimera

Pletnev et al. (2002) used a similar strategy to construct a chimeric virus, WN/DEN4, containing the DEN4 genetic background and the WNV prM and E genes. The chimerization process itself was sufficient to substantially reduce neurovirulence and neuroinvasiveness in mice compared to the WNV parental virus. A targeted deletion of a region of the 3′UTRc not essential for replications produced a more attenuated version of the chimeric virus (WN/DEN4Δ30), which displayed reduced ability to replicate in mouse brain in vivo as well as human and mouse neural cell lines in vitro (Pletnev et al., 2003, 2006). The chimeras induced a moderate to high titre of neutralizing antibodies in rhesus monkeys and prevented viraemia in animals challenged with WNV NY99. The more attenuated virus, WN/DEN4Δ30, was prepared as a clinical vaccine lot and further assessed in pre-clinical trials for safety and immunogenicity in monkeys (Pletnev et al., 2006). A single subcutaneous dose of 10^5 PFU induced consistent neutralizing antibody responses to WNV, although no detectable viraemia was observed in the vaccinated monkeys. Due to the extensive global distribution of DENV, the vaccine interference of prior immunity to DENV was also assessed. The immune response to vaccination with WN/DEN4Δ30 in DEN4 immune monkeys was only slightly lower than that in naïve animals, suggesting that prior infection with DEN4 may not significantly affect vaccine efficacy. However, the effect of immunity generated by exposure to multiple serotypes of DENV, which is common in many endemic areas, has not been addressed. The results of a phase I clinical trial of WN/DEN4Δ30 [referred to by Pletnev et al. (2006)] have yet to be published.

To assess the potential of WNV/DEN4 chimeric viruses to enter into natural transmission cycles, their ability to replicate and produce viraemic responses in natural avian hosts of WNV was investigated in domestic geese, a species highly susceptible to infection with the NY99 strain of WNV. Two-week-old goslings inoculated with 10^4 PFU of WNV/DEN4 or WN/DEN4Δ30 failed to develop detectable viraemia or seroconvert indicating the chimeric viruses were non-infectious in these hosts (Pletnev et al., 2006). Transmission studies with common mosquito vectors of WNV and DENV also revealed that the WN/DEN4 chimeras were unable to infect and spread to the salivary glands of *Aedes aegypti*, the major vector of DENV or *Culex tarsalis*, an efficient vector of WNV in the USA. However, it should be noted that both WN/DEN4 and WN/DEN4Δ30 were as infectious as the parental viruses for *Ae. albopictus*, a mosquito capable of transmitting both WNV and DENV. Nevertheless, the reduced ability of the chimeric viruses to infect and disseminate in mosquito vectors combined with the lack of detectable viraemia in vaccinated primates suggests they are unlikely to be transmitted from vaccinees to mosquitoes.

6.3 DEN2/WNV Chimera

Chimeras have also been generated from the genetic background of DEN2 viruses for assessment as WNV vaccine candidates in mice. Huang et al. (2005) reported the construction of several recombinant viruses, which included the prM and E genes of the NY99 WNV strain spliced into the backbone of the 16681 DEN2 wild-type strain and two attenuated DEN2 viruses that had been selected by serial passage in primary dog kidney (PDK) cells. Incorporation of the WNV prM signal sequence significantly improved viability of the recombinant viruses, as did point mutations in the structural and non-structural genes that were spontaneously generated during transfection and passage in cell culture. In each case, the chimeric viruses were attenuated to at least the same degree as their parental DEN2 when assessed for growth in monkey and mosquito cells in vitro, consistent with their small plaque size. In both cell types, growth kinetics of the chimera was also several orders of magnitude less than WNV NY99. Two of the chimeric clones derived from the attenuated DEN-2 also exhibited reduced neurovirulence in newborn mice, exhibiting 0 and 6% mortality after intracranial inoculation of 10^4 PFU. Four-week-old mice immunized i.p. with a single dose of 10^4 PFU of these clones developed moderate neutralizing antibody titres ($PRNT_{50}$ range of 108–160) and were protected from i.p. challenge with a relatively high dose (10^5 PFU) of WNV NY99 (Huang et al., 2005).

7 Conventional Non-infectious DNA Vaccines

Mammalian expression plasmids expressing the flaviviral structural genes prM and E have proved successful for the induction of immunity to several flaviviruses [reviewed by Chang et al. (2001)]. Cells transfected with these plasmids secrete highly immunogenic virus-like particles (VLPs) that are antigenically indistinguishable from virions and elicit strong neutralizing antibody responses (Chang et al., 2001) (see Fig. 2d). As these vaccines also result in the synthesis of endogenous viral protein, cell-mediated responses are also stimulated. Unmethylated cytosine–guanine dinucleotides (CpGs) found in plasmid DNA enhance the immune response to these vaccines (Davis, 2000).

Researchers at the Centers for Disease Control and Prevention (CDC) in Fort Collins assessed a recombinant plasmid that expressed the WNV prM and E proteins as a vaccine in mice and horses (Davis et al., 2001). A single intramuscular injection of 100 μg in mice or 1 mg in horses induced neutralizing antibody and protected animals from WNV challenge (Davis et al., 2001). In July 2005, the vaccine was licensed to Fort Dodge Laboratories for use in horses, becoming the world's first veterinary DNA vaccine.

Vical Incorporated also secured a license from CDC for the technology used in the vaccine. In a joint study with the National Institute of Allergy and Infectious Diseases (NIAID) that commenced in 2005,

promoters placed into back-to-back orientation directs transcription of two RNAs: (1) capsid-deleted self-replicating RNA and (2) capsid mRNA. The former is packaged into SRIPs by the three structural proteins. PrM–E particles are also secreted from these cells. Secreted SRIPs infect and deliver the replicon RNA to other cells to further stimulate the host immune response by expressing viral non-structural proteins and secreted prM–E. The absence of capsid gene expression in SRIP-infected cells effectively terminates the infectious cycle. (*See Color Plates*)

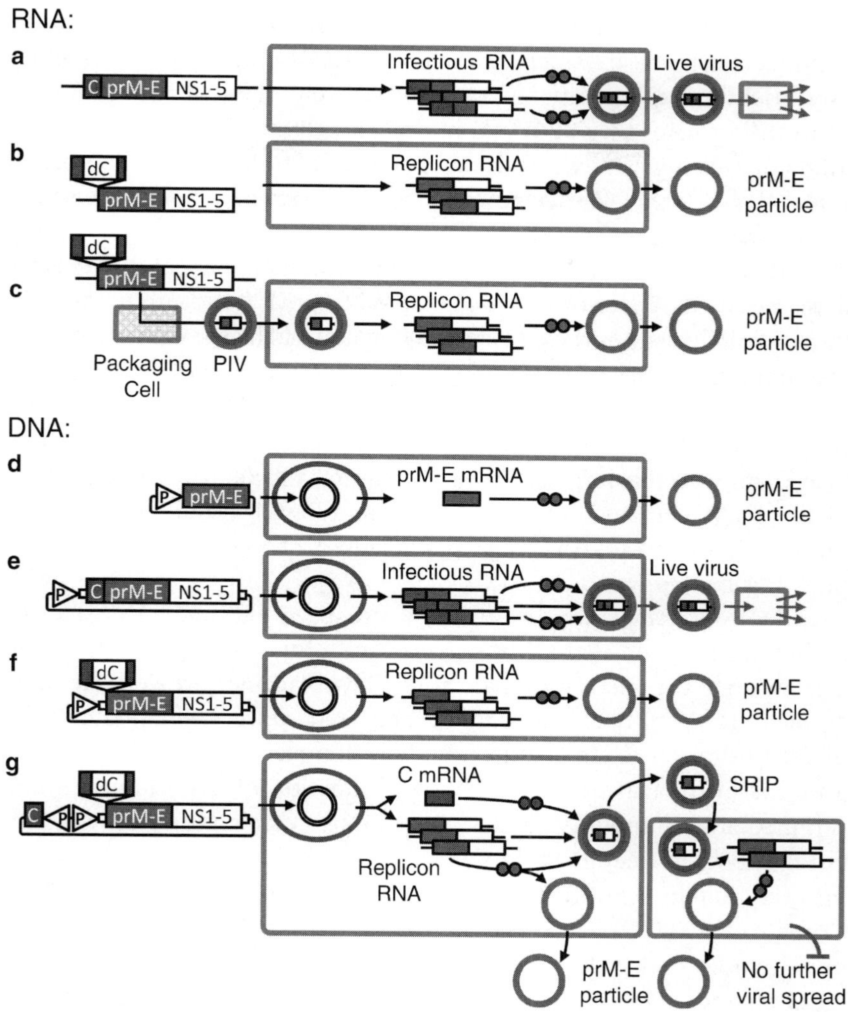

Figure 2. Nucleic acid-based vaccines in development. *RNA:* (**a**) Infectious RNA; the full-length infectious RNA when delivered into cells results in generation of live virus which spreads to other cells. (**b**) Non-infectious RNA: a large internal deletion in the capsid gene in the full-length RNA prevents the packaging of replicon RNA into viral particles. The two structural proteins prM and E form secreted immunogenic prM–E particles, which elicit potent antibody response. (**c**) Pseudo-infectious viruses (PIV): a non-infectious RNA with large deletion in capsid gene is first introduced into a packaging cell expressing structural protein to generate PIVs. PIVs are then used to immunize recipients where they infect cells and deliver the capsid-deleted RNA to produce highly immunogenic secreted prM–E particles. *DNA:* (**d**) prM–E expressing vector: plasmid DNA encoding only the prM and E genes driven by a mammalian expression promoter (P) transcribes prM–E mRNA which leads to the formation and secretion of prM–E particles. (**e**) Infectious DNA: the full-length infectious RNA is transcribed by cellular RNA polymerase II from the mammalian expression promoter and produce infectious virus which spreads to other cells. (**f**) Non-infectious DNA: capsid-deleted self-replicating RNA is transcribed by RNA polymerase II and produces secreted prM–E particles. (**g**) Single-round infectious particles (SRIPs): a single plasmid DNA with two mammalian expression

Vical assessed the vaccine for safety and efficacy in 12 healthy human volunteers in a phase I, open-label clinical trial. Using Vical's proprietary DNA delivery technology, the vaccine was well tolerated, and produced neutralizing responses to WNV in the 11 volunteers who returned for follow-up testing after vaccination. The vaccination schedule comprised three 4 mg doses of vaccine at 1-month intervals via intramuscular needle-free injection.

Researchers from the Johns Hopkins University School of Medicine developed a DNA vaccine expressing the WNV prM–E proteins fused to the transmembrane and carboxyl terminal domains of the lysosome-associated membrane protein (LAMP) (Anwar et al., 2005). This strategy relied on the LAMP sequences directing the WNV antigens to MHC II vesicular compartments of transfected cells, thus enhancing antigen presentation. Consistent with this rationale, mice immunized subcutaneously with 50 μg of this DNA construct induced stronger and longer lasting neutralization titres compared to mice vaccinated with DNA expressing only the WNV prM–E genes (Anwar et al., 2005).

DNA vaccines are an attractive prospect. Plasmid DNA can be highly purified and there is less opportunity for adventitious viruses to contaminate the vaccine preparation. These vaccines are very stable at ambient temperatures and can be delivered in very small quantities under optimized route of injection. The endogenous synthesis of viral antigens in the transfected cell allows induction of both humoral and cell-mediated immunity. Fears that DNA vaccination may induce autoantibodies to human DNA or integration of plasmid sequences into the human genome have so far proven unfounded (Le et al., 2000; Ledwith et al., 2000). Nevertheless, the long-term safety of DNA vaccines must remain under close scrutiny. The large amount of DNA and the requirement for multiple injections to achieve protective immunity in larger animals also remains an unresolved issue.

8 Novel Nucleic Acid-Based Approaches for WNV Vaccine Development

8.1 Infectious Nucleic Acid Vaccines

Delivery of a live attenuated virus via naked nucleic acid (RNA or DNA) combines the advantages of conventional live virus vaccines by mimicking natural infection and inducing long-lasting immunity with those of nucleic acid–based vaccines, i.e., ease of manufacture, stability and purity.

This approach was first demonstrated for flaviviruses using infectious RNA (Fig. 2a) as a vaccine candidate for the tick-borne encephalitis virus (TBEV) (Mandl et al., 1998). Gene gun immunization of mice with gold particles coated with less than 1 ng of viral RNA, in vitro transcribed from a genetically engineered attenuated TBEV DNA clone, initiated virus replication and induced protective immunity against challenge with a highly pathogenic TBEV strain. Coating on gold particles appeared to increase the stability of the otherwise very unstable naked RNA molecules when stored under refrigeration at 4°C. However, rigorous evaluation of long-term stability of this infectious RNA preparation as well as further standardization and quality assurance of RNA preparations are required to estimate the feasibility of this approach.

Use of plasmid DNA for delivery of infectious virus (Fig. 2e) simplifies the vaccine manufacture and eliminates potential concerns over the purity and stability of vaccine preparations. Successful in vivo transcription of infectious flavivirus RNA from a CMV promoter–based plasmid DNA was first demonstrated for the KUNV strain of WNV (Khromykh et al., 2001). Later studies with KUNV infectious DNA pKUN1 and infectious DNA of attenuated lineage 2 WNV clone WN 956 (pCMVWN) demonstrated that this type of DNA vaccine elicited potent protective immune responses against the pathogenic WNV NY99 strain (Hall et al., 2003; Seregin et al., 2006). A single intramuscular immunization of mice with as little as 100 ng of KUNV infectious DNA or 10 ng of W956 infectious DNA vaccine was sufficient to protect against challenge with 100 PFU and 10 LD_{50} of NY99 strain of WNV, respectively (Hall et al., 2003; Seregin et al., 2006). Following immunization with pKUN1 and pCMVWN DNAs, low viral titres could be detected in the blood between 3 and 5 days post-immunization demonstrating that limited virus replication had taken place in muscle cells that had taken up the plasmids. Importantly, the virus isolated from the blood of pKUN1 DNA-immunized mice retained the attenuated phenotype (Hall et al., 2003). Both KUNV and WN 956 DNA vaccines induced potent humoral immune responses, and our most recent studies demonstrated that pKUN1 DNA also induced a strong $CD8^+$ T-cell response to a K^d (or D^d)-restricted CTL epitope in the NS3 protein (Chang et al., 2008).

Thus, the delivery of attenuated WNV strains via an infectious plasmid DNA vector provides an efficient and convenient vaccination strategy. Because viral replication occurs in vivo after injection, thus amplifying immune response, only relatively small doses of DNA are required. However, the same potential safety concerns as discussed above for live attenuated vaccines still remain.

8.2 Non-infectious, Replicating Vaccines

A number of non-infectious but replicating nucleic acid–based flavivirus vaccine candidates have been recently developed. Essentially, this strategy is based on introducing in-frame deletions into the capsid gene to eliminate production of infectious virus but retain the ability of RNA to replicate and produce highly immunogenic secreted prM–E particles. This approach was originally developed for TBEV (Kofler et al., 2004) and later adopted for WNV vaccine development (Mason et al., 2006; Seregin et al., 2006). The vaccines were shown to be effective in mice when delivered as naked RNA (Aberle et al., 2005; Kofler et al., 2004), as VLPs containing capsid-deleted RNA (Mason et al., 2006), and as plasmid DNA encoding capsid-deleted RNA under control of the CMV promoter (Seregin et al., 2006).

8.3 RNA-Based Capsid-Deleted Vaccines

In the original paper by Kofler et al. (2004) a large deletion encompassing residues 29–89 was introduced in the capsid gene of TBEV genomic RNA (Fig. 2b). This deletion includes the first three hydrophobic helices of capsid protein and completely eliminates production of infectious virus (Kofler et al., 2002, 2003). As the production of secreted prM–E particles from this RNA was not efficient, additional point mutations were introduced into the signal sequence for translation of the downstream prM protein to improve signalase cleavage (Lee et al., 2000; Stocks and Lobigs, 1998). These mutations significantly improved production of secreted prM–E particles in RNA-transfected cells (Kofler et al., 2004). When these capsid-deleted naked RNAs were coated onto gold particles and used to immunize mice by gene gun application, all animals seroconverted after two immunizations, developed neutralizing antibodies, and were protected against challenge with 1,000 LD_{50} of the highly virulent TBEV strain Hypr (Kofler et al., 2004). Later studies by the same group demonstrated that two gene gun immunizations of mice with the capsid-deleted TBEV RNA induced humoral and cellular (Th1 and $CD8^{+}$ T-cell response) immune responses comparable to that induced by live vaccines and that even a single immunization with 1 μg of RNA was sufficient to induce a long-lasting (1 year) neutralizing antibody response (Aberle et al., 2005).

8.4 VLP-Based Capsid-Deleted Vaccines

Mason et al. (2006) used a similar approach based on large deletions in the capsid gene to develop VLP-based vaccines against WNV and YFV. Deletions of capsid amino acids 30–101 for WNV and 23–93 for YFV required no additional mutations in the prM signal sequence for efficient production of secreted prM–E particles (Fig. 2c). RNA from both constructs was efficiently packaged into VLPs (referred to as pseudo-infectious viruses or PIV by the authors) by transfecting in vitro transcribed RNA into packaging cells expressing the corresponding structural gene cassettes (Mason et al., 2006) with VLP titers reaching 10^8 mL^{-1}. VLPs were further amplified by additional passage in packaging cell lines. Single immunization of mice with as little as 3×10^4 WNV VLPs elicited neutralizing antibodies and protected mice against challenge with 100 LD_{50} of NY99 strain of WNV (Mason et al., 2006).

8.5 DNA-Based Capsid-Deleted Vaccine

WNV vaccine candidates employing the CMV promoter for generating capsid-deleted RNAs in vivo have also been recently developed (Seregin et al., 2006; Chang et al., submitted). A capsid-deleted vaccine derived from the WNV strain WN 956 pCMVWN(ΔC) was constructed from an infectious cDNA clone by deleting the coding sequences for amino acids 44–59 (Fig. 2f). No infectious virus was produced in transfected cells after 48 h, and no signs of virus infection was detected in mice 2 weeks after intramuscular injection with up to 10 μg of pCMVWN(ΔC) DNA (Seregin et al., 2006). Single intramuscular immunizations with equivalent doses of infectious (pCMVWN) and capsid-deleted pCMVWN(ΔC) DNAs showed that the magnitude of immune response to pCMVWN(ΔC) DNA was three- to sixfold lower compared to animals immunized with pCMVWN DNA. An additional booster immunization with pCMVWN(ΔC) DNA was required to achieve comparable immune responses. Two immunizations with as little as 0.1 μg of pCMVWN(ΔC) DNA resulted in complete protection against challenge with 10 LD_{50} of the NY99 strain of WNV. Although the data on the protective efficacy of pCMVWN(ΔC) DNA after a single immunization were not presented, the complete protection afforded by two i.m. immunizations with low doses of DNA is very encouraging. Some concern, however, remains over the potential for this DNA to produce infectious virus due to the small deletion (16 amino acids) in the capsid gene used in this study. Previous studies with TBEV revealed that spontaneous compensatory mutations in the capsid gene can rescue the defect in virus production caused by capsid deletions of less than 30 amino acids (Kofler et al., 2003). Although no virus was detected in initial tests with pCMVWN(ΔC) DNA, more rigorous evaluation is required to confirm this.

8.6 Replicating DNA Vaccine Producing Single Round Infectious Particles

This WNV DNA vaccine is based on using the KUNV cDNA backbone, with a large deletion (amino acids 18–100) in the capsid gene (Fig. 2g). The plasmid, named pKUNdC/C, also expresses full-length capsid gene under the control of a second CMV promoter which allows production of secreted single-round infectious particles (SRIPs) containing packaged capsid-deleted self-replicating RNA (Chang et al., 2008). The approach thus combines all the advantages of capsid-deleted DNA vaccines with the additional ability to amplify the immune response via in vivo generation and spread of SRIPs. In the immunized host, both initially transfected and SRIP-infected cells contain replicating RNA producing non-structural proteins that induce CTL responses and secrete highly immunogenic prM–E particles that induce neutralizing antibody responses. In vitro experiments showed that SRIPs titres reached ~10^5 particles per microgram of DNA by day 2 post-transfection and maintained this level for up to 5 days. In contrast, control pKUNdC RNA not expressing the full-length capsid gene failed to produce SRIPs. Importantly, passaging of SRIPs collected from pKUNdC/C-transfected cells on Vero cells did not reveal infectious virus, thus ruling out the possibility of recombination. A single immunization of mice with as little as 2 ng of pKUNdC/C DNA by gene gun protected all animals against lethal challenge with WNV NY99. The pKUNdC/C DNA vaccine has also been evaluated for immunogenecity in horses. Two gene gun immunizations with as little as 4 μg of DNA elicited detectable neutralizing antibodies against both KUN and NY99 WNV strains, with additional boosts providing substantial increases in neutralizing antibody titres. These encouraging results warrant further evaluation of the vaccine in clinical trials.

9 Conclusions and Future Directions

The significant morbidity and mortality associated with WNV infection over the last 5 years in North America and the spread of the virus into regions of Central and South America suggest that the disease will continue to be a significant public health problem. Future incursions of virulent strains to other continents and countries such as

Australia, South-East Asia and Western Europe may further increase the incidence of the disease. This trend may be exacerbated in developed countries where a significant increase in the mean age of the population will result in higher rates of severe and fatal disease as the at-risk population expands. Other issues such as global warming and increased travel and trade may further contribute to the increased distribution and prevalence of this vector-borne disease. Thus, the development of an effective vaccine against WNV should be a high priority.

Clearly, the issues that drive the selection and approval of WNV vaccine for human and veterinary use will be safety, efficacy, convenience and expense. Ideally, vaccines that cause no adverse reactions, are not released into natural transmission cycles, and induce long-lived protective immunity from a single dose are highly desirable. While live chimeric WNV vaccines have performed well in clinical trials to date and have fulfilled most if not all of these criteria, the potential for these hybrid viruses to adapt to human or animal hosts or arthropod vectors and initiate natural transmission cannot be ignored, and the risk must be carefully weighed against the benefits of the release of the chimeric virus vaccine for use in the general community. Finally, the expense and logistical issues of manufacture, quality control, storage and transport of live virus preparations also must be considered.

The current consensus that non-infectious DNA vaccines pose no significant risk for use in humans suggests this strategy is a safer and cheaper alternative to live vaccines against WNV. Indeed, the approval of a naked DNA WNV vaccine for veterinary use in 2005 has set a precedent for the future approval for use in humans. Although multiple doses may still be required, this approach avoids many of the problems associated with live viruses discussed above. A significant advance in WNV DNA vaccine technology by the recent development of DNA-based, non-infectious, replicating vaccines will further enhance the potency of this approach, possibly allowing for a single-dose regime. The ease of manufacture and stability of DNA vaccines and the reduced risk of contamination with other pathogens provide a strong incentive to adopt this strategy. The potential for novel, needle-free delivery of DNA is a further attractive option (Kendall, 2006).

References

Aberle JH, Aberle SW, Kofler RM, Mandl CW (2005) Humoral and cellular immune response to RNA immunization with flavivirus replicons derived from tick-borne encephalitis virus. J Virol 79:15107–13

Anwar A, Chandrasekaran A, Ng ML, Marques E, August JT (2005) West Nile premembrane-envelope genetic vaccine encoded as a chimera containing the transmembrane and cytoplasmic domains of a lysosome-associated membrane protein: increased cellular

concentration of the transgene product, targeting to the MHC II compartment, and enhanced neutralizing antibody response. Virology 332:66–77
Arroyo J, Miller C, Catalan J, Myers GA, Ratterree MS, Trent DW, Monath TP (2004) ChimeriVax-West Nile virus live-attenuated vaccine: preclinical evaluation of safety, immunogenicity, and efficacy. J Virol 78:12497–507
Beasley DW, Li L, Suderman MT, Barrett AD (2002) Mouse neuroinvasive phenotype of West Nile virus strains varies depending upon virus genotype. Virology 296:17–23
Borisevich V, Seregin A, Nistler R, Mutabazi D, Yamshchikov V (2006) Biological properties of chimeric West Nile viruses. Virology 349:371–81
Brandler S, Tangy F (2007) Recombinant vector derived from live attenuated measles virus: potential for flavivirus vaccines. Comp Immunol Microbiol Infect Dis 31:271–91
Bray M, Lai CJ (1991) Construction of intertypic chimeric dengue viruses by substitution of structural protein genes. Proc Natl Acad Sci U S A 88:10342–6
Chambers TJ, Nestorowicz A, Mason PW, Rice CM (1999) Yellow fever/Japanese encephalitis chimeric viruses: construction and biological properties. J Virol 73:3095–101
Chang GJ, Davis BS, Hunt AR, Holmes DA, Kuno G (2001) Flavivirus DNA vaccines: current status and potential. Ann N Y Acad Sci 951:272–85
Chang DC, Liu WJ, Anraku I, Clark DC, Pollitt CC, Suhrbier A, Hall RA, Khromykh AA (2008) Single-round injections particles enhance immunogenicity of a DNA vaccine against West Nile virus. Nat Biotechnol 26:571–577.
Chu JH, Chiang CC, Ng ML (2007) Immunization of flavivirus West Nile recombinant envelope domain III protein induced specific immune response and protection against West Nile virus infection. J Immunol 178:2699–705
Davis HL (2000) CpG motifs for optimization of DNA vaccines. Dev Biol (Basel) 104:165–9
Davis BS, Chang GJ, Cropp B, Roehrig JT, Martin DA, Mitchell CJ, Bowen R, Bunning ML (2001) West Nile virus recombinant DNA vaccine protects mouse and horse from virus challenge and expresses in vitro a noninfectious recombinant antigen that can be used in enzyme-linked immunosorbent assays. J Virol 75:4040–7
Despres P, Combredet C, Frenkiel MP, Lorin C, Brahic M, Tangy F (2005) Live measles vaccine expressing the secreted form of the West Nile virus envelope glycoprotein protects against West Nile virus encephalitis. J Infect Dis 191:207–14
Diamond MS, Shrestha B, Mehlhop E, Sitati E, Engle M (2003) Innate and adaptive immune responses determine protection against disseminated infection by West Nile encephalitis virus. Viral Immunol 16:259–78
Hall RA, Khromykh AA (2004) West Nile virus vaccines. Expert Opin Biol Ther 4:1295–305
Hall RA, Khromykh AA (2007) ChimeriVax-West Nile vaccine. Curr Opin Mol Ther 9:498–504
Hall RA, Khromykh AA, Mackenzie JM, Scherret JH, Khromykh TI, Mackenzie JS (1999) Loss of dimerisation of the nonstructural protein NS1 of Kunjin virus delays viral replication and reduces virulence in mice, but still allows secretion of NS1. Virology 264:66–75
Hall RA, Broom AK, Smith DW, Mackenzie JS (2002) The ecology and epidemiology of Kunjin virus. Curr Top Microbiol Immunol 267:253–69
Hall RA, Nisbet DJ, Pham KB, Pyke AT, Smith GA, Khromykh AA (2003) DNA vaccine coding for the full-length infectious Kunjin virus RNA protects mice against the New York strain of West Nile virus. Proc Natl Acad Sci U S A 100:10460–4
Hayes C (1988) West Nile fever. In: Monath TP (ed) The arboviruses: epidemiology and ecology. CRC Press, Boca Raton, FL, pp 59–88
Heinz F, Collett MS, Purcell RH (2000) Family Flaviviridae. In: Van Regenmortel MH, Fauquet CM, Bishop DH (eds)Virus taxonomy, Seventh report of the International Committee for the Taxonomy of Viruses. Academic Press, San Diego, CA, pp 859–878
Huang CY, Silengo SJ, Whiteman MC, Kinney RM (2005) Chimeric dengue 2 PDK-53/West Nile NY99 viruses retain the phenotypic attenuation markers of the candidate PDK-53 vaccine virus and protect mice against lethal challenge with West Nile virus. J Virol 79:7300–10

Iglesias MC, Frenkiel MP, Mollier K, Souque P, Despres P, Charneau P (2006) A single immunization with a minute dose of a lentiviral vector-based vaccine is highly effective at eliciting protective humoral immunity against West Nile virus. J Gene Med 8:265–74

Johnston LJ, Halliday GM, King NJ (2000) Langerhans cells migrate to local lymph nodes following cutaneous infection with an arbovirus. J Invest Dermatol 114:560–8

Johnson BW, Chambers TV, Crabtree MB, Arroyo J, Monath TP, Miller BR (2003) Growth characteristics of the veterinary vaccine candidate ChimeriVax-West Nile (WN) virus in Aedes and Culex mosquitoes. Med Vet Entomol 17:235–43

Jones T (2004) Technology evaluation: ChimeriVax-DEN, Acambis/Aventis. Curr Opin Mol Ther 6:443–50

Jones CT, Patkar CG, Kuhn RJ (2005) Construction and applications of yellow fever virus replicons. Virology 331:247–59

Karaca K, Bowen R, Austgen LE, Teehee M, Siger L, Grosenbaugh D, Loosemore L, Audonnet JC, Nordgren R, Minke JM (2005) Recombinant canarypox vectored West Nile virus (WNV) vaccine protects dogs and cats against a mosquito WNV challenge. Vaccine 23:3808–13

Kendall M (2006) Engineering of needle-free physical methods to target epidermal cells for DNA vaccination. Vaccine 24:4651–6

Khromykh AA, Varnavski AN, Sedlak PL, Westaway EG (2001) Coupling between replication and packaging of flavivirus RNA: evidence derived from the use of DNA-based full-length cDNA clones of Kunjin virus. J Virol 75:4633–40

Kofler RM, Heinz FX, Mandl CW (2002) Capsid protein C of tick-borne encephalitis virus tolerates large internal deletions and is a favorable target for attenuation of virulence. J Virol 76:3534–43

Kofler RM, Leitner A, O'Riordain G, Heinz FX, Mandl CW (2003) Spontaneous mutations restore the viability of tick-borne encephalitis virus mutants with large deletions in protein C. J Virol 77:443–51

Kofler RM, Aberle JH, Aberle SW, Allison SL, Heinz FX, Mandl CW (2004) Mimicking live flavivirus immunization with a noninfectious RNA vaccine. Proc Natl Acad Sci U S A 101:1951–6

Lai CJ, Monath TP (2003) Chimeric flaviviruses: novel vaccines against dengue fever, tick-borne encephalitis, and Japanese encephalitis. Adv Virus Res 61:469–509

Lanciotti RS, Roehrig JT, Deubel V, Smith J, Parker M, Steele K, Crise B, Volpe KE, Crabtree MB, Scherret JH, Hall RA, MacKenzie JS, Cropp CB, Panigrahy B, Ostlund E, Schmitt B, Malkinson M, Banet C, Weissman J, Komar N, Savage HM, Stone W, McNamara T, Gubler DJ (1999) Origin of the West Nile virus responsible for an outbreak of encephalitis in the northeastern United States. Science 286:2333–7

Langevin SA, Arroyo J, Monath TP, Komar N (2003) Host-range restriction of chimeric yellow fever-West Nile vaccine in fish crows (*Corvus ossifragus*). Am J Trop Med Hyg 69:78–80

Ledwith BJ, Manam S, Troilo PJ, Barnum AB, Pauley CJ, Griffiths TGII, Harper LB, Schock HB, Zhang H, Faris JE, Way PA, Beare CM, Bagdon WJ, Nichols WW (2000) Plasmid DNA vaccines: assay for integration into host genomic DNA. Dev Biol (Basel) 104:33–43

Le TP, Coonan KM, Hedstrom RC, Charoenvit Y, Sedegah M, Epstein JE, Kumar S, Wang R, Doolan DL, Maguire JD, Parker SE, Hobart P, Norman J, Hoffman SL (2000) Safety, tolerability and humoral immune responses after intramuscular administration of a malaria DNA vaccine to healthy adult volunteers. Vaccine 18:1893–901

Lee E, Stocks CE, Amberg SM, Rice CM, Lobigs M (2000) Mutagenesis of the signal sequence of yellow fever virus prM protein: enhancement of signalase cleavage in vitro is lethal for virus production. J Virol 74:24–32

Lieberman M, Tesh RB, Clements D (2003) A recombinant subunit West Nile vaccine protects against fatal West Nile encephalitis in hamsters. In: Viral Vaccine Meeting, Barcelona, Spain

Lieberman MM, Clements DE, Ogata S, Wang G, Corpuz G, Wong T, Martyak T, Gilson L, Coller BA, Leung J, Watts DM, Tesh RB, Siirin M, Travassos da Rosa A, Humphreys T, Weeks-Levy C (2007) Preparation and immunogenic properties of a recombinant West Nile subunit vaccine. Vaccine 25:414–23

Liu WJ, Wang XJ, Mokhonov VV, Shi PY, Randall R, Khromykh AA (2005) Inhibition of interferon signaling by the New York 99 strain and Kunjin subtype of West Nile virus involves blockage of STAT1 and STAT2 activation by nonstructural proteins. J Virol 79:1934–42

Liu WJ, Wang XJ, Clark DC, Lobigs M, Hall RA, Khromykh AA (2006) A single amino acid substitution in the West Nile virus nonstructural protein NS2A disables its ability to inhibit alpha/beta interferon induction and attenuates virus virulence in mice. J Virol 80:2396–404

Lorin C, Mollet L, Delebecque F, Combredet C, Hurtrel B, Charneau P, Brahic M, Tangy F (2004) A single injection of recombinant measles virus vaccines expressing human immunodeficiency virus (HIV) type 1 clade B envelope glycoproteins induces neutralizing antibodies and cellular immune responses to HIV. J Virol 78:146–57

Malkinson M, Banet C, Khinich Y, Samina I, Pokamunski S, Weisman Y (2001) Use of live and inactivated vaccines in the control of West Nile fever in domestic geese. Ann N Y Acad Sci 951:255–61

Mandl CW, Aberle JH, Aberle SW, Holzmann H, Allison SL, Heinz FX (1998) In vitro-synthesized infectious RNA as an attenuated live vaccine in a flavivirus model. Nat Med 4:1438–40

Martina BE, Koraka P, van den Doel P, van Amerongen G, Rimmelzwaan GF, Osterhaus AD (2007) Immunization with West Nile virus envelope domain III protects mice against lethal infection with homologous and heterologous virus. Vaccine 26(2):153–7

Mason PW, Shustov AV, Frolov I (2006) Production and characterization of vaccines based on flaviviruses defective in replication. Virology 351:432–43

McDonald WF, Huleatt JW, Foellmer HG, Hewitt D, Tang J, Desai P, Price A, Jacobs A, Takahashi VN, Huang Y, Nakaar V, Alexopoulou L, Fikrig E, Powell TJ (2007) A West Nile virus recombinant protein vaccine that coactivates innate and adaptive immunity. J Infect Dis 195:1607–17

Minke JM, Siger L, Karaca K, Austgen L, Gordy P, Bowen R, Renshaw RW, Loosmore S, Audonnet JC, Nordgren B (2004) Recombinant canarypoxvirus vaccine carrying the prM/E genes of West Nile virus protects horses against a West Nile virus-mosquito challenge. Arch Virol Suppl 18:221–30

Monath TP, Heinz FX (1996) Flaviviruses. In: Fields B, Knipe DM, Howley PM(eds) Fields virology, 3rd edn. Lippincott-Raven, Philadelphia, PA, pp 961–1034

Monath TP, Arroyo J, Miller C, Guirakhoo F (2001) West Nile virus vaccine. Curr Drug Targets Infect Disord 1:37–50

Monath TP, McCarthy K, Bedford P, Johnson CT, Nichols R, Yoksan S, Marchesani R, Knauber M, Wells KH, Arroyo J, Guirakhoo F (2002) Clinical proof of principle for ChimeriVax: recombinant live, attenuated vaccines against flavivirus infections. Vaccine 20:1004–18

Monath TP, Liu J, Kanesa-Thasan N, Myers GA, Nichols R, Deary A, McCarthy K, Johnson C, Ermak T, Shin S, Arroyo J, Guirakhoo F, Kennedy JS, Ennis FA, Green S, Bedford P (2006) A live, attenuated recombinant West Nile virus vaccine. Proc Natl Acad Sci U S A 103:6694–9

Ng T, Hathaway D, Jennings N, Champ D, Chiang YW, Chu HJ (2003) Equine vaccine for West Nile virus. Dev Biol (Basel) 114:221–7

Pletnev AG, Putnak R, Speicher J, Wagar EJ, Vaughn DW (2002) West Nile virus/dengue type 4 virus chimeras that are reduced in neurovirulence and peripheral virulence without loss of immunogenicity or protective efficacy. Proc Natl Acad Sci U S A 99:3036–41

Pletnev AG, Claire MS, Elkins R, Speicher J, Murphy BR, Chanock RM (2003) Molecularly engineered live-attenuated chimeric West Nile/dengue virus vaccines protect rhesus monkeys from West Nile virus. Virology 314:190–5

Pletnev AG, Swayne DE, Speicher J, Rumyantsev AA, Murphy BR (2006) Chimeric West Nile/dengue virus vaccine candidate: preclinical evaluation in mice, geese and monkeys for safety and immunogenicity. Vaccine 24:6392–404

Rice CM, Grakoui A, Galler R, Chambers TJ (1989) Transcription of infectious yellow fever RNA from full-length cDNA templates produced by in vitro ligation. New Biol 1:285–96

Rosas CT, Tischer BK, Perkins GA, Wagner B, Goodman LB, Osterrieder N (2007) Live-attenuated recombinant equine herpesvirus type 1 (EHV-1) induces a neutralizing antibody response against West Nile virus (WNV). Virus Res 125:69–78

Samina I, Havenga M, Koudstaal W, Khinich Y, Koldijk M, Malkinson M, Simanov M, Perl S, Gijsbers L, Weverling GJ, Uytdehaag F, Goudsmit J (2007) Safety and efficacy in geese of a PER.C6-based inactivated West Nile virus vaccine. Vaccine 25:8338–45

Seino KK, Long MT, Gibbs EP, Bowen RA, Beachboard SE, Humphrey PP, Dixon MA, Bourgeois MA (2007) Comparative efficacies of three commercially available vaccines against West Nile Virus (WNV) in a short-duration challenge trial involving an equine WNV encephalitis model. Clin Vaccine Immunol 14:1465–71

Seregin A, Nistler R, Borisevich V, Yamshchikov G, Chaporgina E, Kwok CW, Yamshchikov V (2006) Immunogenicity of West Nile virus infectious DNA and its noninfectious derivatives. Virology 356:115–25

Siger L, Bowen RA, Karaca K, Murray MJ, Gordy PW, Loosmore SM, Audonnet JC, Nordgren RM, Minke JM (2004) Assessment of the efficacy of a single dose of a recombinant vaccine against West Nile virus in response to natural challenge with West Nile virus-infected mosquitoes in horses. Am J Vet Res 65:1459–62

Siger L, Bowen R, Karaca K, Murray M, Jagannatha S, Echols B, Nordgren R, Minke JM (2006) Evaluation of the efficacy provided by a recombinant Canarypox-vectored equine West Nile Virus vaccine against an experimental West Nile virus intrathecal challenge in horses. Vet Ther 7:249–56

Stocks CE, Lobigs M (1998) Signal peptidase cleavage at the flavivirus C–prM junction: dependence on the viral NS2B-3 protease for efficient processing requires determinants in C, the signal peptide, and prM. J Virol 72:2141–9

Tesh RB, Arroyo J, Travassos Da Rosa AP, Guzman H, Xiao SY, Monath TP (2002) Efficacy of killed virus vaccine, live attenuated chimeric virus vaccine, and passive immunization for prevention of West Nile virus encephalitis in hamster model. Emerg Infect Dis 8:1392–7

Turell MJ, Bunning M, Ludwig GV, Ortman B, Chang J, Speaker T, Spielman A, McLean R, Komar N, Gates R, McNamara T, Creekmore T, Farley L, Mitchell CJ (2003) DNA vaccine for West Nile virus infection in fish crows (*Corvus ossifragus*). Emerg Infect Dis 9:1077–81

Watts DM, Tesh RB, Siirin M, Rosa AT, Newman PC, Clements DE, Ogata S, Coller BA, Weeks-Levy C, Lieberman MM (2007) Efficacy and durability of a recombinant subunit West Nile vaccine candidate in protecting hamsters from West Nile encephalitis. Vaccine 25:2913–8

Yamshchikov VF, Wengler G, Perelygin AA, Brinton MA, Compans RW (2001) An infectious clone of the West Nile flavivirus. Virology 281:294–304

Yamshchikov G, Borisevich V, Seregin A, Chaporgina E, Mishina M, Mishin V, Kwok CW, Yamshchikov V (2004) An attenuated West Nile prototype virus is highly immunogenic and protects against the deadly NY99 strain: a candidate for live WN vaccine development. Virology 330:304–12

21. Novel Therapeutics Against West Nile Virus

PEI-YONG SHI

Abstract

No effective therapy is currently available for clinical treatment of flavivirus infections. Recent advances in the structural and molecular biology of flaviviruses have provided new opportunities for the development of antiviral therapies. This chapter summarizes the current status of West Nile virus (WNV) therapeutics. First, strategies for identifying and characterizing small molecular inhibitors are reviewed. These strategies include structure-based rational design, biochemical enzyme-based screening, and reverse genetic system-based screening. Second, known WNV inhibitors are summarized. Both small and macromolecular inhibitors have been identified to inhibit WNV. The macromolecular inhibitors include WNV antibodies, interferon, and nucleic acid-based agents (i.e., antisense oligomer and siRNA). Since the antibody-based therapy is reviewed elsewhere in this book, this chapter emphasizes the nonantibody macromolecular and small molecular inhibitors. Finally, new potential antiviral targets and issues related to WNV therapeutics are discussed.

Keywords

antiviral, siRNA, small molecule inhibitor, replicon, flavivirus

1 Introduction

Flavivirus infection is a major public health threat. Human vaccines for flavivirus infections are currently available only for yellow fever virus (YFV), Japanese encephalitis virus (JEV), and tick-borne encephalitis virus (TBEV). For West Nile virus (WNV), vaccines are only approved for equine use (Davis et al., 2001; Minke et al., 2004; Ng et al., 2003).

Development of a vaccine for dengue virus (DENV) has been challenging, principally because of the need to simultaneously immunize and induce long-lasting protection against all four DENV serotypes; an incompletely immunized individual may be sensitized to dengue hemorrhagic fever or dengue shock syndrome. Because of these complications, it is critical to develop effective therapies for the treatment of flavivirus infections. This chapter focuses on the current development of WNV therapeutics, including (1) strategies for the identification of novel small molecular inhibitors, (2) known WNV inhibitors, and (3) potential new antiviral targets and issues related to the antiviral development. Since members from the *Flavivirus* genus share a common replication strategy, antiviral studies of flaviviruses other than WNV are also selectively discussed.

2 Strategies for the Identification of Novel Small Molecular Inhibitors

Each step of the WNV life cycle is a potential target for antiviral intervention. Three approaches are commonly taken to identify inhibitors, including structure-based rational design, biochemical enzyme-based screening, and reverse genetic system-based screening. Each approach has its own advantages and potential problems. Once combined, these approaches can provide complementary information about individual inhibitors.

2.1 Rational Design

The single open reading frame of the flavivirus RNA genome encodes a long polyprotein that is co- and posttranslationally processed into ten mature proteins by a combination of host proteases and the viral protease, resulting in three structural proteins [capsid (C), membrane (M), and envelope (E)] and seven nonstructural proteins (NS1, NS2a, NS2b, NS3, NS4a, NS4b, and NS5) (Lindenbach et al., 2007). The atomic structures of many flavivirus proteins and domains have been solved. This structural information makes the rational design of inhibitors possible. Among the three structural proteins, the structure of C has been solved by NMR (Ma et al., 2004) and crystallography (Dokland et al., 2004). The structures of the E ectodomain are available in both pre- and postfusion forms (Bressanelli et al., 2004; Modis et al., 2003, 2004; Rey et al., 1995). Moreover, the virion structures of WNV and DENV have been determined by using a combination of cryoelectron microscopy and fitting of the known structure of E into the electron density map (Kuhn et al., 2002; Mukhopadhyay et al., 2003; Zhang et al., 2003c). Among the seven nonstructural proteins, two have enzymatic activities. The N-terminal region of NS3 exhibits a protease activity (with cofactor NS2B), which is responsible for all cleavages of the polyprotein on the cytoplasmic side of the endoplasmic reticulum (ER). The C-terminal portion of NS3 has 5′-RNA triphosphatase, NTPase, and helicase activities. The 5′-RNA triphosphatase activity is required for formation of the cap structure of the viral genome. The NTPase and helicase activities are needed to unwind RNA structure during viral replication. The N-terminal region of NS5 encodes a methyltransferase, which is required to methylate the cap guanine N7 and the ribose 2′-OH of the first nucleotide during cap formation (m7GpppAm). The C-terminal portion of NS5 contains an RNA-dependent RNA polymerase (RdRp). Crystal structures of individual domains of the protease complexed with a Bowman–Birk inhibitor (Murthy et al., 2000) or with the NS2b peptide (Aleshin et al., 2007; Erbel et al., 2006), helicase (Mancini et al., 2007; Mastrangelo et al., 2007b; Wu et al., 2005; Xu et al., 2005), methyltransferase (Assenberg et al., 2007; Egloff et al., 2002, 2007; Mastrangelo et al., 2007a; Zhou et al., 2007), and RdRp (Malet et al., 2007; Yap et al., 2007) have been solved for a number of flaviviruses. Furthermore, the crystal structure of full-length NS3 of DENV-4 was recently resolved (Luo et al., 2008). The structure of the full-length NS5 remains to be determined.

Despite the breakthrough in structure solving of flavivirus proteins, few inhibitors have been developed through rational design. A virtual screening, using DENV-2 NS5 methyltransferase structure, recently revealed a compound that inhibited DENV 2′-O cap methylation with an IC_{50} value of 60 μM (Luzhkov et al., 2007). The study used *S*-adenosylmethionine (SAM), the methyl donor, as a starting molecule to search for analogs that could specifically dock into the SAM-binding pocket of the DENV-2 methyltransferase. Since the SAM molecule bound in the same pocket donates methyl groups to both N7 and 2′-O positions during cap methylations (Dong et al., 2008), the identified compound is expected to inhibit N7 methylation of the viral RNA cap. Given that the flavivirus methyltransferases are highly conserved in structure and sequence, it is likely that this compound would inhibit methyltransferases from flaviviruses other than DENV. However, experiments are needed to verify these speculations and to demonstrate the potency of the compound in a viral infection assay. Along the same line, sinefungin (SIN), a SAM analog in which the transferring methyl group is replaced by an amino group, inhibits both N7 and 2′-O methylations of the WNV RNA cap, with an IC_{50} value of approximately 14 μM (Dong et al., 2008). In BHK cells, SIN inhibits WNV infection with an

IC_{50} of 27 μM, CC_{50} of 4.5 mM, and therapeutic index (TI = CC_{50}/IC_{50}) of 167. The cytotoxicity of SIN is expected, because the compound can also bind to host methyltransferase and suppress cap methylations of cellular mRNA. As the residues that form the SAM-binding pocket between the host and viral methyltransferases are poorly conserved (Martin and McMillan, 2002), the specificity of SIN binding to the viral methyltransferase could be improved to reduce its cytotoxicity.

2.2 Biochemical Enzyme-Based Screening

NS3 and NS5 are key enzymes of the viral replication complex and are ideal targets for antiviral screening. Of the numerous high-throughput screening (HTS) assay formats, the scintillation proximity assay (SPA), a radioactive homogeneous assay technology has been commonly used to screen large compound libraries. The SPA method relies upon the fact that the energy emitted from a radioisotope will only travel a limited distance in an aqueous environment. When a radioisotope-labeled molecule binds to the microsphere, the radioisotope is brought into close proximity to the scintillant and effective energy transfer from the β-particle occurs, resulting in light emission. When the radioisotope remains free in solution, it is too distant from the scintillant; the β-particle dissipates the energy into the aqueous medium and remains undetected. The SPA method has been used to assess DENV-3 RdRp activity (Yap et al., 2007), and can be readily adapted to flavivirus helicases and methyltransferases:

1. In the RdRp assay, a 5′ biotin-labeled oligomer is annealed to an RNA template and extended to incorporate ^{3}H-labeled GTP in the presence of recombinant RdRp. The newly synthesized RNA is then captured onto the streptavidin-coupled SPA beads, and the amount of ^{3}H-GTP incorporation (indicating the RdRp activity) is quantified by scintillation counting.
2. An SPA-formatted methyltransferase assay could be similarly developed. As flavivirus methyltransferase requires specific viral RNA for cap methylation (Dong et al., 2007a), the biotin-labeled RNA substrate should contain the 5′ terminal sequence of the viral genome. Upon transfer of the ^{3}H-labeled methyl group from SAM to RNA cap, the biotin-labeled RNA is captured onto the streptavidin SPA beads and measured through a scintillation counter.
3. For the helicase assay, one strand of the double-stranded RNA substrate (derived from annealing of two RNAs with complementary sequences with 3′ end overhangs) should be labeled with radioactive

nucleotide, while the other RNA strand is coupled with biotin. After unwinding of the double-stranded RNA, the biotinylated RNA could be captured by the streptavidin SPA beads, and counted in a scintillation counter. The helicase activity will be reflected by a reduction of the scintillation signal.

For the NS3 protease assay, a fluorogenic peptide substrate-based assay has been widely used (Mueller et al., 2007; Yusof et al., 2000). This method uses a peptide substrate coupled to a highly fluorescent dye, whose fluorescence is quenched while linked to the uncleaved peptide substrate. Upon protease-mediated cleavage, the fluorescent dye is released from the peptide, thereby producing an increase in fluorescence intensity.

The above assays have made it possible to screen for potential inhibitors of key viral enzymes in an HTS fashion. The major advantage of the enzyme-based assays is that the identified compounds possess known targets. The availability of structures of these proteins will facilitate further improvement of the inhibitors. However, compared to the cell-based screening assays (see next section) the enzyme-based screening does not examine cellular uptake and nonspecific binding of serum proteins to the compounds. The latter factors could dramatically affect the potency of the compounds in cell culture and animals.

2.3 Genetic Cell-Based Screening

Genetic cell-based assays have two major advantages: more than one target and step are analyzed during replication and the uptake of compounds into cells is required, which represents a more authentic therapeutic environment. Inhibitors identified through cell-based assays have had a higher success rate in subsequent animal experiments. Since most flavivirus infections cause cytopathic effects (CPE), the traditional cell-based screening assay involves viral infection of cultured cells and monitoring of inhibition through measurement of CPE (Jordan et al., 2000; Morrey et al., 2002) or quantification of viral RNA by RT-PCR (Shi et al., 2001). Unfortunately, these assays are labor-intensive and lack sensitivity. Thus, while informative, they are not ideal for screening large compound libraries.

To overcome the above problems, three cell-based HTS assays have been developed for WNV drug discovery (Fig. 1) (Puig-Basagoiti et al., 2005). These antiviral assays were optimized in a 96- or 384-well format, validated with known WNV inhibitors, and proven useful in identifying new inhibitor(s) through screening compound libraries:

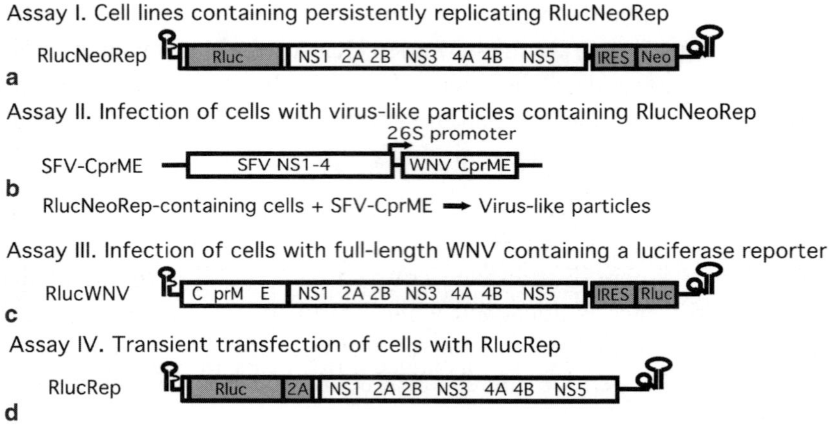

Figure 1. Cell-based assays for development of WNV therapeutics. A *Renilla* luciferase (Rluc) gene was engineered into a replicon or into a full-length viral genome to monitor viral replication. Potential inhibitors could be indicated through suppression of luciferase signals upon compound incubation. (**a**) A reporting cell line assay. The cell line contains a persistently replicating replicon that harbors dual reporter genes, Rluc and neomycin phosphotransferase (Neo, driven by an EMCV IRES), resulting in RlucNeoRep. The assay allows screening inhibitors of all targets involved in viral replication. (**b**) Virus-like particle (VLP) infection assay. WNV structural proteins are provided *in trans* to package luciferase-reporting replicon. The example here uses Semliki forest virus vector (containing viral nonstructural proteins, SFV NS1–4) to express WNV structural proteins C-prm-E under the control of the SFV 26S subgenomic promoter (SFV-CprME). Transfection of RlucNeoRep-containing cells with SFV-CprME packages RlucNeoRep RNA into VLPs. Infections of naïve cells with such VLPs result in the replication of RlucNeoRep, yielding Rluc signals. The VLP-based assay allows screening inhibitors of viral entry and replication. (**c**) A full-length luciferase-reporting WNV infection assay. An EMCV IRES-Rluc was inserted at the 3′ UTR of the genome, resulting in RlucWNV. Infection of cells with RlucWNV generates Rluc signals. This assay allows screening inhibitors of all steps of viral infection cycle, including entry, replication, and assembly. (**d**) A transient luciferase-reporting replicon. A fragment containing Rluc fused with a foot-and-mouth disease virus (FMDV) 2A protease was engineered to replace the viral structural genes, resulting in RlucRep (Lo et al., 2003a). Transfection of cells with RlucRep results in two distinct Rluc peaks: the first peak (at 1–6 h posttransfection) represents input RNA translation, and the second peak (at >10 h posttransfection) represents viral RNA synthesis. The transient RlucRep system is useful to dissect whether a compound inhibits viral translation or RNA synthesis.

1. *Cell lines expressing WNV replicons.* Cell lines have been established that persistently replicate the WNV subgenomic replicon in the absence of structural genes (Lo et al., 2003b). Many of these cell lines harbor replicons that contain a *Renilla* luciferase (Rluc; substituting for viral structural genes) and a neomycin phosphotransferase gene [Neo; driven by an encephalomyocarditis virus internal ribosomal

entry site (EMCV IRES) in the 3′ UTR] (RlucNeoRep, Fig. 1a). Incubation of the RlucNeoRep-containing cells with candidate inhibitors results in a decrease in Rluc activity. The reporting cell lines allow for screening inhibitors at all steps involved in viral translation and RNA replication, but do not interrogate steps of viral assembly and entry. One major advantage of the replicon cell line is that the assay can be performed in a biosafety level 2 (BSL-2) laboratory, as no infectious particles are involved, whereas BSL-3 containment is required for working with the full-length WNV.

2. *Packaged virus-like particles (VLPs) containing replicon RNA*. The luciferase replicon can be packaged into VLPs by supplying structural proteins *in trans* (Khromykh et al., 1998). The structural proteins of WNV can be expressed from a 26S subgenomic promoter from the Semliki forest virus (SFV) expression vector (SFV-CprME, Fig. 1b). Alternatively, the structural proteins could be provided in an inducible expression cell line (Harvey et al., 2004), or in a constitutive expression cell line harboring a noncytopathic Venezuelan equine encephalitis virus (VEEV) replicon (Fayzulin et al., 2006). Compared with the replicon cell line assay, the VLP-infection system allows for screening inhibitors of viral entry as well as translation and RNA synthesis.
3. A full-length WNV with reporter gene activity. An IRES-Rluc fragment was inserted into the 3′ UTR of the WNV genome (RlucWNV, Fig. 1c). Transfection of the RlucWNV RNA into BHK cells produced full-length reporting WNV (Deas et al., 2005). Unfortunately, the reporting virus was not stable; after multiple rounds of infection in cell culture, the engineered IRES-Rluc was deleted from the full-length virus. Nevertheless, an early passage of this reporting virus can be used to screen for inhibitors against all steps of the viral life cycle (Puig-Basagoiti et al., 2005). Recent studies showed that stable reporting YFV could be generated when a heterologous gene was inserted between E and NS1 (Bredenbeek et al., 2006), or when a foreign gene was engineered upstream of the C of the open reading frame (Shustov et al., 2007). These strategies may help to generate WNV that express reporter genes in a more stable fashion.

Besides the above cell-based assays, a transient Rluc-reporting replicon of WNV was developed to analyze mode of action of inhibitors (Lo et al., 2003a). In this replicon, the Rluc was fused in-frame with a foot-and-mouth disease virus (FMDV) 2A protease. The resulting Rluc-2A was engineered to replace the viral structural genes, resulting in the replicon RlucRep (Fig. 1d). Transfection of cells with RlucRep generates two distinct Rluc peaks: one at 1–6 h posttransfection and another at >10 posttransfection. The two Rluc peaks can be used to differentiate between translation of the input RNA and translation of the replicating RNA, respectively. A similar approach was used to develop reporting replicons for other flaviviruses (Alvarez et al., 2005; Jones et al., 2005; Puig-Basagoiti et al., 2006). The transient RlucRep is useful to examine whether a compound inhibits viral translation and/or RNA synthesis.

3 Current WNV Inhibitors

Although a number of compounds have been reported to inhibit WNV enzymes or viral replication in cell culture, few have shown in vivo efficacy. Based on the molecular weights, these antiviral agents can be divided into two categories: macromolecular inhibitors and small molecular inhibitors.

3.1 Macromolecular Inhibitors

3.1.1 Antibody

Antibody-based therapy has yielded the most promising results for treatment of WNV infection (Ben-Nathan et al., 2003; Engle and Diamond, 2003; Julander et al., 2005). The results agreed with previous studies showing that passive administration of monoclonal antibodies prevented and alleviated encephalitis caused by St. Louis encephalitis virus (SLEV) (Mathews and Roehrig, 1984), JEV (Kimura-Kuroda and Yasui, 1988), and YFV (Schlesinger et al., 1985). Humanized monoclonal antibodies against WNV E protein have shown efficacy in mice, even administered as a single dose at day 5 postinfection when WNV has already infected the CNS. These antibodies showed high neutralizing activity in cell culture and provide equivalent protection in mice and hamsters in comparison to gamma-globulin (Gould et al., 2005; Morrey et al., 2006, 2007; Oliphant et al., 2005; Throsby et al., 2006). Intravenous immunoglobulin containing high titers of anti-WNV antibodies appeared effective in patients with WN encephalitis in an open-label study (Shimoni et al., 2001). However, the latter results require further confirmation with appropriate controls. The current status of the antibody-based therapy is detailed elsewhere in this book.

3.1.2 Interferon

Interferon-α-2b is currently under clinical trial for treatment of patients with WNV-mediated meningoencephalitis. Although interferon-α-2b inhibits WNV replication in vitro (Anderson and Rahal, 2002), the effect is dramatically reduced once viral replication has initiated, probably due to the blockage of interferon signaling by several of the nonstructural proteins (Best et al., 2005; Lin et al., 2004; Liu et al., 2005; Munoz-Jordan et al., 2003, 2005). Mice that lack expression of the receptor for interferon-α/β (and are therefore defective in interferon response) exhibited higher mortality, shorter survival time, and altered cellular tropism of infection in comparison with wild-type mice upon WNV inoculation (Samuel and Diamond, 2005). Treatment of WNV-infected primary neurons with interferon-α/β, however, appeared to independently increase the survival of neurons in culture. Significant recovery of neurologic function in five patients with WNV CNS disease was reported when treated with interferon-α-2b soon after symptom onset (Sayao et al., 2004). However, one interferon-α-2b failure case was reported, possibly due to delayed diagnosis and treatment, and other complications (Chan-Tack and Forrest, 2005). More controlled studies are needed to demonstrate the efficacy of the interferon treatment.

3.1.3 Small Peptides

Antiviral peptides were identified from a murine brain cDNA phage display library when screened for binding to the WNV E protein (Bai et al., 2007). The most potent peptide, with a sequence of CDVIALLACHLNT, had an IC_{50} of 2.6 μM in cell culture. The peptides were able to penetrate the blood-brain barrier in mice. Mice challenged with WNV that had been preincubated with the peptide exhibited reduced viremia and mortality compared with the mice challenged with untreated virus. However, the peptide did not elicit any protection when administered intraperitoneally (i.p.) in mice at 3 h post-WNV challenge.

Short peptides representing fragments from the viral E protein have also been screened for inhibition of WNV and DENV (Hrobowski et al., 2005). The underlying rationale was that some peptides may interfere with the intramolecular interactions required for virus-host membrane fusion. The effort resulted in several peptides (representing junction regions between domains I and II) that had IC_{50} values in the 10 μM range. Since flavivirus membrane fusion occurs in endosomes, efficient delivery of the E peptides into cells is required to further improve the efficacy.

3.1.4 Antisense Phosphorodiamidate Morpholino Oligomers

Two phosphorodiamidate morpholino oligomers (PMOs) were reported to potently inhibit WNV infection in cell culture (Deas et al., 2005). One PMO targets the first twenty nucleotides of the WNV genome, another PMO targets the 3′ conserved sequence 1 (CS1) RNA that is required for genome cyclization during viral replication (Khromykh et al., 2001; Lo et al., 2003a). Conjugation of an arginine-rich peptide with the PMO results in more efficient cellular delivery and greater potency. Mode-of-action analyses showed that the PMOs targeting the 5′ end and the 3′ CS1 distinctly suppressed RNA translation and synthesis, respectively. Since the CS1 RNA sequence is conserved among mosquito-borne flaviviruses, the CS1 PMO inhibited a broad spectrum of flaviviruses in cell culture. PMO-resistant viruses can be selected in cell culture (Deas et al., 2007). Viruses resistant to the 5′ end PMO accumulated mutations within the PMO-targeted region, whereas viruses resistant to the 3′ CS1 PMO accumulated mutations downstream of the PMO-targeted region. In vivo analysis showed that the PMOs could partially protect mice from WNV disease, even when administered at day 5 postinfection (Deas et al., 2007). In these studies, viral inoculation and PMO treatment were administered subcutaneously (s.c.) and i.p., respectively. Since the arginine-rich peptide (conjugated to the PMO) is essential for cellular uptake, but is also responsible for toxicity, modification of the peptide composition to reduce its toxicity while maintaining its function is needed to improve the PMO-mediated therapy.

3.1.5 siRNA

Synthetic short interfering RNA (siRNA) inhibits WNV both in cell culture and in mice. The potency of siRNA depends on the routes of viral inoculation and siRNA delivery. In one study, both viral inoculation and siRNA [complexed with transfection reagents cationic lipid formulation (JetSI) and the fusogenic lipid dioleoylphosphatidylethanolamine (DOPE)] were administered intracranially (i.c.) (Kumar et al., 2006). Mice receiving siRNA treatment at 6 h postviral inoculation exhibited 100% survival, whereas the mock-treated WNV-infected mice showed 100% mortality. In another study, pretreatment of mice with siRNA through a hydrodynamic injection (into tail vein) 24 h before i.p. inoculation of WNV yielded partial protection (Bai et al., 2005). In a third cell-based study, WNV replication was greatly reduced when siRNA was introduced by cytoplasmic-targeted transfection prior to, but not after the establishment of viral replication (Geiss et al., 2005). However, when siRNA was transfected through electroporation, suppression of viral replication could be achieved after the establishment of viral replication. The latter results indicate that the transfection reagent is critical for delivering the siRNA into the viral replication compartment.

3.2 Small Molecular Inhibitors

3.2.1 Inhibitors of Nucleoside Triphosphate Synthesis

Inhibitors of nucleotide triphosphate synthesis have been reported to have anti-WNV activities (Anderson and Rahal, 2002; Jordan et al., 2000; Morrey et al., 2002). Among those, ribavirin and mycophenolic acid (MPA) are inhibitors of cellular IMP dehydrogenase (IMPDH), an enzyme that is essential for de novo biosynthesis of guanine nucleotides. Ribavirin is a guanosine analog that competitively inhibits the IMPDH, whereas MPA is a nonnucleoside, noncompetitive inhibitor of IMPDH. Both ribavirin and MPA exhibit a broad spectrum of antiviral activity, primarily through depletion of intracellular GTP pools (Leyssen et al., 2005). Besides inhibition of GTP synthesis, two other antiviral mechanisms have been demonstrated for ribavirin: (1) ribavirin can be incorporated into viral RNA through base-pairing with cytidine or uridine, leading to lethal levels of mutagenesis within the virus population (Crotty et al., 2000; Day et al., 2005) and (2) ribavirin blocks viral cap methylation through binding to a GTP-binding pocket of the DENV-2 NS5 methyltransferase (Benarroch et al., 2004). Although both ribavirin and MPA inhibit WNV in cell culture, treatment of WNV-infected hamsters with ribavirin led to increased mortality (Morrey et al., 2004). Furthermore, an increased mortality was observed among 37 patients who received ribavirin treatment in a WNV outbreak in 2000 (Chowers et al., 2001). In agreement with the WNV results, no clinical benefit was observed when DENV-infected mice and monkeys were treated with ribavirin (Koff et al., 1983; Malinoski et al., 1990). Similar to ribavirin, an increased mortality was observed when WNV-infected mice were treated with MPA (Noueiry et al., 2007). Collectively, the results suggest that ribavirin and MPA are not promising candidates for treatment of WNV infection in vivo.

Several other nucleoside inhibitors were reported to have anti-WNV activity in cell culture, with TIs of >10. The inhibitors are 6-azauridine (IC_{50} 5 μM and CC_{50} 600 μM), 2-thio-6-azauridine (IC_{50} 3 μM and CC_{50} > 3.5 mM), 6-azauridine triacetate (IC_{50} 3.5 μM and CC_{50} 210 μM), pyrazofurin (IC_{50} 6 μM and CC_{50} 110 μM), and cyclopentenylcytosine (IC_{50} 1 μM and CC_{50} 110 μM) (Morrey et al., 2002). Additionally, 5-aza-7-deazaguanosine (ZX-2401) was shown to have a broad-spectrum antiviral activity against viruses in the *Flaviviridae* family, including WNV, hepatitis C replicon, and pestivirus (bovine viral diarrhea virus) (Ojwang et al., 2005). No in vivo efficacy has been reported for these compounds against WNV.

3.2.2 NTPase/Helicase Inhibitors

A series of nucleoside analogs have been identified that inhibit WNV NS3NTPase/helicase:

1. An imidazo[4,5-d]pyridazine nucleoside was reported to inhibit WNV helicase with an IC_{50} of about 30 μM in both enzyme and viral titer reduction assay (Borowski et al., 2002). The antiviral activity of the compound appears to be specific to the WNV NS3, because it did not inhibit hepatitis C virus helicase.
2. A set of ring-expanded heterocyclic nucleoside analogs, which contain either the imidazo[4,5-e][1,3]diazepine ring (Zhang et al., 2003a) or the 6-aminoimdazo[4,5-e][1,3]diazepine-4,8-dione ring systems (Zhang et al., 2003b), inhibited WNV NS3 NTPase/helicase with IC_{50} in low micromolar concentrations.
3. Random screening of a compound library containing a broad range of unrelated small molecules revealed 5,6-dichloro-1-(β-d-ribofuranosyl)benzotriazole which inhibited the WNV helicase with an IC_{50} of about 0.3 μM (Borowski et al., 2003). However, the potency of the compounds described in 2 and 3 has not been examined in cell culture.

3.2.3 Protease Inhibitors

Both small molecular and peptide inhibitors have been reported to inhibit the WNV NS3 protease. Based on the crystal structure of the DENV protease in complex with the mung bean Bowman–Birk inhibitor, five compounds were identified that could mimic the interactions between substrate P1 arginine and protease and therefore inhibit DENV protease activity (Ganesh et al., 2005). Due to the structural conservation among flavivirus proteases, these compounds also inhibited WNV protease, three of which showed IC_{50} of 13–35 μM. With the recent advances in structure solving of flavivirus protease complexes [NS2b-protease-peptide inhibitor (Erbel et al., 2006) and NS2b-protease-aprotinin inhibitor

(Aleshin et al., 2007)], the computational approach could be extended to improve the current compounds or to search for new inhibitors that block the interactions among substrate, NS2b, and protease.

Flavivirus NS3 is a serine protease that belongs to the trypsin superfamily. The cleavage sites of the flavivirus NS3 protease are conserved and contain unique dibasic residues at the P1 and P2 sites (lysine or arginine) followed by a residue with a short side chain at the P1′ position (most commonly glycine, serine, or alanine). Peptides mimicking the conserved cleavage substrate have been explored as possible inhibitors of the flavivirus protease. For WNV, tetrapeptide variants, derived from the benzoyl-norleucine-lysine-arginine-arginine (Bz-nKRR) sequence, inhibited the protease with IC_{50} of 1–200 μM (Knox et al., 2006). Similarly, at micromolar concentrations peptides with α-keto amide backbones inhibited DENV protease (Leung et al., 2001). The potency of the compounds may be improved through structural optimization, based on the recent crystal structure of NS2b-protease-inhibitor (Bz-nKRR) (Erbel et al., 2006). Besides the peptide inhibitors, standard serine protease inhibitors have been tested against flavivirus proteases. Aprotinin inhibits both WNV and DENV proteases with 160 nM of IC_{50} and 26–65 nM of IC_{50}, respectively (Leung et al., 2001; Mueller et al., 2007). The mode of action of aprotinin was recently revealed by the structure of a tertiary complex consisting of WNV NS2b-protease-aprotinin (Aleshin et al., 2007). Aprotinin binds in a substrate-mimetic fashion. In comparison with the structure of the NS2b-protease complex without substrate, the tertiary complex suggests an "induced fit" mechanism of the flavivirus protease, providing an opportunity to develop inhibitors that block the conformational change required for catalytic activity of the protease.

3.2.4 Host Glucosidase Inhibitors

The host glycosylation machinery is required for modifications of flavivirus proteins. Inhibitors of ER α-glucosidase, such as *N*-nonyl-deoxynojirimycin and castanospermine, were reported to inhibit JEV and DENV at low micromolar concentrations in cell culture (Courageot et al., 2000; Wu et al., 2002). The α-glucosidase inhibitors block the trimming step of N-linked glycosylation of prM and E, leading to delayed formation of preM/E heterodimer and nonproductive virion assembly. The *N*-nonyl-deoxynojirimycin compound could partially protect mice from JEV infection at 200 mg/kg/day (Wu et al., 2002). Similarly, castanospermine significantly reduced the mortality of A/J mice infected with DENV-2 at doses of 50 and 250 mg/kg/day. However, castanospermine did not substantially suppress WNV infection in cell culture or in mice (Whitby et al., 2005).

3.2.5 Other Small Molecular Inhibitors

A number of new classes of WNV inhibitors have been identified through HTS using luciferase-reporting replicon cell lines:

1. Triaryl pyrazoline was reported to inhibit WNV without detectable cytotoxicity (IC_{50} 28 μM and $CC_{50} \geq 300$ μM). Besides WNV, this compound also inhibited other flaviviruses (DENV, YFV, and SLEV), an alphavirus (Western equine encephalitis virus), a coronavirus (mouse hepatitis virus), and a rhabdovirus (vesicular stomatitis virus). However, the compound did not suppress an orthomyxovirus (influenza virus) or a retrovirus (HIV-1). Mode-of-action analysis with WNV showed that the compound did not block viral entry or virion assembly, but specifically suppressed viral RNA synthesis (Goodell et al., 2006; Puig-Basagoiti et al., 2006).
2. One pyrozolopyrimidine compound was shown to have antiviral activity (IC_{50} 11 μM and CC_{50} 85 μM) (Gu et al., 2006).
3. Several compounds within the sulfonamide family were recently reported to inhibit WNV in cell culture at the submicromolar range with TIs of >10. Besides WNV, these compounds also inhibit other flaviviruses. One of these compounds was shown to specifically block translation of a YFV replicon, but not an alphavirus (Sindbis virus) replicon or an EMCV IRES-containing mRNA (Noueiry et al., 2007). Experiments are needed to determine the exact target and the in vivo efficacy of the above compounds.

Besides the above novel molecules, minocycline (a tetracycline derivative that inhibits bacterial translation through binding to the 30S ribosomal subunit thus preventing the amino-acyl tRNA from binding to the A site of the ribosome) was recently shown to inhibit WNV infection in cell culture (IC_{50} 18–25 μM and CC_{50} 140 μM). The compound also inhibited WNV-induced apoptosis and suppressed virus-induced activation of c-Jun N-terminal kinase and its target c-jun (Michaelis et al., 2007). Minocycline was previously shown to protect mice from fatal Sindbis virus encephalitis, although it did not inhibit Sindbis replication in cell culture (Irani and Prow, 2007). It will be interesting to examine whether minocycline has anti-WNV activity in animal models.

4 New Antiviral Targets and Challenges on Antiviral Development

Significant progress has been made toward the development of therapeutics against WNV. Among the known inhibitors, antibody-based treatment has so far shown the most promising results. Due to the quasispecies nature of WNV and the error-prone mutation of the viral RdRp, one potential problem of the antibody-based therapy is rapid development of resistant viruses. Therefore, novel classes of inhibitors should be developed in parallel to the antibody-mediated therapy. The establishment of HTS assays has made it possible to screen large compound libraries for inhibitors. One challenge for the development of effective therapeutics of WNV infection is that candidate inhibitors must effectively reach the CNS to clear the virus from infected neurons. Patients with the most severe WNV disease often have immune deficits and present to clinical attention relatively late in their disease course. Ultimately, a combinatory strategy that inhibits viral replication, boosts protective immune responses, and minimizes CNS injury may be more effective than any single antiviral agent.

Improvement of our understanding of the basic virology and pathogenesis of flavivirus will continue to open new avenues for antiviral development. For example, crystallographic studies showed that viral E protein (Bressanelli et al., 2004; Modis et al., 2003, 2004) undergoes a sequential structural change during the fusion-activating transition. Small molecular inhibitors could be developed to block the structural transitions essential for viral/host membrane fusion. The recent characterization of flavivirus NS5 methyltransferase has demonstrated that guanine N7 methylation of the viral RNA cap is essential for WNV replication (Zhou et al., 2007). Since flavivirus methyltransferase specifically methylates the cap structure from viral RNA, the NS5 methyltransferase represents an antiviral target (Dong et al., 2007a). The findings that flavivirus proteins antagonize host innate immune responses have also provided opportunities to develop antiviral therapies (Munoz-Jordan et al., 2003). Inhibitors that block the interferon antagonism of viral protein(s) may allow the innate antiviral response to effectively suppress viral infection and, therefore, reduce viral burden. Finally, host proteins essential for viral replication could potentially be targeted for antiviral therapy. For instance, host Src family kinase c-Yes was found to be critical for maturation of WNV particles (Hirsch et al., 2005). Inhibitors of the c-Src protein kinase were recently shown to potently inhibit four serotypes of DENV, with IC_{50} values in submicromolar range (Chu and Yang, 2007).

In clinical settings, it will be essential to develop a specific and rapid assay for diagnosis of WNV infection prior to initiation of chemotherapy. Upon human infection with WNV after mosquito inoculation, viremia is detected during the first 1–2 weeks postinfection, after which time symptoms develop and antibodies are produced (Prince et al., 2005). Various RT-PCR-based assays have been developed for sensitive detection of the viral RNA (Shi and Kramer, 2003). It is expected that early treatment with antiviral agents will yield better therapeutic outcome than late treatment during the course of infection. Since viremia diminishes when symptoms develop and antibodies are produced, serology is a key diagnostic method for WNV infection (Shi and Wong, 2003). The current serological assays detect WNV-specific antibody against viral structural proteins by enzyme immunoassays (EIA) (Martin et al., 2002; Tardei et al., 2000) or enzyme-linked immunoabsorbent assay for IgG (ELISA) (Davis et al., 2001; Tardei et al., 2000). These assays lack specificity because of cross-reactivity of the structural proteins among flaviviruses. A luminex assay using recombinant NS5 of WNV as an antigen was reported to differentiate WNV infection from infections by other flaviviruses (DENV and SLEV), and also to differentiate between natural WNV infection and flavivirus vaccination (Wong et al., 2003). It remains to be determined whether NS5 from other flaviviruses could be used for serologic diagnosis. Other viral-type specific diagnostic methods are needed not only for antiflavivirus therapy, but also for vaccine development.

References

Aleshin, A., Shiryaev, S., Strongin, A., and Liddington, R. (2007). Structural evidence for regulation and specificity of flaviviral proteases and evolution of the Flaviviridae fold. Protein Sci 16, 795–806.

Alvarez, D. E., De Lella Ezcurra, A. L., Fucito, S., and Gamarnik, A. V. (2005). Role of RNA structures present at the 3′UTR of dengue virus on translation, RNA synthesis, and viral replication. Virology 339, 200–212.

Anderson, J. F., and Rahal, J. J. (2002). Efficacy of interferon alpha-2b and ribavirin against West Nile virus in vitro. Emerg Infect Dis 8, 107–108.

Assenberg, R., Ren, J., Verma, A., Walter, T. S., David Alderton, Hurrelbrink, R. J., Fuller, S. D., Bressanelli, S., Owens, R. J., Stuart, D. I., and Grimes, J. M. (2007). Crystal structure of the Murray Valley encephalitis virus NS5 methyltransferase domain in complex with cap analogues. J Gen Virol 88, 2228–2236.

Bai, F., Wang, T., Pal, U., Bao, F., Gould, L. H., and Fikrig, E. (2005). Use of RNA interference to prevent lethal murine West Nile virus infection. J Infect Dis 191, 1148–1154.

Bai, F., Town, T., Pradhan, D., Cox, J., Ashish, Ledizet, M., Anderson, J. F., Flavell, R. A., Krueger, J. K., Koski, R. A., and Fikrig, E. (2007). Antiviral peptides targeting the West Nile virus envelope protein. J Virol 81, 2047–2055.

Benarroch, D., Egloff, M. P., Mulard, L., Guerreiro, C., Romette, J. L., and Canard, B. (2004). A structural basis for the inhibition of the NS5 dengue virus mRNA 2′-O-methyltransferase domain by ribavirin 5′-triphosphate. J Biol Chem 279, 35638–35643.

Ben-Nathan, D., Lustig, S., Tam, G., Robinzon, S., Segal, S., and Rager-Zisman, B. (2003). Prophylactic and therapeutic efficacy of human intravenous immunoglobulin in treating West Nile virus infection in mice [see comment]. J Infect Dis 188, 5–12.

Best, S. M., Morris, K. L., Shannon, J. G., Robertson, S. J., Mitzel, D. N., Park, G. S., Boer, E., Wolfinbarger, J. B., and Bloom, M. E. (2005). Inhibition of interferon-stimulated JAK-STAT signaling by a tick-borne flavivirus and identification of NS5 as an interferon antagonist. J Virol 79, 12828–12839.

Borowski, P., Lang, M., Haag, A., Schmitz, H., Choe, J., Chen, H. M., and Hosmane, R. S. (2002). Characterization of imidazo[4,5-d]pyridazine nucleosides as modulators of unwinding reaction mediated by West Nile virus nucleoside triphosphatase/helicase: evidence for activity on the level of substrate and/or enzyme. Antimicrob Agents Chemother 46, 1231–1239.

Borowski, P., Deinert, J., Schalinski, S., Bretner, M., Ginalski, K., Kulikowski, T., and Shugar, D. (2003). Halogenated benzimidazoles and benzotriazoles as inhibitors of the NTPase/helicase activities of hepatitis C and related viruses. Eur J Biochem 270, 1645–1653.

Bredenbeek, P., Molenkamp, R., Spaan, W., Deubel, V., Marianneau, P., Salvato, M., Moshkoff, D., Zapata, J., Tikhonov, I., Patterson, J., et al. (2006). A recombinant yellow fever 17D vaccine expressing Lassa virus glycoproteins. Virology 345, 299–304.

Bressanelli, S., Stiasny, K., Allison, S. L., Stura, E. A., Duquerroy, S., Lescar, J., Heinz, F. X., and Rey, F. A. (2004). Structure of a flavivirus envelope glycoprotein in its low-pH-induced membrane fusion conformation. EMBO J 23, 728–738.

Chan-Tack, K., and Forrest, G. (2005). Failure of interferon alpha-2b in a patient with West Nile virus meningoencephalitis and acute flaccid paralysis. Scand J Infect Dis 37, 944–946.

Chowers, M. Y., Lang, R., Nassar, F., Ben-David, D., Giladi, M., Rubinshtein, E., Itzhaki, A., Mishal, J., Siegman-Igra, Y., Kitzes, R., et al. (2001). Clinical characteristics of the West Nile fever outbreak, Israel, 2000. Emerg Infect Dis 7, 675–678.

Chu, J. J., and Yang, P. L. (2007). c-Src protein kinase inhibitors block assembly and maturation of dengue virus. Proc Natl Acad Sci U S A 104, 3520–3525.

Courageot, M. P., Frenkiel, M. P., Dos Santos, C. D., Deubel, V., and Despres, P. (2000). Alpha-glucosidase inhibitors reduce dengue virus production by affecting the initial steps of virion morphogenesis in the endoplasmic reticulum. J Virol 74, 564–572.

Crotty, S., Maag, D., Arnold, J. J., Zhong, W., Lau, J. Y., Hong, Z., Andino, R., and Cameron, C. E. (2000). The broad-spectrum antiviral ribonucleoside ribavirin is an RNA virus mutagen [erratum appears in Nat Med 2001 Feb;7(2):255]. Nat Med 6, 1375–1379.

Davis, B., Chang, G., Cropp, B., Roehrig, J., Martin, D., Mitchell, C., Bowen, R., and Bunning, M. (2001). West Nile virus recombinant DNA vaccine protects mouse and horse from virus challenge and expresses in vitro a noninfectious recombinant antigen that can be used in enzyme-linked immunosorbent assays. J Virol 75, 4040–4047.

Day, C. W., Smee, D. F., Julander, J. G., Yamshchikov, V. F., Sidwell, R. W., and Morrey, J. D. (2005). Error-prone replication of West Nile virus caused by ribavirin. Antiviral Res 67, 38–45.

Deas, T. S., Bennett, C. J., Jones, S. A., Tilgner, M., Ren, P., Behr, M. J., Stein, D. A., Iversen, P. L., Kramer, L. D., Bernard, K. A., and Shi, P.-Y. (2007). In vitro resistance selection and in vivo efficacy of morpholino oligomers against West Nile virus. Antimicrob Agents Chemother 51:2470–2482.

Dokland, T., Walsh, M., Mackenzie, J. M., Khromykh, A. A., Ee, K. H., and Wang, S. (2004). West Nile virus core protein; tetramer structure and ribbon formation. Structure 12, 1157–1163.

Dong, H., Ray, D., Ren, S., Zhang, B., Puig-Basagoiti, F., Takagi, Y., Ho, C., Li, H., and Shi, P. (2007a). Distinct RNA elements confer specificity to flavivirus RNA cap methylation events. J Virol 81, 4412–4421.

Dong, H., Ren, S., Zhang, B., Zhou, Y., Puig-Basagoiti, F., Li, H., and Shi, P. (2008). West Nile Virus methyltransferase catalyzes two methylations of the viral RNA cap through a substrate repositioning mechanism 82:4295–4307.

Egloff, M. P., Benarroch, D., Selisko, B., Romette, J. L., and Canard, B. (2002). An RNA cap (nucleoside-2′-O-)-methyltransferase in the flavivirus RNA polymerase NS5: crystal structure and functional characterization. EMBO J 21, 2757–2768.

Egloff, M. P., Decroly, E., Malet, H., Selisko, B., Benarroch, D., Ferron, F., and Canard, B. (2007). Structural and functional analysis of methylation and 5′-RNA sequence requirements of short capped RNAs by the methyltransferase domain of dengue virus NS5. J Mol Biol.

Engle, M., and Diamond, M. (2003). Antibody prophylaxis and therapy against West Nile virus infection in wild-type and immunodeficient mice. J Virol 77, 12941–12949.

Erbel, P., Schiering, N., D'Arcy, A., Renatus, M., Kroemer, M., Lim, S., Yin, Z., Keller, T., Vasudevan, S., and Hommel, U. (2006). Structural basis for the activation of flaviviral NS3 proteases from dengue and West Nile virus. Nat Struct Mol Biol 13, 372–373.

Fayzulin, R., Scholle, F., Petrakova, O., Frolov, I., and Mason, P. W. (2006). Evaluation of replicative capacity and genetic stability of West Nile virus replicons using highly efficient packaging cell lines. Virology 351, 196–209.

Ganesh, V. K., Muller, N., Judge, K., Luan, C. H., Padmanabhan, R., and Murthy, K. H. (2005). Identification and characterization of nonsubstrate based inhibitors of the essential dengue and West Nile virus proteases. Bioorg Med Chem 13, 257–264.

Geiss, B., Pierson, T., and Diamond, M. (2005). Actively replicating West Nile virus is resistant to cytoplasmic delivery of siRNA. Virol J 2, 53.

Goodell, J. R., Puig-Basagoiti, F., Forshey, B. M., Shi, P. Y., and Ferguson, D. M. (2006). Identification of Compounds with Anti-West Nile Virus Activity. J Med Chem 49, 2127–2137.

Gould, L. H., Sui, J., Foellmer, H., Oliphant, T., Wang, T., Ledizet, M., Murakami, A., Noonan, K., Lambeth, C., Kar, K., et al., (2005). Protective and therapeutic capacity of human single-chain Fv-Fc fusion proteins against West Nile virus. J Virol 79, 14606–14613.

Gu, B., Ouzunov, S., Wang, L., Mason, P., Bourne, N., Cuconati, A., and Block, T. M. (2006). Discovery of small molecule inhibitors of West Nile virus using a high-throughput subgenomic replicon screen. Antiviral Res 70, 39–50.

Harvey, T., Liu, W., Wang, X., Linedale, R., Jacobs, M., Davidson, A., Le, T., Anraku, I., Suhrbier, A., Shi, P., and Khromykh, A. (2004). Tetracycline-inducible packaging cell line for production of flavivirus replicon particles. J Virol 78, 531–538.

Hirsch, A. J., Medigeshi, G. R., Meyers, H. L., DeFilippis, V., Fruh, K., Briese, T., Lipkin, W. I., and Nelson, J. A. (2005). The Src family kinase c-Yes is required for maturation of West Nile virus particles. J Virol 79, 11943–11951.

Hrobowski, Y., Garry, R., and Michael, S. (2005). Peptide inhibitors of dengue virus and West Nile virus infectivity. Virol J 2, 49.

Irani, D., and Prow, N. (2007). Neuroprotective interventions targeting detrimental host immune responses protect mice from fatal alphavirus encephalitis. J Neuropathol Exp Neurol 66, 533–544.

Jones, C., Patkar, C., and Kuhn, R. (2005). Construction and applications of yellow fever virus replicons. Virology 331, 247–259.

Jordan, I., Briese, T., Fischer, N., Lau, J. Y., and Lipkin, W. I. (2000). Ribavirin inhibits West Nile virus replication and cytopathic effect in neural cells. J Infect Dis 182, 1214–1217.

Julander, J. G., Winger, Q. A., Olsen, A. L., Day, C. W., Sidwell, R. W., and Morrey, J. D. (2005). Treatment of West Nile virus-infected mice with reactive immunoglobulin reduces fetal titers and increases dam survival. Antiviral Res 65, 79–85.

Khromykh, A. A., Varnavski, A. N., and Westaway, E. G. (1998). Encapsidation of the flavivirus Kunjin replicon RNA by using a complementation system providing Kunjin virus structural proteins in trans. J Virol 72, 5967–5977.

Khromykh, A. A., Meka, H., Guyatt, K. J., and Westaway, E. G. (2001). Essential role of cyclization sequences in flavivirus RNA replication. J Virol 75, 6719–6728.

Kimura-Kuroda, J., and Yasui, K. (1988). Protection of mice against Japanese encephalitis virus by passive administration with monoclonal antibodies. J Immunol 141, 3606–3610.

Knox, J. E., Ma, N. L., Yin, Z., Patel, S. J., Wang, W. L., Chan, W. L., Ranga Rao, K. R., Wang, G., Ngew, X., Patel, V., et al., (2006). Peptide inhibitors of West Nile NS3 protease: SAR study of tetrapeptide aldehyde inhibitors. J Med Chem 49, 6585–6590.

Koff, W. C., Pratt, R. D., Elm, J. L., Jr., Venkateshan, C. N., and Halstead, S. B. (1983). Treatment of intracranial dengue virus infections in mice with a lipophilic derivative of ribavirin. Antimicrob Agents Chemother 24, 134–136.

Kuhn, R. J., Zhang, W., Rossmann, M. G., Pletnev, S. V., Corver, J., Lenches, E., Jones, C. T., Mukhopadhyay, S., Chipman, P. R., Strauss, E. G.et al., . (2002). Structure of dengue virus: implications for flavivirus organization, maturation, and fusion. Cell 108, 717–725.

Kumar, P., Lee, S. K., Shankar, P., and Manjunath, N. (2006). A single siRNA suppresses fatal encephalitis induced by two different flaviviruses. PLoS Med 3, e96.

Leung, D., Schroder, K., White, H., Fang, N.-X., Stoermer, M., Abbenante, G., Martin, J., PR, Y., and Fairlie, D. (2001). Activity of recombinant dengue 2 virus NS3 protease in the presence of a truncated NS2B co-factor, small peptide substrates, and inhibitors. J Biol Chem 276, 45762–45771.

Leyssen, P., Balzarini, J., De Clercq, E., and Neyts, J. (2005). The predominant mechanism by which ribavirin exerts its antiviral activity in vitro against flaviviruses and paramyxoviruses is mediated by inhibition of IMP dehydrogenase. J Virol 79, 1943–1947.

Lindenbach, B. D., Thiel, H.-J., and Rice, C. M. (2007). Flaviviridae: the virus and their replication, 4th edn. Lippincott Williams & Wilkins, Philadelphia, PA.

Lin, R.-J., Liao, C.-L., Lin, E., and Lin, Y.-L. (2004). Blocking the alpha interferon-induced Jak-Stat signaling pathway by Japanese encephalitis virus infection. J Virol 78, 9285–9294.

Liu, W., Wang, X., Mokhonov, V., Shi, P., Randall, R., and Khromykh, A. (2005). Inhibition of interferon signaling by the New York 99 strain and Kunjin subtype of West Nile virus involves blockage of STAT1 and STAT2 activation by nonstructural proteins. J Virol 79, 1934–1942.

Lo, L., Tilgner, M., Bernard, K., and Shi, P.-Y. (2003a). Functional analysis of mosquito-borne flavivirus conserved sequence elements within 3′ untranslated region of West Nile virus using a reporting replicon that differentiates between viral translation and RNA replication. J Virol 77, 10004–10014.

Lo, L., Tilgner, M., and Shi, P.-Y. (2003b). A potential high-throughput assay for screening inhibitors of West Nile virus replication. J Virol 77, 12901–12906.

Luo, D., Xu, T., Hunke, C., Gruber, G., Vasudevan, S., and Lescar, J. (2008). Crystal structure of the NS3 protease-helicase from Dengue virus. J Virol 82:173–183.

Luzhkov, V., Selisko, B., Nordqvist, A., Peyrane, F., Decroly, E., Alvarez, K., Karlen, A., Canard, B., and Qvist, J. (2007). Virtual screening and bioassay study of novel inhibitors for dengue virus mRNA cap (nucleoside-2′O)-methyltransferase. Bioorg Med Chem 15, 7795–7802.

Malet, H., Egloff, M., Selisko, B., Butcher, R., Wright, P., Roberts, M., Gruez, A., Sulzenbacher, G., Vonrhein, C., Bricogne, G., et al., (2007). Crystal structure of the RNA polymerase domain of the West Nile virus non-structural protein 5. J Biol Chem 282, 10678–10689.

Malinoski, F. J., Hasty, S. E., Ussery, M. A., and Dalrymple, J. M. (1990). Prophylactic ribavirin treatment of dengue type 1 infection in rhesus monkeys. Antiviral Res 13, 139–149.

Ma, L., Jones, C. T., Groesch, T. D., Kuhn, R. J., and Post, C. B. (2004). Solution structure of dengue virus capsid protein reveals another fold. Proc Natl Acad Sci U S A 101, 3414–3419.

Mancini, E., Assenberg, R., Verma, A., Walter, T., Tuma, R., Grimes, J., Owens, R., and Stuart, D. (2007). Structure of the Murray Valley encephalitis virus RNA helicase at 1.9 Angstrom resolution. Protein Sci 16, 2294–2300.

Martin, D., Biggerstaff, B., Allen, B., Johnson, A., Lanciotti, R., and Roehrig, J. (2002). Use of immunoglobulin m cross-reactions in differential diagnosis of human flaviviral encephalitis infections in the United States. Clin Diagn Lab Immunol 9, 544–549.

Martin, J., and McMillan, F. (2002). SAM (dependent) I AM: the S-adenosylmethionine-dependent methyltransferase fold. Curr Opin Struct Biol 12, 783–793.

Mastrangelo, E., Bollati, M., Milani, M., Selisko, B., Peyrane, F., Canard, B., Grard, G., De Lamballerie, X., and Bolognesi, M. (2007a). Structural bases for substrate recognition and activity in Meaban virus nucleoside-2′-O-methyltransferase. Protein Sci 16:1133–1145.

Mastrangelo, E., Milani, M., Bollati, M., Selisko, B., Peyrane, F., Pandini, V., Sorrentino, G., Canard, B., Konarev, P., Svergun, D.et al., . (2007b). Crystal structure and activity of Kunjin virus NS3 helicase; protease and helicase domain assembly in the full length NS3 protein. J Mol Biol 372, 444–455.

Mathews, J. H., and Roehrig, J. T. (1984). Elucidation of the topography and determination of the protective epitopes on the E glycoprotein of Saint Louis encephalitis virus by passive transfer with monoclonal antibodies. J Immunol 132, 1533–1537.

Michaelis, M., Kleinschmidt, M., HW, D., and Cinatl, J. J. (2007). Minocycline inhibits West Nile virus replication and apoptosis in human neuronal cells. J Antimicrob Chemother 60, 981–986.

Minke, J. M., Siger, L., Karaca, K., Austgen, L., Gordy, P., Bowen, R., Renshaw, R. W., Loosmore, S., Audonnet, J. C., and Nordgren, B. (2004). Recombinant canarypoxvirus vaccine carrying the prM/E genes of West Nile virus protects horses against a West Nile virus-mosquito challenge. Arch Virol Suppl, 221–230.

Modis, Y., Ogata, S., Clements, D., and Harrison, S. C. (2003). A ligand-binding pocket in the dengue virus envelope glycoprotein. Proc Natl Acad Sci U S A 100, 6986–6991.

Modis, Y., Ogata, S., Clements, D., and Harrison, S. C. (2004). Structure of the dengue virus envelope protein after membrane fusion. Nature 427, 313–319.

Morrey, J. D., Day, C. W., Julander, J. G., Blatt, L. M., Smee, D. F., and Sidwell, R. W. (2004). Effect of interferon-alpha and interferon-inducers on West Nile virus in mouse and hamster animal models. Antivir Chem Chemother 15, 101–109.

Morrey, J. D., Siddharthan, V., Olsen, A. L., Roper, G. Y., Wang, H., Baldwin, T. J., Koenig, S., Johnson, S., Nordstrom, J. L., and Diamond, M. S. (2006). Humanized monoclonal antibody against West Nile virus envelope protein administered after neuronal infection protects against lethal encephalitis in hamsters. J Infect Dis 194, 1300–1308.

Morrey, J., Smee, D., Sidwell, R., and Tseng, C. (2002). Identification of active antiviral compounds against a New York isolate of West Nile virus. Antiviral Res 55, 107–116.

Morrey, J., Siddharthan, V., Olsen, A., Wang, H., Julander, J., Hall, J., Li, H., Nordstrom, J., Koenig, S., Johnson, S., and Diamond, M. (2007). Defining limits of treatment with humanized neutralizing monoclonal antibody for West Nile virus neurological infection in a hamster model. Antimicrob Agents Chemother 51, 2396–2402.

Mueller, N. H., Yon, C., Ganesh, V. K., and Padmanabhan, R. (2007). Characterization of the West Nile virus protease substrate specificity and inhibitors. Int J Biochem Cell Biol 39, 606–614.

Mukhopadhyay, S., Kim, B. S., Chipman, P. R., Rossmann, M. G., and Kuhn, R. J. (2003). Structure of West Nile virus. Science 302, 248.

Munoz-Jordan, J. L., Sanchez-Burgos, G. G., Laurent-Rolle, M., and Garcia-Sastre, A. (2003). Inhibition of interferon signaling by dengue virus. Proc Natl Acad Sci U S A 100, 14333–14338.

Munoz-Jordan, J. L., Laurent-Rolle, M., Ashour, J., Martinez-Sobrido, L., Ashok, M., Lipkin, W. I., and Garcia-Sastre, A. (2005). Inhibition of alpha/beta interferon signaling by the NS4B protein of flaviviruses. J Virol 79, 8004–8013.

Murthy, H. M., Judge, K., DeLucas, L., and Padmanabhan, R. (2000). Crystal structure of Dengue virus NS3 protease in complex with a Bowman–Birk inhibitor: implications for flaviviral polyprotein processing and drug design. Journal of Molecular Biology 301, 759–767.

Ng, T., Hathaway, D., Jennings, N., Champ, D., Chiang, Y. W., and Chu, H. J. (2003). Equine vaccine for West Nile virus. Dev Biologicals 114, 221–227.

Noueiry, A., Olivo, P., Slomczynska, U., Zhou, Y., Buscher, B., Geiss, B., Engle, M., Roth, R., Chung, K., Samuel, M., and Diamond, M. (2007). Identification of novel small-molecule inhibitors of West Nile virus infection. J Virol 81, 11992–12004.

Ojwang, J., Ali, S., Smee, D., Morrey, J., Shimasaki, C., and Sidwell, R. (2005). Broad-spectrum inhibitor of viruses in the Flaviviridae family. Antiviral Res 68, 49–55.

Oliphant, T., Engle, M., Nybakken, G., Doane, C., Johnson, S., Huang, L., Gorlatov, S., Mehlhop, E., Marri, A., Chung, K., et al. (2005). Development of a humanized monoclonal antibody with therapeutic potential against West Nile virus. Nat Med 11:522–530.

Prince, H. E., Tobler, L. H., Lape-Nixon, M., Foster, G. A., Stramer, S. L., and Busch, M. P. (2005). Development and persistence of West Nile virus-specific immunoglobulin M (IgM), IgA, and IgG in viremic blood donors. J Clin Microbiol 43, 4316–4320.

Puig-Basagoiti, F., Deas, T. S., Ren, P., Tilgner, M., Ferguson, D. M., and Shi, P.-Y. (2005). High-throughput assays using luciferase-expressing replicon, virus-like particle, and full-length virus for West Nile virus drug discovery. Antimicrob Agent Chemother 49, 4980–4988.

Puig-Basagoiti, F., Tilgner, M., Forshey, B., Philpott, S., Espina, N., Wentworth, Goebel, S., Masters, P. S., Falgout, B., Ren, P., et al. (2006). Triaryl pyrazoline compound inhibits flavivirus RNA replication. Antimicrob Agents Chemother 50, 1320–1329.

Rey, F. A., Heinz, F. X., Mandl, C., Kunz, C., and Harrison, S. C. (1995). The envelope glycoprotein from tick-borne encephalitis virus at 2 A resolution. Nature 375, 291–298.

Samuel, M. A., and Diamond, M. S. (2005). Alpha/beta interferon protects against lethal West Nile virus infection by restricting cellular tropism and enhancing neuronal survival. J Virol 79, 13350–13361.

Sayao, A. L., Suchowersky, O., Al-Khathaami, A., Klassen, B., Katz, N. R., Sevick, R., Tilley, P., Fox, J., and Patry, D. (2004). Calgary experience with West Nile virus neurological syndrome during the late summer of 2003. Can J Neurol Sci 31, 194–203.

Schlesinger, J. J., Brandriss, M. W., and Walsh, E. E. (1985). Protection against 17D yellow fever encephalitis in mice by passive transfer of monoclonal antibodies to the nonstructural glycoprotein gp48 and by active immunization with gp48. J Immunol 135, 2805–2809.

Shimoni, Z., Niven, M. J., Pitlick, S., and Bulvik, S. (2001). Treatment of West Nile virus encephalitis with intravenous immunoglobulin. Emerg Infect Dis 7, 759.

Shi, P. Y., and Kramer, L. D. (2003). Molecular detection of West Nile virus RNA. Expert Rev Mol Diagn 3, 357–366.

Shi, P.-Y., and Wong, S. (2003). Serologic diagnosis of West Nile virus infection. Expert Rev Mol Diagn 3, 733–741.

Shi, P. Y., Kauffman, E. B., Ren, P., Felton, A., Tai, J. H., Dupuis, A. P., 2nd, Jones, S. A., Ngo, K. A., Nicholas, D. C., Maffei, J.et al., . (2001). High-throughput detection of West Nile virus RNA. J Clin Microbiol 39, 1264–1271.

Shustov, A., Mason, P., and Frolov, I. (2007). Production of pseudoinfectious yellow fever virus with a two-component genome. J Virol 81, 11737–11748.

Tardei, G., Ruta, S., Chitu, V., Rossi, C., Tsai, T., and Cernescu, C. (2000). Evaluation of Immunoglobulin M (IgM) and IgG enzyme immunoassays in serologic diagnosis of West Nile virus infection. J Clin Microbiol 38, 2232–2239.

Throsby, M., Geuijen, C., Goudsmit, J., Bakker, A. Q., Korimbocus, J., Kramer, R. A., Clijsters-van der Horst, M., de Jong, M., Jongeneelen, M., Thijsse, S.et al., . (2006). Isolation and characterization of human monoclonal antibodies from individuals infected with West Nile Virus. J Virol 80, 6982–6992.

Whitby, K., Pierson, T., Geiss, B., Lane, K., Engle, M., Zhou, Y., Doms, R., and Diamond, M. (2005). Castanospermine, a potent inhibitor of dengue virus infection in vitro and in vivo. J Virol 79, 8698–8706.

Wong, S. J., Boyle, R. H., Demarest, V. L., Woodmansee, A. N., Kramer, L. D., Li, H., Drebot, M., Koski, R. A., Fikrig, E., Martin, D. A., and Shi, P.-Y. (2003). An immunoassay targeting nonstructural protein 5 to differentiate West Nile virus infection from dengue and St. Louis encephalitis virus infections, and form flavivirus vaccination. J Clin Microbiol 41, 4217–4223.

Wu, J., Bera, A., Kuhn, R., and JL, S. (2005). Structure of the Flavivirus helicase: implications for catalytic activity, protein interactions, and proteolytic processing. J Virol 79, 10268–10277.

Wu, S.-F., Lee, C.-J., Liao, C.-L., Dwek, R., Zitzmann, N., and Lin, Y.-L. (2002). Antiviral effects oa an iminosugar derivative on flavivirus infections. J Virol 76, 3596–3604.

Xu, T., Sampath, A., Chao, A., Wen, D., Nanao, M., Chene, P., Vasudevan, S., and Lescar, J. (2005). Structure of the dengue virus helicase/nucleoside triphosphatase catalytic domain at a resolution of 2.4 A. J Virol 79, 10278–10288.

Yap, T., Xu, T., Chen, Y., Malet, H., Egloff, M., Canard, B., Vasudevan, S., and Lescar, J. (2007). Crystal structure of the dengue virus RNA-dependent RNA polymerase catalytic domain at 1.85-angstrom resolution. J Virol 81, 4753–4765.

Yusof, R., Clum, S., Wetzel, M., Murthy, H. M., and Padmanabhan, R. (2000). Purified NS2B/NS3 serine protease of dengue virus type 2 exhibits cofactor NS2B dependence for cleavage of substrates with dibasic amino acids in vitro. J Biol Chem 275, 9963–9969.

Zhang, N., Chen, H. M., Koch, V., Schmitz, H., Liao, C. L., Bretner, M., Bhadti, V. S., Fattom, A. I., Naso, R. B., Hosmane, R. S., and Borowski, P. (2003a). Ring-expanded ("fat") nucleoside and nucleotide analogues exhibit potent in vitro activity against flaviviridae NTPases/helicases, including those of the West Nile virus, hepatitis C virus, and Japanese encephalitis virus. J Med Chem 46, 4149–4164.

Zhang, N., Chen, H. M., Koch, V., Schmitz, H., Minczuk, M., Stepien, P., Fattom, A. I., Naso, R. B., Kalicharran, K., Borowski, P., and Hosmane, R. S. (2003b). Potent inhibition of NTPase/helicase of the West Nile Virus by ring-expanded ("fat") nucleoside analogues. J Med Chem 46, 4776–4789.

Zhang, Y., Corver, J., Chipman, P. R., Zhang, W., Pletnev, S. V., Sedlak, D., Baker, T. S., Strauss, J. H., Kuhn, R. J., and Rossmann, M. G. (2003c). Structures of immature flavivirus particles. EMBO J 22, 2604–2613.

Zhou, Y., Ray, D., Zhao, Y., Dong, H., Ren, S., Li, Z., Guo, Y., Bernard, K., Shi, P., and Li, H. (2007). Structure and function of flavivirus NS5 methyltransferase. J Virol 81, 3891–3903.

Index

Printed in the United States of America